Organic Farming

As the demand for healthy and sustainable food options increases, organic farming is becoming a viable alternative to conventional farming practices that traditionally rely heavily on synthetic inputs. *Organic Farming: A Comprehensive Guide to Sustainable Agriculture* examines a wide range of topics related to organic farming, including soil health, organic fertilizers, biodiversity, biotechnological interventions, microbial inoculants and bio stimulants, genome editing, as well as certification and marketing. This book serves as an important and timely resource for those interested in sustainable and organic farming practices, including farmers, students, researchers, and policymakers. It

- Offers practical advice and strategies for farmers and policymakers looking to transition to or promote sustainable and organic farming practices.
- Presents the latest biotechnological interventions for organic farming.
- Emphasizes the importance of organic farming for a sustainable future and highlights the challenges and opportunities facing the organic farming industry.

Organic Farming
A Comprehensive Guide to Sustainable Agriculture

Edited by Zaffar Bashir, Rohitashw Kumar,
and Mehrun Nisa

CRC Press
Taylor & Francis Group
Boca Raton London New York

CRC Press is an imprint of the
Taylor & Francis Group, an **informa** business

Designed cover image: Shutterstock

First published 2025
by CRC Press
2385 NW Executive Center Drive, Suite 320, Boca Raton FL 33431

and by CRC Press
4 Park Square, Milton Park, Abingdon, Oxon, OX14 4RN

CRC Press is an imprint of Taylor & Francis Group, LLC

ISBN: 978-1-032-79313-9 (hbk)
ISBN: 978-1-032-79314-6 (pbk)
ISBN: 978-1-003-49146-0 (ebk)

DOI: 10.1201/9781003491460

Typeset in Times
by Apex CoVantage, LLC

Contents

Acknowledgements

I sincerely acknowledge the efforts of Dr Zaffar Bashir, Assistant Professor, Microbiology, Center of Research for Development and PG Programme in Microbiology, University of Kashmir, Srinagar and Mrs Mehrun Nisa, Research Scholar, University of Kashmir, Srinagar for their support and guidance to bring this manuscript in final refined shape, constructive criticism and utmost cooperation at every stage during this work.

I am grateful to Er Munjid Maryam, who put great efforts and worked day and night to bring this manuscript in a refined shape. I am grateful to my students, Er Zeenat Farooq, Er Tanzeel Khan, Er Dinesh Vishkarma, Er Faizan Masoodi, Er Noureen, Er Muneeza, Er Nuzhat, Er Adil, and Er Mahrukh for providing all necessary help to write this book.

I am highly obliged to Hon'ble Vice Chancellor Professor N A Ganai for his support and affection and encouragement always for innovation. I wish to extend my sincere thanks to College of Agricultural Engineering and Technology, SKUAST-Kashmir for their support and encouragement.

I express my regards and reverence to my late parents as their contribution in whatever I have achieved till date is beyond expression. It was their love, affection, and blessed care that has helped me to move ahead in my difficult times and complete my work successfully. I thank my family members, who have been a source of inspiration always. I thankfully acknowledge the contribution of all my teachers since schooldays, for showing me the right path at different steps of life.

I consider it a proud privilege to express my heartfelt gratitude to ICAR–All India Coordinated Research Project on Plastic Engineering in Agriculture Structures & Environment Management, for providing a facility to carry out this project at SKUAST-Kashmir, Srinagar.

I sincerely acknowledge with love, the patience and support of my wonderful wife Reshma. She has loved and cared for me without ever asking anything in return, and I am thankful to God for blessing me with her. She has spent the best and the worst of times with me, but her faith in my decisions and my abilities has never wavered. I would also like to thank my beloved daughter Meenu, son-in-law Vineet Ravish, and son Vineet for making my home lovely with their sweet activities.

Finally, I bow my head before the almighty God, whose divine grace gave me the required courage, strength, and perseverance to overcome various obstacles that stood in my way.

Rohitashw Kumar
Associate Dean
College of Agricultural Engineering and Technology
SKUAST-Kashmir, Srinagar

Preface

Growth in agricultural productivity is vital for sustaining economic wellbeing and food security which are being threatened by climate change. Soil and water are the most precious gifts of nature. The prosperity and history of a nation depend, to a great extent, on these resources and their management. Organic farming is quickly gaining popularity as researchers around the world become more aware of the negative impacts of conventional farming practices on the environment, human health, and animal welfare. Organic farming involves using natural processes and resources to produce crops and livestock without the use of synthetic chemicals or genetically modified organisms. As the demand for healthy and sustainable food options increases, organic farming is becoming a viable alternative to conventional farming practices that rely heavily on synthetic inputs. The book's proposal on organic farming is an important contribution to the literature on sustainable agriculture. It covers a wide range of topics related to organic farming, including soil health, organic fertilizers, marketing organic products and certification, biodiversity, biotechnological interventions, microbial inoculants and bio stimulants, genome editing, precision agriculture, plant-microbe interactions, synthetic biology, non-GMO biotechnology, bioremediation, and bioprospecting. This book is accessible to readers of all backgrounds, including farmers, students, researchers, and policymakers. The themes covered provide a wealth of information on organic farming that is not easily available elsewhere. The book emphasizes the importance of organic farming for a sustainable future and highlights the challenges and opportunities facing the organic farming industry. The comprehensive guide to organic farming practices and innovations presented in this book makes it a must-read for anyone interested in sustainable agriculture. It offers practical advice and strategies for farmers and policymakers looking to transition to or promote sustainable and organic farming practices. Overall, the book's proposal on organic farming is an important and timely resource for anyone interested in sustainable and organic farming practices. The book introduces the importance of organic farming for a sustainable future and highlights the challenges and opportunities facing the organic farming industry. It will help with teaching and research in soil, environment, and sustainable agriculture. It is a newly emerging field in conservation agriculture and climate change and has potential to attract a lot of attention.

This book, with 17 chapters, delves into the principles and practices of organic farming, offering a thorough exploration of sustainable agriculture to guide and inspire readers committed to nurturing both the land and their communities.

Chapter 1 deals with an introduction to organic farming. Organic farming has grown in importance over the last 20 years, particularly in developing countries where there is an increasing need for organic goods on the global market. Although organic farming produces fewer harvests than conventional agriculture, it is more profitable, is environmentally friendly, and delivers food that is healthy and free of pesticides. Nevertheless, due to the large cost difference between organic and conventional farming, organic farming remains a marginal choice. Ensuring that wholesome, sustainable food is available at the proper prices, places, and hours is the current topic of conversation. In addition to the necessity of strategic planning, organic growers and the organic sector at large need to promote innovation and in-depth education.

Chapter 2 deals with marketing organic products and organic certification. Organic farming, a traditional practice of our ancestors, avoids pesticides and chemical fertilizers, benefiting health and the environment. In India, the organic food market is rapidly expanding, driven by consumer trust in the safety of organic products and a preference for sustainable practices. However, high costs, limited awareness, and challenges in certification hinder broader adoption. To enhance organic farming's potential, government support for lowering prices, better education for consumers, and simplified certification processes for small farmers are crucial. Growing demand and strategic marketing can boost organic food's future in India.

Chapter 3 highlights the trend and innovations in organic farming. The future of organic farming promises exciting growth through the integration of technology, environmental care, and a commitment to both human and planetary health. Blending tradition with innovation, organic farming is poised to go beyond sustainability, fostering regeneration and resilience. As demand for healthier, safer food rises, organic farming is gaining popularity, becoming synonymous with food safety and environmental preservation. Its global market is rapidly expanding, offering benefits for consumer health, ecological well-being, and economic growth through increased income opportunities.

Chapter 4 explores the role of organic farming in promoting biodiversity. By aligning with natural processes, organic farming enhances biodiversity through practices that improve soil health, increase crop diversity, and create habitats for wildlife. Studies show that organic farming supports significantly higher biodiversity levels compared to conventional methods in most cases. Small farms, with their diverse field edges, further boost species richness. Ultimately, organic farming helps preserve genetic diversity, ecosystems, and the well-being of various species, contributing to a more sustainable planet.

Chapter 5 discusses organic fertilizers, which are derived from natural sources such as plant or animal waste. These fertilizers provide essential nutrients, improve soil structure, and enhance long-term fertility, unlike chemical fertilizers which harm the environment. Organic fertilizers, such as compost and manure, release nutrients slowly and help maintain soil health by boosting its physical, chemical, and biological properties. They increase water retention, aid nutrient absorption from deeper soil layers, and promote sustainable agriculture. Organic fertilizers are a key factor in improving agricultural productivity and soil fertility.

Chapter 6 highlights the vital role of soil microorganisms, including bacteria, fungi, and protozoa, in maintaining soil health. These microbes decompose organic matter, recycle nutrients, and produce growth regulators that enhance plant development. They also play a key role in disease control and improved crop productivity. With growing concerns about the environmental impact of chemical fertilizers, optimizing the relationship between plants and soil microbes is crucial for sustainable agriculture.

Chapter 7 explores the role of silicon solubilizing bacteria (SSBs) in organic agriculture and their impact on crop productivity. The use of silicon-based fertilizers has increased due to the benefits of silicon supplementation, but challenges remain regarding the efficient dissolution and optimal dosage of these fertilizers. The chapter examines the current understanding of silicon absorption and deposition in plants, the role of soil pH, and the contributions of substances like Indole-3-Acetic Acid (IAA) and organic acids produced by SSBs in enhancing silicon solubilization. It also discusses ongoing advancements and technological progress in this area.

Chapter 8 focuses on the critical role of zinc (Zn) as a micronutrient essential for various plant functions, including photosynthesis, root and shoot development, and stress protection. Zinc deficiency is widespread, with regions like Jammu and Kashmir exhibiting significant shortages. While zinc supplements can address deficiencies, they often become fixed in the soil, reducing their effectiveness. Synthetic fertilizers also pose risks to plant and soil health. The chapter highlights the potential of plant growth-promoting microbes in solubilizing zinc and addressing deficiencies through innovative microbial formulations. It also discusses the application of microbial enzymes in bioremediation and the need for future studies to optimize microbial solutions for long-term soil health.

Chapter 9 examines the role of plant growth-promoting rhizobacteria (PGPR) in improving crop productivity and resilience. As agriculture faces environmental challenges and strives for sustainable practices, PGPR offer a promising alternative to chemical fertilizers and pesticides. These beneficial microbes enhance plant health through various mechanisms, such as producing growth regulators like IAA and solubilizing essential nutrients including phosphorus, potassium, zinc, and silicates. PGPR not only reduce reliance on synthetic inputs but also contribute to soil health and biodiversity, supporting environmentally friendly and sustainable agricultural methods.

Chapter 10 explores the potential of biocontrol agents, specifically *Trichoderma* spp. and *Pseudomonas* spp., as eco-friendly alternatives to chemical fertilizers. Despite their promising benefits, such as disease suppression and plant growth promotion through mechanisms like nutrient solubilization, these agents are underutilized due to limited field application and commercialization efforts. Challenges include insufficient farmer knowledge and the need for improved extension services. Advances in genetic engineering and molecular techniques offer new opportunities to enhance the effectiveness and production of these biocontrol agents. To realize their full potential, increased focus on mass production, understanding their mechanisms, and evaluating environmental factors are essential.

Chapter 11 explores green manuring, vermicomposting, and composting as sustainable practices for enhancing soil health. Green manuring involves planting cover crops that are later incorporated into the soil to improve its fertility and structure, reduce erosion, and decrease dependence on synthetic fertilizers. Vermicomposting converts organic waste into nutrient-rich vermicompost using earthworms, offering a high-quality, natural alternative to chemical fertilizers and boosting soil health. Composting turns organic waste into valuable compost, enhancing soil fertility and promoting environmentally friendly waste management. Together, these practices provide effective solutions for improving soil quality, increasing crop productivity, and advancing sustainable agriculture.

Chapter 12 addresses the impact of synthetic fertilizers and pesticides on soil health and microbiology. As the global population grows, the pressure to increase food production has led to widespread use of chemical inputs to boost agricultural output. However, this approach has significantly harmed soil health, leading to severe pollution, loss of biodiversity, and disruption of microbial communities. Excessive use of artificial fertilizers and pesticides has resulted in soil degradation, negatively affecting nutrient content and microbial diversity. While these chemicals can sometimes serve as carbon sources for microorganisms, they often cause long-term damage, reducing soil quality and crop productivity. Organic practices and biocontrol agents present viable alternatives, promoting soil fertility and sustainable agriculture. Future research should focus on understanding how different agricultural practices affect soil ecosystems and microbial responses to chemical inputs, aiming for more sustainable and productive farming systems.

Chapter 13 explores the role of biotechnological interventions in enhancing organic farming practices. Traditional intensive farming methods, reliant on chemicals, have led to biodiversity loss, disrupted soil microflora, and toxic contamination of food chains, necessitating a shift towards more sustainable approaches. Organic farming, which emphasizes biodiversity, natural pest control, and soil fertility through methods like crop rotation, offers significant advantages over conventional methods, including improved yield and quality of produce. The integration of biotechnology in organic farming—through bioformulations like biopesticides, biofertilizers, and bioherbicides—can further advance these benefits, though it also presents challenges that need addressing. Key actions for achieving sustainable agriculture include developing climate-resilient crops, identifying beneficial plant metabolites, optimizing biopesticide formulations, and implementing advanced monitoring technologies like artificial intelligence. These approaches aim to improve soil health, enhance productivity, and ensure environmental and food safety, highlighting the need for a combined strategy of organic farming practices, biotechnological innovation, and real-time monitoring for future agricultural sustainability.

Chapter 14 delves into the complex interactions between physical, chemical, and biological factors that influence water movement in organic soils. Soil organic matter (SOM) is pivotal in enhancing soil water management by increasing water retention, reducing runoff, improving aeration, and structuring the soil. Unlike mineral soils, organic soils exhibit diverse porosities, leading to higher preferential flow, hydrophobicity, and varying expansion and contraction behaviors. The lower bulk density of organic soils facilitates faster water movement and greater porosity, while the presence of SOM promotes aggregation and further expands porosity. Peatlands, as significant water reserves, excel in water absorption but often require artificial drainage for optimal use. Agriculture's role in draining organic soils has contributed to significant greenhouse gas emissions and land subsidence. Understanding these processes is essential for managing organic soils effectively for conservation and agricultural purposes. Continued research and monitoring are vital to adapt to environmental changes and ensure sustainable use of these valuable soils.

Chapter 15 explores various composting structures and their roles in improving waste management and soil health. Composting offers a safer, more sustainable alternative to improper waste disposal, enhancing human and environmental health by reducing reliance on chemical fertilizers. This process not only enriches soil structure and nutrient availability but also integrates plant-derived biocides to foster organic farming. Composting methods include the aerated static-pile system, enclosed system, and windrow system, each chosen based on operational site, climate, waste characteristics, and available bulking materials. To ensure effectiveness, composts must be tested for diseases and maturity before field application. Additionally, there is a need for continued research to speed up the composting process and improve technology. Raising awareness about these composting technologies is crucial for their widespread adoption by farmers, promoting better waste management and sustainable agriculture practices.

Chapter 16 discusses the application of IoT technology in vermicomposting, emphasizing its role in overcoming practical challenges and enhancing the composting process. The chapter introduces an advanced IoT-based vermicompost monitoring system that includes functions for data recording, monitoring, logging, reporting, visualization, and alerting on critical compost parameters. This system addresses limitations of current solutions by reducing costs, eliminating the need for specialized infrastructure, and simplifying procedures. Additionally, the chapter evaluates the most effective sensor technologies for optimizing vermicomposting operations.

Chapter 17 explores the dual role of plastics in agriculture, highlighting both their benefits and environmental challenges. Plastics enhance agricultural productivity and resource efficiency, but their persistence in the environment, including the accumulation of macro-, micro-, and nanoplastics in soils, poses significant ecological risks. Traditional plastics, such as polyethylene and polypropylene, are resistant to degradation, leading to long-term pollution. Biodegradable polymers offer a potential solution, yet their effectiveness and safety must be rigorously tested to ensure they decompose properly in field conditions without leaving harmful residues. Sustainable plastic use in agriculture requires improved collection, reuse, recycling processes, and the development of non-toxic additives. Global initiatives, like the Plastics Treaty, emphasize the need for comprehensive strategies to manage plastic waste while preserving its agronomic benefits, ensuring a balance between sustainability and agricultural productivity.

Zaffar Bashir

Rohitashw Kumar

Mehrun Nisa

About the Editors

Dr. Zaffar Bashir is currently working as microbiology faculty at university of Kashmir, UT J&K, India. He has experience of fourteen months of research and more than six years of teaching experience in different subjects of Microbiology. His research interests include biofertilizers, biopesticides, bioremediation, microbial ecology, organic farming, soil microbiology, plant pathology, Extremophiles, and elemental profiling of seeds. Dr Zaffar is involved in teaching post graduate courses such as, introductory microbiology, agriculture microbiology, food microbiology, environmental microbiology, molecular microbiology, industrial microbiology, medical microbiology, virology, bacteriology, mycology, phycology, extremophiles, laboratory course and environmental biotechnology etc at university level. His editorial activities and scientific memberships include publishing research and review articles in national and international journals and as a reviewer. Dr. Zaffar has published various research and review papers, book chapters of his own field. Dr. Zaffar was honored with three best presentation awards in national and international conferences and recently got best thesis award in an international conference. He has supervised project work of many M.Sc Microbiology students of university of Kashmir. Dr. Zaffar submitted some of the germ plasm lines of Buckwheat (*Fagopyrum* spp.) in NBPGR -New Delhi, with unique traits. Dr. Zaffar also submitted 17 microbial sequences in GenBank (NCBI).

Dr. Rohitashw Kumar (B.E., M.E., and Ph. D.) is Associate Dean and Professor in College of Agricultural Engineering and Technology, Sher-e-Kashmir University of Agricultural Sciences and Technology of Kashmir, Srinagar, India. He worked as Professor Water Chair (Sheikh-ul-Alam Shiekh Nuruddin Water Chair), Ministry of Jal Shakti, Government of India, at the National Institute of Technology, Srinagar (J&K) during 2018–2021. He is also Professor and Head, Division of Irrigation and Drainage Engineering. He obtained his Ph.D. degree in Water Resources Engineering from NIT, Hamirpur, and Master of Engineering Degree in Irrigation Water Management Engineering from MPUAT, Udaipur. He got, Gold medal award-2022, leadership award in 2020, Special Research award in 2017 and Student Incentive Award-2015 (Ph.D. Research) from the Soil Conservation Society of India, New Delhi. He also got the first prize in India for best M. Tech thesis in Agricultural Engineering in 2001. He has published over 150 papers in peer-reviewed journals, more than 30 popular articles, 10 books, 4 practical manuals, and 40 book chapters. He has guided two Ph. D students and 18 M. Tech in soil and water engineering. He has handled more than 15 research projects as a principal or co-principal investigator. Since 2011, he has been Principal Investigator of ICAR- All India Coordinated Research Project on Plastic Engineering in Agriculture Structural and Environment Management.

Mehrun Nisa completed M.Sc Botany from Hemvati Nandan Bahuguna Garhwal University, Uttrakhand (Central University), India and M.Phil. from Bhagwant University, Ajmer, Rajasthan, India. Mehrun Nisa is currently pursuing Ph. D Botany at university of Kashmir, UT Jammu and Kashmir, India. Her research interests include plant pathology, microbial ecology, post-harvest diseases of apricot/ walnut and soil microbiology. Her editorial activities include publishing research and review articles in national and international journals. Mehrun Nisa published various research papers, review papers, and book chapters.

Contributors

Irfan Ahmad
Center of Research for Development and PG programme in Mmicrobiology
University of Kashmir
Srinagar (Jammu and Kashmir)

Qurut-Ul- Ain
Center of Research for Development and PG Programme in Microbiology
University of Kashmir
Srinagar, J&K, India

Rizwana Ali
Center of Research for Development (CORD) and PG Programme in Microbiology
University of Kashmir
Srinagar, J&K, India

Arbeena
Center of Research for Development and PG Programme in Microbiology
University of Kashmir
Srinagar, J&K, India

Zaffar Bashir
Center of Research for Development and PG Programme in Microbiology
University of Kashmir
Srinagar, J&K, India

Mehnoor Mohmad Bhat
Center of Research for Development and PG Programme in Microbiology
University of Kashmir
Srinagar, (Jammu and Kashmir)

Bilal Ahmad Dar
Department of Botany
University of Kashmir
Srinagar, J&K, India

Ufaq Fayaz
Center of Research for Development (CORD) and PG Programme in Microbiology
University of Kashmir
Srinagar, J&K, India

Mudasir Gani
Sher-e-Kashmir University of Agricultural Sciences & Technology –Kashmir
Shalimar, Srinagar, Jammu & Kashmir, India

Burhan Hamid
Center of Research for Development and PG Programme in Microbiology
University of Kashmir
Srinagar, J&K, India

Asma Jan
Center of Research for Development and PG Programme in Microbiology
University of Kashmir
Srinagar, J&K, India

Gulzar Ahmad Khan
Central Sericultural Research & Training Institute Pampore
Jammu & Kashmir, India

Tanzeel Khan
College of Agricultural Engineering and Technology
Sher-e-Kashmir University of Agricultural Sciences & Technology of Kashmir
Shalimar, Srinagar, Jammu & Kashmir, India

Noureen Khursid
College of Agricultural Engineering and Technology
Sher-e-Kashmir University of Agricultural Sciences & Technology of Kashmir
Shalimar, Srinagar, Jammu & Kashmir, India

Rohitashw Kumar
College of Agricultural Engineering and Technology
Sher-e-Kashmir University of Agricultural Sciences & Technology of Kashmir
Shalimar, Srinagar, Jammu & Kashmir, India

Shagufta Majeed
Center of Research for Development
University of Kashmir
Srinagar (Jammu and Kashmir)

Shugufta Majeed
Center of Research for Development and PG Programme in Microbiology
University of Kashmir
Srinagar, J&K, India

Munjid Maryam
College of Agricultural Engineering and Technology
Sher-e-Kashmir University of Agricultural Sciences & Technology of Kashmir
Shalimar, Srinagar, Jammu & Kashmir, India

Nifa Mehraj
College of Agricultural Engineering and Technology
Sher-e-Kashmir University of Agricultural Sciences & Technology of Kashmir
Shalimar, Srinagar, Jammu & Kashmir, India

Mehrun Nisa
Department of Botany
University of Kashmir
Srinagar, J&K, India

Gousia Qadir
Cluster University Srinagar
(Jammu and Kashmir)

Rukhsana Qadir
Department of Botany
University of Kashmir
Srinagar, J&K, India

Mahrukh Qureshi
College of Agricultural Engineering and Technology
Sher-e-Kashmir University of Agricultural Sciences & Technology of Kashmir,
Shalimar, Srinagar, Jammu & Kashmir, India

Saniya Rafiq
Center of Research for Development (CORD) and PG Programme in Microbiology
University of Kashmir
Srinagar, J&K, India

Gulab Khan Rohela
Central Sericultural Research & Training Institute (CSR&TI), CSB
Pampore -192 121, Jammu & Kashmir, India

Bisma Rashi
Division of Basic Sciences and Humanities, FOA, Wadura
Sher-e-Kashmir University of Agricultural Sciences and Technology of Kashmir
Srinagar, J&K, India

Rehana Rasol
Division of Soil Science
Sher-e-Kashmir University of Agricultural Sciences and Technology of Kashmir
Srinagar, J&K, India

Farahanaz Rasool
Directorate of Extension
Sher-e- Kashmir University of Agricultural Sciences & Technology- Kashmir,
Shalimar, Srinagar (Jammu and Kashmir)

Rehana Rasool
Division of soil sciences
SKUAST-Kashmir
Srinagar (Jammu and Kashmir)

Plabani Roy
Central Sericultural Research & Training Institute Pampore
Jammu & Kashmir, India

Syam, S
Central Sericultural Research & Training Institute Pampore
Jammu & Kashmir, India

Pawan Saini
Central Sericultural Research & Training Institute (CSR&TI), CSB
Pampore -192 121, Jammu & Kashmir, India

Saima Shafi
Division of Basic Sciences and Humanities
Sher-e-Kashmir University of Agricultural Sciences and Technology of Kashmir
Srinagar, J&K, India

Parvaze Ahmad Wani
Center of Research for Development (CORD) and PG Programme in MicrobiologyUniversity of Kashmir
Srinagar, J&K, India

1 Introduction to Organic Farming

Challenges and Opportunities for Development of Plant Organic Farming Research

Saima Shafi and Zaffar Bashir

1.1 INTRODUCTION

Organic farming, also known as ecological farming or biological farming, is an agricultural system that emphasizes the use of organic fertilizers such as compost manure, green manure, biofertilizers, biopesticides, ash, blood meal, and bone meal. It places importance on techniques like crop rotation and companion planting. Companion planting refers to planting different crops together to enhance growth, deter pests, and improve soil quality. The concepts of organic agriculture were developed in the early 1900s by visionaries such as Sir Albert Howard, F.H. King, and Rudolf Steiner. Agricultural development policies in developing countries should focus on increasing land productivity, reducing costs, and improving product efficiency without causing harm to humans or the environment. Organic farming systems have gained attention over the past decade due to their potential for environmental protection, resource conservation, and improved food quality. They are a societal need, benefiting both consumers and farmers. Organic farming can help transform rural agriculture into sustainable agriculture, recoup conversion costs, and maintain soil sustainability (Asif et al., 2015). India, home to 30% of global organic producers, accounts for only 2.59% of the total organic cultivation area. The majority of the farming community is resource-poor, making organic farming beneficial for small and scattered landholders. The excessive use of energy and natural resources to meet market demand is leading to overuse of resources (Aktar-Uz-Zaman et al., 2017). The Green Revolution system can expand agricultural output capacity by improving soil fertility, using modern machinery, and increasing fertilizers and plant protection chemicals. This could lead to significant economic benefits at the farm level, supplying food for the world's growing population. Farmers now have a stable income source due to high yields and high-quality products, while consumers can

DOI: 10.1201/9781003491460-1

purchase a wide range of agricultural products at competitive prices. However, the luxury issue in consumerist culture is the overproduction of goods like sugar, wheat, milk, and meat. These intensive agriculture methods are the primary source of agricultural foods in industrialized nations. The overuse of synthetic chemicals has led to resistant weeds, fungi, and insects, polluting the environment and contributing to the greenhouse effect and global warming (Christov et al., 2014).

Sustainable development is a concept defined by the World Commission on Environment and Development, aiming to meet present needs without compromising future generations' ability to meet their own needs. It focuses on reducing the environmental impact of intensive agricultural practices. This system, which includes social, economic, and environmental aspects, represents the next evolutionary step in agricultural production (Graham et al., 2017). The social component focuses on establishing a high standard of living for rural populations, while the economic component focuses on efficient resource use and the development of the rural economy's competitiveness. Environmental considerations include long-term conservation of production resources. Sustainable agriculture is based on technology that increases output while minimizing environmental and human impact. Organic farming is a sustainable method that preserves natural resources and has minimal environmental impact (Hijbeek et al., 2018). It produces food while preserving land, water, energy, and biodiversity, meeting the criteria of sustainable agriculture. Banjara and Poudel argue that organic farming significantly impacts the socioeconomic standing of Nepalese farmers and strengthens the bond between people and their environment. They emphasize the importance of involving all relevant parties, including public and commercial sectors, individual farmers, and consumers, for the sustainability of organic agriculture (Muller, 2017). Testing the efficacy of this model is crucial, and proper resource management and understanding of ecological and biological processes are essential. Organic farming should also be run to maximize farm profits by ensuring market stability.

1.2 PROSPECTS AND CONSTRAINTS FOR THE ADVANCEMENT OF RESEARCH ON PLANT ORGANIC FARMING

1.2.1 Biodiversity

Organic farming methods may reduce environmental negatives, according to a meta-analysis of research comparing organic and conventional farming in Europe. Organic farming techniques often have positive environmental effects per unit of land, but not always per product unit. Differences between the two methods include organic matter in soil, nitrogen leaching, nitrous oxide emissions, energy usage, and land utilization (Stagnari, 2017). Most research shows that organic farming has less impact on the ecosystem compared to conventional farming.

Organic farming outperforms conventional farming in terms of environmental benefits, including reduced use of harmful chemicals, water usage, and trophic chain dispersion. A life cycle analysis of winter wheat production showed organic farming practices performed 8.5 and 5.9 times better than conventional methods (Ryffel, 2017).

The European Union's Biodiversity Strategy for 2020 aims to combat the decline of plant biodiversity and ecosystem deterioration. The strategy aims to increase the EU's contribution to raising awareness of biodiversity loss and strengthen the role of Europe's agriculture, forestry, and fisheries sectors in conservation and sustainable use. Crop diversity has decreased by 75% over the past 100 years, with 90% in the EU. The EU's biodiversity and climate change targets are at risk due to degradation and loss of natural capital. The 2010 biodiversity target was not met, and only 17% of species and habitats examined under the Habitats Directive are considered in good quality. The strategy aims to prevent invasive alien species from endangering biodiversity and strengthen the EU's agriculture, forestry, and fisheries sectors' role in biodiversity conservation (Jahrl et al., 2016).

EU regulation 834/2007 outlines the goals of organic farming, which aim to establish a sustainable governance structure; sustain soil, water, plants, and animals' health; and contribute to high biological diversity. Organic farming preserves or enriches biodiversity at the field, farm, and ecosystem levels, as per regulatory demands. Conventional farming methods can increase biodiversity when managed appropriately, but organic farming systems often have more floral and faunal richness (Stein-Bachinger et al., 2019). The environment around agricultural land has the capacity to improve biodiversity in agricultural regions. Biodiversity declines due to intensive farming systems disrupting natural ecosystems. Low-input and organic farming can preserve biodiversity and control pests using natural methods. Studies show increased biodiversity on organic farms improves insect pest management (Dimambro, 2018). In areas where conventional agriculture is common, organic farming can reverse the reduction of habitual species. Organic farming has been found to significantly increase biodiversity compared to conventional methods. A meta-analysis of data from organic farming zones over a 30-year period showed a 30% increase in species richness, with the impact varying depending on the crop and organism type (Lichtenberg, 2017). Organic farms generally have greater predator population densities than conventional ones, but insect damage to many crops is typically not higher in well-manicured organic fields. Some scientists suggest that organic farming may benefit intensively managed agricultural regions, but small-scale landscapes with distinct non-crop biotopes may not benefit (Wilhelm, 2016). Further research is needed to fully understand the impact of organic farming on biodiversity, but several studies suggest that this influence is variable at the farm level. Figure 1.1 depicts the point of interest with relevance to change in biodiversity with the implementation of organic farming practices. Table 1.1 shows biodiversity impact assessment for organic management techniques. In general, organic farming results in biodiversity that is 10.5%–30% greater than conventional farming (Armengot, 2013).

1.2.2 Soil Fertility

Organic farming systems can significantly contribute to the world's growing population while preserving soil fertility/soil health and structure. Conservation agriculture, which focuses on crop rotation, permanent soil cover, and minimal mechanical soil disturbance, is being promoted to increase agricultural production, profitability, and

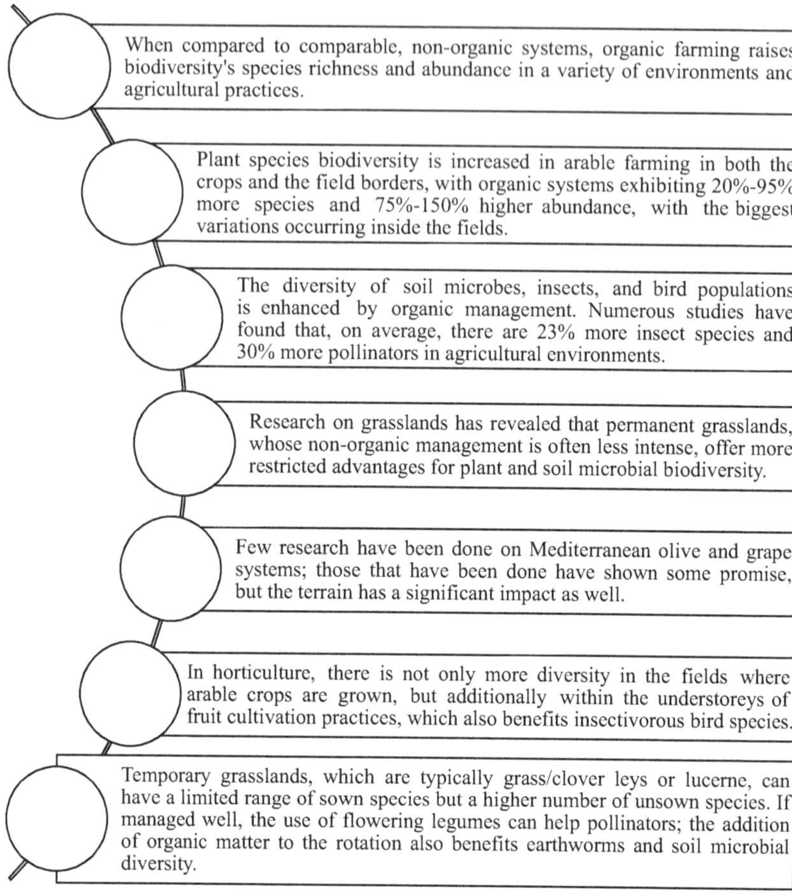

FIGURE 1.1 The figure depicts the point of interest with relevance to change in biodiversity with the implementation of organic farming practices.

food security (Cong et al., 2016). Studies have shown that conservation agriculture benefits agroecosystems by reducing soil erosion, organic carbon loss, water usage efficiency, and improved nutrient cycling, and reduced greenhouse gas emissions. However, soil in organic fields often has a high concentration of organic matter due to a cascade effect caused by microorganisms breaking down organic residues (Bashir et al., 2017, 2018a, 2018b, 2019; Doernberg et al., 2016). Researchers need to intensify their studies on humus content, microorganism variety, and soil structure to gain a deeper understanding of these processes. Soil metagenomics has been used to investigate the complexity of soil microbiology under various agricultural situations and soil types. The taxonomic makeup and microbial architectures of agricultural soils grown under conventional and no-till systems differ significantly, suggesting that changes in the microbiota of the same soil types might be caused by tillage system type and agronomic usage (Hemmerling et al., 2016). The microbiota of

TABLE 1.1
Biodiversity Impact Assessment for Organic Management Techniques

Practice	Impact on Biodiversity	Key Findings	References
Use of mixed farming practices	Moderate	In conjunction, the agriculture and livestock industries may use manure on land to minimize the amount of artificial fertilizer used and increase system variety.	Altieri & Nicholls, 2017
Rotations	High	A variety of crop varieties that are intended to lessen disease and insect loads and increase the diversity of root architecture in order to improve soil health	
Manure and compost fertilizers	Low	Natural manure and compost instead of synthetic NPK fertilizers	Hijbeek et al., 2018
Ruminants predominantly forage fed	Moderate	Presupposes ruminant diets based on forage (such as grass and clover), meaning that little or no grain or complex feed is fed	DEFRA, 2017
Mechanical weed control	Moderate	Weed removal using soil surface cultivation, such as finger weeding, inter-row cultivation, and spring-tine harrowing	Pesticide Action Network, 2017
The use of leys in arable rotations	Moderate	Introducing a grass or legume ley for one to three years into the arable rotation in order to increase biological activity and soil fertility	Sanders & Hess, 2019.
Encouraging natural predators (outdoors)	High	Agricultural areas dedicated to conservation in order to support natural predators; this includes protecting and preserving wildlife areas (such as species-rich field margins and beetle banks) in order to support predators that are necessary for pest management.	Vogeler et al., 2017
Use of green waste compost	Low	Green waste from gardens and landscaping is composted and added to fields as a beneficial soil supplement that lowers the need for fertilizer, improves soil health, and reduces landfill trash.	
Animals provided with maximum possible outdoor access year-round	None	All year-round access to the outdoors for cattle, including sacrifice fields and loafing yards	Muller, 2017
Diverse/herbal sward	High	Utilizing a variety of mixed species in grasslands, such as legumes, ryegrass, timothy, chicory, cocksfoot, fescue, clovers, lucerne, and sainfoin, among others	Lampkin & Sanders, 2021
Increasing use of legumes in crop rotations	None	Legumes come in two primary varieties: perennial forages used as green manure and seed/grain legumes. This assessment's effects are concentrated in the first category, which is crop rotation with grains and seed legumes.	Stagnari, 2017
Novel forages	Moderate	Lucene, sainfoin, and chicory are examples of feed crops that are planted in pasture areas on animal ranches.	Hodgson et al., 2010
Undersowing of leys in crops	High	Allowing the ley to form while the primary crop is still in the ground by seeding a ley sward either during or after the sowing of a crop, often grain	Wilkins, 2008

conventionally farmed soils is more copiotrophic when there is more nutrient amendment, while uncultivated soils are more oligotrophic. Conventional tillage is more likely to cause copiotrophy, which lowers the stability of soil organic matter and increases nutrient availability (Jarosch et al., 2017). Soil metagenomics studies have shown that differences in microbiota between tillage and no tillage are more significant and have less impact on crop management. This suggests that the microbiota of tillage-affected conventionally farmed soils has a higher number of bacteria engaged in xenobiosis, carbon and nitrogen cycle, phosphorus, potassium solubilization and residue breakdown (Bashir et al., 2018, 2019b, 2021a; Bashir, 2021b; Panda & Patra, 2017; Hamid & Bashir, 2019a). Crop rotation, primarily intended to disrupt the plant-pathogen cycle, has minimal effect on microbiota diversity. Soil metagenomics could play a significant role in organic agriculture by providing insights into the biology of pathogen-suppressive soils and potentially paving the way for their creation in the future. By using natural, non-synthetic products like composting, mulching, intercropping, and biological weed, disease, and pest management, the biological qualities of the soil are restored (Tsvetkov et al., 2017).

Crop rotation is crucial for maintaining the balance of the microbial community in organically managed farms. By boosting the populations of beneficial species grazing microbial films and promoting soil nutrient mineralization, soil fertility is enhanced. Mycorrhizal fungi are essential for increasing nutrient absorption, biocontrol, and microbial ecology in organic agricultural approaches. A symbiotic relationship between fungi and vascular plants' roots is the most prevalent type of mutualistic relationship, with up to 90% of plant species on Earth capable of forming a mycorrhizal association (Wu, 2017). Organic farming promotes a symbiosis between plants and fungi, protecting them from biotic and abiotic stressors while increasing the root's absorption surface area. This benefits the plant by providing water and nutrients, and supplying sugars, a carbon source produced by photosynthetic processes, to the fungus. Arbuscular mycorrhizal (AM) also enhances agroecosystem sustainability by preserving and enhancing microbial ecology and soil structure. Organic management increases the variety of arbuscular mycorrhizal fungi (AMF) assemblages, making AMF communities more diverse and richer, resembling wild grasslands. As the conversion to organic management grows, mycorrhiza richness increases. A higher frequency of crop rotation using a grass-clover blend as a fodder crop may account for the beneficial effects of organic management on AMF diversity. Adding legumes to crop rotations improves soil characteristics. However, some writers argue that variations in AMF richness and community structure are more noticeable at the end of the growing season, suggesting organic farming may favor AMF with longer life cycles (Yang et al., 2015). Organic farming research faces challenges in identifying markers associated with induced resistance, developing predictive models for these interactions, and understanding the interactions between soil, plant, and mycorrhiza under organic management. Practical implications for efficient biological control are also being considered. A multidisciplinary study comparing conventional and organic tomato agro-ecosystems revealed that organic soil has higher microbial activities, suppressing root pathogens due to microbial antagonistism (Palm et al., 2014). Nitrogen content is a key factor influencing yield in organic farming, with leguminous plants, crop wastes, manure, and compost being the sole sources of nitrogen. Integrated nutrient management, which combines organic, inorganic, and biological nutrient sources, may increase soil nutritional content (Mikha et al., 2013). However, synthetic nitrogen fertilizers can lower soil fertility due to the loss of organically bonded nitrogen and organic carbon. Badgley et al. (2007) assessed the nitrogen content of leguminous cover crops, finding that enough nitrogen could be fixed by these crops, eliminating the need for synthetic fertilizer.

Transitioning from conventional to organic farming techniques requires a period for soil ecosystems to adjust to changing conditions. Mulvaney et al. (2009) suggest a gradual switch from synthetic nitrogen fertilizers to crop rotations typical of organic farming and legume species for long-term sustainability. Organic farms typically have a 20%–25% lower productivity than conventional systems, which require higher synthetic nitrogen fertilizers. However, crop yield loss can be compensated by adding the same percentage of arable land. After a few years, the yield difference between organic and conventional systems is less noticeable, especially in industrialized nations where organic farming is common and technical advancements allow for various farm systems (Graham et al., 2017). In emerging nations, the yield difference is substantial due to low purchasing power and limited access to natural resources. The yield difference is less noticeable in industrialized nations where organic farming is customary and technical advancements allow for various farm systems.

1.2.3 Plant Breeding

In order to close technical discrepancies in organic cultivation, processing, and commercialization, advancements in research and development are essential (Wickson et al., 2016). In order to benefit all stakeholders in sustainable agriculture systems, Siderer et al. (2005) recommend that studies take into account farming practices in addition to food composition and nutritional impacts. Future organic techniques may be designed with the importance of farmer experience in mind (Bradshaw, 2017). Figure 1.2 depicts the aims of organic plant breeding, and Table 1.2 shows an overview of the varying criteria indulged with concerning plant breeding strategies for successful organic farming. Utilizing cutting-edge techniques and technology in organic farming requires a dynamic interplay between variety and innovation in the viewpoints of scientists, farmers, institutions, and social networks (Pacifico & Paris, 2016). Promoting a daring attitude toward creativity and an environment of intense exploration and collaboration for novel ideas and practices is crucial for organic producers and the organic industry as a whole. To create organic genotypes with increased production, efficient resource usage, and little environmental effect, clever methods are required (Messmer et al., 2015). Nowadays, high-yielding crop varieties produced for use in the conventional high-input industry account for 95% of organic output in modern countries (Wilkins, 2008). This prevents the expression of yield potential and resistance to diseases in low-input farming. High-yielding crops grown as monocultures by conventional agriculture frequently lead to the extinction of regional cultivars whose traits make them ideal for low-input and organic farming methods. Some may adjust to varying environmental circumstances and management techniques, which has led breeders to create new initiatives for organic methods. Organic seed rules should not exclude landraces and local variants with desirable

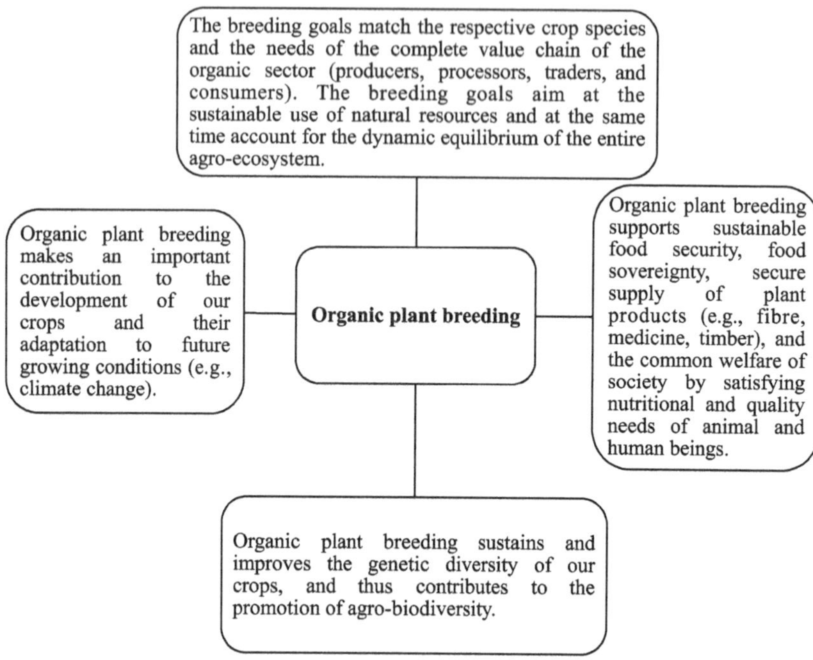

FIGURE 1.2 Aims of organic plant breeding.

features. Breeding can improve the genetic variety in nitrogen usage efficiency of crops such as potatoes, wheat, and barley; agronomic methods can also boost this efficiency (Lombardo & Zelasco, 2016). The advantageous relationship between soil and plant microorganisms may account for the productivity of recently developed crops at reduced nitrogen contents in organic fields. It is unclear if the goal of organic agriculture is to achieve yields that are similar to conventional farming or just greater than they are currently (Zou et al., 2017). However, focusing plant breeding efforts on this sector can help close the production disparities amongst conventional and organic agricultural systems. The breeding objectives of conventional and organic agriculture are similar in that they both seek increased production, tolerance to both abiotic and biotic stresses, and increased resource efficiency. Plant breeders frequently have to cope with the genotype-by-environment interaction; yet, if properly utilized, this interaction can lead to advancements in crop growth. Organic farming may be viewed as a distinct habitat that incorporates essential features and selection techniques, with a heavy emphasis on local adaptability. Although the breeding objectives of conventional and organic agricultural systems are similar, their priority may vary because conventional agriculture lacks some desirable features. Efficiency in using nutrients, competitiveness against weeds, symbiotic relationships with soil microorganisms, and resistance to mechanical weed control are all crucial characteristics for organic farming (Quiedeville et al., 2017). Cereals also exhibit genetic diversity relevant to weed competitiveness, and features such as allelopathy and early vigor can be valuable in breeding for increased weed suppression. Encouraging

TABLE 1.2
Overview of the Varying Criteria Indulged with Concerning Plant Breeding Strategies for Successful Organic Farming

Criteria	Objectives
Breeding strategies	To take into consideration the plant-environment interaction, speed up the selection gain, and take advantage of potential epigenetic effects, the selection environment is set up in line with organic farming practices. This indicates that organic agricultural practices are used during the selecting process.
	Breeders of organic plants are only allowed to create organic kinds using genetic material free of genetic engineering byproducts.
	Additional methods of selection (such as the examination of natural chemicals or molecular markers for diagnostic reasons) can be employed in addition to phenotypic selection in the field.
Socio-economic criteria	It is recommended that genetic resources be exchanged and that living things, their metabolites, gene sequences, and breeding techniques not be patented.
	To enable producers and consumers to select varieties based on their values, the breeding process, the starting material, and the applied breeding techniques will be made public (e.g., clear declaration of varieties derived from mutation breeding).
	Programs for participatory breeding that involve farmers, processors, merchants, and consumers are encouraged.
	It is hoped that a variety of autonomous breeding programs and breeders with various crop varieties would boost agricultural diversification.
Ethical criteria	Technical or physical intrusion into the plant genome is avoided (e.g., by transfer of isolated DNA, RNA, or proteins, or through induced mutagenesis), as the genome is recognized as an indivisible organism.
	It is forbidden to technically or physically infiltrate an isolated cell on growth media, since the cell is revered as an inseparable functional unit (e.g., digestion of the cell wall, death of the cell nucleus by cytoplast fusions).
	It is necessary to preserve a variety's capability for species-specific reproduction, therefore methods (such as Terminator technology) that limit the germination potential of crops grown from seeds should be avoided.
	The standards for breeding operations are based on the health, environment, justice, and caring ideals of organic farming.
	Non-hybrid types will be created to supplement the already popular hybrids and allow farmers the option to cultivate their own seeds (farmers' privilege).
	The fusing of pollen and egg cells occurs inside plant-specific crossing barriers, creating genetic diversity. It is not recommended to force somatic cells to hybridize, for example, by cell fusions.
	A variety needs to be useful for both seed propagation and future agricultural enhancement. This indicates that patenting is prohibited, breeders' exemptions and farmers' rights are legitimately provided, and crossing capacity is unrestricted by technological methods (e.g., by utilizing male sterility without the option of restoration).

innovation and strengthening collaboration between agriculture, research, and industry at the regional, national, and European levels necessitate membership in the European Innovation Partnership (EIP). In order to address obstacles with organic regulations, research and innovation are prioritized in the EU Commission's Action Plan for the Future of Organic Production in the European Union. The production of protein crops might benefit from renewed investment if the gap is closed (DEFRA, 2017). This would increase the stability of yields and product quality and increase the profitability of protein crops for both growers and the supply chain. If research focuses on greater sustainability, it may also result in advances to animal husbandry, nutrition, breeding, and feed efficiency in organic agriculture.

1.3 CAN ORGANIC FARMING FEED THE WORLD?

The potential of organic agriculture to feed the world is sometimes questioned, yet conventional approaches are failing because of problems with social organization, food distribution, and issues related to racism, poverty, and gender. There may be production losses while converting farmland from inorganic to organic farming, which would reduce the amount of food accessible to the underprivileged and drive up food prices. Organic farming is profitable and sustainable in spite of these difficulties. But, as switching to organic farming may affect food distribution and equality concerns, it is crucial to take into account both economic and social ramifications. There are several issues involved in development of organic sector in India (Garibaldi et al., 2017).

1.3.1 Yield during Conversion from Conventional to Organic Farming

Organic farming is becoming more and more popular among farmers in Punjab, Haryana, and Eastern Uttar Pradesh as a result of worries about the chemical-based agricultural system. Some farmers think chemical agriculture can lower health risks, while others think it can keep output levels stable. Because of artificial fertilizers and the restoration of soil fertility, yields in irrigated farms may decline during the conversion phase. Yields following the conversion, however, can be on par with or even greater than in traditional farming. Rainfed farmland yields considerably less, though. Although there are few studies that compare yield, some suggest that organic farms are more productive than conventional farms. It is unusual for conversion from conventional low-external input systems to provide poorer yields (Hodgson et al., 2010). Yields, however, can decrease in the first several years following conversion until the fertility and natural tilth of the soil return. Following that, yields might level off or increase, according to how well organic management works and how good the organic fertilizers are. When the best practices are applied during conversion, organic farming may be economically competitive with conventional farming. This may be accomplished in part by assessing the information requirements of organic farmers and customizing information distribution methods.

For commodities like rice, organic cultivation is less cost-effective, but in the long run, it has greater chances to reduce both financial expenses and environmental harm. It keeps the ecosystem balanced so that people, animals, plants, and crops may all coexist together. Reducing the usage of pesticides may boost net returns and lower input costs, which directly helps the economy (Pesticide Action Network, 2017). By providing protection against unfavorable price fluctuations in a particular item, crop variety on organic farms can also have a positive economic impact. Rather than the higher price of organic products, farmers that practice organic farming are generally motivated by lower input costs and yields that are comparable to those of conventional. After 25 years of research in Switzerland, Mader et al. (2002) reported sustained yields in temperate regions without the use of agrochemicals. Nonetheless, the majority of agricultural experts contend that some crops cannot provide large quantities of food without agrochemicals, and that farmyard waste and other biomass materials would not be accessible without chemical fertilizers. The nation may experience food insecurity as a result of organic farming. Because they use fewer inputs and their costs and revenue are distributed more evenly, organic farmers require less borrowing money. Lenders, on the other hand, discriminate against them, which is seen as an economic disadvantage. Over time, organic farming provides more benefits than conventional farming, such as increased yield security, greater yields, and less reliance on outside inputs, which lessens the vulnerability of low-income households to crises. These points are especially important when it comes to marginal places. Researchers Padel and Lampkin (2007) studied conventional and organic agricultural techniques. Padel discovered that although organic farms' net revenue per unit area was equivalent, their market value of crops per unit area was 11% lower. While both studies indicated that the market value of crops generated per unit area was equivalent for both systems, Willer et al. (2015) observed that organic farms usually had greater net returns and lower production costs. Two studies comparing cash grain farms in Washington found that organic farms had a 38% higher net income per unit area compared to conventional farms. However, a follow-up study found that organic farms had a 22% higher net return than conventional farms. Berardi's 1978 study compared 10 organic and 10 conventional farms in New York and Pennsylvania for returns from wheat production. When cash operating costs were included, organic farms had higher returns. However, when land and unpaid family labor costs were included, conventional farms had a higher average net return. The studies had limitations, including a small sample size, inability to conduct statistical tests, and not including the livestock enterprise, which may be essential for optimal economic performance of organic farms. In a 1984 Regenerative Agriculture Association poll, 213 farmers claimed that, when they started farming with less bought inputs, their net income either climbed or stayed the same, while 12% reported a drop (Willer & Lernoud, 2015). The poll indicates a high level of satisfaction with the financial performance of low-input farming, notwithstanding the possibility that the sample is not typical of organic farmers.

1.3.2 Soil Fertility

Two decades of rapid destruction of fertile agricultural soils in India have been caused by the green revolution paradigm, which replaced the nutrients cycle with chemical fertilizers and focused on producing marketable agricultural commodities. However, the Punjab experience demonstrates that soil fertility cannot be restored by nitrogen, phosphorus, and potassium (NPK) from factories, and agricultural productivity requires returning biological products to the soil. Technologies cannot substitute nature's ecological processes without destroying the basis of production, and markets cannot provide the only measure of output and yields. The idea that soil fertility is created in chemical factories and that agricultural yields are determined by commodities that are sold is a result of the green revolution. Millets, which have large yields from adding organic matter to the soil, have supplanted nitrogen-fixing crops like pulses (Wang et al., 2021). The internal inputs of biological products that sustain soil fertility were disregarded in the green revolution miracle's cost-benefit analysis. Future-Based Management has long been the main strategy for restoring lost soil organic matter (SOM), a sign of the health and potential for productivity of the soil. In addition to providing crops with nutrients and enhancing the texture of the soil in drylands, organic manures also serve as mulches, shielding crops from the effects of extreme heat, enhancing seed germination, boosting water retention capacity, and fostering the ideal microclimate for the growth of beneficial soil microbes. Soils grown organically have a higher capacity to prevent soil deterioration and are more adapted to handle drought and nutrient loss (Wang et al., 2021). The current agricultural system mainly concentrates on growing plants, ignoring the significance of nourishing the soil. Plants that fix nitrogen, such as soy beans, can be planted to offset the export of elements with seeds and enable the use of less fertilizer in an organic system. Farmers are logical decision-makers who strive for net gain by balancing expenses and possible rewards. What was deemed wasteful and unproductive agriculture has been turned into sustainable and productive agriculture by the green revolution? The people who caused the issue and prioritized the market over the life of the soil are not the ones who can provide a solution to the crisis of soil health. Using the market as the organizational basis for agriculture will not lead to the repair and recovery of soil health. Recuperation instead consists in relearning how to renew naturally and in respecting the soil's right to share its output in order to keep it healthy (Aulakh et al., 2022).

1.3.3 Livestock

A vital component of agriculture, livestock also create jobs and revenue in rural regions. By generating manure and affecting the availability of soil organic carbon, it directly contributes. Household income is indirectly impacted by livestock. Utilizing animals and crops together in mixed farming makes it possible to use animal manure to improve the fertility of the soil. Farmers are aware of the advantages of manure and its ability to contribute more to the upkeep of soil fertility. In order to keep the soil fertile, provide draught power, and feed households, livestock is a necessary component (Slagboom et al., 2016). A more closed nutrient management system has resulted from the weakening of the historical ties that once connected cattle and forests. A household's ability to invest in productivity-boosting strategies by buying off-farm inputs is increased by livestock revenue. The landholdings of marginal, tiny, and semi-medium farmers are insufficient for the household's needs year-round, and raising cattle provides these smallholders with an option. The rise of animal stocks during droughts is slower than that of milch animal stocks, indicating a quick shift in the livestock sector (Mie et al., 2017). This is because of things like growing automation, shrinking common property resources, and shrinking farm size. The availability of manure is affected by this. Local animal breeds should be preserved through integrated farming and native land-use systems as they are essential to livelihoods and sustainable agriculture. According to a jury of "Prajateerpu" farmers, the corporate agriculture outlined in "Vision 2020" would worsen the loss of livestock biodiversity. They demand that government assistance, suitable training, and research be provided in order to bring livestock back into agriculture. Reduced availability of water and fodder, together with widespread practices in animal breeding, are contributing to a decline in livestock genetic diversity, draught power, natural fertilizers, livelihoods, and family assets. Reusing other available nutrients, recycling crop leftovers, and improving the integration of crop and animal systems may all increase agricultural output. Overall, there should be more promotion of a farming system that maximizes the revenue and employment possibilities for small farms. Comprehending current farming systems is essential for large research initiatives. The inadequate power supplied by animals in rainfed ecosystems must be addressed, and the needs for draught power must be determined and met. The self-generating business of raising livestock lowers the volatility and unpredictability of agricultural revenue. Dairy and poultry farming combined with agriculture can increase the average net income of small farmers. Reduced soil nitrogen availability and a closed nutrient management system are the results of decreasing ties between livestock and forests (Van Wagenberg et al., 2017). In summarization maintaining soil fertility, supplying food and draught power, and boosting agricultural production all depend on livestock.

1.3.4 Institution/Certification

Various organizational structures, such as those run by businesses, non-governmental organizations, the government, and self-help groups, promote organic farming in India. Despite the success of organic farming, it is challenging to provide an unambiguous structure for its further advancement. Smallholders find it expensive to participate in organic farming because of institutional and financial barriers that farmers in developing and transitional nations must overcome in order to become accredited organic producers. Uncertain standards, laborious paperwork, the absence of one point of contact certification body, and costly certification have not inspired farmers in places such as Chhattisgarh (Altieri & Nicholls, 2017). The low export volume of the state also makes certification worthwhile for both farmers and consumers. There is still a sizable portion of the organic agricultural community that is excluded and does not get paid more for their goods. A support framework must be established, specifically for resource-limited small farmers, to effectively move from non-organic to organic farming. The conversion phase, which is vital to neutralizing chemical contaminants left in soil by agricultural practices, is the interval between beginning organic management and certification. Unlike conventional agriculture, sustainable farm management does not rely on a unified plan. Annual crops typically take 24 months to convert, but perennials can take as long as 36 months to mature. Depending on the ecological circumstances of the farm and the agricultural method employed during the pre-conversion phase, the certifying authority may decide to shorten or lengthen this duration (Bernacchia et al., 2016). Wage prices may rise with the use of organic procedures, and obtaining knowledge and accreditation from a recognized body may be expensive. If there are no other options, such as internal control systems and small farmers' organization accreditation, this might be very difficult for small farmers. In emerging and transitional economies, there are three types of certification programs: Participatory Guarantee System (PGS), Internal Control System (ICS)-certified small-scale farmers, and third-party certification for people. Originating in the poor world, where organic growers oversee and maintain quality assurance systems, the PGS is a participative certification program. Every variation is unique to certain markets, governments, geographies, and communities. For organic agriculture to continue spreading throughout India, there must be strong organizational backing. An enabling situation for small farms, group certification, an easy-to-understand and affordable certification procedure, technological help for record keeping, and internal control systems are examples of appropriate institutional support.

1.3.5 Ecology

A viable approach to ensuring the ecological stability and financial stability of the nation's millions of small-scale farmers is organic biodiverse farming. It protects fish, animals, soil nutrients, energy, pollution, and the availability of food for future generations. But there are obstacles to conversion and a dearth of reliable evidence about the advantages of organic farming for policymakers. Because of the rise in chemicals like pesticides and fertilizers since the 1960s, the use of biochemical innovations in agriculture has a detrimental influence on the environment. Concerns over these compounds' effects on the environment have arisen as a result. It was realized in the 1980s that sustainable development required the use of alternative farming methods. People have suffered for a long time from ailments like leukemia and epilepsy as a result of the ongoing usage of pesticides (Yue et al., 2017).

In order to support SAD within the larger context of health and the environment, deliberate efforts must be undertaken to identify workable substitutes for pesticides. Policies encouraging environmentally friendly agricultural methods should be put into place, and the financial and environmental effects of crop regulations on pesticides reduction should be carefully considered. Though it's not always practical to use the instruction on their primary agricultural area, trained farmers are becoming more and more interested in ecological agriculture. On homestead property, on the other hand, where there are no outside influences and market forces to govern the situation, farmers are progressively using this strategy. Although farmers are becoming more conscious of the issue, a full transition from inorganic to organic farming will take time (Sanders & Hess, 2019). Farmers who employ organic agriculture suffer because pesticides from farmers upstream contaminate their farms in states like Chhattisgarh, wherein organic agriculture is prevalent. This is particularly relevant for medicinal herbs, which are used in medications that may save lives and other health items, giving them a higher sensitivity index. The agricultural community must be jointly mobilized and made more conscious in order for the sustainable farming movement to succeed in getting rid of inorganic methods. Consumer demand for safe, hazard-free food and expanding knowledge of the health risks connected with agrochemicals and environmental conservation are the main drivers of increasing interest in alternative kinds of agriculture (Reganold & Wachter, 2016). In both industrialized and developing nations, consumer demand for organic food is rising at an average yearly growth rate of 20%–25%, making organic agriculture one of the most environmentally friendly production practices. Because organic agriculture can be combined with integrated agricultural systems and has potential environmental advantages, it might be viewed as a development tool for developing nations such as India. An analysis comparing crop output under the Inorganic Farming System (IFS) and the Organic Farming System (OFS) revealed a progressive decrease in production costs in the OFS. Assigning monetary values to enhanced soil health, less pollution, increased resilience, and decreased greenhouse gas emissions is difficult, though (Rahmann et al., 2017).

1.3.6 Marketing

Compared to traditional marketing, organic marketing is far more involved and calls for the creation of expansive marketplaces and distribution networks as well as cautious market selection. It demands specific knowledge, expertise, and experience—all of which disorganized farmers frequently lack. Since the export market receives more than 80% of the nation's organic production, the local market is underdeveloped. The challenges farmers encounter when switching to organic practices are exacerbated by the absence of domestic marketing outlets (Nuttavuthisit & Thøgersen, 2017). Understanding markets, business or operational structure, communication and transportation networks, and a suitable policy environment are all necessary for small manufacturers to get access to markets. Small farmers require increased managerial capability, as well as improved access to finance and education (Nuttavuthisit & Thøgersen, 2017). Cooperatives and organizations play a key role in ensuring that new goods and procedure standards are met and that supply chain exclusion is prevented. It is also essential to have professional expertise in technical production elements and marketing. Value addition in the chain of supply may be captured by bolstering small farmer groups and offering technical support to raise productivity and enhance produce quality (Bruschi et al., 2015).

1.3.7 Policy Support

India's policymakers have historically prioritized outside solutions to farmers' problems, which has increased reliance on outside inputs—which are more expensive, harmful to the environment, and economically unproductive than alternatives that save resources. New policies need to concentrate on creating circumstances based on locally accessible resources as well as local expertise and knowledge in order to solve this. In order to support farmers' assessments and express their structured demands, policy makers should form partnerships and conversations with other stakeholders. This would enable quick feedback and alternative policy adaptation. The main goal of agricultural policy should be to empower experts and the general public to maximize the use of biological and social resources. India is far behind other countries in embracing the practice of organic farming, despite the efforts of certain NGOs. Government authorities should encourage producer and consumer groups, grow the market for organic goods, streamline the procedure for obtaining certification, and lower certification prices in order to promote organic agriculture. To raise knowledge among farmers and consumers, an awareness campaign highlighting the advantages of organic farming over conventional farming practices is necessary (Huber et al., 2017). Despite providing farmers with an alternate approach, organic farming is not included in the National Agricultural Policy. However, because organic food certification requires a lot of organizational, regulatory, control, and paperwork work, it might be difficult. The government of Chhattisgarh encourages the growth of horticultural and agricultural products as well as medicinal, aromatic, and coloring plants. Some state government organizations that support organic farming of medicinal, agricultural, horticultural, and aromatic crops include the Chhattisgarh Vanoushadhi Board, the Departments of Horticulture and Agriculture, and the Chhattisgarh State Minor Forest Produce Federation (Banjara & Poudel, 2017).

Even in the absence of direct government effort, the government may still be vital to organic farming since non-governmental organizations do not have the commercial acumen needed for successful marketing. In these kinds of circumstances, cooperation among NGOs and governments may prove beneficial. In order to reach organic markets, contract farming agreements between private enterprises and organic farmers need to implement efficient marketing tactics. Conflicts of interest, however, might exist between the interests of farmers and profit-driven parties. Therefore, for resource-poor and disorganized farmers, a legal structure enforcing contracts and offering reliable arbitration is essential. Farmers' groups are helpful in protecting their interests, but they need a lot of help in the beginning, throughout operations, during training, and during marketing. The state's government or the non-governmental sector can provide this support (Bazzani et al., 2017). Karnataka, Punjab, Madhya Pradesh, Arunachal Pradesh, and Meghalaya have all adopted these programs. The "Uttarakhand organic" campaign advocates for the use of the organic model as a farming technique and a crucial component of initiatives for rural development. Development of the home market is also emphasized. It is too early to comment on the initiative, but if it is done well, it might serve as a model for state-driven organic growth in India.

1.4 STANDARDS AND REGULATIONS IN ORGANIC FARMING

With the number of organic farmers and production regions in the EU increasing dramatically over the past 20 years, the cultivation of organic produce has become a global phenomenon. Twenty-seven percent of all the world's organically agricultural land is currently in Europe, where organic farming got its start in the 1980s. The area used for organic production in the EU, Norway, and Switzerland is 2.4%; several nations have exceeded this limit. As a result, there is now more focus on the manufacturing, processing, and monitoring of food safety through the use of specialized guidelines, protocols, and risk management initiatives. In 2002, the International Federation of Organic Agriculture Movements (IFOAM) published a fundamental guideline for organic farming. In Europe, organic agriculture is regulated under EC

Council Regulation (EEC) No. 2092/91. National production standards and organic agricultural rules are currently in place in every EU member state, and European legislation is in line with IFOAM's fundamental requirements (IFOAM, 2017). EU regulations must be met by organic products entering the EU, and third-country certification—which includes accreditation, audit traits, yearly inspections, material lists, specified conversion times, and sustainable farm plans—is necessary. Switzerland, Australia, Israel, and New Zealand all have certifications but also their unique set of rules. The FiBL assessment on organic laws and regulations states that as of 2016, there were 87 nations globally having organic standards. Since its adoption in 1991, Regulation (EEC) 2092/91 has undergone several changes, such as the development of a community emblem and standards for animal production. The EU strictly defines what constitutes organic food and farming, and it places a premium on the quality of products that are provided for sale. By being registered as a collective trade mark with the European Office for Harmonization in the Internal Market and a select few adjacent nations, such as Switzerland and Norway, the EU organic logo is safeguarded (Ilukpitiya and Khanal, 2016). The acquis communautaire's laws pertaining to the organic cultivation of agricultural goods are compiled into organic rules, which are applicable to all EU member states. They include regulations that apply to the manufacturing, processing, marketing, and trading of both raw and processed goods. They encompass agricultural and animal products intended for human use as well as animal feed. Each member state must appoint a Competent Authority in accordance with Regulation 2092/91 in order to authorize and monitor certifying organizations, carry out processes for import inspection and certification, and generally regulate all operations pertaining to organic cultivation, processing, and marketing. Numerous nations have created their own initiatives to support organic farming, providing a portion of the costs associated with certification and inspection. Since 1996, manufacturers in Denmark have been eligible for free certification. However, because of changes in mainstream agricultural institutions, there are worries that the natural farming community in Central and Eastern European nations is losing influence over the growth of the organic industry (Lampkin & Sanders, 2021). There is enormous potential in organic agriculture to build sustainable agricultural systems, but no one strategy will guarantee the world's food security. It is necessary to combine organic farming practices with other cutting-edge agricultural methods, however implementing these methods is fraught with difficulties.

1.5 LOOKING TO THE FUTURE

Numerous elements of organic agriculture are included, such as crop husbandry, livestock, and horticulture, along with related farm operations. There is still more work to be done, but India has developed and executed a number of programs to support organic agriculture, which have increased acreage and exports over the years many times over. The following are the key suggestions for increasing the use of organic farming throughout the nation:

- Growth in organic farming is limited by a number of factors, including a decrease in yield in the first years of conversion, a lack of organic manures on the farm to meet nutrient demands, a gradual absorption of nutrients from organic manures resulting in an imbalance between demand for crops and soil supply, challenges managing the bulky compost, and inadequate infrastructure for certification and marketing. Thus, it's important to have a comprehensive plan to deal with each of these problems (Nene, 2017).

- Lack of a consistent and dependable availability of certified supplies (such as seeds, bioagents, fertilizers, and manures) and the financially feasible marketing of organic agricultural products are the primary issues facing organic producers. Therefore, it is necessary to construct stable and dependable input-output linkages in prospective organic clusters (Ponisio et al., 2015). To increase capacity utilization and ensure sustainability, the public and private organic input manufacturing units developed under the nation's development plans should be connected to appropriate marketing channels. It is imperative that certified organic supply marketing channels be established in order to support the nation's organic farming industry.
- Potential organic agricultural zones must be designated as "Special Organic Agriculture Systems Zones" and classified as "Special Economic Zones" in order to capitalize on both home and foreign export markets.
- The foundation of India's organic farming industry is the widespread use of crop + animal farming systems. This is a fantastic chance to set up integrated organic farming systems in all of the specialty markets, which may function as a unit for research and demonstration. Depending on the niche, a cluster of villages has to be encouraged to use organic agricultural practices (Ramesh et al., 2010).
- To ensure secure food security for the nation, the "Towards Organic" method should be used rather than quickly moving from inorganic to organic in high-intensity agricultural regions. This strategy will help to boost the efficiency with which fertilizers and water are used while also reducing the initial substantial loss of yield during the conversion phase. To achieve the intended results, government programs for water management, plant protection, and integrated nutrient management must be combined.
- The national organic production rules are equivalent to those of the European Union along with other major nations. It benefits the export. It is necessary to study the domestic standard for sustainable farming and certification, which also adheres to export norms. A method that moves "forward toward organic" and incorporates integrated crop-management techniques can lead to "safe food for all." The production methods of a combined strategy with recommended optimum use of micronutrients in the type of synthetic fertilizers can be taken into consideration by the domestic standard. Nonetheless, the control of weeds and pests should adhere to export standards. This advice is further supported by the claim that plants receive nutrients in forms of inorganic matter from a variety of sources, including organic ones (Panwar et al., 2021).
- Along with boosting farm households' income and building climate resilience, "certified organic farming," which combines tradition, innovation,

and science in the de facto organic areas (hills) and rainfed/dryland regions, will contribute to secure food security in the future. This strategy will benefit the health of people, animals, and ecosystems as well. Thus, organic farming needs to be encouraged for certain crops and specialized markets (Yadav et al., 2009).

- The nation's organic products should be produced, processed, and marketed via Farmer Producer Organizations (FPOs). In actuality, organic advertising will greatly benefit from ties to a guaranteed market. Anand pattern, which proved effective in the dairy industry, needs to be investigated for the country's organic farming growth.
- For the South Asia Association for Regional Cooperation (SAARC) countries to share technologies and reap the benefits of complementarity, networking among academic institutions, research centers, markets, certifying bodies, and non-governmental organizations (NGOs) is crucial. For the SAARC countries to implement organic agriculture, a reliable web-enabled information system is crucial.

1.6 CONCLUSION

Organic farming has grown in importance over the last 20 years, particularly in developing countries where there is an increasing need for organic goods on the global market. Although organic farming produces fewer harvests than conventional agriculture, it is more profitable, environmentally friendly, and delivers food that is healthy and free of pesticides. Nevertheless, due to the large cost difference between organic and conventional farming, organic farming remains a marginal choice. Making ensuring that wholesome, sustainable food is available at the proper prices, places, and hours is the current topic of conversation. In addition to the necessity of strategic planning, organic growers and the organic sector at large need to promote innovation and in-depth education.

REFERENCES

Aktar-Uz-Zaman, M., Tuhina-Khatun, M., & Hanafi, M. M. (2017). Genetic analysis of rust resistance genes in global wheat cultivars: An overview. *Biotechnology & Biotechnological Equipment*, *31*, 431–445.

Altieri, M. A., & Nicholls, C. I. (2017). The adaptation and mitigation potential of traditional agriculture in a changing climate. *Climatic Change*, *140*(1), 33–45.

Armengot, L. (2013). Weed harrowing in organically grown cereal crops avoids yield losses without reducing weed diversity. *Agronomy for Sustainable Development*, *33*(2), 405–411.

Asif, M., Yang R-C., & Navabi, A. (2015). Mapping QTL, selection differentials, and the effect of Rht-B1 under organic and conventionally managed systems in the Attila £ CDC Go spring wheat mapping population. *Crop Science*, *55*, 1129.

Aulakh, C. S., Sharma, S., Thakur, M., & Kaur, P. (2022). A review of the influences of organic farming on soil quality, crop productivity and produce quality. *Journal of Plant Nutrition*, *45*(12), 1884–1905.

Badgley, C., Moghtader, J., & Quintero, E. (2007). Organic agriculture and the global food supply. *Renewable Agriculture and Food Systems*, *22*(2), 86–108.

Banjara, R. K., & Poudel, M. (2017). Sustainable model of organic agriculture: A case study of Nepalese farmers. *Journal of Advanced Academic Research*, *3*(1), 142–163.

Bashir, Z. (2021). characterization of potassium solubilizing bacteria from rhizospheric soils of cherry (*Prunusavium*). *Research Journal of Agricultural Sciences*, *12*(2), 494–496.

Bashir, Z., Hamid, B., Zargar, M. Y., & Kundoo, A. A. (2021). Morphological and biochemical characterization of potassium solubilizing bacteria from Rhizospheric soils of Pinus (*Pinus wallichiana*). *Journal of Research and Development*, *21*, 58–63.

Bashir, Z., Zargar, M. Y., Baba, Z. A., & Mohiddin, F. A. (2018). Solubilization of Potassium containing mineral by bacteria from Kashmir iris (*Iris kashmiriana*) Rhizosphere. *The Pharma Innovation Journal*, *l7*(4), 747–750.

Bashir, Z., Zargar, M. Y., Baba, Z. A., Mohiddin, F. A., Peer, F. A., Mir, S. A., & Najar, G. R. (2018a). Isolation and characterization of phosphorus solubilizing bacteria (PSB) from Rhizospheric soils of Apple (*Malus Domestica*). *Journal of Pharmacognosy and Phytochemistry*, *7*(2), 2997–3000.

Bashir, Z., Zargar, M. Y., Bhat, T. A., Kousar, S., Baba, A., & Mohiddin, F. A. (2018b). Solubilization of phosphorus containing mineral by bacteria from Rhizospheric region of Walnut (*Juglans regia*). *International Journal of Current Microbiology and Applied Sciences*, *7*(8), 2391–2398.

Bashir, Z., Zargar, M. Y., Mohiddin, F. A., Baba, Z. A., & Hamid, B. (2019a). Isolation and biochemical characterization of phosphorus solubilizing bacteria from Rhizospheric region of Apricot and Peach. *Journal of Research and Development University of Kashmir, 19*, 972–5407.

Bashir, Z., Zargar, M. Y., Mohiddin, F. A., Kousar, S., Husain, M., & Rasool, F. (2017). Phosphorus solubilizing microorganisms: Mechanism and diversity. *International Journal of Communication Systems*, *5*(5), 666–673.

Bashir, Z., Zargar, M. Y., & Vishwakarma, D. K. (2019b). Potassium solubilizing microorganisms for sustainable agriculture. *Applied Agricultural Practices for Mitigating Climate Change*, *2*, 17–28.

Bazzani, C., Caputo, V., & Nayga, R. (2017). Revisiting consumers' valuation for local versus organic food using a non-hypothetical choice experiment: Does consumers' personality matter? *Food Quality and Preference*, *62*, 144–154.

Berardi, G. M. (1978). Organic and conventional wheat production: Examination of energy and economics. *Agro-Ecosystems*, *4*, 367–376.

Bernacchia, R., Preti, R., & Vinci, G. (2016). Organic and conventional foods: Differences in nutrients. *Italian Journal of Food Science*, *28*(4), 565–578.

Bradshaw, J. E. (2017). Plant breeding: Past, present and future. *Euphytica, 213*, 60.

Bruschi, V., Shershneva, K., & Dolgopolova, I. (2015). Consumer perception of organic food in emerging markets: Evidence from Saint Petersburg, Russia. *Agribusiness*, *31*(3), 414–432.

Christov, N. K., Ilchovska, D., & Hristov, K. N. (2014). Chemical mutagenesis, mutation breeding and quantitative genetic analyses of maize mutants: From theory to practice. In N. B. Tomlekova, M. I. Kozgar, & M. R. Wani (Eds.), *Mutagenesis: Exploring genetic diversity of crops* (169–196). Wageningen Academic Publishers.

Cong, R., Li, H., & Zhang, Z. (2016). Evaluate regional potassium fertilization strategy of winter oilseed rape under intensive cropping systems: Large-scale field experiment analysis. *Field Crops Research*, *193*, 34–42.

DEFRA. (2017). Ecological focus areas: Features on farms in England 2015/16.

Dimambro, M. (2018). Countryside stewardship organic management and conversion options: A scoping study to establish a monitoring protocol. Literature review. *CAWR & Coventry*. https://pure.coventry.ac.uk/ws/portalfiles/portal/40416921/Dimambro_Rayns_Steiner_and_Carey_2018.pdf

Doernberg, A., Zasada, I., & Bruszewska, K. (2016). Potentials and limitations of regional organic food supply: A qualitative analysis of two food chain types in the Berlin metropolitan region. *Sustainability*, *8*(11), 11–25.

Garibaldi, L. A., Gemmill-Herren, B., & D'Annolfo, R. (2017). Farming approaches for greater biodiversity, livelihoods, and food security. *Trends in Ecology & Evolution*, *32*(1), 68–80.

Graham, R. F., Wortman, S. E., & Pittelkow, C. M. (2017). Comparison of organic and integrated nutrient management strategies for reducing soil N2O emissions. *Sustainability*, *9*, 510.

Hamid, B., & Bashir, Z. (2019a). Potassium solubilization microorganisms: An Alternative technology to chemical fertilizers. *Journal of Research and Development University of Kashmir*, *19*, 972–5407.

Hemmerling, S., Asioli, D., & Spiller, A. (2016). Core organic taste: Preferences for naturalness-related sensory attributes of organic food among European consumers. *Journal of Food Products Marketing*, *22*, 824–850.

Hijbeek, R., ten Berge, H. F. M., Whitmore, A. P., Barkusky, D., Schröder, J. J., & van Ittersum, M. K. (2018). Nitrogen fertiliser replacement values for organic amendments appear to increase with N application rates'. *Nutrient Cycling in Agroecosystems*, *110*(1), 105–115.

Hodgson, J. A., Kunin, W. E., Thomas, C. D., Benton, T. G., & Gabriel, D. (2010). Comparing organic farming and land sparing: Optimizing yield and butterfly populations at a landscape scale. *Ecology Letters*, *13*(11), 1358–1367.

Huber, B., Schmid, O., & Moller, C. (2017). Standards and regulations: Overview. In: *FiBL&IFOAM—Organics International* (pp. 150–157). The world of organic agriculture, FiBL&IFOAM.

IFOAM. (2017). *Principles of organic agriculture* (pp. 4–9). Organics International.

Ilukpitiya, P., & Khanal, P. (2016). Consumer perception of organic food and product marketing. *Organic Farming for Sustainable Agriculture*, *9*, 315–324.

Jahrl, I., Moschitz, H., & Stolze, M. (2016). Growing under the common agricultural policy: The institutional development of organic farming in Central and Eastern European countries from 2004 to 2012. *International Journal of Agricultural Resources, Governance and Ecology*, *12*(4), 357–380.

Jarosch, K., Oberson, A., & Emmanuel, F. (2017). Phosphorus (P) balances and P availability in a field trial comparing organic and conventional farming systems since 35 years. In *Proceedings of the 19th EGU General Assembly (EGU2017); 2017 Apr 23–28* (Geophysical Research Abstracts, Vol. 19, p. 15377). European Geosciences Union.

Lampkin, N., & Sanders, J. (2021). Policy support for organic farming in the European Union: 2011, 2015 and 2019. *Project Report*. Thünen Institute.

Lichtenberg, E. M. (2017). A global synthesis of the effects of diversified farming systems on arthropod diversity within fields and across agricultural landscapes. *Global Change Biology*, *23*(11), 4946–4957.

Lombardo, L., & Zelasco, S. (2016). Biotech approaches to overcome the limitations of using transgenic plants in organic farming. *Sustainability*, *8*, 497.

Mader, P., Fliebbach, A., & Dubois, D. (2002). Soil fertility and biodiversity in organic farming. *Science*, *296*, 1696–1697.

Messmer, M., Wilbois, K. P., & Baier, C. (2015). *Plant breeding techniques: An assessment for organic farming* (2nd ed.). Research Institute of Organic Agriculture (FiBL).

Mie, A., Andersen, H. R., Gunnarsson, S., Kahl, J., Kesse-Guyot, E., Rembialkowska, E., Quaglio, G., & Grandjean, P. (2017). Human health implications of organic food and organic agriculture: A comprehensive review. *Environmental Health*, *16*, 22.

Mikha, M. M., Vigil, M. F., & Benjamin, J. G. (2013). Long-term tillage impacts on soil aggregation and carbon dynamics under wheat-fallow in the central great. *Soil Science Society of America Journal*, *77*, 594–605.

Muller, A. (2017). Strategies for feeding the world more sustainably with organic agriculture. *Nature Communications, 8*(1).

Mulvaney, R. L., Khan, S. A., & Ellsworth, T. R. (2009). Synthetic nitrogen fertilizers deplete soil nitrogen: A global dilemma for sustainable cereal production. *Journal of Environmental Quality, 38*(6), 2295–2314.

Nene, Y. L. (2017). A critical discussion on the methods currently recommended to support organic crop farming in India. *Asian Agri–History, 21*(3), 267–285.

Nuttavuthisit, K., & Thøgersen, J. (2017). The importance of consumer trust for the emergence of a market for green products: The case of organic food. *Journal of Business Ethics, 140*(2), 323–337.

Pacifico, D., & Paris, R. (2016). Effect of organic potato farming on human and environmental health and benefits from new plant breeding techniques. Is it only a matter of public acceptance? *Sustainability, 8*(10), 1054.

Padel, S., & Lampkin, N. (2007). The development of governmental support for organic farming in Europe. In W. Lockeretz (Ed.). *Organic farming: An international history* (pp. 93–122). CAB International.

Palm, C., Blanco-Canqui, H., & DeClerck, F. (2014). Conservation agriculture and ecosystem services: An overview. *Agriculture, Ecosystems & Environment, 187*, 87–105.

Panda, R., & Patra, S. (2017). Depletion and contribution pattern of available potassium in Indian coastal soils under intensive cropping and fertilization. *International Journal of Pure & Applied Bioscience, 5*(2), 1144–1152.

Panwar, A. S., Shamim, M., Ravisankar, N., Ansari, M. A., Singh, R., Prusty, A. K., & Noopur, K. (2021). Influence of long term fertilization with organic sources on crop productivity and soil quality in rice–wheat system under arid and sub humid conditions. *Indian Journal of Fertilizers, 17*(6), 544–554.

Pesticide Action Network. (2017). *Alternative methods in weed management to the use of glyphosate and other herbicides*. file:///C:/Users/youhe/Downloads/kdoc_o_00042_01.pdf

Ponisio, L. C., M'Gonigle, L. K., Mace, K. C., Palomino, J., de Valpine, P., & Kremen, C. (2015). Diversification practices reduce organic to conventional yield gap. *Proceedings of Royal Society B, 282*, 20141396. https://doi.org/10.1098/rspb.2014.1396

Quiedeville, S., Barjolle, D., & Mouret, J. C. (2017). Ex-post evaluation of the impacts of the science-based research and innovation program: A new method applied in the case of farmers' transition to organic production in the Camargue. *Journal of Innovation Economics & Management, 1*(22), 145–170.

Rahmann, G., Ardakani, M. R., & Barberi, P. (2017). Organic Agriculture 3.0 is innovation with research. *Organic Agriculture, 7*(3), 169–197.

Ramesh, P., Panwar, N. R., Singh, A. B., Ramana, S., Yadav, S. K., Shrivastava, R., & Subba Rao, A. (2010). Status of organic farming in India. *Current Science, 98*(9), 1190–1194.

Reganold, J. P., & Wachter, J. M. (2016). Organic agriculture in the twenty-first century. *National Plants, 2*, 15221.

Ryffel, G. I. (2017). Have a dream: Organic movements include gene manipulation to improve sustainable farming. *Sustainability, 9*, 392.

Sanders, J., & Hess, J. (2019). Leistungen des ökologischen Landbaus fur Umwelt und Gesellschaft. 2. überarbeitete und ergänzte Auflage. *Thünen Report 65*. Johann Heinrich von Thünen-Institute.

Siderer, Y., Maquet, A., & Anklam, B. (2005). Need for research to support consumer confidence in the growing organic food market. *Trends in Food Science and Technology, 16*, 332–343.

Slagboom, M., Kargo, M., Edwards, D., Sørensen, A. C., Thomasen, J. R., & Hjortø, L. (2016). Organic dairy farmers put more emphasis on production traits than conventional farmers. *Journal of Dairy Science, 99*(12), 9845–9856.

Stagnari, F. (2017). Multiple benefits of legumes for agriculture sustainability: An overview. *Chemical and Biological Technologies in Agriculture*, *4*(2). https://doi.org/10.1186/s40538-016-0085-1

Stein-Bachinger, B., Haub, A., & Gottwald, F. (2019). Biodiversität'. In J. Sanders & J. Hess. *Leistungen des ökologischen Landbaus für Umwelt und Gesellschaft* (pp. 97–131). Thünen Institut. https://literatur.thuenen.de/digbib_extern/dn060722.pdf

Tsvetkov., I., Georgieva, L., & Tsvetkova, D. (2017). Benefits of the mycorrhizal fungi *Glomus* spp. for grapevine nutrient uptake, biocontrol and microbial ecology. *Journal of Mountain Agriculture on the Balkans*, *20*(1), 227–250.

Van Wagenberg, C. P. A., de Haas, Y., Hogeveen, H., van Krimpen, M. M., Meuwissen, M. P. M., van Middelaar, C. E., & Rodenburg, T. B. (2017). Animal board invited review: Comparing conventional and organic livestock production systems on different aspects of sustainability. *Animals*, *11*(10), 1839–1851.

Vogeler, I., Ronaldo, V., & Rogerio, C. (2017). Potential benefits of diverse pasture swards for sheep and beef farming. *Agricultural Systems, Elsevier*, *154*(C), 78–89.

Wang, M., Xu, S., Yang, J., Xu, L., Yu, Q., & Xie, X. (2021). The effect of organic and conventional management practices on soil macropore structure in greenhouse vegetable production. *European Journal of Soil Science*, *72*(5), 2133–2149.

Wickson, F., Binimelis, R., & Herrero, A. (2016). Should organic agriculture maintain its opposition to GM? New techniques writing the same old story. *Sustainability*, *12*(8), 1105.

Wilhelm, B. (2016). Funktionen und Regulierung von Ackerwildkräutern. In B. Freyer (Ed.), *Ökologischer Landbau: Grundlagen, Wissensstand und Herausforderungen* (pp. 486–503). Wiley.

Wilkins, R. J. (2008). Eco-efficient approaches to land management: A case for increased integration of crop and animal production systems. *Philosophical Transactions of the Royal Society B: Biological Sciences*, *363*(1491), 517–525.

Willer, H., & Lernoud, J. (2015). The world of organic agriculture. *Statistics and emerging trends. FiBLIFOAM report*. Research Institute of Organic Agriculture (FiBL), Frick and IFOAM-Organics International.

Wu, Q. S. (2017). *Arbuscular mycorrhizas and stress tolerance of plants*. Springer.

Yadav, D. S., Kumar, V., & Yadav, V. (2009). Effect of organic farming on productivity, soil health and economics of rice (*Oryza sativa*)–wheat (*Triticum aestivum*) system. *Indian Journal of Agronomy*, *54*(3), 267–271.

Yang, Y. Y., Song, H., & Scheller, A. (2015). Community structure of arbuscular mycorrhizal fungi associated with *Robinia pseudoacacia* in uncontaminated and heavy metal contaminated soils. *Soil Biology and Biochemistry*, *86*, 146–158.

Yue, Q., Xu, X., & Hillier, J. (2017). Mitigating greenhouse gas emissions in agriculture: From farm production to food consumption. *Journal of Cleaner Production*, *149*, 1011–1019.

Zou, M., Lu, C., & Zhang, S. (2017). Epigenetic map and genetic map basis of complex traits in cassava population. *Scientific Reports*, *7*, 41232.

2 Marketing Organic Products and Organic Certification

Bisma Rashi and Zaffar Bashir

2.1 INTRODUCTION

One of the many tactics used to achieve the goals of sustainable agriculture is organic farming. Many of the methods utilised in organic agriculture are not novel to other systems of agriculture, such as the conventional farming methods employed in historic countries like India. Examples of these practises include intercropping, mulching, and integrating livestock and crops. However, soil health is the primary goal of organic farming, which is based on a number of laws and certification schemes that forbid the use of almost all synthetic inputs. Numerous studies have demonstrated the detrimental impacts of contemporary farming methods on the environment, farm productivity, and the health of all living things. Although the term "organic farming" has many definitions, most people concur that the United States Department of Agriculture's (USDA) is the most thorough and rigorous. "A system that is designed and maintained to produce agricultural products by the use of methods and substances that maintain the integrity of organic agricultural products until they reach the consumer," is how it is defined.

Farmers in developed and developing nations alike can trade using organic agriculture. Few farmers will take risks because of a lack of financial support, technical information, and market information altering their production process. Therefore, having current understanding of possibilities (market requirements) and trends of the organic market is essential for significant stakeholders (e.g., NGOs, farmer organisations, traders, exporters, etc.) who promote organic farming in developing countries. India serves as one example, as the nation is home to a substantial number of small-scale farmers who practise conventional farming practices and minimally rely on agricultural inputs. Farmer support in these areas is provided by NGOs that promote organic farming and other groups. As organic agriculture has come to be recognised as a viable option for rural development, connections have been forged across all continents. Even though there is still scepticism surrounding the movement (Kirchmann & Ryan, 2004; Trewavas, 2001; Trewavas, 2004), organic farming has strong marketing appeal, positive growth projections, and suggestions that it is now an "industry" rather than just a "movement" have been made (Cornish & Stewart, 2002).

It is believed that our ancestors practiced organic farming in the distant past. Food products produced without using chemical fertilisers or pesticides—both of which

DOI: 10.1201/9781003491460-2

are detrimental to human health—are the primary components of organic farming. Going back to tradition, organic farming was a pillar of the great Indian civilisation until British intervention into the nation. India subsequently implemented chemical farming as a result of the serious food shortage that occurred there in the early 1960s. The Indian government had to acquire food from other countries as a result. The Green Revolution got underway. In terms of health, people started to gradually get used to eating organic food. Many farmers modified their farming practices in response to the rising demand for organic products both domestically and globally. Recent data for the financial year 2022 shows that India consumed over 2% of the net area, and the global market for food that is organic is sizable. In the US and the United Kingdom, where it has been widely practiced for an extended period of time, organic agriculture is well known. Asia is regarded as the marketplace of the future, per a recent study. India leads the world in organic farming, according to the 2018 World of Organic Agricultural Report. It claims that for the previous five years, the Indian government has been actively supporting organic farming. Among the many projects, the Indian government has worked on the well-known Mission Organic Value Chain Development for the North Eastern Region (MOVCDNER) and Paramparagat Krishi Vikas Yojana (PKVY).

The PKVY, a thorough section included in the National Mission of Sustainable Agriculture, focuses on Soil Health Management. Organic farming is encouraged under PKVY through PGS certification and the implementation of the organic compounds village by cluster concept.

The states of Arunachal Pradesh, Assam, Manipur, Meghalaya, Mizoram, Nagaland, Sikkim, and Tripura participated in an initiative called Organic Value Chain Development for North Eastern Region. The strategy aims to foster certified organic farming in an economic mode to facilitate the interaction between producers and consumers and encourage the growth of value along the whole chain. The first steps in this process are the components, seedlings, certification, and building of facilities for operations related to acquiring, integrating, manufacturing, marketing, and branding.

Chemical fertilisers have an impact on organic farming, and agrochemical-based cultivation has been connected to water contaminants (Misra et al., 1991). The impact of organic farming on production levels, high costs, and inadequate funding is the main focus of the study (Elayaraja & Vijai, 2020a, 2020b). The regions with the largest percentages of organic agricultural land, according to current trends, are Oceania (37%), Europe (24%), and Latin America (20%). Oceania and Europe had the highest percentages, according to the comparison. India is presently ranked tenth out among the top ten nations in the world in regard to the total amount of land suitable for organic farming. Ten percent of the total certified area—0.5 million hectares—is cultivable, with the remaining 90% being wild and forested areas used to gather small-scale forest products. APEDA (2013) reports that 5.21 million hectares are fully certified organic. India produced 1.34 million metric tonnes of all kinds of food goods types, such as fruits, oilseeds, sugarcane, cotton, basmati rice, pulses, tea, spices, coffee, and their value-added products. In addition to goods connected to the edible sector, other goods are produced, such as nutrient-dense foods and organically woven cotton (APEDA, 2013). Out of all the states, Madhya Pradesh has the largest percentage of certified organic land, followed by Uttar Pradesh and Rajasthan. Madhya Pradesh (1.1 MHA, or 52%) has the largest proportion of land in India devoted to organic farming, followed by Maharashtra (0.96 MHA, or 33.6%), Orissa (0.67 MHA, or 9.7%), Uttarakhand, and Sikkim. To produce organic food, farmers employ a range of methods. It is regarded as one of the best strategies for implementing eco-friendly behaviours. In 162 countries, organic farming is conducted. Germany's regime is one of the leading nations worldwide for organic farming, and its products rule the international market. A few Asian countries with developed economies are Singapore, Hong Kong, and Japan. Emerging economies such as India are becoming more visible on the global scene and are being given more consideration. Another claim is that Denmark and Switzerland have the highest per-capita densities. India is a developing country with 21 agro-ecological zones, which leads to an excess of food products that are organic. The respondents claimed that India's temperate climate and rich soil are ideal for organic farming.

In today's business environment, ecological and social concerns are important marketing factors. Earth's temperature is rising mostly due to global warming, which is caused by burning coal, petroleum, and other types of air pollutants. The Intergovernmental Panel on Climate Change (IPCC) claims that there will probably be a 1.1–6.4 degree Celsius increase in global temperature "between" 1990 and 2100. As a result, there is an increasing global consumer interest in environmental protection, which has given rise to organic marketing—the practise of offering environmentally and socially conscious products and services. Societies are starting to take environmental responsibility more seriously. In order to change their behaviour towards societal concerns, business organisations do so. Numerous governments across the globe have recognised the significance of organic marketing initiatives and are working to enact regulations that will safeguard their nations' natural resources, public health, and sustainability for future generations. Therefore, implementing organic marketing in daily life is the social duty of all parties involved, including businesses, organisations, retail retailers, intermediaries, cooperative farming societies, and consumers. They try to transition from the traditional approach to the organic approach, which is essential for both the environment and society to keep developing and growing.

The agricultural sector is, as everyone knows, the foundation of India's economic growth. Low prices for crops, inadequate storage facilities, poor supply chain management, expensive chemical fertiliser costs, labour shortages, the movement of minimal workers from rural to urban areas, a lack of government assistance for agriculture, trouble repaying loans, and other issues are just a few of the difficulties that farmers face in the agricultural sector. These issues drive farmers to consider suicide. Farmers begin to lose faith in agriculture as a result, looking instead to the industrial sectors for a living. The ingestion of food products containing hazardous pesticides, fertilisers, and preservative chemicals represents a serious risk to human health. Customers who are aware of this are choosing organic food items over conventional ones because of their long-term benefits. Organic food products fall into a number of categories, including pulses and food grains, fruits and vegetables, dairy products, drinks, sweeteners, honey, jaggery, meat, fish, and poultry products. The marketplace for food that is organic is anticipated to expand at a 25% annually between 2015 and 2022, as per TechSci research report. But organic farming gives farmers an opportunity to overcome these problems. These days, maintaining a healthy lifestyle while protecting the environment and people's health requires using organic products. In addition to their obvious environmental benefits, organic products have special benefits for manufacturers, consumers, and the surroundings. The advantages help the world market for organic food, which is expanding quickly in both developed and developing countries. Eco-friendly farming methods and natural resources are used in the production of organic food products. This farming method's main goal is to cultivate the land and enhance agricultural products in a way that releases carbon dioxide into the atmosphere while maintaining the soil's vitality of vitamins to agricultural products for raised production that is sustainable in a pollution-free, ecologically friendly environment via biological resources, including crop, animal, and farm wastes, as well as aquatic waste products, in addition to beneficial microbes (biofertilisers). From harvesting to shipping goods from the farm to markets, organic farming follows less-pesticide, less-chemical, less genetically modified organisms, and preservative-free practices at every turn. The field of organic agriculture has seen tremendous growth in recent years. During the subsequent 20 years, the number of organic growers increased dramatically from about 0.3 million in 1998 to 10 million hectares of total production. An estimated 2.8 million organic growers farmed 71.5 million hectares in 2018. The two continents with the highest numbers of organic producers are Asia (1.4 million) and Africa (850,000); North America, Europe, Latin America and the Caribbean, and Oceania follow in order of importance. Oceania has 35.9 million hectares under organic cultivation, the most of any continent despite having the fewest organic producers (FiBL and IFOAM—Organics International, 2020). Even with its rapid expansion, only 1.5% of all agricultural land globally is used for organic farming. The possibility of future growth in organic agriculture is increased by this small area of land devoted to organic farming. For instance, according to FiBL and IFOAM—Organics International 2020, 1.25 million hectares of organic agriculture were added to the European continent in just one year, representing an 8.7% growth. An additional sign of the expansion of organic agriculture is the rising demand for organic food. In accordance, the United States contributes to the world with regard of economic utilisation of organic food, coming in second with 8 billion euros followed by Germany with 4 billion euros (Pandey et al., 2019). According to the FIBL and IFOAM—Organics International report (2020), China ranked fourth in the world in 2018. Organic agriculture is described as "a method for farming that sustains the good condition of soils, ecosystems, and people" by IFOAM—Organics International (2008). Rather than utilising ingredients that are harmful, it depends on ecological processes, biological diversity, and cycles that are tailored to the environment. Furthermore, IFOAM—Organics International (2008) promoted organic agriculture as a blend of "innovation, tradition, and research that will enhance our common surroundings and encourage equitable interactions as well as a positive standard of their lives for all involved."

2.2 FARMING IN AN ORGANIC WAY

According to the USDA study team, organic farming is a system that tries to minimise the application of chemical inputs (including fertilisers as well as pesticides, hormones, additives to feed, etc.) and instead relies as much as possible on rotations of crops, residue from crops, animal manure, elsewhere organic debris, mineral-grade rock additives, as well as biological systems for nutrient and plant protection mobilisation. "Organic agriculture is holistic production management system which promotes and enhances agro eco system health; including biodiversity, biological cycles, and soil biological activity" is a commonly accepted definition of organic agriculture. Enhancing soil biological activity, increasing the biological diversity, and preserving the soil's long-term fertility are the objectives of an organic agriculture system. It also aims to recycle plant and animal wastes and restock the earth's soil with nutrients while consuming fewer non-renewable resources and managing agricultural products with care, using specific processing techniques to maintain the product's vital characteristics and organic compounds reliability throughout its life. This can be established by any existing farm via a transformation period, the duration which is determined by site-specific elements like the land's past, the type of farm animals to be brought up, and the crops that need to be grown.

2.2.1 Environment and People Gain Long-Term Benefits from Organic Farming

Encouraging organic farming allows farmers to reap the rewards of increased crop yields, increased energy, and lower emissions of greenhouse gases and consequences of global warming; conserve water; localise sales to cut down on long-distance transportation costs; do away with pesticides; improve nutrition; save organic seeds; create jobs; and move towards sustainability.

2.2.2 Organic Agriculture Benefits

- It raises the soil's fertility over time.
- It manages illnesses and pests without endangering the environment.
- It guarantees the safety and cleanliness of water.
- It assists farmers in making better use of their current resources, thereby reducing the amount of money they need to spend on inputs.
- It produces superior crops to be offered for sale at a fair price, as well as wholesome food and animal feed.

2.2.3 Organic Product: What Is It?

- No chemicals are used in the production, processing, packing, or sale of organic products. The category of organic products includes the following food categories: Veggies and fruits; dairy and milk merchandise; cereals and pulses; meats, poultry, eggs, and seafood; honey; baked goods; baby food; medicines and herbs; beverages; desserts; and other items.
- Organic farming combines cutting-edge scientific research and technology with practical, time-tested farming techniques. It eliminates the need to revert to farming practises from hundreds of years ago and uses organic pesticides and manures for fruits and vegetables as well as animals.
- Converting conventional land to organic requires a six-year process to meet National Organic Standards Board (NOSB) requirements.

- Because growing food organically has real costs associated with it, it seems that organic food costs more than regular food grown conventionally (higher labour, seed, fertiliser, and manure costs). No government subsidies are given as needed.

2.2.4 Features of an Organic Product

- Naturally grown products (organic farming).
- Items that are biodegradable, recyclable, and reusable.
- Items made of natural materials.
- Non-toxic products made with recycled materials.
- Product ingredients listed under authorised organic.
- Items that don't harm the environment.
- Goods which won't be put through animal testing.
- Items with environmentally friendly packaging, such as refillable or reusable containers.

2.3 GUIDELINES FOR ORGANIC PRODUCTS

The ecology principle of sustainability depends on residing ecological systems and agriculture. The equitable principle concerns shared surroundings and life prospects. A care philosophy emphasises the need to practise organic agriculture responsibly and with caution in order to preserve the environment for present and future generations; the one and indivisible planets, animals, humans, and soil all benefit from this principle of health.

It's crucial to read the labels before deciding to purchase organic goods. Using the USDA Organic seal is not mandatory. The following symbols in fact are found on labels:

a) "100% organic" denotes 100% organic cultivation.
b) "Organic" denotes ingredients that are a minimum of 95% organic.
c) "Made with organic ingredients" indicates that the substances being used are at least 70% organic.

2.3.1 Organic Products' Benefits

In actuality, non-organic farmers save over 25% on production costs. In just five years, it can increase crop yields five times while reducing soil erosion by up to 50%. Consisting of natural ingredients devoid of artificial flavours, preservatives, and hazardous chemicals, organic food may lower the chance of cancer, heart attacks, and strokes.

2.3.2 The Reasons for Utilising Organic Products

An organic product is free of chemicals, highly nutritious, naturally occurring, kind to the environment and animals, and contains no genetically modified organisms

(GMOs). It has been approved by competent boards. It is of superior quality, tastes better, and feels fresher.

2.3.3 Justifications for the Higher Price of Organic Products

Taking into account social, political, and environmental factors, the price of organic goods is actually quite reasonable; however, we must ascertain who and what bears the additional cost of production. Nearly everywhere in the world provides subsidies for conventional farming. Conversely, organic farming is heavily favoured. Aside from this, the biggest expenses associated with organic farming are those related to consultants, certification, inspection, etc. It should have certification from affiliated bodies with the NPOP, USDA, and NOSB.

2.4 THE ORGANIC FOOD SECTOR

India is currently at the cusp of an organic revolution, and although the country's organic food industry is still in its infancy, it has grown steadily and quickly over the last several years. Due to changes in consumer spending and behaviour, the nation's expanding global organic food market is anticipated to expand at the fastest rate.

2.4.1 Products and Food Category: Organic

The organic food categories and products are illustrated in Table 2.1.

2.5 CONSCIOUSNESS ABOUT HEALTH

Health consciousness has a big influence on consumer attitudes (Liu, 2007). The main reason consumers buy organic food is because of its health benefits (Shepherd et al., 2005). Numerous studies demonstrate that consumers' propensity to purchase

TABLE 2.1
Organic Food Categories and Items

Category of Organic Foods	Organic Food Items
Cereals made organic	Corn, rice, wheat, or maize
Whole grains and pulses	Black and red grammes
Fruit organic produce	Walnut, cashew nut, orange, pineapple, passion fruit, banana, and mango
Oils and seeds with organic content	Groundnut, castor, sunflower, mustard, cotton seed, and soybean
Vegetable organic produce	Onion, tomato, garlic, potato, and Brinjal
Spices and herbs made organic	Mace, chilli, peppermint, cardamom, turmeric, black pepper, white pepper, tamarind, ginger, clove, cinnamon, nutmeg, and amla
Additional	Cotton, textiles, sugar, tea, coffee, and jiggery

organic foods is significantly influenced by health-related factors (Grossman, 1972; Kim & Chung, 2011). Customers may view the products they purchase as investments in themselves as individuals because one of the primary factors impacting these individuals may be the decline in their health (Grossman, 1972). According to Bourn and Prescott (Bourn & Prescott, 2002), the higher nutritional content of organic foods gives them a competitive advantage over conventional foods.

2.6 CUSTOMER EXPERTISE

The theory of reasoned action (TRA), which examines the motivational influences on consumer behaviour, advances our knowledge of how consumer behaviour develops (Fishbein & Ajzen, 1975). A foundation for forecasting consumer attitudes and behaviour is provided by TRA (Sheppard et al., 1988). Liu (2007) provides additional evidence that the best theory for forecasting consumer behaviour regarding organic foods is TRA. Customers want to fulfil their needs and desires and be informed about the products they are purchasing. As a result, knowledge is crucial in influencing how consumers behave when purchasing food. According to Sapp (Sapp, 1991; Richa & Deepak, 2016), gaining knowledge necessitates a cognitive learning process. Depending on their level of expertise, consumers have different purchase intentions (Chiou, 1998). It is impossible to divorce consumers' purchases of organic goods from their awareness and comprehension of organic foods (Gotschi et al., 2010; Saleki et al., 2012). According to recent research, consumers worldwide lack the awareness and knowledge that European consumers have regarding organic food. Because the organic food market is more developed than the others, there is a greater awareness of organic food (Compagnoni et al., 2000; Demeritt, 2002; Paul & Rana, 2012). Research has also revealed that there are significant differences in consumers' views of what constitutes "organic." In one study, for instance, participants believed that produce from organic farms was grown without the use of growth regulators, fertilisers, or pesticides (Jolly et al., 1989). Besides this, participants within a related UK research conducted by Hutchings and Greenhalgh (Hutchins & Greenhalgh, 1997) believed that "organic" farming was chemical-free and naturally grown. Respondents also believed that organic food is not produced in an intensive manner. An understanding of and familiarity with organic foods are crucial when making purchasing decisions as consumers. Smith and Paladino (2010) investigated how customers' decisions to buy organic food were positively impacted by being informed regarding social and environmental issues, according to research on the variables influencing consumer choices (Smith & Paladino, 2010). Still, as we can see, consumer awareness of organic foods varies greatly. Though they may believe that food that is organic is natural, and healthy, people may hold this belief because they believe organic foods to be free of chemical pesticides and fertilisers.

2.7 FOOD THAT IS ORGANICALLY AVAILABLE

One aspect that promotes the purchase of organic foods is availability (Davies et al., 1995). Makatouni (Makatouni, 2002) brought attention to the possibility that limited access to organic foods could serve as a deterrent to their consumption. In a 2005 study, Tarkiainen and Sundqvist (Tarkiainen & Sundqvist, 2005) demonstrated that the availability of organic foods positively impacted consumers' decisions to buy. According to the Young et al. Survey (Young et al., 2010), consumers favour products that are simple to obtain. They are therefore hesitant to invest time in looking for organic products. Still, organic food has been gaining popularity lately, and grocery stores across the country are carrying it on their shelves. Because of improved advertisement of such goods by major supermarkets and specialty shops, consumers can now find organic foods more readily (Dettmann & Dimitri, 2007).

2.8 THE PURPOSE OF BUYING AND THE ACT OF BUYING BEHAVIOUR

The concept of planned behaviour states that a person's response is influenced by their intentions and their perception of behavioural control. Sheppard et al.'s (1988) study demonstrated the relationship between choices and actions in a number of purchasing-related categories. Ajzen (1991) asserts that intentions or willingness can precisely forecast actual purchasing behaviour. Studies conducted by Tanner and Wolfing (2003) and Vermeir and Verbeke (2008) revealed discrepancies between the actual and stated purchasing behaviour of clients who displayed positive attitudes. Hughner et al. (2007) discovered that while the majority of consumers thought buying organic food was a good idea, very few actually did. Researchers think there is a relationship between attitudes and behaviours based on the previously mentioned data. This is consistent with the work that Wheale and Hinton (2007) conducted. Opinions regarding foods produced organically may have a positive and significant impact on consumer behaviour (Voon et al., 2011). This suggests that buying food that is organic is a consequence of having a particular shopping goal. The discrepancy between consumers' actual buying habits for foods that are organic and their favourable attitudes towards them is known as the attitude-behaviour gap. This disparity implies that favourable reviews of organic goods may not always translate into a purchase. This disparity may be influenced by numerous factors. There are many factors that can cause a disparity among what clients think that they are going to spend money on and what they actually do, including price, availability, and social influence.

2.9 SOCIODEMOGRAPHIC ELEMENTS

Beyond attitudes, there are other factors that affect behaviour. Voon et al. (2011) discovered, for instance, that sociodemographic variables have an impact on consumers' purchase decisions. Gender is one important factor. According to Lockie et al. (2004), women tend to have more favourable opinions about organic foods than do men. According to Stobbelaar et al. (2007), teenage girls are more inclined than males to prefer organic products. It has been shown that purchasing organic food is influenced by age. Age-related health concerns may make older people more likely to purchase organic food, according to research by Misra et al. (1991). However, younger consumers are more willing to pay prices that are more than 6% higher to ensure that food products are free of pesticides, according to research results by Cranfield and

Magnusson (2003). Furthermore, Rimal et al. (2005) discovered that older people are less likely than younger people to purchase organic foods. Nonetheless, because they think organic foods are more significant, women and younger people buy them (Govidnasamy & Italia, 1990). Income is another important aspect of customer demographics that is thought to have a significant impact on how frequently people purchase organic food. According to two studies by Govidnasamy and Italia (1990) as well as Maria et al. (2001), households with higher incomes are more likely to buy organic products. Similarly, Voon et al.'s (2011) study discovered a favourable correlation between purchasing organic food and household income. Additionally, moms with more disposable income in their 30s to 45s are more likely to purchase organic food (Dettmann & Dimitri, 2007). The consumption of organic food was found to positively correlate with education, according to studies by Cunningham (2002) and O'Donovan and McCarthy (2002). This is also supported by the research of Dettmann and Dimitri (2007). They discovered that people who are more educated are more likely to buy organic food. Aryal et al. (2009) also came to this conclusion. They provided an example of how education can also have an impact on consumers' decisions to buy organic products.

2.10 ORGANIC CERTIFICATION

Manufacturers of organic agricultural products, including food, must go through the certification process. Generally speaking, retailers, restaurants, farmers, food processors, seed suppliers, and other businesses directly involved in the food production industry are all eligible to become certified. While requirements vary by country, they typically include a collection of production guidelines that cover the following areas: growing, processing, packing, storing, and shipping:

- Staying away from synthetic chemicals and genetically modified organisms (fertiliser, chemical pesticides, antibiotics, food additives, etc.).
- Using agricultural land that has not been treated with chemicals for an extended length of time (often three years or more).
- Maintaining meticulous written records of manufacturing and sales (an audit trail); physically separating non-certified from organic items; and carrying out regular on-site inspections.

Within certain nations, the government regulates certification, and it is against the law to use the term "organic" for commercial purposes. The same government rules pertaining to agriculture, food safety, and other matters that affect non-certified producers also apply to certified organic producers.

The most common tool for evaluating regulatory compliance and assuring consumers that a process, product, or service complies with standards is certification (Nassar, 2003). FAO (2012) states that organic certification indicates respect for particular production and setup guidelines. Verification and standardisation form the cornerstone of various mechanisms for certification. There is agreement on what constitutes an organic claim on a product based on the parameters outlined by the rules. FAO (2012) and IFOAM—Organics International 2019 define verification as the truthfulness test that is contingent upon the fulfillment of the relevant standards. Direct consumer-producer interactions, techniques attesting to group production, third-party stores that purchase directly from producers, and, fourth, businesses hired to conduct this kind of verification but unrelated to the manufacturing, sorting, or transportation process are some of the methods that can be used for verification (FAO and IFOAM—Organics International, 2019). The packaging of goods whose conformance has been assessed and endorsed by certified producers may bear a seal, enabling customers to recognise products that adhere to organic standards. Certification procedures ensure that producers and processors can positively differentiate their products while safeguarding consumers against potential fraud (FAO, 2012). Customers pay more attention to certifications and their seals, which have come to represent quality and ethics, not just in North America and Europe but on every other continent as well (Raynolds et al., 2007). The scope of action established by ISO 17065 and ISO 19011, which define agricultural and livestock product and food certifiers, must be followed. Both adhere to the normative rules of Codex Alimentarius, which establish the requirements for best practices in ensuring food safety. According to Botonaki et al. (2006), conformity assessment programmes are crucial for guaranteeing the quality of food and fostering more equitable and environmentally balanced social relationships. TPC stands for third-party certification, the most widely used certification methodology (Hatanaka et al., 2005; Kaufmann & Vogl, 2018). According to ISO 17065, TPC's primary attributes are competence, independence or impartiality, and accreditation. Using this method, producers pay independent, private certifiers to attest to their organic compliance. According to Fouilleux and Loconto (2016), there were 85 countries having 549 TPC certifiers registered by 2012. Some TPC features restrict smallholder farmers' ability to be certified, even with the regulatory structure that lends the TPC legitimacy (FAO, 2007). These characteristics are more pronounced in the Global South. TPC's three primary obstacles are its concentration on lengthy distribution routes, frequently for export, the high level of standardisation, and the costs associated with administrative overhead (Kaufmann & Vogl, 2018; Smith & Grasseni, 2020). To incorporate smallholder farmers in TPC programmes, IFOAM—Organics International created the Internal Control System (ICS), a form of group certification. This is justified by the fact that farmers can split expenses and engage experts to assist with administrative and technical problems if they complete the certification procedure in groups. According to Cavallet et al. (2018), the primary target market for ICS is smallholder farmers looking for export markets. Projections from FiBL and IFOAM—Organics International 2020 indicate that about all certified goods are cocoa and coffee, with an estimated 2.6 million smallholder farmers having received ICS certification. ICS performance and compliance are evaluated by a recognised certifying organisation via TPC. An alternate technique for conformance assessment has grown in acceptance over the past ten years. PGS is commonly used for providing national markets, interruptions, and to establish relationships between farmers and consumers. It puts manufacturers, customers, and participants at the helm of the certification process (Cavallet et al., 2018). As stated on page 15, Smith and Grasseni (2020) observed that PGS "is considered a means for integrating the abilities and experience of food suppliers with the various skills and abilities of buyers, to connect each other's information and networks together." As of October 2019, 223 PGSs were sent throughout 76 economies, with engagement from 567,142 producers, most of whom were based in Asia, FiBL and IFOAM—Organics International (2020) state as much. The 8,168% increase in PGS membership over the ten-year period from 2010 to 2019 is noteworthy. Following Brazil, Bolivia, Chile, Costa Rica, Ecuador, Mexico, Paraguay, Uruguay, FiBL, and IFOAM—Organics International 2020 recognised PGS as a national organic quality assurance system. Brazil was the first nation to formally recognise PGS as being similar to TPC, according to Hirata et al. (2019). The two primary features of this system, according to Smith and Grasseni (2020) and Hirata et al. (2019), are solidarity and control over society. A PGS's guiding principles are generally social supervision, accountability, and shared responsibility; however, the actions of the government and other organisations influence the structure and operations of a PGS (Turner et al., 2004; Hirata et al., 2019; Smith & Grasseni, 2020). The subject of organic accreditation has gained prominence in the discourse around the structure of sustainable agrifood systems. For this reason, it is imperative to understand the protocols that certifying bodies for organic production systems adhere to globally.

2.10.1 The Reason for Certification

Organic certification helps to meet the expanding demand for organic food worldwide. Its goals are to guard against fraud and ensure quality. Certification is a means for organic producers to identify suppliers of goods that have been authorised intended for application in approved operations. "Certified organic" as is just like "low fat," "100% whole wheat," or "no artificial preservatives" in terms of providing consumers with a sense of product assurance. The primary goals of certification are to control and streamline the consumer sale of organic goods. Each certification body has its own service marks that it uses to brand itself to consumers. For example, in order to provide producers with a competitive advantage, a certifier might highlight the high degree of consumer acceptance of its logo. The majority of certifying organisations use organic standards that satisfy the minimal requirements set forth by the national government.

2.10.2 The Procedure for Certification

To get their farm certified, farmers typically have to complete several extra tasks on top of their regular farming duties:

- Study the officially recognised organic standards, which outline exactly what may and cannot be utilised for every facet of farming, such as transportation, storage, and sales.
- Regulation: Farm structures and production practices need to follow the guidelines; this could entail, among other things, changing the structures or locating new suppliers.
- A substantial amount of evidence is required, encompassing the findings of water and soil analysis, as well as history of the farm and its current configuration.

- Planning: Every aspect from seed to sale must be covered in a written annual production plan, which should include information on seed suppliers, crop and field locations, fertilisation and pest control procedures, harvesting techniques, storage locations, and more.
- Inspection: Included in the process are an oral interview, a written record review, and a yearly physical on-farm inspection requirement.
- Fee: Growers are required to pay a fee to the certifying body in order to facilitate annual quality surveillance and a mark that is recognised as a mark of quality in the market.
- Record-keeping: Daily written records of all farming and marketing operations must be kept on file and accessible for review at all times.

In addition, specific tests (such as tests on soil, water, or plant tissue) could be asked for, as well as short-notice or surprise inspections.

For a farm to be certified for the first time, the soil needs to meet the basic requirements of not having been exposed to prohibited materials (synthetic compounds, and so on) for an array of years. During this time—usually three years—a traditional farm is required to follow organic guidelines. We call this the state of being in transition. Crops in transition are not regarded as entirely organic. A farm that is currently producing without the use of chemicals can get certified right away.

The certification process is similar for non-farm operations. The emphasis is on handling and processing conditions, ingredients, and other inputs. A transportation company would have to provide information about how its cars, warehouses, shipping containers, and other equipment are used and maintained. A restaurant would have an inspection of its facilities and have the organic certification of its suppliers confirmed.

2.10.3 Authenticity and Product Labeling

In today's consumer market, being able to label a food product "organic" offers a significant marketing benefit. The goal of certification is to safeguard customers from term abuse and facilitate the purchase of organic products. But the organic labeling that certification itself makes possible typically needs to be explained. Three levels of organics are defined by organic legislation in many countries. Products bearing the label "100% organic" are created solely using certified organic materials and techniques. Goods labeled as "organic" must have a minimum of 95% organic content. Both could exhibit an organic seal as well. "Made with organic ingredients" designates a third category, which must have at least 70% organic ingredients. The certifying authority's logo may also be displayed on products to indicate their approval. Products that have less than 70% organic content are only allowed to list this information as an ingredient; they cannot be advertised to consumers. Figure 2.1 shows different organic logos for all nations.

2.10.4 Worldwide Certification

Governments in certain nations create and regulate organic standards. The US, the EU, and Japan have strict organic laws that limit referring to certified producers as

FIGURE 2.1 Organic logos for a few of the nations.

"organic" only. There might or might not be government regulations in nations without organic laws, and private businesses and non-profits handle certification.

The EU-Eco-regulation of 1992 gave member states of the EU comprehensive organic legislation. The national level is where certification is managed.

- The Soil Association and Organic Farmers and Growers are the two largest organisations in the UK that manage organic certification. Regulations are imposed on all certification bodies by the UK's Organic Food Standards Register (UKROFS), which is governed by EU legislation.
- In Sweden, the private company KRAV is in charge of organic certification.

October 2002 saw the establishment of the National Organic Program (NOP), a federal law in the United States. It limits the implementation of the term "organic" to producers who are certified organic, excluding producers who sell less than $5,000 a year; these growers are excluded from formal application processes but must nonetheless conform and be subject to a documents audit upon request. State, nonprofit, and private organisations have been given permission by the US Department of Agriculture (USDA) to oversee certification.

A national organic standard has been published by the Canadian government, but it is merely a guideline; legislation is currently being developed. Organisations in the private sector offer certification.

In April 2001, the Japanese Agricultural Standard (JAS) became a legally binding decree in Japan. In November 2005, the Ministry of Agriculture made updates to this and required all JAS certifiers to reapply for accreditation.

Since there are no national criteria for organic goods in Australia, the agency in charge of overseeing organic certification is the Australian Quarantine and Inspection Service (AQIS). Since the government only becomes engaged in certifying organic products when a product that is organic is exported, AQIS now acts as the primary certification agency.

The retail public is most impacted by this, despite the lack of a framework for policing the promotion of organic products marketed in Australia. Commercial purchasers that view this as an issue have accepted the global export network as the accepted norm and are prepared to pay a premium for produce sourced from farms recognised by national programmes. From 2,345 accredited operators in 2004 to seven AQIS-approved certifying organisations in 2006, the number of organisations that can issue Organic Produce Certificates increased. Japan is Australia's largest supplier of organic vegetables by weight, making up 33.59% of all imports into the country. Second place goes to the UK (17.51%), then to France (10.51%), and New Zealand (10.21%). The most well-known organic chemicals farmers' association in the nation, Biological Farmers Australia, is the parent organisation of Australian Certified Organic, the largest certifier of organic items in the nation.

The two standards that are recognised in China by the China Green Food Development Center are AA and A. The latter is stricter as the former does allow limited usage of synthetic chemicals for agriculture.

The Ministry of Commerce's Agricultural Processed Foods Export Development Authority (APEDA) is in charge of overseeing organic certification for export in

India. There are currently no national guidelines in India for organic produce. Eleven certification bodies are currently permitted to carry out the National Programme for Organic Production's (NPOP) 13 certification processes. This mainly affects the retail public, even though there is no system in place to monitor the labeling of organic produce sold within India. Commercial purchasers who find this problematic simply come to terms with the export system as the accepted norm and are prepared to pay more for goods produced by NPOP-certified farms.

Global equivalency negotiations are presently underway, and agreements have already been made to standardise certification between countries, facilitating trade. The International Federation of Organic Agriculture Movements (IFOAM), Ecocert, and the Organic Crop Improvement Association (OCIA) certifying agencies are members of the global organisation. Organic products meant for export are often approved by organisations in importing nations in the absence of formal agreements between states; these agencies may set up long-term operations abroad specifically for this purpose. India's NPOP organic certification system was accorded equivalence by the European Union in 2006. Furthermore, it has been recognised for compliance assessment by USDA's NOP (Figure 2.1).

2.10.5 Certifications and Government Regulations

When certifying organic products in India, APEDA (Agricultural and Processed Food Products Export Development Authority) adheres to the NPOP. The purpose of NPOP was to establish national guidelines for organic goods. In 2000, the Government of India (GOI) established the National Standards of Organic Products (NSOP), which required all products labeled as "organic" to go through an obligatory inspection process and be certified by a nationally recognised certification authority. In support of the NPOP, the "India Organic" logo was unveiled on January 26, 2002.

2.11 EXAMINING AS WELL AS VERIFYING NATURALLY ORIGINAL PRODUCTS

The National Steering Committee of Organic Products (NSCOP), operating under the auspices of the Indian government's Ministry of Commerce, devised and executed the NPOP. NSCOP is in charge of creating national standards for organic products in addition to national accreditation rules and programmes. The International Federation of Organic Agriculture Movements (IFOAM) recommendations, the Codex Alimentarius Commission, and EU legislation served as the foundation for the development of the NPOP and NAP standards.

2.11.1 Organic Channels of Distribution

Between the initial creator and the final buyer, there are intermediaries such as wholesalers, retailers, and brokers. The term "middleman" describes those people or organisations that facilitate, negotiate, or sell while acting as agents and brokers, such as outlets, businesses, dealers, and brokers.

- **Broker:** A middleman whose role it is to connect sellers and buyers. The broker is involved in the finances and risk assessment but does not carry inventory.
- **Facilitator:** A middleman who helps with distribution but does not negotiate a purchase or sale or claim title to the goods.
- **Merchant:** A middleman who purchases, seises ownership of, and resells goods.
- **Producer:** A person, an organisation, a cooperative farming society, or a business that creates the products for the market.
- **Retailer:** A company that offers products to customers directly.
- **Sales Agent:** A middleman who looks for customers and bargains on behalf of producers.
- **Sales Force:** A team of individuals employed by the business to market its goods and services.
- **Distributors or Wholesalers:** A company that sells goods to customers who resell them or use them for other commercial purposes.

2.11.2 Segmenting the Organic Market

There are two categories within the Indian organic market.

- Organised Sector: This industry is supported by major brands and uses conventional business channels. In urban India, this became increasingly significant.
- Unorganised Sector: Players in this sector are either unbranded or uncertified, and their focus is primarily on faith. This gained significant traction in rural India.
- The organic product market is divided into three segments based on market demands and needs:
    - Open Market: This category consists of small mom-and-pop shops, wholesalers, and conventional retailers.
    - Mainstream: Stand-alone stores and large national and regional retail chains and outlets are included in this category.
    - Food Service/Institutional Buyers: This market consists of lodging facilities, eateries, caterers, and other establishments.

2.11.3 Mix for Organic Marketing

A lot of green marketing strategies are integrated into larger plans. Similar to traditional marketing strategies, green marketing strategies rely on the primary 4Ps. Businesses and farmers can expand their market shares and boost their profits by entering new markets and introducing innovative ideas such as eco-friendly products. The four Ps of green organic marketing are similar to the four Ps of conventional marketing—place, promotion, pricing, and product— however, they differ only a little.

- **Organic Product:** Utilises "organic," additionally its claim to be chemical-free is made. Eco-friendly products are generally less expensive, use less water, and require less natural resources; their procurement occurs across multiple locations. The item is recyclable and comes in good packaging. The product is put through quality testing at every point in the value chain. Energy- and water-efficient products reduce their negative impact on the environment while also saving money.

- **Organic Price:** The market for organic foods is highly sensitive to price. It balances the needs of people, the environment, and business in a way that protects worker, farmer, and community health and promotes effective productivity. It can be made more valuable by altering its look, enhancing its features, customising it, etc.
- **Organic Place:** Organic place reduces its carbon footprint by controlling logistics to minimise transportation emissions. For instance, mango juice can be licensed for domestic production rather than being sold as an import to India. It keeps products from being shipped long distances, saves money on shipping, and—above all—reduces the carbon emissions that come from ships and other forms of transportation. The products are sold through two channels. The product isn't suitably offered in the marketplace.
- **Organic Promotion:** The lady of the house has the biggest influence on it, along with customer purchasing behaviour. Therefore, they should be targeted, and the main tactic will be mouth publicity. In order to promote sustainability, promotional materials, signage, white papers, websites, videos, and presentations must be designed with people, the environment, and financial gain in mind.

2.12 MARKET STUDY ON ORGANIC PRODUCTS: A SWOT ANALYSIS

A SWOT analysis is given in Table 2.2.

TABLE 2.2
SWOT Analysis of Market Study on Organic Products

Strengths	Weaknesses
□ Safe food items	□ Insufficient awareness among agriculturalists
□ Enhanced nutrition and high quality; enhanced soil health; and environmental sustainability	□ Insufficient knowledge in remote regions
□ Keeps traditional varieties alive	□ Reduced government support
□ Adequate weather conditions	□ Expensive and labour-intensive
□ Young clientele	□ Productivity gaps
□ Double advantages	□ Difficult access to the products in the city
□ A little less expensive than a specialty store for "healthy nutrition"	□ Imprecise demand estimate and inadequate marketing
□ The company has a robust farmer base and sourcing, an efficient state-of-the-art lab for assurance of quality and research and development, and its own production and packaging operation.	

(Continued)

TABLE 2.2 (Continued)
SWOT Analysis of Market Study on Organic Products

Opportunities	Threats
□ The expanding middle class	□ Lack of infrastructure
□ Urban drift is spreading, modern shopping is growing	□ Lack of certification bodies
□ Affluent income is rising, health awareness is rising	□ Substitutions
□ Significant food and fertiliser subsidies are being reduced	□ Competition
□ Make significant profits from exports	□ Expensive and complicated certification process
□ There is a sizable unexplored market.	□ The implementation of genetically modified crops
□ A growing group of individuals who are concerned about their health	□ Low understanding of organic inputs
□ The absence of organic food suppliers	□ The authenticity of organic products through certification
	□ People's perspectives
	□ Developing a novel idea for food products that provide people with the necessary health and other advantages

2.13 CONCLUSION

Organic farming is believed to be one of the long-standing customs of our forefathers. Food products grown organically are generated without the use of pesticides or chemical-based fertilisers, both of which have negative health effects. Going back to the past, organic farming was the main source of prosperity for the great Indian civilisation up until British intervention. In India, the market for organic food is expanding. It began manufacturing chemical-free organic food products. Additionally, much more food that is organic will be exported. It will be recognised as an item that is sold on a worldwide scale. Individual characteristics are very significant. Indian consumers value and believe that organic food is safe, so they prefer it. Sustainable environmental practices are yet another reason why people favour organic food products. According to this study, most people who use organic products prefer them because of their perceived health benefits and because they firmly believe that organic products are pesticide-free. However, some people question whether organic food is actually pesticide-free and think that organic food products are being overused. The government should address the high cost of organic food products, as even those with modest incomes may have a preference for them. Organic farming has more potential in the future, and in the near future, there may be a rise in the demand for organic food. Because certain pollutants aggravate health conditions, eating organic food is essential for maintaining good health. India is quickly becoming one of the biggest possible markets for foods made organically. The organic food market is developing at the highest rate in the world, right now, in India. Consumers are aware that using compounds in food production increases the risk of disease; these chemicals are also bad for the natural world, farmers, and people's health. Consequently, there is a greater demand for organic food. When it pertains to organic products, consumers are highly price-sensitive. Businesses are required to disclose to them the value added over the cost for which they sell organic products. The primary issues are an absence of scientific knowledge, misleading marketing, inadequate consumer knowledge of the product, financial constraints, and a lax enforcement of the law. Adequate consumer education campaigns are necessary to attain environmental sustainability in the field of organic foods through marketing strategies. Marketers must find ways to improve product performance and build customer loyalty. There are plenty of opportunities for organic marketing, but they may also encounter many obstacles because a sizable portion of the population is not aware of or willing to learn about the issues surrounding natural marketing campaigns. Furthermore, there is a growing demand for education and support to help organic farmers obtain certification for their products. One of the problems is that many producers and producer organisations in the nation do not hold certification under either of the two certification schemes. PGS-India is presently being simplified to make it more user-friendly for small producers (Business Standard). Since certified organic produce can only be sold at premium organic product prices, small and medium-sized organic producers would benefit from greater awareness of the streamlined certification systems. It is the responsibility of Jaivik Kheti to fully inform these organic farmers and producing organisations about simplified certification procedures and straightforward marketing. The steps taken by the Thiruvalluvar Uzlavar and Vashista Uzlavargal organisations to assist small organic farmers in certifying and marketing their produce through direct and digital markets must be highlighted.

REFERENCES

Ajzen, I. (1991). The theory of planned behavior. *Organizational Behavior and Human Decision Processes, 50*(2), 179–211.

APEDA. (2013). *Annual report 2013–14.* https://www.apeda.gov.in/apedawebsite/Annual_Reports/ANNUAL_REPORT_ENGLISH_2013-14.pdf

Aryal, K. P., Chaudhary, P., Pandit, S., & Sharma, G. (2009). Consumers' willingness to pay for organic products: A case from Kathmandu Valley. *Journal of Agriculture and Environment, 10*, 12–22.

Botonaki, A., Polymeros, K., Tsakiridou, E., & Mattas, K. (2006). The role of food quality certification on consumers' food choices. *British Food Journal, 108*, 77–90.

Bourn, D., & Prescott, J. (2002). A comparison of the nutritional value, sensory qualities, and food safety of organically and conventionally produced foods. *Critical Reviews in Food Science and Nutrition, 42*(1), 1–34.

Cavallet, L. E., Canavari, M., & Fortes, P. (2018). Participatory guarantee system, equivalence and quality control in a comparative study on organic certifications systems in Europe and Brazil. *Revista Ambiente & Água, 13*(4), 1–14.

Chiou, J. S. (1998). The effects of attitude, subjective norm, and perceived behavioural control on consumers' purchase intentions: The moderating effects of product knowledge and attention to social comparison information. *Proceedings of the National Science Council, 9*(2), 298–308.

Compagnoni, A., Pinton, R., & Zanoli, R. (2000). Organic farming in Italy. In S. Graf & H. Willer (Eds.). *Organic agriculture in Europe: Current status and future prospects of organic farming in twenty-five European countries*. Bad Durkheim.

Cornish, P. S., & Stewart., T. E. B. (2002). Certification-case studies with Australian market gardeners. In R. Thompson (Ed.). *Cultivating Communities. Proceedings of the 14th IFOAM Organic World Congress, Victoria, Canada* (p. 222). Canadian Organic Growers.

Cranfield, J. A., & Magnusson, E. (2003). Canadian consumers' willingness-to-pay for pesticide-free food products: An ordered probit analysis. *International Food and Agribusiness Management Review, 6*(4), 14–30.

Cunningham, R. (2002). *Who is the organic consumer?* Paper presented at Growing Organic Conference.

Davies, A., Titterington, A., & Cochrane, C. (1995). Who buys organic food? A profile of the purchasers of organic food in Northern Ireland. *British Food Journal, 97*(10), 17–23.

Demeritt, L. (2002). *All things organic: A look at the organic consumer*. The Hartman Group.

Dettmann, R., & Dimitri, C. (2007). Who's buying organic vegetables? Demographic characteristics of US consumers. *Journal of Food Distribution Research, 16*(1), 49–62.

Elayaraja, M., & Vijai, C. (2020a). Organic farming in India: Benefits and challenges, *European Journal of Molecular & Clinical Medicine, 7*(11), 3021–3029.

Elayaraja, M., & Vijai, C. (2020b). Organic food in India: Health and environmental advantages and disadvantages. *Sambodhi Journal, 45*(3), 688–695.

FAO. (2007). *Organic certification schemes: Managerial skills and associated costs*. FAO.

FAO. (2012). *Organic agriculture and the law*. FAO.

FiBL & IFOAM—Organics International. (2020). Organic producers and other operator types. In H. Willer, B. Schlatter, J. Travnicek, L. Kemper, & J. Lernoud (Eds.), *The word organic agriculture: Statistics & emerging*. FAO.

Fishbein, M., & Ajzen, I. (1975). *Belief, attitude, intention, and behavior: An introduction to theory and research*. Addison Wesley Publishing Company.

Food and Agriculture Organization of the United Nations (FAO) & IFOAM—Organics International. (2019). Why invest in Participatory Guarantee Systems? *Opportunities for organic agriculture and PGS for sustainable food systems* (p. 61). FAO/IFOAM.

Fouilleux, E., & Loconto, A. (2016). Voluntary standards, certification, and accreditation in the global organic agriculture field: A tripartite model of techno-politics. *Agriculture and Human Values, 34*, 1–14.

Gotschi, E., Vogel, S., Lindenthal, T., & Larcher, M. (2010). The role of knowledge, social norms, and attitudes toward organic products and shopping behavior: Survey results from high school students in Vienna. *The Journal of Environmental Education, 41*(2), 88–100.

Govidnasamy, R., & Italia, J. (1990). Predicting willingness to pay a premium for organically grown fresh produce. *Journal of Food Distribution Research, 30*(2), 44–53.

Grossman, M. (1972). *The demand for health: A theoretical and empirical investigation*. Columbia University Press.

Hatanaka, M., Bain, C., & Busch, L. (2005). Third-party certification in the global agrifood system. *Food Policy, 30*, 354–369.

Hirata, A., Rocha, L., Assis, T., Souza-Esquerdo, V., & Bergamasco, S. (2019). The contribution of the participatory guarantee system in the revival of agroecological principles in Southern Minas Gerais. *Sustainability, 11*(17), 1–15.

Hughner, R., McDonagh, P., Prothero, A., Shultz, J., & Stanton, J. (2007). Who are organic food consumers? A compilation and review of why people purchase organic food. *Journal of Consumer Behavior, 6*(2/3), 94–110.

Hutchins, R. K., & Greenhalgh, L. A. (1997). Organic confusion: Sustaining competitive advantage. *British Food Journal, 99*(9), 336–338.

IFOAM—Organics International. (2008). *Definition of organic agriculture*. IFOAM General Assembly.

Jolly, D. A., Schutz, G. H., Diaz–Knauf, K. V., & Johal, J. (1989). Organic foods: Consumer attitudes and use. *Food Technology, 43*(11), 60–63.

Kaufmann, S., & Vogl, C. (2018). Participatory guarantee systems (PGS) in Mexico: A theoretic ideal or everyday practice? *Agriculture and Human Values, 35*, 457–472.

Kim, H. Y., & Chung, J. E. (2011). Consumer purchase intention for organic personal care products. *Journal of Consumer Marketing, 28*(1), 40–47.

Kirchmann, H., & Ryan, M. H. (2004). Nutrients in organic farming—Are there advantages from the exclusive use of organic manures and untreated minerals? *International Crop Science*, 89–116. https://doi.org/10.1007/978-1-4020-9316-65

Liu, M. E. (2007). *US college students' organic food consumption behavior* [PhD diss., Texas Tech University].

Lockie, S., Lyons, K., Lawrence, G., & Grice, J. (2004). Choosing organics: A path analysis of factors underlying the selection of organic food among Australian consumers. *Appetite, 43*(2), 135–146.

Makatouni, A. (2002). What motivates consumers to buy organic food in the UK? Results from a qualitative study. *British Food Journal, 104*(35), 345–352.

Maria, L., Jill, M., & Ron, M. (2001). Assessing consumer preferences for organic, eco-labeled, and regular apples. *Journal of Agricultural and Resource Economics, 26*(2), 404–416.

Misra, S. K., Huang, C. L., & Ott, S. L. (1991). Consumer willingness to pay for pesticide-free fresh produce. *Western Journal of Agricultural Economics, 16*(2), 218–227.

Nassar, A. (2003). Certifcação no Agribusiness. In D. Zylbersztajn & R. F. Scare (Eds.). *Gestão da Qualidade no Agribusiness: Estudos e casos* (pp. 30–46). Atlas.

O'Donovan, P., & McCarthy, M. (2002). Irish consumer preference for organic meat. *British Food Journal, 104*(3/4/5), 353–370.

Pandey, D., Kakkar, A., Farhan, M., & Khan, T. (2019). Factors influencing organic foods purchase intention of Indian customers. *Organic Agriculture, 9*, 357–364.

Paul, J., & Rana, J. (2012). Consumer behaviour and purchase intention for organic food. *Journal of Consumer Marketing, 29*(6), 412–422.

Raynolds, L., Murray, D., & Heller, A. (2007). Regulating sustainability in the coffee sector: A comparative analysis of third party environmental and social certification initiatives. *Agriculture and Human Values, 24*, 147–163.

Richa, M., & Deepak, S. (2016). An analysis of factors affecting growth of organic food. *British Food Journal, 118*(9), 2308–2325.

Rimal, A., Moon, W., & Balasubramanian, S. (2005). Agro-biotechnology and organic food purchase in the United Kingdom. *British Food Journal, 107*, 84–97.

Saleki, Z. S., Seyedsaleki, S. M., & Rahimi, M. R. (2012). Organic food purchasing behavior in Iran. *International Journal of Business & Social Science, 3*(13), 278–285.

Sapp, S. G. (1991). Impact of nutritional knowledge within an expanded rational expectations model of beef consumption. *Journal of Nutrition Education and Behavior, 23*(5), 214–222.

Shepherd, R., Magnusson, M., & Sjoden, P. O. (2005). Determinants of consumer behavior related to organic foods. *Ambio, 34*(4/5), 352–359.

Sheppard, B. H., Hartwick, J., & Warshaw, P. R. (1988). The theory of reasoned action: A meta-analysis of past research with recommendations for modifications and future research. *Journal of Consumer Research, 15*(3), 325–343.

Smith, R., & Grasseni, C. (2020). Ambivalent solidarities: Food governance reconfigurations in Croatia and Italy. *Anthropology Today, 36*.

Smith, S., & Paladino, A. (2010). Eating clean and green? Investigating consumer motivations toward the purchase of organic food. *Australasian Marketing Journal, 18*(2), 93–104.

Stobbelaar, D. J., Casimir, G., Borghuis, J., Marks, I., Meijer, L., & Zebeda, S. (2007). Adolescents' attitudes toward organic food: A survey of 15 to 16-year-old schoolchildren. *International Journal of Consumer Studies, 31*(4), 349–356.

Tanner, C., & Wolfing, K. S. (2003). Promoting sustainable consumption: Determinants of green purchases by Swiss consumers. *Psychology & Marketing, 20*(10), 883–902.

Tarkiainen, A., & Sundqvist, S. (2005). Subjective norms, attitudes, and intentions of Finnish consumers in buying organic food. *British Food Journal, 107*(11), 808–822.

Trewavas, A. (2001). Urban myths of organic farming. *Nature, 410*, 409–410.

Trewavas, A. (2004). A critical assessment of organic farming-and-food assertions with particular respect to the UK and the potential environmental benefits of no-till agriculture. *Crop Protection*, *23*(9), 757–781.

Turner, N., Angus, J., McIntyre, L., Robertson, M., Borrell, A., & Lloyd, D. (2004). New directions for a diverse planet. In *Proceedings of the 4th International Crop Science Congress*. The Regional Institute.

Vermeir, I., & Verbeke, W. (2008). Sustainable food consumption among young adults in Belgium: Theory of planned behavior and the role of confidence and values. *Ecological Economics*, *64*(3), 542–553.

Voon, J. P., Ngui, K. S., & Agrawal, A. (2011). Determinants of willingness to purchase organic food: An exploratory study using structural equation modelling. *International Food and Agribusiness Management Review*, *14*(2), 103–120.

Wheale, P., & Hinton, D. (2007). Ethical consumers in search of markets. *Business Strategy and the Environment*, *16*(4), 302–315.

Young, W., Hwang, K., McDonald, S., & Oates, C. J. (2010). Sustainable consumption: Green consumer behaviour when purchasing products. *Sustainable Development*, *18*(1), 20–31.

3 The Future of Organic Farming
Trends and Innovations

Rukhsana Qadir, Bilal Ahmad Dar, Burhan Hamid, Zaffar Bashir, and Mehrun Nisa

3.1 INTRODUCTION

Organic farming has undergone a significant transformation since its initiation as a countercultural movement in the mid-20th century. Organic farming was primarily perceived as a specialized approach to agriculture and has evolved into a global phenomenon, acknowledged for its value by both consumers and farmers. Amidst global challenges like climate change, biodiversity decline, and the necessity for sustainable food systems, organic farming is emerging as a hopeful solution. It transcends being merely a production method, representing a sustainability philosophy that involves active engagement with nature, particularly focusing on biodiversity conservation and environmental sustainability. Recognized as a substitute and justifiable approach (Eyhorn et al., 2019) contributing to the shift towards a (khaki) economy (Aceleanu, 2016), organic farming emphasizes sustainability, environmental health, and biodiversity while rejecting synthetic pesticides, fertilizers, and genetically modified organisms (Murmu et al., 2022).

Adopting traditional farming techniques, organic farmers depend on natural processes to enhance soil fertility, control pests, and manage weeds. This method underscores the utilization of organic matter, crop rotation, and companion planting to maintain soil health and minimize the reliance on external inputs. By eschewing synthetic chemicals, organic farming strives to reduce environmental impact, ensuring water quality and safeguarding biodiversity. Additionally, it often champions animal welfare by prohibiting antibiotics and growth hormones in livestock. The all-encompassing unbiased of organic farming is to launch a balanced and harmonious system fostering ecological resilience, producing nutritious crops while respecting the delicate equilibrium of nature. With a growing consumer emphasis on health and sustainability, organic farming has gained popularity. Organic agriculture, serving as both an attitude and a farming system, strives to produce uncontaminated and nutritious food, prioritizing human health (Henning et al., 1991). The ecological compensations of organic farming include the maintenance of soil fertility, carbon dioxide packing, reduction of fossil fuel usage, preservation of landscapes, and conservation of biodiversity (Gomiero et al., 2011). Guided by the principles of health,

DOI: 10.1201/9781003491460-3

ecology, fairness, and care outlined by IFOAM (IFOAM, 2018), organic agriculture establishes the foundation for its growth and development (Gamage et al., 2023).

Despite critiques from scholars highlighting inefficiencies in diet making (Kirchmann et al., 2016; Pickett, 2013), there is widespread recognition of organic farming's potential to address various sustainability goals—environmental, economic, and social—as evident in hearsays and academic studies (Reganold & Wachter, 2016; Saffeullah et al., 2021). Arguments propose that organic farming has the capacity to encounter the sustenance difficulties of a growing population without necessitating additional land, concurrently mitigating environmental impacts through societal shifts such as reduced consumption of animal products and decreased food waste (Muller et al., 2017). Moreover, the adoption of state-of-the-art practices can enhance carbon-based farming yields and facilitate adaptation to climate change (Purnhagen et al., 2021; Reganold & Wachter, 2016). This chapter explores the trends and innovations influencing the future trajectory of organic farming. Conscious alternative to conventional agriculture. Organic farming is a sustainable and holistic approach to agriculture that prioritizes environmental conservation, biodiversity, and soil health. It relies on natural processes and avoids synthetic pesticides and herbicides and genetically modified organisms. Organic farmers use crop rotation, composting, and natural predators to manage pests and enhance soil fertility. The goal is to create a balanced ecosystem that fosters healthy crops while minimizing negative impacts on the environment. Organic farming promotes ethical and transparent practices, providing consumers with food that is free from synthetic chemicals and genetically modified ingredients (Figure 3.1).

3.2 INNOVATIONS IN ORGANIC FARMING

Scientific innovations collectively contribute to the ongoing evolution of organic farming, making it more resilient, productive, and aligned with ecological principles.

3.2.1 Pest Management

In the realm of organic farming, a comprehensive or whole-farm approach is adopted, heavily relying on ecological processes and biodiversity within the agroecosystem. The escalating demand for organic products underscores the necessity for sustainable pest management practices. Rather than resorting to synthetic chemicals, farmers are increasingly turning towards biological control methods. Preventive measures employed for pest control in organic farming include crop rotation, soil and nutrient management, sanitation, and cultural practices to enhance crop health (Headrick, 2021). Non-synthetic controls encompass the use of lures, traps, and repellents. Lure utilization typically involves the use of semiochemicals or food attractants to attract precise classes or sophisticated taxonomic groups (e.g., fruit flies) for trapping and elimination (Yousef et al., 2018). These lures may also attract natural enemies (Lucchi et al., 2017). Pheromone mating disruption, a successful technique for lepidopteran pests in agriculture for several decades, is well-documented (Kehat et al., 1995; Lykouressis et al., 2005; Orkun et al., 2005; Vetter et al., 2006). Azadirachtin and oils are among the top four insecticidal products (Wilson et al., 2019). Entomopathogenic

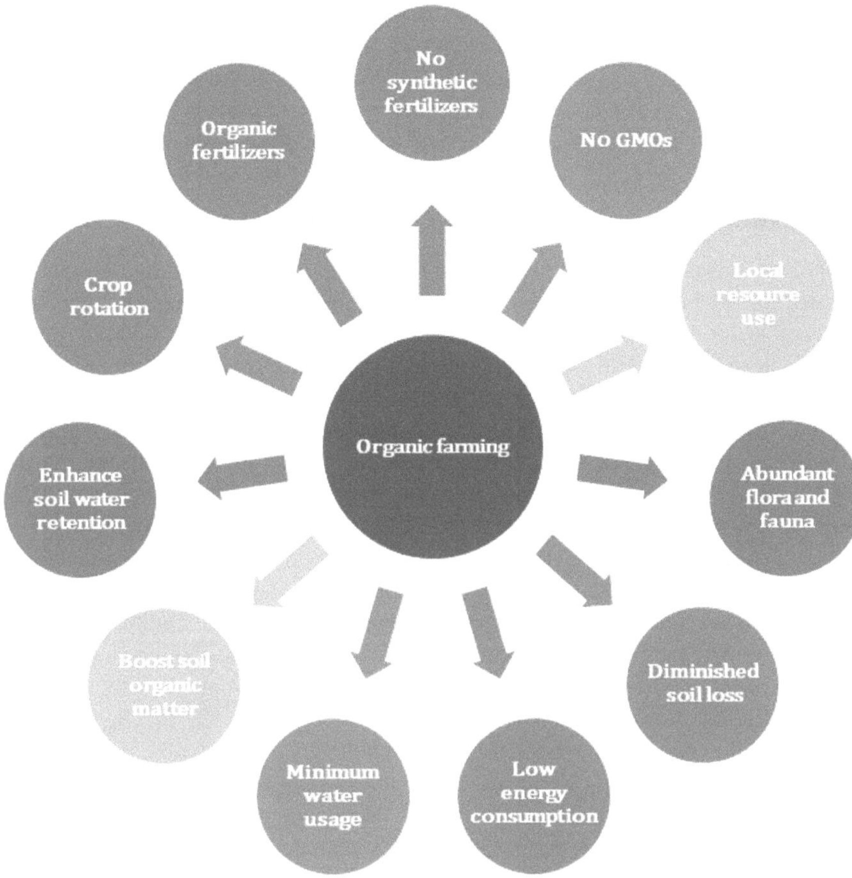

FIGURE 3.1 Organic farming processes: key practices and impacts.

nematodes (EPN) are gaining popularity as an alternative to chemical control due to their unique biology and easy application (Stevens & Lewis, 2017). Beneficial insects, such as ladybugs and predatory mites, are employed to maintain a balance between predator and prey populations, aligning with the principles of organic farming. Innovations like precision agriculture, which utilizes data and technology for real-time identification and treatment of pest issues, are anticipated to revolutionize organic pest management.

Furthermore, the majority of strategies, principles, and elements associated with Integrated Pest Management (IPM) are in accordance with organic farming systems (Edwards-Jones & Howells, 2001). Integrated Pest Management techniques, which involve the strategic deployment of natural predators and pest-resistant crops, are gaining prominence. Successful IPM programs in the milieu of organic agriculture typically entail the monitoring of crops for pests, precise identification of pests, establishment of economic thresholds, implementation of integrated pest control tactics, and maintenance of records for assessment. Factors that create an unsuitable habitat for vermin and ailments encompass resource constraint, rivalry, parasitism, and predation (Drinkwater et al., 1995). These factors play a pivotal role in maintaining the equilibrium of the agroecosystem and suppressing detrimental pests. Additionally, the diversity of fauna and flora significantly contributes to pest and disease management within the organic farming system (Hajjar et al., 2008; Westerman et al., 2003).

3.2.2 Smart Farming: Harnessing Technology for Sustainability

Technology serves as a tool to enhance production levels, and its integration into organic farming practices is not a compromise of principles but a means to improve efficiency and sustainability. Various technological advancements, such as drones, precision farming, and IoT applications, empower farmers to monitor and manage crops in real time. Multispectral cameras on drones provide valuable data for informed decision-making on resource allocation, pest control, and irrigation. Precision agriculture (PA) is a data-driven strategy that witnesses, trials, and evaluates the distinct requirements of individual fields and crops (Monteiro et al., 2021). Aligned with the "produce more with less" principle (Bucci et al., 2018), precision agriculture, alternatively termed precision or smart farming, plays a role in addressing the increasing food demand while prioritizing sustainability through precise and resource-efficient production management (Research European Commission, 2017). The Internet of Things (IoT) is a transformative know-how enabling remote connectivity for smart farming (Patil & Kale, 2016). Real-time data facilitated by IoT empower organic growers to make knowledgeable decisions vis-à-vis irrigation, pest control, and nutrient management. Sensor technologies cover a wide array of agricultural aspects, including cattle crusade, micro-climate data assortment, and even precise pH level measurements of the soil. These advancements contribute to the growth of agricultural ecosystems in a self-sufficient and sustainable manner.

3.2.3 Vertical Farming: Reimagining Agricultural Space

Conventional industrial agricultural methods are diminishing soil quality at a pace surpassing the natural replenishment process. This heightened erosion, combined with extensive freshwater consumption, has resulted in a lessening of arable land and heightened stress on existing water reservoirs (Pimentel, 2013). In rejoinder to these global encounters of dwindling resources, vertical farming emerges as a transformative solution. Vertical farming (VF) involves stacking layers of crops in controlled environments to optimize space, reduce water consumption, and minimize the reliance on pesticides. Vertical farming not only addresses the limitations of traditional agriculture but also presents an innovative approach to resource efficiency. By maintaining precise control over environmental conditions, vertical farming significantly reduces resource consumption while allowing for increased production throughout different times. This not only improves yield (Benke & Tomkins, 2017) but also brings agricultural activities closer to urban centres, thereby reducing the environmental impact associated with food transportation.

A crucial aspect of vertical farming is the monitoring and control of carbon dioxide levels. Nondispersive infrared (NDIR) CO_2 sensors show a vivacious role in stalking and regulating the circumstances within vertical ranches, ensuring optimal growth and resource utilization. This technology represents a noteworthy advancement in sustainable and efficient farming, offering a promising alternative to conventional agriculture.

3.2.4 Regenerative Agriculture: Nurturing the Soil Ecosystem

Regenerative agriculture encompasses farming structures and arena practices that aim to minimize soil disorder, incorporating sundry revolutions and cover crops, and integrating grazing livestock (Giller et al., 2021). Its primary objectives are the diminution of greenhouse gas emissions, the enhancement of soil carbon content, the improvement of soil health and biology, the promotion of farm-scale nutrient use competence (NUE), and the support of biodiversity and associated ecosystem amenities (Magistrali et al., 2023). Moving beyond mere sustainability, regenerative agriculture focuses on the restoration and rejuvenation of soil, recognizing its pivotal role as the cornerstone of a resilient food system. Various techniques, including cover cropping, rotational grazing, and agroforestry, are employed in regenerative agriculture to contribute to soil health, carbon sequestration, and overall ecosystem restoration. This holistic approach ensures the continued productivity of the land for future generations. Both carbon-based and non-organic regenerative farming systems promote comprehensive approaches to managing soil, crops, and livestock. These approaches encompass the utilization of resilient and weed-resistant plant varieties, diverse crop rotations incorporating nitrogen-fixing legumes, consistent application of animal manure or compost derived from organic waste, the conservation and creation of biodiversity zones within farms, limited tillage, sustained vegetation cover through intercropping, companion planting, or cover crops, and the amalgamation of crop and livestock production through outdoor grazing or foraging (Holka et al., 2022; Magistrali et al., 2023; Ozlu et al., 2022; Rempelos et al., 2021). It is essential to underscore a notable distinction: organic farms are required to adhere to legally mandated farming standards and undergo regular inspections by certified entities to label and market their products as organic; in contrast, there are presently no legally binding canons for regenerative farming arrangements (Magistrali et al., 2023; Rempelos et al., 2021). Consequently, regenerative and organic creation approaches generate unique soil corporal, biochemical, besides biotic conditions in comparison to hard farming systems (Rempelos et al., 2020, 2023).

3.2.5 Biological Disease Management

In organic farming, biological disease management signifies an environmentally sustainable strategy that utilizes living organisms to regulate and suppress plant diseases. Aligned with the principles of organic farming, which prioritize natural and holistic solutions over synthetic chemicals, this approach employs beneficial microbes, such as specific microorganisms (bacteria, fungi, actinomycetes), to combat plant pathogens. One method within this framework is Anaerobic Soil Disinfestation (ASD), which entails incorporating fresh carbon-based material into saturated soil under impermeable lithe for a duration of three to six weeks, contingent on external temperatures (Lamers et al., 2004; Momman, 2008). The carbon source promotes the propagation of bacteria, depleting accessible oxygen and enabling anaerobic bacteria to decompose, effectively managing various soil-borne plant pathogens and nematodes. This method induces a swing in bacteriological community composition even though preserving microbial diversity (Birkhofer et al., 2008). Organic farmers also seek to enhance bacterial populaces and variety on plant exteriors through the application of compost extracts, commonly referred to as composters (Kasselaki et al., 2011; Litterick et al., 2004). Additionally, extracts from composted olive wastes have proven effective against tuber blight (Patsaki et al., 2005). Direct application of biocontrol agents, such as *Gliocladium*, *Trichoderma*, *Streptomyces*, *Pseudomonas*, and *Bacillus* species, is another biological control method primarily used against soil-borne plant pathogens and foliar diseases (Caldwell et al., 2015). The yeast *Aureobasidium* is employed as a controller of fire blight (Holb et al., 2015; Tamm et al., 2015).

In total, to microorganisms, extracts derived from diverse plant species can be employed to combat plant pathogens and insect vectors. Commercial products for organic agriculture include yucca and citrus extracts obtained through water extraction or other solvents. However, scientific substantiation supporting the efficacy of numerous plant extracts is currently limited. For instance, horsetail excerpts have been pragmatic to control diseases such as peach leaf curl and powdery mildew, but outcomes may vary. Extracts from horseradish roots, dock, nettle, chives, garlic, and other herbs, spices, and medicinal plants are under examination for their impacts on innumerable plant diseases. While certain products, such as lecithin extracted from soybeans, have received approval from organic regulatory agencies, the overall endorsement of plant-based antifungal agent remains constrained (Datnoff et al., 2007; Dagostin et al., 2011; Finckh & Van Bruggen, 2015; Keqiang & Van Bruggen, 2001; Koch & Groot, 2015; Slusarenko et al., 2008).

3.3 SUSTAINABLE AGRICULTURE AND AGROECOLOGY

Given the forecasted global population upsurge to 9.7 billion by the year 2050, considerable pressure is placed on our food systems. In response to this challenge, carbon-based farming emerges as a promising avenue for achieving sustainable agriculture. Grounded in principles that prioritize soil health, minimize chemical inputs, and foster biodiversity, organic farming is positioned to play a pivotal role in addressing the pressing need to combat climate change and preserve natural resources. Functioning as a holistic approach to agriculture, organic farming aims for long-term sustainability and reduced environmental impact, and gives priority to the well-being of the planet, farmers, and consumers. It serves as a crucial component in the broader initiatives to construct a more supportable and robust food system. At the core of the concept of agricultural sustainability is the principle that present needs should be fulfilled without jeopardizing the capacity of forthcoming cohorts to encounter their individual requirements. In the context of agriculture, sustainability encompasses various dimensions (Malik et al., 2020). Sustainable agriculture practices, while producing crops or livestock, avoid causing harm to human or environmental systems. This philosophy and farming system involve steering clear of detrimental significances for soil, water, biodiversity, nearby or downstream resources, and the well-being of those working or residing on the farm or in the surrounding area (Etingoff, 2016). Maintainable agriculture embraces practices such as permaculture, agroforestry, cohesive farming, manifold cropping, and yield alternation. It is essential to cultivate adaptable commercial methods and farming methods within the framework of supportable food systems (FAO, 2018), contributing to the enduring viability of the human population. The positive aspects of organic farming are in line with sustainable agricultural practices. Strategies such as Integrated Pest Management (IPM), yield alternation, establishing nitrogen-fixing ground casings, using mulch, and adopting no-till or reduced-tillage agronomic methods contribute to enhanced soil health and decreased pesticide consumption (Merah et al., 2021). Furthermore, organic farming advocates for the cautious application of fertilizers, both organic and synthetic, guided by soil samples, promoting precision farming. The development of novel irrigation schemes also optimizes harvest or meadow yield while minimizing water depletion, destruction, and salinization (Baker et al., 2015; Chaichi et al., 2018; Merah & Djazouli, 2016; Merah et al., 2021; Salim et al., 2020). These practices highlight the significance of pulverized protection, mulches, and limited insecticide practice to avert soil attrition through promoting biodiversity conservation.

Carbon-based farming, recognized as a thriving sector, not only offers many vigor assistances but also addresses key environmental concerns. Consumer preferences are shifting towards organic products, with people demonstrating a willingness to pay higher prices due to perceived health benefits. The increasing awareness and commitment to environmental sustainability have further contributed to source maintenance and a positive influence on the demand for organic products.

3.4 ORGANIC CROP NUTRITION

On a global scale, regulations governing organic agriculture prohibit the utilization of artificial goods such as fertilizers and scientifically manufactured agrochemicals. Consequently, carbon-based farmers employ various practices to augment soil productiveness, including the application of organic modifications, crop revolutions, and cover crops. These agronomical practices aim to enhance the biological, chemical, and physical properties of the soil, while also providing essential nutrients to plants (Tejada et al., 2016). Commonly utilized methods involve fertilization through visceral compost, composted crop deposits, and leguminous floras as main and transitional crops. Despite the widespread adoption of these practices, there is presently no universally applicable instruction overseeing the usage of organic fertilizers globally. Moreover, the implementation of an effective crop rotation design, incorporating concealment crops and (khaki) manures, along with the application of approved inorganic and organic fertilizers, holds significance (Durán-Lara et al., 2020; Reeve et al., 2016). These activities contribute to an upsurge in soil organic matter, enhance the bustle of advantageous invertebrates and microbes, improve soil bodily properties, boost plant nutrient availability, reduce disease risks, and foster overall crop health (Brust, 2019; Reeve et al., 2016). Biological nitrogen fixation facilitated by legumes, in contrast to chemically synthesized nitrogen-bearing fertilizers, constitutes a crucial aspect of fertility management (Durán-Lara et al., 2020; Möller, 2018). Fertilizers of animal origin, such as fowls, brute, ewes, and livestock guano, plus urine, are extensively employed in organic production, with their availability and utilization contingent on regional factors like quantity, price, transportation, and handling (Garg et al., 2005; Guo et al., 2019). Composting emerges as an effective approach for handling organic solid waste, transforming into a humified, constant, odourless, and pathogen-free substantial that enhances degraded soils (Lavado et al., 2012; Orden, 2018; Prono et al., 2010). Recognized for its profitability, environmental friendliness, and ease of handling, manure is a high-quality artefact commonly utilized in biological gardening (Tognetti et al., 2007; Bian et al., 2019). It is imperative to note that while compost provides crops with certain nutrients, primarily nitrogen, additional organic or approved mineral fertilizers are often incorporated. Vermicompost, a product in increasing demand as an organic fertilizer, involves the stabilization of organic matter through earthworms. Notably, vermicompost tends to have higher levels of macronutrients and micronutrients compared to traditional compost, offering soluble forms of essential elements. Its properties contribute to enhance soil fertility, ensuing in improved crop yields (Lim et al., 2015). Another waste management technique employed in organic farming is biochar, a carbon-rich charcoal derived through the warm air treatment (pyrolysis) of agronomic rests and animate biomass. Biochar serves as a sustainable and efficient source of organic and mineral nutrients, playing a role in enhancing soil health, addressing low fertility, improving crop yields, immobilizing heavy metals, and alleviating plant stress. Moreover, it contributes to the retention of ammonia, ammonium, and nitrates, aiding in carbon sequestration and mitigating global warming (Adekiya et al., 2019; Ni et al., 2020; Xiao et al., 2014; Zhu et al., 2017). Over the past 25 years, there has been a growing focus on bio stimulants and biofertilizers within both the organized community and the agrochemical engineering. Positioned by mode of agroecological solutions for fertility challenges, abiotic stress tolerance, and the enhancement of organic food production quality, these products serve as complementary elements in integrated crop nutrition, particularly in organic farming (Chawla & Sadawarti, 2020; Santini et al., 2021). Bio stimulants, well-defined fertilizer products designed to stimulate plant nutrition progressions irrespective of their nutrient gratified, and biofertilizers, containing living organisms that enhance the primary nutrient supply to crops, are pivotal components. Natural substances like betain, chitin, humic and fulvic acids, vegetal hydrolysed proteins, phenolic compounds, and seaweed extracts fall under the category of plant bio stimulants. Biofertilizers contributing to plant growth include nitrogen-fixing biofertilizers, orthophosphate biofertilizers, potassium biofertilizers, biofertilizers for micronutrients, plant growth-promoting rhizobacteria, and muck (Aloo et al., 2020; Bashir et al., 2017, 2018a, 2018b, 2019). Both bio stimulants and biofertilizers (N_2 fixers, P and K solubilizers) are expected to show a vital role in future supportable food creation, especially in organic crop nutrition, though the importance of a comprehensive global regulatory framework is underscored (Bashir et al., 2018, 2019b, 2021a; Bashir, 2021; Hamid & Bashir, 2019a).

3.5 BIODIVERSITY AND CONSERVATION

The heightened and broadened development of contemporary farming practices presents significant coercions to comprehensive biodiversity. During the final quarter of the 20th century, there were substantial let-downs in the array and richness of farmland in various countries, prompting concerns about the sustainability of prevailing intensive farming practices. Intensive land use, marked by a substantial influx of nutrients and insecticides (Isbell et al., 2013; Meehan et al., 2011), limited crop rotations (Kremen et al., 2002; Robinson & Sutherland, 2002), diminished set-asides (Flade, 2012; Herzon et al., 2011; Tscharntke et al., 2011), and the perpetuation of perpetual grassland (Eglington, 2009), coupled with the elimination of landscape elements (Fabian et al., 2013), efficient harvesting techniques, and seed cleaning (McCracken & Tallowin, 2004), as well as the enhancement or preservation of expansively utilized habitats (Berg et al., 2004; Uchida & Ushimaru, 2014), constitute key factors contributing to the decline of species and their habitats (Benton et al., 2003; Haber, 2014; Habel et al., 2019; Uchida & Ushimaru, 2014). The decline in biodiversity has far-reaching consequences on successive elements of the food shackle (Heldbjerg et al., 2016; Krauss et al., 2011). Additionally, the superiority and functionality of cultivated land as a territory diminish due to deteriorated reproduction conditions, including the loss of appropriate upbringing sites in compact foliage (Eglington, 2009; Quinn et al., 2017). In response to this ongoing loss of biodiversity, farming systems deemed 'sustainable,' such as organic farming, have been acknowledged as potential solutions and receive substantial support through subsidy payments under national government legislation (Hole et al., 2005). Numerous assessments and meta-analyses conducted over the past three decades consistently highlight the positive impacts of organic farming on both above- and belowground biodiversity compared to orthodox agribusiness (Bengtsson et al., 2005; Hole et al., 2005; Mondelaers et al., 2009; Rahmann, 2011; Tuck et al., 2014). These impacts encompass various facets, such as the richness and activity of soil microbes (Lori et al., 2017), the variety of arthropods (Lichtenberg et al., 2017), cultivated plant species (Campiglia et al., 2018; Rydberg & Milberg, 2000), and avian populations in agricultural fields (Chamberlain et al., 1999). Qualitative evaluations conducted by Hole et al. (2005) and Rahmann (2011) emphasize the positive repercussions of organic farming on biodiversity. Collectively, there is ample evidence indicating numerous beneficial impacts of organic farming on local biodiversity (Bengtsson et al., 2005; Hole et al., 2005; Tuck et al., 2014) and ecosystem services, including natural pest control, pollination, and carbon sequestration (Gattinger et al., 2012) (Figure 3.1). However, it is important to acknowledge potential harmful paraphernalia on harvest vintages (Seufert & Ramankutty, 2017; Wittwer et al., 2021). Moreover, the extent of organic farming possessions on different taxa and various bionetwork service areas can vary significantly (Ostandie et al., 2021; Wittwer et al., 2021).

3.6 FUTURE PROSPECTS OF ORGANIC FARMING

Organic farming, a long-standing agricultural tradition deeply rooted in many rural and farming communities over the centuries, has experienced a transition toward conventional farming practices, driven by the advent of modern techniques and the growing demands of a burgeoning population. This transformation entails embracing synthetic fertilizers, chemical pesticides, and genetic modification methods. Even in emergent states such as India, there is a growing desire for organically cultivated goods, driven by an enhanced awareness of food safety and quality. The biological process, which eschews chemical pesticides, has a significant impact on soil health, offering substantial prospects for revenue cohort (Bhardwaj & Dhiman, 2019). Different areas derive advantages from naturally occurring organic nutrient sources that facilitate organic farming (Adolph & Butterworth, 2002; Deshmukh & Babar, 2015; Reddy, 2010). Historically, numerous nations possess firmly established agricultural systems, resourceful farmers, expansive arid lands, and limited reliance on chemical fertilizers and pesticides. Farmers, equipped with deep insights drawn from experience, extensive observation, resilience, and traditional practices, play a key role in nourishing soil fertility and controlling pests. Their effective approaches contribute to the strengthening of biological creation and successive monetary progress. India, currently the largest global carbon-based producer, holds the eighth position of carbon-based cultivated land in 2017 (Adolph & Butterworth, 2002; Sharma & Goyal, 2000; Willer & Lernoud, 2017, 2019; Willer & Lernoud, 2019). In the realm of carbon-based farming, innovative apparatuses have emerged, such as the amalgamation of mycorrhizal fungi and nano-bio stimulants for environmentally friendly agrarian efficiency. Sensor technology and spatial geodata enable more conscious cultivation area mapping, while 3D printers aid smallholders. The Bee Scanning App represents an advancement for beekeepers in combating the Varroa destructor parasite mite and supports population modelling and breeding programs (Nova-Institute GmbH, 2018). The Inhana Rational Farming Technology, based on the principle of 'Element Energy Activation,' offers an inclusive carbon-based method for environmentally and frugally justifiable yield creation, grounded in modern scientific knowledge (Barik & Sarkar, 2017). It focuses on energizing the soil system by restoring soil-plant-microflora dynamics and the plant system by reviving nutrient use competence and higher plant resistance against vexation/ailment contamination. Despite India's traditional roots in organic farming, the influx of techniques and pesticides led to its marginalization. However, the snowballing demand for organic foodstuffs in India and globally, driven by awareness of food quality, soil health improvement, and pest control, positions organic farming as an alternative system. Advocates argue that dry land holds great potential for organic agriculture due to its suitability for low-input farming and biodiversity utilization. Carbon-based farming, with its emphasis on upholding and enlightening loam strength while avoiding impurities and relying on home-grown efforts and labour, has to augment the economic and ecological health of dry lands and their inhabitants. However, obstacles to the advancement of organic farming persist. Obstacles include a lack of knowledge, marketing challenges, shortages of biomass, inadequate supporting infrastructure, elevated input expenses, issues related to marketing organic inputs, the absence of a cohesive agricultural policy, a scarcity of financial backing, diminished yields, an incapacity to fulfil export demands, and the non-existence of quality standards for bio-manures, and the influence of political and social factors. The primary constraint lies in the failure of government policies to decisively promote organic agriculture.

3.7 CONCLUSION

The prospect of organic farming unfolds as an exhilarating mosaic, intricately composed of technological advancements, environmental stewardship, and a collective dedication to sustaining both humanity and the Earth. Moving ahead, the harmonious blend of tradition and technology is set to propel organic farming into uncharted territories, ensuring that our agricultural methods transcend mere sustainability, embracing regeneration, resilience, and a seamless connection with the intricate tapestry of life. Organic farming not only yields more nourishing and secure food but is also witnessing a surge in popularity as consumers increasingly seek the perceived health and safety benefits of organic produce. Consequently, organic food is becoming synonymous with food safety, ensuring a secure journey from farm to plate. In assessment of organic farming, the carbon-based farming process stands out as more environmentally friendly, nurturing soil health and preserving environmental integrity, thus contributing to the well-being of patrons. Notably, the global carbon-based yield bazaar is currently experiencing unparalleled growth. In essence, organic agriculture serves as a holistic catalyst for promoting the vigor of a homeland's consumers, the biological well-being of the realm, and comprehensive pecuniary development through income generation.

REFERENCES

Aceleanu, M. I. (2016). Sustainability and competitiveness of Romanian farms through organic agriculture. *Sustainability*, *8*, 245.

Adekiya, A. O., Agbede, T. M., Aboyeji, C. M., Dunsin, O., & Simeon, V. T. (2019). Effects of biochar and poultry manure on soil characteristics and the yield of radish. *Scientia Horticulturae, 243*, 457–463.

Adolph, B., & Butterworth, J. (2002). *Soil fertility management in semi-arid India: Its role in agricultural systems and the livelihoods of poor people*. Natural Resources Institute.

Aloo, B. N., Makumba, B. A., & Mbega, E. R. (2020). Plant growth promoting rhizobacterial biofertilizers for sustainable crop production: The past, present, and future. *Preprints*, 2020090650.

Baker, B. P., Cooley, D., Futrell, S., Garling, L., Gershuny, G., Green, T. A., & Young, S. L. (2015). *Organic agriculture and integrated pest management: Synergistic partnership needed to improve the sustainability of agriculture and food systems*. https://organicipmwg.files.wordpress.com/2015/07/white-paper.pdf

Barik, A., & Sarkar, N. (2017). Organic farming in India: Present status, challenges and technological break through. In *3rd International Conference on Bio-Resource and Stress Management, Jaipur*.

Bashir, Z. (2021). Characterization of potassium solubilizing bacteria from Rhizospheric soils of Cherry (Prunusavium). *Research Journal of Agricultural Sciences*, *12*(2), 494–496.

Bashir, Z., Hamid, B, Zargar, M. Y., & Kundoo, A. A. (2021). Morphological and biochemical characterization of potassium solubilizing bacteria from Rhizospheric soils of Pinus (*Pinus wallichiana*). *Journal of Research and Development*, *21*, 58–63.

Bashir, Z., Zargar, M. Y., Baba, Z. A., & Mohiddin, F. A. (2018). Solubilization of potassium containing mineral by bacteria from Kashmir iris (*Iris kashmiriana*) Rhizosphere. *The Pharma Innovation Journal*, *17*(4), 747–750.

Bashir, Z., Zargar, M. Y., Baba, Z. A., Mohiddin, F. A., Peer, F. A., Mir, S. A., & Najar, G. R. (2018a). Isolation and characterization of phosphorus solubilizing bacteria (PSB) from Rhizospheric soils of Apple (*Malus Domestica*). *Journal of Pharmacognosy and Phytochemistry*, *7*(2), 2997–3000.

Bashir, Z., Zargar, M. Y., Bhat, T. A., Kousar, S., Baba, Z. A., & Mohiddin, F. A. (2018b). Solubilization of phosphorus containing mineral by bacteria from Rhizospheric region of Walnut (*Juglans regia*). *International Journal of Current Microbiology and Applied Sciences*, *7*(8), 2391–2398.

Bashir, Z., Zargar, M. Y., Mohiddin, F. A., Baba, Z. A., & Hamid, B. (2019). Isolation and biochemical characterization of phosphorus solubilizing bacteria from Rhizospheric region of Apricot and Peach. *Journal of Research and Development University of Kashmir*, *19*, 972–5407.

Bashir, Z., Zargar, M. Y., Mohiddin, F. A., Kousar, S., Husain, M., & Rasool, F. (2017). Phosphorus solubilizing microorganisms: Mechanism and diversity. *International Journal of Communication Systems*, *5*(5), 666–673.

Bashir, Z., Zargar, M. Y., & Vishwakarma, D. K. (2019b). Potassium solubilizing microorganisms for sustainable agriculture. *Applied Agricultural Practices for Mitigating Climate Change*, *2*, 17–28.

Bengtsson, J., Ahnström, J., & Weibull, A. C. (2005). The effects of organic agriculture on biodiversity and abundance: A meta-analysis. *Journal of Applied Ecology*, *42*(2), 261–269.

Benke, K., & Tomkins, B. (2017). Future food-production systems: Vertical farming and controlled-environment agriculture. *Sustainability: Science, Practice and Policy*, *13*, 13–26.

Benton, T. G., Vickery, J. A., & Wilson, J. D. (2003). Farmland biodiversity: Is habitat heterogeneity the key? *Trends in Ecology & Evolution*, *18*, 182–188.

Berg, C., & Naturschutz und Geologie Mecklenburg-Vorpommern. Landesamt für Umwelt. (2004). *Die Pflanzengesellschaften Mecklenburg-Vorpommerns under Gefährdung: Text band*. Weissdorn-Verlag.

Bhardwaj, M., & Dhiman, M. (2019). Growth and performance of organic farming in India: What could be the future prospects? *Journal of Current Science*, *20*, 1–8.

Bian, B., Hu, X., Zhang, S., Yang, Z., Yang, W., & Zhang, L. (2019). Pilot-scale composting of typical multiple agricultural wastes: Parameter optimization and mechanisms. *Bioresource Technology*, *287*, 121482.

Birkhofer, K., Bezemer, T. M., Bloem, J., Bonkowski, M., Christenen, S., & Dubois, D. (2008). Long-term organic farming fosters below and aboveground biota: Implications for soil quality, biological control and productivity. *Soil Biology and Biochemistry*, *40*, 2297–2308.

Brust, G. E. (2019). Management strategies for organic vegetable fertility. In B. Biswas & S. A. Micallef (Eds.). *Safety and Practice for Organic Food* (pp. 193–212). Academic Press.

Bucci, G., Bentivoglio, D., & Finco, A. (2018). Precision agriculture as a driver for sustainable farming systems: State of art in literature and research. *Calitatea*, *19*, 114–121.

Caldwell, B., Rosen, E. B., Sideman, E., Shelton, A. M., & Smart, C. D. (2015). *Resource guide for organic insect and disease management* (pp. 1–201). New York State Agricultural Experiment Station.

Campiglia, E., Radicetti, E., & Mancinelli, R. (2018). Floristic composition and species diversity of weed community after 10 years of different cropping systems and soil tillage in a Mediterranean environment. *Weed Research*, *58*(4), 273–283.

Chaichi, W., Djazouli, Z., Zebib, B., & Merah, O. (2018). Effect of vermicompost tea on Faba bean growth and yield. *Compost Science & Utilization*, *26*(4), 279–285.

Chamberlain, D. E., Wilson, J. D., & Fuller, R. J. (1999). A comparison of bird populations on organic and conventional farm systems in southern Britain. *Biological Conservation*, *88*(3), 307–320.

Chawla, R., & Sadawarti, R. K. (2020). Effect of bio-fertilizers and organic manures on growth, yield and fruit quality of fruit crops. *Plant Arch*, *20*, 3767–3768.

Dagostin, S., Schärer, H. J., Pertot, I., & Tamm, L. (2011). Are there alternatives to copper for controlling grapevine downy mildew in organic viticulture? *Crop Protection*, *30*, 776–788.

Datnoff, L. E., Elmer, W. H., & Huber, D. M. (2007). *Mineral nutrition and plant disease* (p. 278). APS Press.
Deshmukh, M. S., & Babar, N. (2015). Present status and prospects of organic farming in India. *European Academic Research, 3*, 4271–4287.
Drinkwater, L. E., Letourneau, D. K., Workneh, F., Van Bruggen, A. H. C., & Shennan, C. (1995). Fundamental differences between conventional and organic tomato agroecosystem in California. *Ecological Applications, 5*, 1098–1112.
Durán-Lara, E. F., Valderrama, A., & Marican, A. (2020). Natural organic compounds for application in organic farming. *Agriculture, 10*, 41.
Edwards-Jones, G., & Howells, O. (2001). The origin and hazard of inputs to crop protection in organic farming systems: Are they sustainable? *Agricultural Systems, 67*, 31–47.
Eglington, S. (2009). Understanding the causes of decline in breeding wetland bird numbers in England. *British Trust for Ornithology, 562*, 124.
Etingoff, K (Ed.). (2016). *Sustainable agriculture and food supply: Scientific, economic, and policy enhancements*. CRC Press.
Eyhorn, F., Muller, A., & Reganold, J. P. (2019). Sustainability in global agriculture driven by organic farming. *Nature Sustainability, 2*, 253–255.
Fabian, Y., Sandau, N., Bruggisser, O. T., Aebi, A., Kehrli, P., Rohr, R. P., & Bersier, L. F. (2013). The importance of landscape and spatial structure for hymenopteran-based food webs in an agro-ecosystem. *Journal of Animal Ecology*, 1203–1214.
FAO. (2018). Transforming food and agriculture to achieve the SDGs: 20 interconnected actions to guide decision-makers. *Technical Reference Document*.
Finckh, M. R., & Van Bruggen, A. H. C. (2015). General introduction. In M. R. Finckh, A. H. C. van Bruggen & L. Tamm (Eds.), *Plant diseases and their management in organic agriculture* (pp. 3–11). APS Press.
Flade, M. (2012). Von der energiewende zum biodiversitats-desaster-zur lage des vogelschutzes in Deutschland. *RSCH, 23*.
Gamage, A., Gangahagedara, R., Gamage, J., Jayasinghe, N., Kodikara, N., Suraweera, P., & Merah, O. (2023). Role of organic farming for achieving sustainability in agriculture. *Farming System, 1*(1), 100005.
Garg, V., Chand, S., Chhillar, A., & Yadav, A. (2005). Growth and reproduction of Eisenia Foetida in various animal wastes during vermicomposting. *Applied Ecology and Environmental Research, 3*, 51–59.
Gattinger, A., Muller, M., Haeni, C., Skinner, A., Fliessbach, N., Buchmann, P., Mäder, M., Stolze, P., Smith, N. E. H., & Scialabba, H. (2012). Enhanced top soil carbon stocks under organic farming. *Proceedings of the National Academy of Sciences, 109*, 18226–18231.
Giller, K. E., Hijbeek, R., Andersson, J. A, & Sumberg, J. (2021). Regenerative agriculture: An agronomic perspective. *Outlook Agricultural, 50*, 13–25.
Gomiero, T., Pimentel, D., & Paoletti, M. G. (2011). Environmental impact of different agricultural management practices: Conventional vs. organic agriculture. *Critical Reviews in Plant Sciences, 30*, 95–124.
Guo, Z., Zhang, J., Fan, J., Yang, X., Yi, Y., Han, X., Wang, D., Zhu, P., & Peng, X. (2019). Does animal manure application improve soil aggregation? Insights from nine long-term fertilization experiments. *Science of the Total Environment, 660*, 1029–1037.
Habel, J. C., Ulrich, W., Biburger, N., Seibold, S., & Schmitt, T. (2019). Agricultural intensification drives butterfly decline. *Insect Conservation and Diversity, 12*(4), 289–295.
Haber, W. (2014). *Landwirtschaft und Naturschutz*. John Wiley.
Hajjar, R., Jarvis, D. I., & Gemmill-Herren, B. (2008). The utility of crop genetic diversity in maintaining ecosystem services. *Agriculture, Ecosystems & Environment, 123*, 261–270.
Hamid, B., & Bashir, Z. (2019a). Potassium solubilization microorganisms: An Alternative technology to chemical fertilizers. *Journal of Research and Development University of Kashmir, 19*, 972–5407.

Headrick, D. (2021). The future of organic insect pest management: Be a better entomologist or pay for someone who is. *Insects*, *12*(2), 7.

Heldbjerg, H., Fox, A. D., Levin, G., & Nyegaard, T. (2016). The decline of the Starling Sturnus vulgaris in Denmark is related to changes in grassland extent and intensity of cattle grazing. *Agriculture, Ecosystems & Environment*, *230*, 24–31.

Henning, J., Baker, L., & Thomassin, P. J. (1991). Economics issues in organic agriculture. *Canadian Journal of Agricultural Economics*, *39*, 877–889.

Herzon, I., Ekroos, J., Rintala, J., Tiainen, J., Seimola, T., & Vepsäläinen, V. (2011). Importance of set-aside for breeding birds of open farmland in Finland. *Agriculture, Ecosystems & Environment*, *143*(1), 3–7.

Holb, I., Heijne, B., & Tamm, L. (2015). Organic apple disease management, in plant diseases and their management in organic agriculture. In M. R. Finckh, A. H. C. van Bruggen & L. Tamm. APS Press, pp. 319–334.

Hole, D. G., Perkins, A. J., Wilson, J. D., Alexander, I. H., Grice, P. V., & Evans, A. D. (2005). Does organic farming benefit biodiversity? *Biological Conservation*, *122*(1), 113–130.

Holka, M., Kowalska, J., & Jakubowska, M. (2022). Reducing carbon footprint of agriculture-Can organic farming help to mitigate climate change? *Agriculture*, *12*, 1383.

IFOAM (International Movement of Organic Agriculture Movements). (2018). *Definition of organic agriculture*. https://www.ifoam.bio/en/organic-landmarks/definition-organic-agriculture

Isbell, F., Reich, P. B., Tilman, D., Hobbie, S. E., Polasky, S., & Binder, S. (2013). Nutrient enrichment, biodiversity loss, and consequent declines in ecosystem productivity. *Proceedings of the National Academy of Sciences*, *110*, 11911–11916.

Kasselaki, A. M., Goumas, D., Tamm, L., Fuchs, J., Cooper, J., & Leifert, C. (2011). Effect of alternative strategies for the disinfection of tomato seed infected with bacterial canker (Clavibacter michiganensis subsp. michiganensis). *NJAS: Wageningen Journal of Life Sciences*, *58*, 145–147.

Kehat, M., Anshelevich, L., Harel, M., & Dunkelblum, E. (1995). Control of the codling moth (*Cydia pomonella*) in apple and pear orchards in Israel by mating disruption. *Phytoparasitica*, *23*, 285–296.

Keqiang, C., & Van Bruggen, A. H. (2001). Inhibitory efficacy of several plant extracts and plant products on Phytophthora infestans. *Journal of Hebei Agricultural University*, *24*, 90–96.

Kirchmann, H., Kätterer, T., Bergström, L., Börjesson, G., & Bolinder, M. A. (2016). Flaws and criteria for design and evaluation of comparative organic and conventional cropping systems. *Field Crops Research*, *186*, 99–106.

Koch, E., & Groot, S. P. C. (2015). Health management for seeds and other organic propagation material. In M. R. Finckh, A. H. C. van Bruggen & L. Tamm (Eds.). *Plant Diseases and their Management in Organic Agriculture* (pp. 189–203). APS Press.

Krauss, J., Gallenberger, I., & Steffan-Dewenter, I. (2011). Decreased functional diversity and biological pest control in conventional compared to organic crop fields. *PloS One*, *6*(5), e19502.

Kremen, C., Williams, N. M., & Thorp, R. W. (2002). Crop pollination from native bees at risk from agricultural intensification. *Proceedings of the National Academy of Sciences of the United States of America*, *99*, 16812–16816.

Lamers, J. G., Wanten, P., & Blok, W. J. (2004). Biological soil disinfestation: A safe and effective approach for controlling soilborne pests and diseases. *Agroindustria*, *3*, 289–291.

Lavado, R. S., Compost, O. D., Compostaje, P., & USO, P. (2012). In Compostaje En la Argentina: Experiencias en la Argentina: Experiencias de Producción, Calidad y USO. In M. J. Mazzarino & P. Satti (Eds.). *Orientación Gráfica Editora: Buenos Aires, Argentina*, pp. 3–12.

Lichtenberg, E. M., Kennedy, C. M., Kremen, C., Batary, P., Berendse, F., Bommarco, R., & Crowder, D. W. (2017). A global synthesis of the effects of diversified farming systems on arthropod diversity within fields and across agricultural landscapes. *Global Change Biology*, *23*(11), 4946–4957.

Lim, S. L., Wu, T. Y., Lim, P. N., & Shak, K. P. Y. (2015). The use of vermicompost in organic farming: Overview, effects on soil and economics. *Journal of the Science of Food and Agriculture*, *95*, 1143–1156.

Litterick, A. M., Harrier, L., Wallace, C. A., & Wood, M. (2004). The role of uncomposted materials, composts, manures, and compost extracts in reducing pest and disease incidence and severity in sustainable temperate agricultural and horticultural crop production—A review. *Critical Reviews in Plant Sciences*, *23*, 453–479.

Lori, M., Symnaczik, S., Mäder, P., De, Deyn, G., & Gattinger, A. (2017). Organic farming enhances soil microbial abundance and activity-A meta-analysis and meta-regression. *PloS One*, *12*(7), e0180442.

Lucchi, A., Loni, A., Gandini, L. M, Scaramozzino, P., Ioriatti, C., Ricciardi, R., & Schearer, P. W. (2017). Using herbivore-induced plant volatiles to attract lacewings, hoverflies and parasitoid wasps in vineyards: Achievements and constraints. *Bulletin of Insectology*, *70*, 273–282.

Lykouressis, D., Fantinou, A., Toutouzas, S., Perdikis, D., & Samartzis, D. (2005). Management of the pink bollworm Pectinophora gossypiella (Saunders) (Lepidoptera: Gelechiidae) by mating disruption in cotton fields. *Crop Protection*, *24*, 177–183.

Magistrali, A., Cooper, J., George, D., & Standen, J. (2023). Identifying and implementing regenerative agricultural practices in challenging environments: Experiences of farmers in the North of England. *BBSRC Project Report No. PR. 640–09*.

Malik, D. S., Sharma, A. K., Sharma, A. K., Thakur, R., & Sharma, M. (2020). A review on impact of water pollution on freshwater fish species and their aquatic environment. *Advances in Environmental Pollution Management: Wastewater Impacts and Treatment Technologies*, *1*, 10–28.

McCracken, D. I., & Tallowin, J. R. (2004). Swards and structure: The interactions between farming practices and bird food resources in lowland grasslands. *Ibis*, *146*, 108–114.

Meehan, T. D., Werling, B. P., Landis, D. A., & Gratton, C. (2011). Agricultural landscape simplification and insecticide use in the Midwestern United States. *Proceedings of the National Academy of Sciences*, *108*, 11500–11505.

Merah, O., & Djazouli, Z. E. (2016). Toxicity evaluation of Dittrichia Viscosa L's Aqueous extracts in combination with bio-adjuvant silene Fuscata on Chaitophorus Leucomelas Koch. (Hom. Aphididae) and on Biocenotic Resumption of Functional Groups. *Jordan Journal of Agricultural Sciences*, *405*(3977), 1–18.

Merah, O., Djazouli, Z. E., & Zebib, B. (2021). Aqueous extract of Algerian nettle (*Urtica dioica* L.) as possible alternative pathway to control some plant diseases. *Iranian Journal of Science and Technology, Transactions A: Science*, *45*, 463–468.

Möller, K. (2018). Soil fertility status and nutrient input–Output flows of specialised organic cropping systems: A review. *Nutrient Cycling in Agroecosystems*, *112*, 147–164.

Momman, N. (2008). Biological soil disinfestation (BSD) of soilborne pathogens and its possible mechanisms. *Japan Agricultural Research Quarterly*, *42*, 7–12.

Mondelaers, K., Aertsens, J., & Van Huylenbroeck, G. (2009). A meta-analysis of the differences in environmental impacts between organic and conventional farming. *British Food Journal*, *111*(10), 1098–1119.

Monteiro, A., Santos, S., & Gonçalves, P. (2021). Precision agriculture for crop and livestock farming-brief review. *Animals (Basel)*, *11*(8), 2345.

Muller, A., Schader, C., El-Hage Scialabba, N., Brüggemann, J., Isensee, A., Erb, K. H., Smith, P., Klocke, P., Leiber, F., Stolze, M., & Niggli, U. (2017). Strategies for feeding the world more sustainably with organic agriculture. *Nature Communications*, *8*(1), 1–13.

Murmu, K., Murmu, P., Das, A., & Bandopadhyay, S. P. (2022). Organic agriculture: As a climate change adaptation and mitigation strategy. *Zeichen Journal*, *8*(3), 171–187.

Ni, J. J., Bordoloi, S., Shao, W., Garg, A., Xu, G., & Sarmah, A. K. (2020). Two-year evaluation of hydraulic properties of biochar-amended vegetated soil for application in landfill cover system. *Science of the Total Environment*, *712*, 136486.

Nova-Institute GmbH. (2018). *High-tech strategies for small farmers and organic farming* [Press Release].

Orden, L. (2018). Evaluación del proceso de compostaje de residuos sólidos orgánicos: Respuesta agronómica de su utilización en un cultivo de cebolla (*Allium cepa L.*). Universidad nacional del sur.

Orkun, B., Kovanci, C. S., Walgenbach, J. F., & Kennedy, G. G. (2005). Comparison of mating disruption with pesticides for management of oriental fruit moth (Lepidoptera: *Tortricidae*) in North Carolina Apple Orchards. *Journal of Economic Entomology*, *98*, 1248–1258.

Ostandie, B., Giffard, O., Bonnard, B., Joubard, S., Richart, C. D., & Thiéry, A. (2021). Rusch multi-community effects of organic and conventional farming practices in vineyards. *Scientific Reports*, *11*, 11979.

Ozlu, E., Arriaga, F. J., Bilen, S., Gozukara, G., & Babur, E. (2022). Carbon footprint management by agricultural practices. *Biology*, *11*, 1453.

Patil, K. A., & Kale, N. R. (2016). A model for smart agriculture using IoT. In *Proceedings of the 2016 International Conference on Global Trends in Signal Processing, Information Computing and Communication, Jalgaon, India, 22–24 December* (pp. 543–545). IEEE.

Patsaki, E. E., Bourbos, V. A., Bbalis, C., & Barbbopoulou, E. A. (2005). Use of liohumus extract for ecological control of potato late blight, in Researching Sustainable Systems. In U. Köpke, U. Niggle, D. Neuhoff, P. Cornish, W. Lockeretz & H. Willer. *First Scientific Conference of the International Society of Organic Agriculture Research (ISOFAR)* (pp. 549–551). Adelaide.

Pickett, J. A. (2013). Food security: Intensification of agriculture is essential, for which current tools must be defended and new sustainable technologies invented. *Food and Energy Security*, *2*, 167–173.

Pimentel, D. (2013). Soil erosion threatens food production. *Agriculture*, *3*, 443–463.

Prono, A. R., Martín, C. A., & Mazzarino, M. J. (2010). Efectos de La Relación C/N y El Contenido de Humedad Sobre La Eficiencia Del Compostaje En Reactores an Escala Banco-Laboratorio. In *Proceedings of the Actas XXII Congreso Argentino de la Ciencia Del Suelo, Rosario*. Ibero-American Society of Physics and Environmental Chemistry.

Purnhagen, K. P., Clemens, S., Eriksson, D., Fresco, L. O., Tosun, J., Qaim, M., Visser, R., Weber, A., Wesseler, J., & Zilberman, D. (2021). Europe's farm to fork strategy and its commitment to biotechnology and organic farming: Conflicting or complementary goals? *Trends Plant Science*, *26*(6), 600–606.

Quinn, J. E., Awada, T., Trindade, F., Fulginiti, L., & Perrin, R. (2017). Combining habitat loss and agricultural intensification improves our understanding of drivers of change in avian abundance in a North American cropland anthrome. *Ecology and Evolution*, *7*(3), 803–814.

Rahmann, G. (2011). Biodiversity and organic farming: What do we know? *TI Agriculture and Forestry Research*, *3*, 189–208.

Reddy, S. B. (2010). Organic farming: Status, issues and prospects-A review. *Agricultural Economics Research Review*, *23*, 343–358.

Reeve, J. R., Hoagland, L. A., Villalba, J. J., Carr, P. M., Atucha, A., Cambardella, C., Davis, D. R., & Delate, K. (2016). Organic farming, soil health, and food quality: Considering possible links. *Advances in Agronomy*, *137*, 319–367.

Reganold, J. P., & Wachter, J. (2016). Organic agriculture in the twenty-first century. *Nature Plants*, *2*, 15221.

Rempelos, L., Almuayrifi, M. S. B., Baranski, M., Tetard-Jones, C., Barkla, B., Cakmak, I., Ozturk., L., Cooper, J., Volakakis, N., & Hall, G. (2020). The effect of agronomic factors on crop health and performance of winter wheat varieties bred for the conventional and the low input farming sector. *Field Crops Research, 254*, 107822.

Rempelos, L., Baranski, M., Wang, J., Adams, T. N., Adebusuyi, K., Beckman, J. J., Brockbank, C. J., Douglas, B. S., Feng, T., & Greenway, J. D. (2021). Integrated soil and crop management in organic agriculture: A logical framework to ensure food quality and human health? *Agronomy, 11*, 2494.

Rempelos, L., Wang, J., Sufar, E. K., Almuayrifi, M. S. B., Knutt, D., Leifert, H., Leifert, A., Wilkinson, A., Shotton, P., & Hasanaliyeva, G. (2023). Breeding bread-making wheat varieties for organic farming systems: The need to target productivity, robustness, resource use efficiency and grain quality traits. *Foods, 12*, 1209.

Research European Commission. (2017). *Precision farming: Sowing the seeds of a new agricultural revolution*. The Community Research and Development Information Service (CORDIS).

Robinson, R. A., & Sutherland, W. J. (2002). Post-war changes in arable farming and biodiversity in Great Britain. *Journal of Applied Ecology, 39*(1), 157–176.

Rydberg, N. T., & Milberg, P. (2000). A survey of weeds in organic farming in Sweden. *Biological Agriculture & Horticulture, 18*(2), 175–185.

Saffeullah, P., Nabi, N., Liaqat, S., Anjum, N. A., Siddiqi, T. O., & Umar, S. (2021). Organic agriculture: Principles, current status, and significance. In K. R. Hakeem, G. H. Dar, M. A. Mehmood & R. A. Bhat (Eds.). *Microbiota and Biofertilizers, 18*(2), 173–182.

Salim, D., De Caro, P., Merah, O., & Chbani, A. (2020). Control of post-harvest citrus green mold using Ulva lactuca extracts as a source of active substances. *International Journal of Bio-resource and Stress Management, 11*(3), 287–296.

Santini, G., Biondi, N., Rodolfi, L., & Tredici, M. R. (2021). Plant biostimulants from cyanobacteria: An emerging strategy to improve yields and sustainability in agriculture. *Plants, 10*, 643.

Seufert, V., & Ramankutty, N. (2017). Many shades of gray-The context-dependent performance of organic agriculture. *Science Advances, 3*(3), e1602638.

Sharma, A. K., & Goyal, R. K. (2000). Addition in tradition on agroforestry in arid zone. *LEISA-India, 2*, 19–20.

Slusarenko, A. J., Patel, A., & Portz, D. (2008). Control of plant diseases by natural products: Allicin from garlic as a case study. *European Journal of Plant Pathology, 121*, 313–322.

Stevens, G., & Lewis, E. (2017). Status of entomopathogenic nematodes in integrated pest management strategies in the USA. In M. Mahfouz, M. Abd-Elgawad, T. H. Askary & J. Coupland (Eds.). *Biocontrol Agents: Entomopathogenic and Slug Parasitic Nematodes* (pp. 289–311). CAB International.

Tamm, L., Speiser, B., & Holb, I. (2015). Direct control of airborne diseases. In M. R. Finckh, A. H. C. van Bruggen & L. Tamm (Eds.). *Plant Diseases and their Management in Organic Agriculture* (pp. 205–216). APS Press.

Tejada, M., Rodríguez-Morgado, B., Gómez, I., Franco-Andreu, L., Benítez, C., & Parrado, J. (2016). Use of biofertilizers obtained from sewage sludges on maize yield. *European Journal of Agronomy, 78*, 13–19.

Tognetti, C., Mazzarino, M. J., & Laos, F. (2007). Improving the quality of municipal organic waste compost. *Bioresource Technology, 98*, 1067–1076.

Tscharntke, T., Batáry, P., & Dormann, C. F. (2011). Set-aside management: How do succession, sowing patterns and landscape context affect biodiversity? *Agriculture, Ecosystems & Environment, 143*(1), 37– 44.

Tuck, S. L., Winqvist, C., Mota, F., Ahnström, J., Turnbull, L. A., & Bengtsson, J. (2014). Land-use intensity and the effects of organic farming on biodiversity: A hierarchical meta-analysis. *Journal of Applied Ecology, 51*(3), 746–755.

Uchida, K., & Ushimaru, A. (2014). Biodiversity declines due to abandonment and intensification of agricultural lands: Patterns and mechanisms. *Ecological Monographs*, *84*(4), 637–658.

Vetter, R. S., Millar, J. G., Vickers, N. J., & Baker, T. C. (2006). Mating disruption of carob moth, Ectomyelois ceratoniae, with a sex pheromone analog. *Southwestern Entomologist*, *31*, 33–47.

Westerman, P. S., Wes, J. S., Kropff, M. J., & Van der Werf, W. (2003). Annual losses of weed seeds due to predation in organic cereal fields. *Journal of Applied Ecology*, *40*, 824–836.

Willer, H., & Lernoud, J. (2017). The world of organic agriculture. *Statistics and Emerging Trends*. FiBL & IFOAM-Organics International.

Willer, H., & Lernoud, J. (2019). The world of organic agriculture. *Statistics and Emerging Trends*. Research Institute of Organic Agriculture (FiBL), Frick and IFOAM-Organics International.

Wilson, H., Bodwitch, H., Carah, J., Daane, K. M., Getz, C. M., Grantham, T. E., & Butsic, V. (2019). First known survey of cannabis production practices in California. *California Agriculture*, *73*, 119–127.

Wittwer, S. F., Bender, K., Hartman, S., Hydbom, R. A. A., Lima, V. L., Nemecek, F., Oehl, P. A., & Olsson, O. (2021). Organic and conservation agriculture promote ecosystem multifunctionality. *Science Advances*, *7*.

Xiao, X., Chen, B., & Zhu, L. (2014). Transformation, morphology, and dissolution of silicon and carbon in rice straw-derived bio chars under different pyrolytic temperatures. *Environmental Science & Technology*, *48*, 3411–3419.

Yousef, M., Aranda-Valera, E., & Quesada-Moraga, E. (2018). Lure-and-infect and lure-and-kill devices based on *Metarhizium brunneum* for spotted wing Drosophila control. *Journal of Pest Science*, *91*, 227–235.

Zhu, X., Chen, B., Zhu, L., & Xing, B. (2017). Effects and mechanisms of biochar-microbe interactions in soil improvement and pollution remediation: A review. *Environmental Pollution*, *227*, 98–115.

4 Biodiversity in Organic Farming

The Role of Organic Farming in Promoting Biodiversity

Saniya Rafiq, Parvaze Ahmad Wani, Burhan Hamid, Mehrun Nisa, and Zaffar Bashir

4.1 INTRODUCTION

The use of biological fertilizers and pest control obtained from plant and animal waste is known as organic farming. Organic farming was founded to combat the damage that chemical pesticides and artificial fertilizers were causing to the environment. Visionaries like F.H. King, Rudolf Steiner, and Sir Albert Howard established the ideas behind organic agriculture in the early 1900s. Sir Albert Howard forwarded the uptake of sustainable farming practices in the West after gaining knowledge about them in India. Organic farming was further forwarded through various publications and texts by personalities such as J.I. Rodale and his son Robert. A sharp rise in the demand for organic food was reported in the 1960s with the publication of Rachel Carson's *Silent Spring*, which emphasized the harm brought by pesticides to the environment.

The United States Department of Agriculture (USDA) defines organic farming as "a system that primarily avoids and minimizes the use of synthetic inputs" (e.g., fertilizers, pesticides, hormones, feed additives, etc). Crop rotations, crop leftovers, animal manures, off-farm organic waste, mineral-grade rock inputs, and biological nutrient mobilization techniques are among the potent ways that organic farming promises plant protection.

There are at least 3.1 million organic farmers managing 72.3 million hectares of cropland throughout 187 countries (Willer et al., 2021). In 2019, the global trade in organic food and beverage products exceeded 106 billion euros, indicating the rising interest of modern consumers in adopting environmentally sensitive and health conscious eating practices. Approximately, almost 75% of global production in agriculture is derived from certified organic farming practices. It is among the world's most widespread and economical agricultural systems. In 2019, the world's largest producer of organic foods was India (Willer et al., 2021). In that year, there were more than 1.4 million organic food producers in India, more than in the other nine largest countries in the world put together (Willer et al., 2020).

In the past couple of decades, agriculture has been focused on industrial and highly intensive agricultural methods in order to provide enough food for the world's

DOI: 10.1201/9781003491460-4

population—a goal that has not been achieved yet. These kinds of agricultural methods also have detrimental effects on the environment, resulting in reduction of biodiversity and farm bird index, where a decline has been put on record in Slovenia since 2008. The result was equally catastrophic for biodiversity as many farmers increased their yield and became highly industrialized, while those who were unable to do so were diminished and commonly forced to leave their land.

The emphasis on the sustainability and minimal environmental impact of organic farming methods resulted in their increasing popularity. In order to promote biodiversity, organic farming is essential, as this chapter manifests. We will discuss all the ways that organic farming practices support biodiversity protection and enhancement, from soil health to agro ecosystem management. In order to illustrate the different ways that organic agriculture contributes to the preservation of diverse ecosystems and the health of our world, this chapter examines the complex relationship between organic agriculture and biodiversity. This chapter will explore the essential role of biodiversity in agricultural systems and ecological systems.

Organic farming which emphasizes environmentally friendly practices enhances and preserves biodiversity. This chapter explores the complex interrelationship between biodiversity and organic farming, highlighting the different ways that organic agriculture supports the preservation of diverse ecosystems and the health of our world. The basic significance of biodiversity in ecological systems and organic agriculture is discussed in detail in this chapter. It addresses several ecosystem services that biodiversity provides, such as soil health, pollination, insect control, and nutrient cycling. The chapter also examines the negative effects of biodiversity loss, including decreased agricultural productivity and increased susceptibility to illnesses and pests.

With the aim of ensuring that there is enough food for everyone, agriculture has focused on industrial and extremely intense agricultural methods over the past several decades. This goal has not been met yet. These farming practices also have a number of detrimental environmental repercussions such as decline in biodiversity, as seen by Slovenia's farm bird index since 2008. A large number of farmers maximized their working and were highly mechanized, while those who could not stay updated to the times were disempowered and occasionally pressured to give up on their land, which had a dreadful impact on biodiversity. Meeting the growing demand for food in a way that is socially and environmentally sustainable over an extended period of time is crucial.

Farming systems can be designed both as productive as well as also maintaining or improving the requirements of ecosystem services (e.g., pollination services, carbon sequestration, energy efficiency, lowering potential global warming, biodiversity, soil quality, nutrient management, water-holding capacity, control of weeds, diseases and pests, resistance and resilience to crop productivity and climate change, and so on) and agro ecosystem resilience and sustainability.

4.2 CONVENTIONAL FARMING VS ORGANIC FARMING

The main difference between conventional and organic farming is that the conventional farming utilizes chemical agents to resist pesticides and weedicides while organic farming does not allow for any chemical interference. Conventional farming uses a large number of chemical pesticides and fertilizers to maximize yields per acre. With the aim of reducing their harmful influence on the environment, chemical fertilizers and pesticides are not used in organic agricultural practices. Comparatively, by applying livestock dung soil fertility is increased (Lori et al., 2017). Organic farming lowers the risk for aquatic environmental toxicity and soil erosion in contrast to conventional farming (Wittwer et al., 2021).

Moreover, organic farming increases revenue per hectare and generates plenty of microorganisms as well as boosts soil biodiversity. However, lower yields can be possible (–22%) depending on the crop, denoting an exchange between agricultural productivity and environmental preservation (Wittwer et al., 2021).

Organic farming does not utilize synthetic chemicals, genetically modified organisms (GMOs), and antibiotics in cattle, resulting in the production of safer and healthier food. Figure 4.1 depicts the use of various types of organic constituents for augmenting organic farming.

In conventional agriculture, the utilization of synthetic chemicals and genetically modified organisms (GMOs) may give rise to concern regarding the safety of food and its long-term health impacts.

Organic farming outperforms conventional farming in terms of energy productivity. In contrast to conventional farming, where the nitrogen fertilizer only accounts for 41% of energy expenses, organic systems utilize 45% less energy overall. Organic foods are currently in demand resulting in their increased cost. Furthermore, the cost of production for organic products is not the only element that influences their price as additional considerations include environmental and food security.

Organic farming generates less pollution compared to conventional farming in terms of its impact on the environment and changes in the climate. Since organic farming builds persistency within the agricultural ecosystem, it makes the system more potent to the adverse consequences of climate change. It develops efficient and ecologically friendly farming techniques that prevent soil erosion and are resistant towards changes in temperature and drought. In addition, organic farming also inspires conservation efforts, restoration projects, and sustainable and ecologically friendly management. Organic farming has fewer economic demands as compared to conventional farming. Moreover, in response to the unpredictable effect of

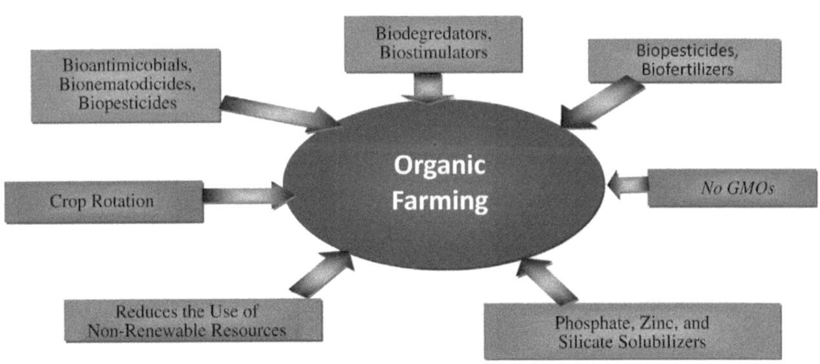

FIGURE 4.1 Use of various types of organic constituents for augmenting organic farming.

climate change, organic farming provides assistance to communities and farmers. Furthermore, organic farming satisfies a wide range of conditions found for effective adaptation strategies (Muller, 2009; Murmu et al., 2022).

Researchers have been interested in organic farming practices because organic farming posse's potential effects on biodiversity. Now let's explore the intriguing connection between organic farming and the environment:

4.3 PRINCIPLES OF ORGANIC FARMING

4.3.1 The Principle of Health

Organic farming maintains and improves the health of humans, animals, and plants. According to this theory, healthy crops grow in healthy soil, which benefits both human and animal health. Organic farming is especially necessary to generate beneficial and high quality food that improves health and well-being. In order to obtain this, pesticides, fertilizers, pharmaceuticals, and food additives that may be hazardous to the health of animals should not be used in organic farming.

4.3.2 The Principle of Ecology

Organic farming ought to include living ecological systems and cycles as the basis of its foundation; these should be cooperated with, repeated, and sustained. Organic farming must create habitats, maintain genetic and agricultural diversity, and implement a well-designed farming system, in order to achieve ecological balance. The environment, which includes the air, water, biodiversity, ecosystems, climate, and landscapes must be protected and improved by people who deal in production and utilization of organic products.

4.3.3 The Principle of Fairness

Relationships that guarantee justice for all people including farmers, laborers, processors, distributors, traders, and consumers should be the foundation of organic farming. It ought to reduce social costs, promote food sovereignty, and minimize poverty. Figure 4.2 depicts roles of organic farming in agriculture, environment, and human health.

4.3.4 The Principle of Care

Organic farming must be managed with comprehension and responsibility in order to protect the environment, and health of present and future generations, and their well-being. Decision making should be clear, and it must involve participation from all parties involved, taking their needs and values into consideration.

4.4 CONSTITUENTS OF ORGANIC FARMING

Organic farming has become currently an important tool for the improvement of soil fertility as use of synthetic elements for improvement in soil fertility can show

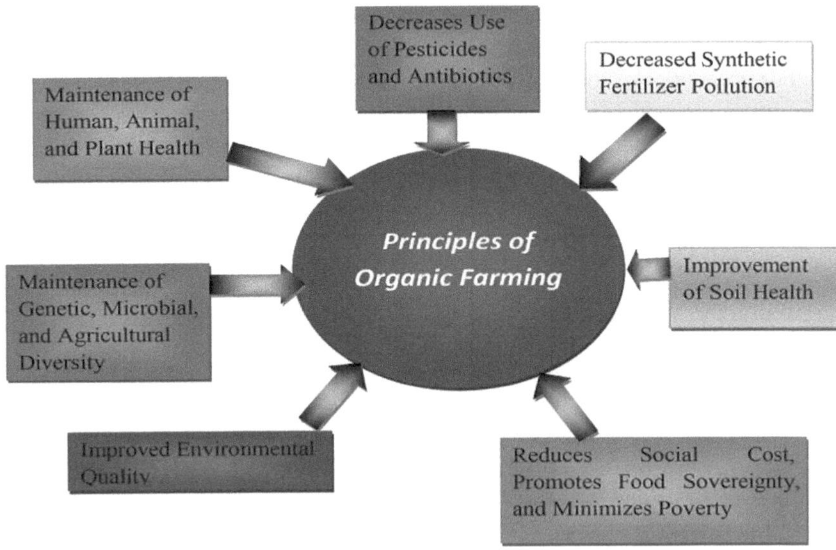

FIGURE 4.2 Roles of organic farming in agriculture, environment, and human health.

toxicity to human and other population. Organic farming constituents which can improve soil fertility for improved crop yield include use of vermin-culture, use of biological agents (microbes), alternate use of crops, biofertilizers, biological agents for disease control, crop residues, farm yard manure, biogas slurry etc (Raj et al., 2024).

4.4.1 Crop Rotation

To achieve good growth in agriculture, it is necessary to alternately use different crops on the same plot of land year by year. Rotation of crops will improve soil fertility and will manage diseases by rotation of the hosts (crops) for different types of pests. Since there will be no specific host available on the yield for specific pests at a specific time, these pests and pathogens will be starved of the nutrients required for their survival during that specific period, and thus crop rotation can manage disease control which can save millions of dollars. Rotation can also improve soil fertility on the field. If alternately legume crops are used, this can provide nitrogen to the soil by fixing it from the atmosphere through nodulation which thus can improve soil fertility through nitrogen fixation which can be used by the next season crop and can improve their yield (Raj et al., 2024).

4.4.2 Use of Organic Matter and Crop Residue

Organic matter and crop residues are good sources of nutrients such as nitrogen, from legume crops such as chickpea, lentil, greengram, cowpea etc, phosphorus from both legume and non-leguminous crops and other types of nutrients from cereals, oil seeds etc. These nutrients present in good quantity can improve soil fertility by tenfold which can improve crop yield in agriculture and thus improve the economy of various populations. Sources of organic matter can be the dead plants, animals, human refuse etc which upon degradation by different types of microbes such as bacteria and fungal species in particular can yield available nutrients for crops to be taken for their nourishment and yield improvement (Raj et al., 2024). Crop residues if tilled in the same field after their harvest can act as good sources of nutrients for the next year's crop upon their degradation by microbes (Raj et al., 2024).

4.4.3 Farm Yard Manure, Green Manuring, and Compost

Sources of farm yard manure and compost include decomposed grass in combination with urine and animal dung and decomposed leftover crops, grasses, stubble, vegetable refuse, and refuse from humans and animals; it also includes industrial and agricultural waste which has been decomposed by either microbes such as bacteria and fungus or by using vermin-culture. In the process of decomposition, complex carbohydrates have been converted to simple carbohydrates to be used by plants for their growth (Raj et al., 2024).

In green manuring, un-decomposed green parts of various plants and green grasses are added to soil and then they are ploughed to be mixed with the soil so they can become food for certain types of microbes which decompose them to increase fertility of the soil and make available nutrients such as phosphorus and nitrogen to be taken up by plants for their growth and development. Crops which are used as green manure in agriculture for increasing fertility of the soil include leguminous crops such as cowpea, cluster bean, sun hemp; fodder crops such as berseem, senji etc (Raj et al., 2024).

4.4.4 Different Types of Biofertilizers and Biopesticides

Different types of biofertilizers such as symbiotic, non-symbiotc, and phosphatic fertilizers are being used in agricultural practices as an organic matter to improve fertility for the enhancement of crop productivity. Symbiotic and non-symbiotic fertilizers used in agriculture include nitrogen-fixing microorganisms such as *Rhizobium*, *Bradyrhizobium*, *Azorhizobium*, *Agrobacterium*, *Azotobacter*, *Sinorhizobium* etc which can fix nitrogen from the atmosphere to be converted into organic form (Ammonia) which can enhance fertility of the soil (Wani et al., 2007). Nitrogen fixation can be achieved by both symbiotic and non-symbiotic bacteria by the formation of nodules on the root system of the legume crops by symbiotic bacteria whereas free-living nitrogen-fixing bacteria can also fix nitrogen without their association with any type of plants. Blue green algae have also been found to be involved in nitrogen fixation to be utilized by the plants. Phosphate solubilizing microorganisms have also been reported to increase fertility of the soil by solubilizing the unavailable phosphate into available form by the release of organic acids. P and K solubilizing microorganisms involved in phosphate and potassium solubilization including species of *Bacillus*, *Pseudomonas*, *Micrococcus*, *Aspergillus*, *Rhizopus*, *Penicillium* etc are prominent species reported for phosphate/potassium solubilization (Bashir et al., 2017; Bashir, Zargar, Baba, et al., 2018a; Bashir, Zargar, Bhat, et al., 2018b, 2019; Bashir, Zargar, Baba, & Mohiddin, 2018; Bashir et al., 2019b; Bashir et al., 2021; Bashir, 2021; Hamid & Bashir, 2019a; Wani & Khan, 2010, 2013). Biopesticides in the form of microbes have also been used to control different types of diseases in agriculture. They have also been reported to act as biofertilizers for crop enhancement.

4.5 ROLE OF ORGANIC FARMING IN SUSTAINABLE AGRICULTURE

4.5.1 Environmental Benefits

4.5.1.1 Protection of Soil and Conservation of Water

Organic farming improves soil and water health by preserving the natural composition of soil in terms of the mineral status, and biodiversity of different types of beneficial microbes and removes disease causing pathogens by the release of antimicrobials from different beneficial microbes. Dependence on the use of synthetic pesticides and fertilizers has reduced as a result of the use of natural growth regulators both bacterial and fungal species. These groups of microorganisms improve soil and plant health by fixing nitrogen through the nodules, solubilizing phosphate and releasing growth regulators. The priority in organic agricultural practices is given to conservation of water. Farmers can reduce water usage through crop rotation, mulching, and by natural irrigation practices to ensure that this precious resource is managed efficiently and sustainably.

4.5.1.2 Storage of Carbon and Safer Production of Crops

Climate change mitigation greatly benefits from organic farming. Ground storage of carbon and restoration of healthier soil with higher organic matter content reduce greenhouse gas emissions.

As organic foods do not contain any hazardous chemical residues, both farmers and customers can grow and consume them with greater safety. Organic farming is a great method to improve diets and lower the risk of health problems by producing healthy food free from pesticide, such pulses and daily cereals.

4.5.1.3 Extended Sustainability and Eco-Friendly

Long-term goals are prioritized above immediate profits in sustainable organic agricultural methods. To ensure that the land stays fruitful for future generations, organic farming focuses on crop rotation, natural insect management using eco-friendly microbes such as phosphate solubilizers and nitrogen-fixing organisms. Those microbes should also have capacity to remove pollutants such as metals, pesticides, and hydrocarbons and also should decompose organic waste so that they are available for the crops to enhance their yield (Wani & Khan, 2013; Wani et al., 2023a, 2024). Organic farming promotes the health of ecosystems and biodiversity. Avoiding the usage of dangerous chemicals reduces pollution and aids in the preservation of wildlife, so that agriculture and the natural world coexist peacefully (Wani et al., 2022).

4.6 ORGANIC FARMING FOR IMPROVED BIODIVERSITY

The variety and variability of life on earth is referred as biodiversity or biological diversity. Variations at the genetic (genetic variability), species (species diversity), and ecological (ecosystem diversity) levels are considered as biodiversity. Organic farming can provide a high standard of biodiversity, which gives the human race a natural wealth. The majority of organic farms use species like microorganisms which act as biofertilizers along with biopesticides. These microbes can enhance the nutritional status of the soil in the form of N, P, and K which are freely available to the plants for them to sustain their crop productivity. These microbes also act as biopesticides by controlling the negative effect of pests by releasing certain antimicrobial agents such as secondary metabolites as well as biosurfactants which can inhibit the growth. Such deleterious microbes thus can enhance crop yield that contribute to agricultural sustainability by using less human products (such as pesticides and fertilizers) (Gani et al., 2021; Wani et al., 2023b).

Agricultural practices such as crop rotation, minimum tillage practices, strip cropping, manuring by organic matter, supplementation of soil with nutrients through organic sources, and minimum usage of pesticides, synthetic fertilizers, and fungicides will improve overall soil health which improves crop production for sustainable agriculture (Kavitha & Chandran, 2017).

4.6.1 Management of Production through Disease Control

Use of organic farming will increase productivity of agriculture by controlling various types of pests and weeds which is an important practice for the development of sustainable agriculture. This is achieved by using minimum inputs which can damage agriculture. There should be minimal usage of agrochemicals and synthetic fertilizers, using natural biological agents. There should be maximum use of natural agriculture practices such as ploughing, peddling, and leveling so that there is minimal growth of weeds. Use of uncontaminated seeds, preservation of water, cultivation in an appropriate season, and use of biocontrol and plants which have natural pest repellent capacity are being encouraged for sustainable agriculture for increased crop production and ecofriendly agriculture (Gamage et al., 2023).

4.6.2 Organic Farming in Richness of Specific Soil Microbes

Organic practices possess a significant favorable effect on soil bound organisms such as symbiotic and non-symbiotic nitrogen-fixing microorganisms (Wani & Khan, 2010, 2013). They also have beneficial effect on phosphate and potassium solubilizing microbes such as *Bacillus*, *Pseudomonas*, *Mesorhizobium*, *Serretia* etc (Bashir et al., 2017, 2018a, 2018b; Wani et al., 2007, 2009). These nitrogen-fixing phosphate solubilizing microbes have been reported to augment the yield of crops such as chickpea, green gram, lentil, cowpea etc by increasing their nitrogen and other elemental concentrations (Wani et al., 2007). The reduced use of pesticides and herbicides sustains these soil microbes, while the use of natural fertilizers such as manure boosts their numbers.

4.6.3 Organic Farming for Enhanced Species, Plant, Floral, and Capital Richness

Organic farming tends to promote a wider variety of species. Almost all noncrop and every naturally occurring species prefer organic farming over conventional farming. Birds, butterflies, earthworms, beetles, spiders, and mammals are some of these species. Organic farms usually have about 30% more related species.

There is a higher diversity of plants and flowers on organic farms. Insects, butterflies, earthworms, and several bird species are all included in this diversity. Comparing organic farming practices to conventional farming practices, there is an average 30% increase in biodiversity.

The natural capital that humans can benefit from is the biodiversity promoted by organic farming. Organic farming practices help to reduce the risks associated with low yields and promote agricultural sustainability by lowering the need for synthetic products such as pesticides and fertilizers.

4.6.4 Role of Microbes as Decomposers in Organic Farming

On comparison with organic farming, conventional farming has detrimental effects on the population of microbial species in the form of bacteria and fungi. Conventional practices can decrease the population of microbes by starving them of the necessary nutrients such as nitrogen and phosphorus which can decrease their population. If microbial population is more in number they can be helpful in bioremediation of pollutants in the form of pesticides, hydrocarbons, and metals (Gani et al., 2021; Wani et al., 2024). They can also be helpful in the decomposition of complex carbohydrates such as cellulose, hemicelluloses etc by the release of organic acids such as acetic acid, lactic acid, butyric acid etc. It is confirmed from different studies that the effects of organic farming on species richness in soil animals are of a considerable significance for decomposition of organic matter such as decomposition of cellulose, hemicelluloses etc which improves soil fertility through availability of various nutrients for improved crop fertility. Tuck et al. described various animals such as annelids (earthworms and pot worms) and nematodes (roundworms) as important taxonomic categories that have important function of decomposition of organic matter. Other studies have described certain trophic groups of nematodes, pot worms (Enchytraeidae), and earthworms (Lumbricidae) as species which are less subjected to various in farming practices (Birkhofer et al., 2008). In another study (Birkhofer et al., 2008, and Diekötter et al., 2010), there is no significant difference in the population of soil arthropods (meso and macrofauna), when various types of crops are being grown in organic farming compared to other farming practices where there is significant decrease in their population which can affect the productivity of various crops in agriculture.

4.7 SOCIO-ECONOMIC BENEFITS OF ORGANIC FARMING

4.7.1 Improved Health of Population

Conventional agriculture practices generally use pesticides and synthetic fertilizers for increased crop productivity. These pesticides and synthetic fertilizers can improve crop productivity, but these inputs can deposit in the soil and accumulate in various parts of crops. Through leaching they can reach the water bodies, and their accumulation in crops from soil ultimately can reach human beings through the food chain. Upon reaching human beings they have adverse effects on them by damaging various parts of the body and can cause various diseases (Tiwari, 2023). But if farms use organic farming instead of pesticides for increasing the crop productivity there will be no side effects to the human population and thus no disease, and there will be improvement in the health of humans (Tiwari, 2023).

4.8 ECONOMIC BENEFIT OF ORGANIC FARMING

The economic benefit of organic farming is due to less use of pesticides and synthetic fertilizers which can enhance the cost of final products, but if organic manure is used instead of the synthetic ones, there will be less cost for the production of the agriculture produce, thus organic produce can have more economic benefit to the general population (Tiwari, 2023).

4.8.1 Organic Farming Benefit to Community

While performing organic farming practices for the production of agricultural crops, there is involvement of the local population and produce is sold in the local market which benefits local population, and thus it can improve the socio economic culture of the local population and can also improve and preserve traditional and local agricultural practices (Tiwari, 2023). Organic farming can benefit society in a number of ways from environmental and economic points of view and an improved relationship among society, but it can also have a challenging effect on the population. In certain cases if the organic matter is not decomposed by various groups of organisms, there will be no availability of nutrients for increased crop productivity and thus there will be economic loss to the farmer and ultimately population will be affected (Tiwari, 2023).

4.9 SUSTAINABILITY IN ORGANIC FARMING

Emission of greenhouse gases will decrease as farming systems adopt organic farming as a viable option to other forms of farming. Organic farming not only may be beneficial for the local people but may also increase yield of the agricultural produce without having any toxic effect to the society. Researchers from the University of California have found organic farming increases the yield in agriculture along with a 58% increase of polyphenols which are known as antioxidants to remove the toxic effect of reactive oxygen species. It is also concluded that decreases in the population of pests have also been reported, otherwise they may have damaged the crops and may have reduced the content of polyphenols which may have acted as a defense mechanism to protect the plants against pests.

If farmers use less synthetic fertilizers for production of agriculture there will be less environmental damage in terms of pollution to the soil and water bodies which may endanger aquatic, wildlife, animal as well as human life. There is no such danger involved with the organic farming for cultivation of different types of crops (Table 4.1). Organic farming serves the purpose of protecting the soil from the

TABLE 4.1
Organic Farming and the Sustainable Development Goals (SDGs)

SDGs	Themes	Functions
1.	Eliminate poverty	Offer wages to organic farmers; low-cost financing for qualified farmers; sustainable production (livestock and crops); increased revenue from premiums on organic farming products; and labor-intensive nature
2.	End hunger, achieve food security, improve nutrition, promote sustainable agriculture	A diversified cropping system to mitigate risks of crop failure, more nutritious and safe food, improved productivity and sustainability of productive systems, and helps protect genetic resources
3.	Ensure healthy lives and promote well-being for all at all ages	Improved health and healthy lifestyles
4.	Ensure inclusive and equitable quality education for all	Effect of spending more money on children's education
5.	Attain gender parity and empower all women and girls	By giving women greater employment options and allowing labor-intensive jobs to be locally safe, it will prevent women from moving to cities in search of employment and enhance family well-being
6.	Make sure that everyone has access to water and sanitation and that it is managed sustainably	Less fertilizer leaching, which lessens the indirect effects of increased access to clean water and sanitation (reducing the pollution of nitrates, phosphates, heavy metals, and agrochemicals) and chemical fertilizers and pesticides that pollute water bodies
7.	Make sure everyone has access to modern, affordable, dependable, and sustainable energy	Integrated farming practices that apply organic principles
8.	Encourages equitable, long-term, sustainable economic growth, complete and fruitful employment, and decent labor for all	A sustainable living and a safer, healthier work environment are offered by organic agriculture.
9.	Develop inclusive and sustainable industrialization, build industrialization, build robust infrastructure, and encourage innovation	Modernizing organic farming might give farmers access to facilities and training in areas like marketing, certification, traceability, harvesting, and postharvest technologies.
10.	Minimize disparities both within and across nations	Small-scale organic producers' integration into international markets. It will guarantee patterns of production and consumption that are sustainable.
11.	Make cities and human settlements inclusive, safe, resilient, and sustainable	As ethical consumerism has grown, so has consumer support for crops grown in socially and environmentally responsible ways.
12.	Assure patterns of production and consumption that are sustainable	Agriculture supported by the community decreases waste and encourages recycling
13.	Act quickly to mitigate the effects of climate change	Practices in organic farming help to slow down climate change.
14.	Seas, oceans, and marine resources should be preserved and responsibly used for sustainable development.	Cut down on water contamination
15.	Stop, reverse, and encourage the sustainable use of terrestrial ecosystems; manage forests responsibly; prevent desertification	Organic methods encourage the replenishment of soils with carbon, the enhancement of damaged areas, and the production of nutritious food

pollution created by using synthetic fertilizers and chemical pesticides. For this purpose, famers now use organic matter, green manuring, compost, crop rotation, and microorganisms in organic farming. The main targets of sustainable development goals (SDGs) are eliminating world hunger, achieving food security, and promoting sustainable agriculture. Table 4.1 depicts organic farming and the Sustainable Development Goals (SDGs). When the SDGs are taken as a whole, agriculture is contributing to all 15 sustainable goals either without deviation or indirectly.

Organic farming is one of the most environmentally friendly and quickly growing sustainability movements. It is often referred to as "regenerative farming"; this idea is the foundation of a growing number of sustainable food and beverage projects. Organic farming goals are to reduce or get rid of synthetic fertilizers, to increase agricultural yields and productivity, promote biodiversity, protect water resources, and isolate carbon in the soil. Organic farming methods are most commonly associated with it (Pretty et al., 2018). The implementation of integrated pest management (IPM) programs to reduce pesticide consumption is one of the benefits of organic farming when it comes to sustainable agricultural practices. Crop rotation, planting cover crops or nitrogen-fixing ground coverings, and using mulch can lend a helping hand to improve soil health. Using no till or reduced tillage agricultural methods can also give assistance to decrease soil compaction. Soil samples can be used to detect the amounts of synthetic and organic fertilizer to be applied to elevate the use of organic fertilizers over synthetic fertilizers and avoid the unrestrained use of synthetic fertilizers. All of these practices aid in improving soil health (Baker et al., 2015; Merah et al., 2021; Salim et al., 2020). Along with focusing on the biodiversity protection, this also gives an assurance to utilize mulches and ground cover, and diminishes the use of pesticides, in order to prevent soil erosion. With many health advantages and significant environmental benefits, organic farming is thought to be a flourishing industry. Consumers are more concerned about them and are willing to pay a higher premium for them. Ecological sustainability and environmental consciousness were prioritized, resulting in resource conservation and good effects from purchasing organic products.

Obstacles are preventing organic production from proceeding, even though organic farming is currently a global trend in agriculture. The costs faced by organic farmers vary greatly from those of conventional farms. Compared to conventional farmers, organic farmers invest less on raw materials, but they pay much more for labor and feed.

The field of organic farming has many challenges. Despite the apparent health benefits of less chemicals in the food production process, organic products are more prone to degradation for a number of causes.

Due to the change in temperature during shipment and their usually shorter shelf life, organic products should be consumed sooner to ensure food safety and appeal. The main problem with growing organic food also includes time. Organic fruits and meats generally need to reach the market sooner, however this isn't always the case. Invasion of pests, which are as old as farming practices themselves, are another significant problem in organic farming. If pests like insects and rodents are not addressed, crops may be destroyed. For a very long time, people have been making use of pesticides as an effective means of controlling them. However, because they are not obtained from natural sources and thus harm the environment, various chemicals are prohibited in organic farming.

Organic farmers pay significantly more for labor and feed, but they also spend less on inputs overall because organic farming takes a systems approach rather than depending on pesticides or fertilizers to control fertility, weeds, and pests. Protecting biodiversity, overseeing intricate grazing systems, and maintaining animal health without the use of antibiotics are all requirements for organic farmers. The output of organic grains is falling, despite the great returns.

4.10 CHALLENGES AND FUTURE PERSPECTIVE

Organic farming presents certain difficulties even if it is a promising strategy for enhancing biodiversity. Lower yields from organic farming may cause problems for feeding the world's expanding population. Thus, increasing organic systems' production and efficiency will be a problem in the future.

In conclusion, by adopting methods that are in tune with nature, organic farming significantly contributes to the promotion of biodiversity. A viable solution to the twin problems of food supply and biodiversity protection is organic farming. Organic agricultural practices are crucial in promoting biodiversity because they prioritize soil health, enhance crop diversity, and provide environments that are conducive to animals.

Creating more productive, sustainable farming systems that preserve or improve the provision of ecosystem services, such as biodiversity, will be a challenge in the future. In 80% of situations, organic farming was shown to have a significantly higher level of biodiversity than conventional farming; in 16% of cases, the differences were unclear, and in 4% of comparisons, where seven to ten biodiversity indicators were considered, there was a decrease in biodiversity. Indirectly, small farms in particular might benefit from biodiversity. Due to their smaller land parcels, these farms typically feature more field edges, which are comparatively species-rich.

In the end, it contributes to a healthier and more sustainable earth by preserving genetic variety, ecosystems, and the welfare of numerous species.

REFERENCES

Baker, B., Cooley, D., Futrell, S., Garling, L., Gershuny, G., Thomas, A., Moyer, J., Rajotte, E., Seaman, A., & Young, S. (2015). Organic agriculture and integrated pest management: Synergistic partnership needed to improve the sustainability of agriculture and food systems. *Organic and IPM: Synergistic Partnership*, *49*, https://doi.org/10.13140/RG.2.1.3616.8085F.Z

Bashir, Z. (2021). characterization of potassium solubilizing bacteria from Rhizospheric Soils of Cherry (*Prunusavium*). *Research Journal of Agricultural Sciences*, *12*(2), 494–496.

Bashir, Z., Hamid, B., Zargar, M. Y., & Kundoo, A. A. (2021). Morphological and biochemical characterization of potassium solubilizing bacteria from Rhizospheric Soils of Pinus (*Pinus wallichiana*). *Journal of Research and Development*, *21*, 58–63.

Bashir, Z., Zargar, M. Y., Baba, Z. A., & Mohiddin, F. A. (2018). Solubilization of potassium containing mineral by bacteria from Kashmir iris (*Iris kashmiriana*) Rhizosphere. *The Pharma Innovation Journal*, *17*(4), 747–750.

Bashir, Z., Zargar, M. Y., Baba, Z. A., Mohiddin, F. A., Peer, F. A., Mir, S. A., & Najar, G. R. (2018). Isolation and characterization of phosphorus solubilizing bacteria (PSB) from Rhizospheric Soils of Apple (*Malus Domestica*). *Journal of Pharmacognosy and Phytochemistry*, *7*(2), 2997–3000.

Bashir, Z., Zargar, M. Y., Bhat, T. A., Kousar, S., Baba, Z. A., & Mohiddin, F. A. (2018b). Solubilization of phosphorus containing mineral by bacteria from Rhizospheric region of Walnut (*Juglans regia*). *International Journal of Current Microbiology and Applied Sciences*, *7*(8), 2391–2398.

Bashir, Z., Zargar, M. Y., Mohiddin, F. A., Baba, Z. A., & Hamid, B. (2019). Isolation and biochemical characterization of phosphorus solubilizing bacteria from Rhizospheric region of Apricot and Peach. *Journal of Research and Development University of Kashmir*, *19*, 0972–5407.

Bashir, Z., Zargar, M. Y., Mohiddin, F. A., Kousar, S., Husain, M., & Rasool, F. (2017). Phosphorus solubilizing microorganisms: Mechanism and diversity. *International Journal of Communication Systems*, *5*(5), 666–673.

Bashir, Z., Zargar, M. Y., & Vishwakarma, D. K. (2019b). Potassium solubilizing microorganisms for sustainable agriculture. *Applied Agricultural Practices for Mitigating Climate Change*, *2*, 17–28.

Birkhofer, K., Bezemer, T., & Bloem., J. (2008). Long-term organic farming fosters below- and aboveground biota: Implications for soil quality, biological control and productivity. *Soil Biology and Biochemistry*, *40*, 2297–2308.

Diekötter, T., Wamser, S., Wolters, V., & Birkhofer, K. (2010). Land-scape and management effects on structure and function of soil arthropod communities in winter wheat. *Agriculture, Ecosystems and Environment*, *137*, 108–112.

Gamage, A., Gangahagedara, R., Gamage, J., Jayasinghe, N., Kodikara, N., Suraweera, P., & Merah, O. (2023). Role of organic farming for achieving sustainability in agriculture. *Farming System*, *1*, 100005.

Gani, G., Asif, M., Wani, P. A., Malik, M. A., Dar, Z. M., Masood, A., & Shafi, S. (2021). Chlorpyrifos degradation, biocontrol potential and antioxidant defense activation under pesticide stress by rhizosphere bacteria isolated from Rhizosphere of peach (*Prunus persica*) plants. *Chemistry and Ecology*. https://doi.org/10.1080/02757540.2021.1985475

Hamid, B., & Bashir, Z. (2019a). Potassium solubilization microorganisms: An alternative technology to chemical fertilizers. *Journal of Research and Development University of Kashmir*, *19*, 972–5407.

Kavitha, V., & Chandran, K. (2017). Organic farming in conserving bio diversity in India—A review. *Agricultural Reviews*, *38*(4), 316–320.

Lori, M., Symnaczik, S., Mäder, P., Deyn, G., & Gattinger, A. (2017). Organic farming enhances soil microbial abundance and activity—A meta-analysis and meta-regression. *PloS One*, *12*, e0180442.

Merah, O., Djazouli, Z. E., & Zebi, B. (2021). Aqueous extract of Algerian Nettle (*Urtica dioïca L.*) as possible alternative pathway to control some plant diseases, Accepté pour publication dans. *Iranian Journal of Science and Technology, Transactions A: Science*, *45*, 463–468.

Muller, A. (2009). Benefits of organic agriculture as a climate change adaptation and mitigation strategy in developing countries. *Environment for Development*, 1–19.

Murmu, K., Das, P., Sarkar, A., & Bandopadhyay, P. (2022). Organic agriculture: As a climate change adaptation and mitigation strategy. *Zeichen Journal*, *8*(3), 171–187.

Pretty, J., Benton, T., Bharucha, Z., Dicks, L., Flora, C., Godfray, C., Goulson, D., Hartley, S., Lampkin, N., Morris, C., Pierzynski, G., Prasad, P. V. V., Reganold, J., Rockstrom, J., Smith, P., Thorne, P., & Wratten, S. (2018). Global assessment of agricultural system redesign for sustainable intensification, *Nature Sustainability*, *1*, 441–446. https://doi.org/10.1038/s41893-018-0114-0

Raj, J., Jat, S., Kumar, M., & Yadav, A. (2024). The role of organic farming in sustainable agriculture. *Advances in Research*, *25*(3), 128–136.

Salim, D., Caro, P., Merah, O., & Chbani, A. (2020). Control of postharvest citrus green mold using Ulva lactuca extracts as a source of active substances. *International Journal of Biodiversity Science & Management*, *11*(3), 287–296, https://doi.org/10.23910/1.2020.2107

Tiwari, A. K. (2023). The role of organic farming in achieving agricultural sustainability: Environmental and socio-economic impacts. *Acta Biology Forum*, *2*(2), 29–32.

Wani, P. A., Abiodun, A. A., Olusesi, Y. K., Rafi, N., Wani, U., Oluwaseun, O. I., & Sirajudeen, A. A. O. (2022). Hydrocarbon degradation and metal remediation by hydrocarbon-utilising and metal-tolerant Klebsiella pneumoniae YSA-9 isolated from soil contaminated with petroleum. *Chemistry and Ecology*, *38*(8), 744–759.

Wani, P. A., Ebudola, A. N., Olusebi, Y. K, Rafi, N, Fawzhia, S. J. O., & Oluwaseun, O. I. (2023a). Hydrocarbon utilizing and metal tolerant bacteria simultaneously degrade hydrocarbons and detoxify metals in petroleum contaminated soil. *Geomicrobiology Journal*, *40*(4), 372–381.

Wani, P. A., & Khan, M. S. (2010). Bacillus species enhance growth parameters of chickpea (*Cicer arietinum* L.) in chromium stressed soils. *Food and Chemical Toxicology*, *48*, 3262–3267.

Wani, P. A., & Khan, M. S. (2013). Nickel detoxification and plant growth promotion by multi metal resistant plant growth promoting *Rhizobium* species RL9. *Bulletin of Environmental Contamination and Toxicology*, *91*, 117–124.

Wani, P. A., Khan, M. S., & Zaidi, A. (2007). Synergistic effects of the inoculation with nitrogen fixing and phosphate solubilizing rhizobacteria on the performance of field grown chickpea. *Journal of Plant Nutrition and Soil Science*, *170*, 283–287.

Wani, P. A., Olusebi, Y. K., & Rilwan, B. A. (2023b). Remediation of hydrocarbons and metals by hydrocarbon utilizing Pseudomonas taiwanensis YSA-17 isolated from soil contaminated with petroleum. *Journal of Basic Microbiology*, *63*, 1426–1439.

Wani, P. A., Olusebi, Y. R., Wani, U., Rafi, N., & Khan, M. S. A. (2024). Role of chromium reductases, antioxidants and biosorption against oxidative damage of metals by *bacillus cereus*. *Journal of Basic Microbiology*, *64*(3), 1–11. https://doi.org/10.1002/jobm.202300589

Wani, P. A., Zaidi, A., & Khan, M. S. (2009). Chromium reducing and plant growth promoting potential of *Mesorhizobium* species under chromium stress. *Bioremediation Journal*, *13*(3), 121–129.

Willer, H., Schlatter, B., Trávníček, J., Kemper, L., & Lernoud, J. (2020). *Organic agriculture worldwide 2018: Current statistics*. Research Institute of Organic Agriculture (FIBL), Frick.

Willer, H., Trávníček, J., & Schlatter, S. (2021). The world of organic agriculture. *Statistics and Emerging Trends*, Research Institute of Organic Agriculture FiBL, and IFOAM-Organic International.

Wittwer, A., Bender, F., Hartman, K., Hydbom, S., Lima, R. A., & Loaiza, V. (2021). Organic and conservation agriculture promote ecosystem multifunctionality. *Science Advances*, *7*, eabg6995.

5 Organic Fertilizers
Types and Benefits of Organic Fertilizers

Zaffar Bashir, Arbeena, Qurut-Ul- Ain, Bisma Rashi, and Mehrun Nisa

5.1 INTRODUCTION

A fertilizer is a material that is given to soil or plants to supply needed nutrients, promote plant development, and increase crop output. Nitrogen, phosphorus, and potassium are examples of nutrients. Fertilizers are manmade or naturally occurring substances that provide specific nutrients required for plant growth. Fertilizers are essentially specialized plant food. They are available as microscopic grains or pellets. Fertilizers contain specific nutrients that plants require in order to develop more quickly and larger, which is critical in farming and gardening. Organic fertilizers are made from natural materials like plant or animal waste. Compost, manure, bone meal, and other organic debris are examples of these. These fertilizers slowly release nutrients, enhancing soil structure and promoting long-term soil fertility. Previously, farmers used chemical fertilizers to boost crop yield, but now they prioritize natural fertilizers above chemical fertilizers. Chemical fertilizers have had such a negative impact on the land in this generation that crop yield has increased while disease and pest outbreaks have increased in cereals, which is why most farmers have now switched to natural fertilizers instead of chemical fertilizers. Research conducted by various scholars has demonstrated that the use of organic fertilizer is among the best ways to enhance the quality of agricultural land and achieve its sustainable use (Zhang et al., 2021).

Natural fertilizers are characteristic fertilizers that contain carbon and are inferred from characteristic assets. Fertilizers, which are substances that give supplements and advance development, can advantage both plants and soil. Common natural fertilizers incorporate creature squander, plant-based fertilizers such as compost, and mineral sources such as meat preparing squander, slurry, guano, excrement, and bio solids. There are various non-chemical or abiotic fertilizer choices that compare to Natural Horticulture Standards. Illustrations incorporate creature squander, peat, agrarian plant squander, and handled sewage slime (Dittmar et al., 2000; Penhallegon, 2003). Almost all ranchers utilize fertilizer and other normal natural fertilizers (Kannan et al., 2001). Horse, poultry, shelled nut husk, and calf fertilizer were frequently utilized alone or in combination. A few ranchers utilize angle excrement to fertilize particular crops. Fertilizer is regularly connected earlier to planting. Most ranchers

 DOI: 10.1201/9781003491460-5

are new with natural fertilizer handling (Guene, 2002). Fertilizer determination is impacted by trim sort, assortment, and season. It is additionally influenced by soil chemistry and past perceptions and encounters of the ranchers (Niassy et al., 2010). Numerous thinks about have been conducted to decide the agronomic potential of natural fertilizer on different trim frameworks. Agriculturists will gain more cash as rural efficiency rises (Ayeni & Adetunji, 2010; Zake et al., 2005).

Natural agribusiness energizes supplement reusing to diminish the sum of supplements imported to the cultivated land. Natural squander application in soil could be an appropriate strategy for keeping up soil natural matter, moving forward soil ripeness and providing supplements required by plants (Davarinrjad et al., 2004). Natural fertilizers are natural materials that are less hurtful to the environment than chemical fertilizers (Dharma, 1996).

5.1.1 Improved with Plant Supplements

It is comprises around 15 million tons of metropolitan squander, 20 million tons of edit buildups, and 275.5 million tons of creature fertilizer (El-Shemy & Aly, 1997). The angling industry produces a part of squander in angle markets and preparing plants (López-Mosquera et al., 2011). Angle remains have customarily been utilized as fertilizer due to their high nutritive esteem (fundamentally N and P) and quick deterioration. After handling, about 75% of the entire weight of the fish was created as strong squander within the shape of intestine, head, skin, bones, blades, and outlines. Angle squanders high in nitrogen, potassium, phosphorus, and follow minerals (Ghaly & Alhattab, 2013) can be utilized to form an assortment of nutritive and nonnutritive products. Organic fertilizers were characterized as materials with a particular chemical composition and a high expository esteem that supply plant supplements in a promptly accessible shape (Penhallegon, 2003). Natural fertilizers are those inferred from creature, human, or vegetable matter (e.g., compost, excrement). Natural fertilizers are made from normal crude materials, such as our biodegradable damp suit. Compost is ordinarily made by breaking down biodegradable squanders. The study revealed that the natural product peelings, extra nourishments, and indeed natural product juices are cases of these squanders. Natural fertilizers are advantageous to the soil. It makes the soil available and perfect for planting. The biggest source of green fertilizer is natural excrement (barring green fertilizer). Depending on the species, huge sums of supplements are displayed in creature fertilizers. As a result, they are one of the foremost imperative natural fertilizer sources in agrarian areas. Fertilizer moves forward soil richness by giving supplements such as nitrogen that soil organisms can utilize. When excrement is connected to soil, it increases its water holding capacity or structure. Synthetic and natural fertilizers are nearly continuously utilized in cultivating, depending on soil chemistry, perception, rancher involvement, season, and trim type (Niassy et al., 2010). Advantages of organics**:** (a). Organic manure provides all the nutrients that are required by plants but in limited quantities. (b). It helps in maintaining C: N ratio in the soil and also increases the fertility and productivity of the soil. (c). It improves the physical, chemical, and biological properties of the soil. (d). It improves both the structure and texture of the soils.

(e). It increases the water holding capacity of the soil. (f). Due to an increase in the biological activity, the nutrients that are in the lower depths are made available to the plants. (g). It acts as mulch, thereby minimizing the evaporation losses of moisture from the soil. Various organic fertilizers are used in the agricultural sector:

5.2 COMPOST

Composting is a biological process that turns organic wastes into a homogenous, plant-available material in a sanitary manner under aerobic (oxygen-containing) conditions with suitable moisture and temperature. Composting is defined by the United Nations Environment Program (UNEP) as the biological decomposition of biodegradable solid waste under predominantly aerobic conditions to a state that is sufficiently stable for nuisance-free storage and handling and is mature enough for safe use in agriculture (Amadi et al., 2012). Compost is created when organic matter is broken down by microbes over time. Compost preparation requires the use of organic components that decompose naturally. Compost is an inexpensive, readily available, and simple-to-make material that can significantly enhance the quality of cropland and soil. The soil's structural integrity is enhanced by compost. Enhancing drainage and lowering erosion, it lets more air into the soil. By retaining more water, compost prevents the soil from drying out during dry spells. Thus, compost is a good organic fertilizer because it contains nutrients as well as organic matter. It enriches the soil with nutrients, increases its ability to retain water, and aids in maintaining a healthy tilth, which promotes greater aeration for seed germination and plant root development (Edwards & Hailu, 2011). The application of compost, which is made primarily from biomass wastes, is one effective technique to raise the Soil Organic Matter content. Table 5.1 shows the chemical composition of organic matter, % (dry weight basis). Nonetheless, the key determinants of SOM-enrichment are the type, quantity, and degree of compost humification, the characteristics of the soil (such as its type and amount of clay), and management practices. Bouajila and Sanaa (2011)

TABLE 5.1
Chemical Composition of Organic Matter

Fractions	% (Dry Weight Basis)
Sugars, starches, amino acids, urea	2–30
Ammonium salts (hot/cold water soluble), fats, oils, waxes, (ether/alcohol soluble)	1–15
Proteins	5–40
Hemicelluloses	10–30
Cellulose	15–60
Lignin	5–30
Mineral matter (ash)	5–30

Source: FAO Soil Bulletin (1987)56

reported that application of manure and household wastes compost resulted in significant increase of organic carbon, with the compost treatment being the most efficient. There are occasions when rock phosphate, *Azotobacter*, *Acetobacter*, and PSB are added to the compost in order to enrich it. According to Jacob and Banerjee (2017), incubation of conventional compost with bacteria such as Aulosira, *Azosprillum*, and *Azotobacter* for six weeks resulted in an enhancement of the N, P, and K content by 2.76, 3.70, and 2.20 times faster. Compost can therefore be made more useful to agriculture and the environment by adding beneficial organisms such as microbes that encourage plant growth (Sousa et al., 2018). In addition to boosting crop production, adding rock phosphate-enriched compost (rock phosphate plus chicken litter plus *Pseudomonas* sp.) improved the chemical and biological characteristics of the soil in comparison to the control (Billah et al., 2020).

5.3 VERMICOMPOST

Vermicompost is a rich organic material that maintains soil health and increases crop output by fostering microorganisms, plant growth, and nutrients. Compared to inorganic fertilizer, which encourages greenhouse gas emissions from agricultural land, this organic fertilizer is involved in the control of such emissions. The high nutrient content of vermicompost, which includes micronutrients, phosphate, potassium, and nitrogen, is well known. Research continuously demonstrates that adding vermicompost to soil improves its nutritional content, which in turn increases plant output and growth. It enhances soil structure by increasing aggregate stability and promoting a well-aerated soil matrix. However, because vermicompost-tea naturally contains a high concentration of carboxylesterase enzyme potentials, it is very effective at breaking down pharmaceutical residues and organopesticide residues found in contaminated soil, such as diclofenac, triclosan, and ibuprofen (Haque & Biswas, 2021). In the soil, vermicompost fosters a vibrant and varied microbial life. Vermicomposting is a labor-intensive process and difficult to separate earthworms from other materials, but it has been demonstrated to be an effective way to treat biomass waste and increase the nutrient content of the end product. In light of recent occurrences, earthworms are often recommended for vermicomposting problems in order to remove harmful pollutants from soil. When it comes to their environment, earthworms are more tolerant than other living things like bacteria, archaea, algae, and fungi. One example of this is the acidity of the soil. The overall health of the soil ecosystem, disease prevention, and nutrient cycling are all positively impacted by the presence of beneficial microbes. According to specific research, vermicompost has the ability to manage various soil-borne infections through bio control capabilities. The interplay between plant pathogens and the advantageous bacteria in vermicompost helps to decrease illness.

5.4 MANURE

Manure is an organic substance that is used to fertilize soil; it is frequently made from animal feces. Its advantages include strengthening soil structure, increasing water retention, and supplying vital minerals for plant growth. Furthermore, manure

TABLE 5.2
Average Nutrient Content of Manure

Manure	Percentage Content		
	N	P_2O_5	K_2O
Animal refuse	0.3–0.4	0.1–0.2	0.1–0.3
Cattle dung, fresh	0.4–0.5	0.3–0.4	0.3–0.4
Horse dung, fresh	0.5–0.5	0.4–0.6	0.3–1.0
Poultry manure, fresh	1.0–1.8	1.4–1.8	0.8–0.9
Sewage sludge, dry	2.0–3.5	1.0–5.0	0.2–0.5
Sewage sludge, activate dry	4.0–7.0	2.1–4.2	0.5–0.7
Cattle urine	0.9–1.2	trace	0.5–1.0
Horse urine	1.2–1.5	trace	1.3–1.5
Human urine	0.6–1.0	0.1–0.2	0.2–0.3
Sheep urine	1.5–1.7	trace	1.8–2.0
Ash, coal	0.73	0.45	0.53
Ash, household	0.5–1.9	1.6–4.2	2.3–12.0
Ash, wood	0.1–0.2	0.8–5.9	1.5–36.0
Rural compost, dry	0.5–1.0	0.4–0.8	0.8–1.2
Urban compost, dry	0.7–2.0	0.9–3.0	1.0–2.0
Farmyard manure, dry	0.4–1.5	0.3–0.9	0.3–1.9
Filter-press cake	1.0–1.5	4.0–5.0	2.0–7.0
Rice hulls	0.3–0.5	0.2–0.5	0.3–0.5
Groundnut husks	1.6–1.8	0.3–0.5	1.1–1.7
Banana, dry	0.61	0.12	1.00
Cotton	0.44	0.10	0.66

encourages healthy microbial activity in the soil, which helps to create a more productive and healthier environment for plants. Bouajila and Sanaa (2011) reported that application of manure resulted in significant increase of organic carbon, with the compost treatment being the most efficient. Table 5.2 displays average nutrient content of manure.

Manure offers several benefits in agriculture:

- Enhanced Nutrient Content: Nutrients that are critical for plant growth, such as potassium, phosphate, and nitrogen, are abundant in manure. With a balanced nutrition source, it functions as a natural fertilizer.
- Soil Structure Improvement: Manure adds more organic matter to the soil, which improves its structure. This fosters a healthier environment for plant roots by enhancing soil aeration, drainage, and water retention.
- Microbial Activity: Beneficial bacteria proliferate in the soil when manure is added. These microorganisms support the cycling of nutrients, the breakdown of organic materials, and the general health of the soil.

- Decreased Requirement for Artificial Fertilizers: Manure can lessen the need for synthetic fertilizers, which may have an adverse effect on the environment, by adding nutrients to the soil. More environmentally friendly farming methods are supported by this change.
- Water Retention: The use of manure improves the soil's structure, which increases its ability to retain water. Because it makes it easier for plants to get water during dry spells, this is especially helpful in places with inconsistent rainfall.
- Biodiversity and Sustainable Agriculture: An ecosystem that is healthier and more varied is supported by the microbial diversity that manure fosters in the soil. Manure is a sustainable agricultural method that promotes soil fertility over the long run and environmental responsibility.

5.5 NITROGEN-FIXING MICROORGANISMS

Nitrogen is essential for all living things because it is a major part of amino acids, which are the building blocks of proteins and of nucleic acids such as DNA, which transfers genetic information to subsequent generations of organisms. Our atmosphere has 78% nitrogen gas and this atmospheric nitrogen is molecular dinitrogen, a relatively nonreactive molecule that is metabolically useless as it cannot be utilized directly by the plants or animals. The living organisms, especially bacteria, convert nitrogen gas of the atmosphere or air into compounds of nitrogen, which can be used by the plants. The process of converting atmospheric nitrogen into ammonia or related nitrogenous compounds in soil, which plants require, is known as nitrogen fixation. Recent studies on nitrogen-fixing organisms have revealed a number of advantages, especially in relation to agriculture:

- Increased Availability of Nitrogen: Certain plants develop symbiotic partnerships with nitrogen-fixing bacteria, such as those in the Rhizobiaceae family, which help convert atmospheric nitrogen into a form that plants can use. As a result, the soil has better access to nitrogen, which encourages plant growth.
- Decreased Reliance on Fertilizers Using Synthetic Nitrogen: The need of artificial nitrogen fertilizers may be decreased by bacteria that fix nitrogen. This is important since using synthetic fertilizers excessively might cause pollution of the water supply and other environmental problems.
- Increase Soil Fertility and Crop Yield: By increasing the amount of nutrients present, nitrogen-fixing bacteria improve the fertility of the soil. The general health of the soil may benefit from enhanced soil structure and nutrient cycling as a result. According to research, using microorganisms that fix nitrogen can increase crop yields. These bacteria assist plant growth and development, leading to increased yields, by increasing nitrogen availability.

It is noteworthy that the efficacy of organisms that fix nitrogen might differ based on several parameters such as crop type, soil properties, and agricultural practices. Continuous investigation endeavors to enhance the utilization of nitrogen-fixing bacteria in farming to optimize their advantages while reducing any possible drawbacks.

5.6 PHOSPHATE SOLUBILIZING MICROORGANISMS (PSMS)

Phosphorus (P) is one of the main macronutrients that limit plant growth and are necessary for healthy plant growth, especially in tropical regions. Nitrogen is the most essential macronutrient for plants, with phosphorus coming in second. But because P is fixed as insoluble phosphates of iron, aluminum, and calcium in the soil, the amount of soluble P that is available for plants to consume is restricted (Sharma et al., 2013). Phosphate solubilizing microorganisms (PSMs) are a class of helpful bacteria that hydrolyze both organic and inorganic phosphorus molecules from insoluble substances. Strains from fungal genera (Penicillium and Aspergillus), actinomycetes, arbuscular mycorrhizal (AM), and bacterial genera (*Bacillus*, *Micrococcus, Serratia Pseudomonas* and *Rhizobium*) are notable among these PSMs (Bashir et al., 2017, 2018b, 2018c, 2019a). A certain PSM's capacity to solubilize phosphate can be evaluated using the solubilization index (SI), which is the ratio of the colony diameter to the total diameter, or clearance zone. As described by several studies, phosphate SI can be determined by the following formula:

$$SI = \frac{\text{colony diameter} + \text{halo zone diameter}}{\text{colony diameter}}$$

Bacteria produce organic acids, which lowers the pH in the root zone, allowing trapped forms of phosphate such as $Ca_3\ (PO4)_2$ in calcareous soils to be freed. In addition to delivering the accessible cumulated phosphate (through solubilization), phosphorus biofertilizers also help to improve organismal N_2-fixation and make Zn, Fe, and other minerals present due to plant growth-promoting compounds synthesis. To enhance phosphorus uptake efficiency, PSB play an important role in supplying phosphate to plants, which is an environment friendly and sustainable approach. The bacterial isolates inoculated into plates with sterilized Pikovskaya medium containing tricalcium phosphate and incubated at 30°C for 72h. Formation of a clear zone around the colony indicated the phosphate solubilization by the bacteria. Figure 5.1 shows the on plate assay of P solubilization on Pikovskaya's agar. The isolate which forms a clear zone was capable of solubilizing the phosphorus.

PSMs reveal significant insights into their roles in nutrient cycling, plant growth promotion, and sustainable agriculture. It is believed that using PSMs to regulate phosphorus is an environmentally beneficial method. These microbes support sustainable farming methods that have a smaller negative impact on the environment by promoting natural phosphate solubilization processes (Bashir et al., 2017).

5.7 POTASSIUM SOLUBILIZING MICROORGANISMS

Potassium (K) is considered as an important macronutrient and is an essential component of a plant nutrition package limiting crop yield and crop quality and performs important biological functions to maintain plant growth. Potassium is present as an

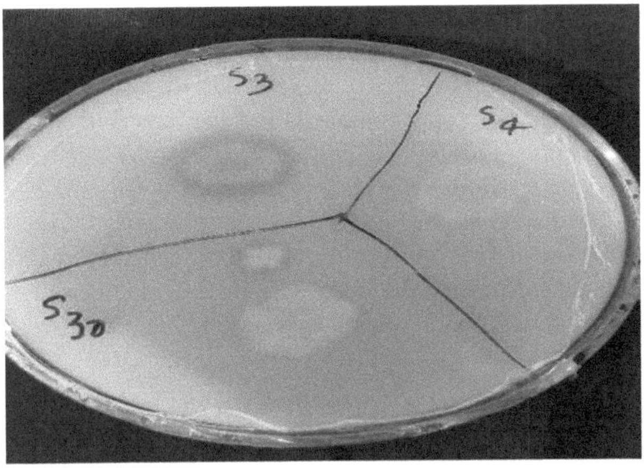

FIGURE 5.1 Isolate S_3 shows the P solubilization on Pikovskaya's agar.

abundant element in soil, but only a small percentage, ranging from 1% to 2%, is available for plant uptake (Singh et al., 2021). The remaining potassium is bound to other minerals, rendering it unavailable for plant absorption. Potassium exists in soil in different forms such as mineral potassium, non-exchangeable potassium, exchangeable potassium, and solution potassium (Yahaya et al., 2023). A group of microorganisms called potassium solubilizing microorganisms (KSMs) has been recognized for their role in converting fixed potassium into forms accessible to plants, facilitating its absorption by the plants (Das et al., 2022). Various microorganisms with the capability to solubilize potassium, such as *Bacillus* sp. and *Pseudomonas* sp. (Bashir et al., 2018a, 2019b, 2021; Bashir, 2021; Hamid & Bashir, 2019a; Liu et al., 2012; Soumare et al., 2023), have been reported for releasing accessible forms of potassium, from the minerals containing potassium in the soil. Potassium solubilizing microorganisms (KSMs) including both fungal and bacterial species assist the plant growth by solubilization of insoluble forms of potassium. Potassium solubilizing microorganisms are widespread in distribution, and their abundance varies from one soil to another. Microorganisms in the rhizosphere play a major role in solubilization of locked form of soil minerals in the soil. This microbial community is used as potassium source (K based biofertilizers) in organic agriculture (Bashir et al., 2021; Bashir, 2021).

5.8 ZINC SOLUBILIZING MICROORGANISMS

Zn is one of the fundamental trace elements required for plants. It is also an important metal constituent compulsory for all the enzymatic activities of plants (Rani et al., 2020). The Zn deficiency would lead to physiological strain due to which different enzymes are not able to perform their function properly (Alloway, 2009). Amongst all micronutrients, Zn is needed by the majority of proteins. Microorganisms can be employed to solubilize insoluble forms of zinc and make plants healthy with strong yield and output in order to address zinc shortage (Fasim et al., 2002). It is well known that a variety of microorganisms are crucial to the solubilization of potassium (K), phosphorus (P), iron (Fe), and zinc (Zn) in soil (Bashir et al., 2018a, 2019b, 2021; Bashir, 2021; Hamid & Bashir, 2019a). Most active rhizospheric fungi and bacteria in solubilizing insoluble zinc compounds include *Aspergillus niger, A. nomius,* and *A. oryzae, Bacillus subtilis, Bacillus thuringiensis, Thiobacillus thioxidans, Pseudomonas striata, Pseudomonas fluoregnence, B. cenocepacia, Cynobacteria, Acinetobacter, S. marcescens, S. liquefaciens, Bacillus megaterium* were isolated from a zinc mining site in Thailand (Anitha et al., 2015; Dhaked et al., 2017; Kalpana et al., 2020; Vandana & Manu, 2021). This zinc solubilizing microbial community is used as Zn based biofertilizers in organic agriculture.

5.9 SILICATE SOLUBILIZING MICROORGANISMS

Silicon (Si), an essential second plentiful element in the earth's crust, is abundant in the soil typically as insoluble silicate forms (Bist et al., 2020). However, plants can uptake Si only in the soluble form of monosilicic acid. Silicon plays important roles such as reduction of metal toxicity, increased development in plants, formation of soil, and attributes in several chemical reactions (Gerard et al., 2008). Production of monosilicic acid by rock-weathering mostly depends on microbial activities. Numerous studies have predicted the role of Indole-3-Acetic Acid (IAA) and organic acids like malic acid, oxalic acid, fumaric acid, citric acid, gluconic acid, etc produced by silicon solubilizing bacteria (SSB) in silicate solubilization (Naureen et al., 2015; Vasanthi et al., 2018). This microbial community is used as silicon based biofertilizers in organic agriculture sector.

5.10 PLANT GROWTH-PROMOTING RHIZOBACTERIA (PGPR)

Sustainable agriculture is very essential in today's world because it has the possibility to satisfy our future agricultural demands, something that conventional agriculture will not be able to achieve. Soil microorganisms with beneficial impact on plant development and health constitute an intriguing alternative to standard agricultural practice. PGPR (plant growth-promoting rhizobacteria) are naturally occurring soil bacteria that penetrate plant roots aggressively and promote plant growth. PGPR are found in plant roots and increase plant production and immunity; however, new research from multiple groups has discovered that PGPR can also cause so-called induced systemic tolerance to salt and drought. PGPR may help boost nutrient absorption, minimizing the requirement for fertilizers and decreasing phosphate build nitrate in agricultural soils. Multidisciplinary approaches are used in scientific studies to better understand PGPR adaptation, effects on induced systemic resistance, bio control of plant pathogens, plant physiology and growth, biofertilization, and potential green alternatives for plant productivity, viability of inoculating, plant microorganism interactions, and root colonization mechanisms. Plant growth-promoting rhizobacteria (PGPR) are a diverse group of naturally occurring soil bacteria that improve plant production and immunity. PGPR must be able to colonize the root surface; they must also be able to live, reproduce, and compete with other micro biota, at least for the period needed to express their plant growth promotion or protective activities

(Ahemad & Kibret, 2014). The specific methods through which PGPR enhance plant development are unknown, although they are known to impact plant development in a variety of ways, both direct and indirect. PGPR can directly facilitate the growth and development of plants by increasing the availability and uptake of nutrients which take place by some mechanisms such as biological N_2 fixation, solubilization of inorganic phosphate, and mineralization of organic phosphate and other nutrients; production of siderophores and synthesis of plant growth regulators like auxin, cytokinins, gibberellins (GA), abscisic acid (ABA), and ethylene (Berger et al., 2015). The technique through which PGPR prevents or neutralizes the negative effects of phytopathogens and conditions on plants is known as indirect stimulation. This includes the production of repressive substances (such as antibiotics), the induction of systemic resistance in plants, the creation of iron-limiting conditions through the production of phytopathogen-controlling siderophores, and the active protection of plant growth under environmental stress (abiotic stress). A specific PGPR can influence plant growth and development through one or a combination of methods; it may also employ distinct features for growth promotion at different stages of the plant's life cycle. Several research projects are currently underway to better understand the diversity and relevance of soil PGPR communities, as well as their roles in agricultural sustainability. Plant growth-promoting microorganisms have become a very appealing technology and study field in agriculture as well as phytoremediation. PGPR and their interactions with plants are economically exploited because they have a positive influence on crop production by promoting plant growth, increasing crop yields, decreasing the expense of chemical fertilizers, and being less hazardous to the environment. PGPR are used as inoculant for biofertilization, bio stimulation, bio control, and bioremediation (Bhattacharyya & Jha, 2012). In vitro tests, greenhouse settings, and in some cases, field experiments have shown that PGPR promotes plant development in a variety of agronomically significant crops such as maize, wheat, barley, peas, canola, soy, potatoes, lentils, radicchio, tomato, and cucumber (Berger et al., 2018). Plant growth yields a wide range of results. Increased germination rates, larger biomass, improved yields, increased leaf area, chlorophyll and nutritional content, drought tolerance, and disease resistance are just a few examples (Lucy et al., 2004). When applied under different circumstances or cropping systems, the diversity in the efficiency of biological inoculants is a cause for worry. A variety of factors, including soil pathogens, low soil pH, and little rainfall throughout the growing season, contribute to low PGPR colonization and limit their impact on plant development. Agricultural methods like fertilizer application may not appear to be as harmful, but they cause significant changes in soil characteristics, which can damage the rhizosphere. PGPR functions are additionally influenced by climatic fluctuations, soil type, nutrient, organic matter, and moisture content. In fact, in a study with wheat and a *Pseudomonas* sp., results suggested that the less fertile the soil, the greater the plant growth stimulation by the PGPR. Interactions between plants and microbes play a significant role in agricultural output. Beneficial plant-associated bacteria play an important role in sustaining and/or boosting plant health and development in both controlled and wild ecosystems (Dawwam et al., 2013). PGPR efficiency, on the other hand, is mostly determined by native biotic and abiotic variables. Indigenous isolates may be more preferable than non-indigenous bacterial flora in the selection of PGPR for inoculation of agricultural plants because they have evolved to their environment and can be more competitive. It is generally understood that PGPR that increase growth in one plant species may not always have the same impact on other plant species (Batista et al., 2018). Some bacteria have broad growth-promoting effects on a number of plant species, whereas others have significant host-plant selectivity and colonize just one or a few plant species (Long et al., 2008). Many studies have found that crops inoculated with PGPR isolated from the same plant species can occasionally have a favorable PGPR impact (Lucy et al., 2004). As a result, considering the variety of the host plants is one of the most successful ways for selecting efficient PGPR. Plant growth-promoting rhizobacteria have traditionally been utilized to help plants develop under typical and special abiotic stress conditions, such as phosphorus shortage or saline stress. Plants, on the other hand, are subjected to several abiotic stressors at the same time. Furthermore, earlier research has found that numerous PGP features are likely to benefit plant growth under a variety of unfavorable conditions (Batista et al., 2018). Apart from choosing efficient PGPR for the creation of inoculants for agricultural production, choosing suited cultivars that benefit from interaction with these bacteria is an essential agronomical consideration. Indeed, early research revealed that the amount of the bacteria's beneficial effects on plant development varied depending on the cultivar of the host plant (Egamberdieva et al., 2010). The extensive use of agrochemicals, pathogenic attacks, and many other stresses has declined the crop yield tremendously which led to huge loss in agriculture productivity worldwide. So, there is a dire need to find an alternative for escalating problems of food security and sustainability in agriculture. Under such scenario, the climate-smart agriculture practices that avail bacterial usage open the avenues in fulfilling the incessant demand for food for global population. The usage of PGPR might help reduce the harmful environmental effects of chemical fertilizers, insecticides, and herbicides in the future as they play a tremendous role in the plant defense mechanism. Plant growth-promoting rhizobacteria is also a wonderful solution for farmers facing the new challenges of contemporary agriculture, since major environmental and social concerns have arisen as a result of agricultural industrialization, which was prompted by the need to feed a growing population.

5.10.1 Mechanisms Utilized by PGPR

Plant growth-promoting chemicals are produced by an expanding number of bacterial species. Table 5.3 reflects plant growth promoting substance released by PGPR. PGPR functions may be classified as direct and indirect. In direct mechanism, PGPR are described as biofertilizers that produce organic molecules that enhance plant growth or boost soil nutrient absorption. In indirect mechanism, bio control techniques include antibiotic synthesis, Fe chelators (also known as siderophores), and exterior cell wall disintegrating enzymes (e.g., chitinase and glucanase) that may hydrolyze the pathogen (i.e. fungus) cell wall. According to this scenario, a number of PGPR, most commonly from the genera *Agrobacterium*, *Bacillus*, *Paenibacillus*, and *Pseudomonas*, have been shown to reduce PPN population density through parasitism, the production of hydrolytic enzymes and antinematicidal metabolites (direct antagonism), and the induction of systemic resistance.

TABLE 5.3
Growth-Promoting Substances Secreted by Specific PGPR (Indirect Antagonism) (Ben & Kamilova, 2009)

PGPR	Plant Growth-Promoting Substances	References
Bacillus spp., *Pseudomonas* spp.	P solubilization, IAA production	Bashir et al., 2018a
Rhizobium spp. (pea)	Siderophores, hydrogen cyanide, ammonia, exo-polysaccharides, Indole-3-acetic acid	Ahemad & Khan, 2012
Mesorhizobium spp.	Indole-3-acetic acid, siderophores, hydrogen cyanide, exo-polysaccharides, ammonia	Ahmad et al., 2011
Rhizobium spp. (lentil)	Indole-3-acetic acid, siderophores, hydrogen cyanide, ammonia, exo-polysaccharides	Babu et al., 2015
Pseudomonas spp.	A3R3 Indole-3-acetic acid, siderophores	Asad et al., 2019
Psychrobacter spp.	SRS8 Heavy metal mobilization	Manoj et al., 2020
Bradyrhizobium spp.	Indole-3-acetic acid, siderophores, exo-polysaccharides, hydrogen cyanide, ammonia	Meena et al., 2017
Pseudomonas aeruginosa	4EA Siderophores	Teja et al., 2013

5.10.2 PGPR as Bio Control

Diseases destroy around 40% of the potential worldwide agricultural output even before harvest, and another 20% is harmed after harvest. Even more concerning is the fact that around 125,000–130,000 metric tons of pesticides are used to exterminate pests every year in some regions of Africa (Asogwa et al., 2009). Although regulatory bodies such as the National Environmental Standards and Regulations Enforcement Agency (NESREA), the National Agency for Foods and Drugs Control (NAFDAC), and the Nigeria Stored Products Research Institute (NSPRI) are working to reduce pesticide use, the implementation and enforcement of prescribed management systems, as well as legislative will, are still lacking (Nnedinma, 2016). A possible solution is provided by PGPR. Around 1,400 bio pesticide products are on the market globally, and the number is growing every day (Arora et al., 2016). According to Kaur et al. (2019) bio control may be defined as the employment of microorganisms to limit the occurrence and frequency of plant diseases (Kaur et al., 2019). When compared to traditional chemical insecticides, these bacteria provide better selectivity and lower toxicity. PGPR native to the soil and rhizosphere play an important role in phytopathogen management because they may contain or suppress a wide range of fungal, bacterial, nematode, and viral diseases. The use of PGPR as a bio control agent is increasing nowadays in the world as the field trials have been mostly inconsistent. Significant management of phytopathogens has been documented mostly in laboratory and greenhouse research. The eradication and prevention of phytopathogen proliferation are two important consequences of microbial inoculants on plants (mostly fungi) (Babalola et al., 2012). In Ramyasmruthi et al. (2012), 18 possible PGPR isolates were identified from the rhizosphere of various Solanaceous crops. Antifungal activity of isolates against selected soil-borne phytopathogens indicated by dual plate assay: *Alternaria alternate* OTA36, *Alternaria brassicola* OCA1, *Alternaria brassiceae* OCA3, *Collectotrichum gleosporidose* OGC1. Seed bacterization with chosen PGPR (*Pseudomonas*) decreased illness incidence by diverse pathogens by over 50%. There have been a number of inhibiting compounds identified. Pseudomonas species generate phenazine, pyoluteorin, oomycin A, pyrrolnitrin, tensin, tropolone, and cyclic lipopeptides, as well as *Bacillus*, *Streptomyces*, and *Stenotrophomonas* species generate zwittermicin A, oligomycin A, kanosamine, 2, 4-diacetylphloroglucinol (DAPG), amphisin, and xanthobac- cin. Specificity and activity of these various antibiotics have been widely investigated, and a number of these bio control strains are now commercially accessible. The necessity for these stresses to be applied continuously and persistently is, however, a typical challenge. Antibiotic resistance may develop among some plant infections. Some scientists have utilized hydrogen cyanide-producing strains as sole or augmented bio control strains as a prophylactic precaution (Devi et al., 2007). Because hydrogen cyanide alone may not have enough activity to operate as a bio control agent, this technique is effective. Nonetheless, it appears to interact with bacteria that have antibiotic genes encoded. Plant root colonizers get a competitive advantage by producing siderophores. By sequestering Fe^{3+} across the rhizosphere, siderophore producers benefit from the growth of disease-fighting organisms, making nutrients easily available for PGPR. Bacterial siderophores differ in their ability to sequester iron, and because fungal siderophores lack strong affinity, they frequently deny disease-causing fungus critical nutrients/elements. In rare situations, PGPR also takes Fe from heterologous siderophores made by co-residents, depriving them of a critical nutrient. For example, Pseudomonas spp. can utilize siderophores released by several bacteria and fungus species to increase iron availability in its environment. In a modified succinic acid medium, strains of *P. fluorescence* and *P. putida* produce significant amounts of hydroxamate-type siderophore. Other soil bacteria, such as *Azotobacter vinelandii* and *B. cereus*, also generate similar compounds, which can be used as competent plant growth promoters to boost agricultural output. *Bacillus megaterium* from the root area can also create siderophores, which help to increase plant development and minimize disease severity. *P. fluorescence* strains have recently been used as seed inoculants on agricultural plants to promote growth and increase productivity of a variety of crops. The antagonistic potential of rhizoplane fungus and bacteria has been found to improve plant development (Beneduzi et al., 2012). On the other hand, determining the features of hostile activity has been challenging. By fighting for available iron with resident microorganisms, PGPR exhibit their ability to promote growth. These processes are critical for antagonistic bacteria to work through induced resistance, and novel inoculants containing resistance-inducing and antagonistic PGPR can be developed as an environmentally benign strategy to lower plant illnesses and enhance agricultural yield. Furthermore, microorganisms that create a wide range of structural, intracellular, and extracellular polysaccharides have been found. Exopolysaccharides (EPS) are involved in biofilm formation. They help to capture the free phosphorous from the insoluble in soils as supplements. They also help the plants to receive the nutrients needed for proper growth and development.

These EPS also protect plants from disease-causing bacteria that aren't local to the area. Most PGPR create antifungal chemicals (lysing enzymes or hydrogen cyanides) that prevent disease-causing fungus from surviving. This provides an alternate technique for treating a variety of illnesses caused by fungi. Bio control agents, on the other hand, still need an appropriate delivery system (Qurashi & Sabri, 2012).

5.10.3 Application of PGPR in Agriculture

Microbial inoculants have been extensively studied, and the application of beneficial microorganisms has been used to develop agricultural systems (Babalola & Glick, 2010), resulting in increased crop development and a reduction in plant pests and disease-causing organisms. Plants' inherent ability to resist illness has been boosted by specialized, well-maintained inoculants, indicating its potential as a bio control tool (González-Chang et al., 2016). In nonchemical agriculture, biofertilizers, which are living organisms integrated into plants, seed, or even soil, are frequently promoted as an alternative to manmade chemical fertilizers. These fertilizers improve plant nutrition by leveraging microorganisms' natural ability to break down, solubilize, and mobilize nutrients (Sattar et al., 2019), lowering fertilizer costs and use (Singh & Gupta, 2018). In most industrialized nations, the adoption of the aforementioned microbial-based agricultural enhancements is rapidly expanding (Timmusk et al., 2017). Thus, in an area where agriculture is important to the economy, the use of PGPR is a promising alternative to traditional agricultural practices. Furthermore, local farmers may be hesitant to adopt a technology whose benefits and impacts are only seen after long-term usage (Igiehon & Babalola, 2018) unless the government provides financial support and commitment. With rising environmental concerns and a recognition that the span of large-scale chemical usage in the environment must come to an end, PGPR provide an appealing alternative that includes the prospect of establishing more sustainable agricultural practices. The discovery of several pathways involved in plant-rhizosphere microbe interaction has opened up new avenues for developing agricultural yield-improving techniques. Along with this, biotechnology can be used to enhance PGPR-positive strains in order to develop transgenic strains with diverse modes of action. Before we can assume that using PGPR as biofertilizers would result in a long-term boost of host plant development, we need to learn more about plant-microbe interactions in the Rhizosphere. Beneficial bacterial strain combinations that interact synergistically are now being developed, and various recent researches in the field of inoculation technology reveal a promising trend. PGPR are great model systems for biotechnologists, since they may offer them with novel genetic ingredients and bioactive compounds with a wide range of applications in agriculture and environmental sustainability.

Within the world, two sorts of fertilizers are utilized: normal and engineered. Other soil added substances, such as liming materials, are not essentially considered fertilizers. Fertilizers, both normally happening and fabricated, are accessible. Potassium (K), phosphorus (P), and nitrogen (N) are three essential macronutrients that are greatly advantageous in most present day rural strategies, with micronutrient aides such as soil shake being utilized on event. Agronomists utilize an assortment of strategies to apply these fertilizers, counting fluid, dry, pelletized, hand-tool strategies, and overwhelming agrarian hardware. Conventional fertilizing strategies included creature fertilizer, compost, mined minerals, excrement, edit revolutions, or mechanical human-nature buildup (for illustration, blood feast from creature butcher or angle handling squander). With advancements in plant nourishment within the late 1800s, an engineered fertilizer industry emerged. These improvements cleared the way for large-scale mechanical farming with high yields. N-fixing chemical forms created amid WWII, such as the Haber handle, fueled an increment within the industry. This was a noteworthy component of the so-called Green Insurgency within the half of the twentieth century, with nitrogen fertilizer utilization expanding by 800% between 1961 and 2019. The utilization of manufactured fertilizers in horticulture has far-reaching natural results (Chen et al., 2019). For illustration, fungicides can crush soil micro biomes and cause poison buildup in biological systems. Characteristic gas fracking, for instance, which is utilized within the Haber process, has a backhanded effect on engineered fertilizers. Mechanical horticulture on an expansive scale has been connected to living space corruption and biodiversity weight. The Economic Advancement Objectives are based on building up an economical nourishment generation framework or a great worldwide climate (MacFeely, 2020). To address these issues, most administrative point is to move rural homes towards regenerative or sustainable rural frameworks that utilize way better soil administration (such as no-till cultivating), less manufactured fertilizers, and way better water administration (i.e., no-till cultivating).

5.11 BLOOD MEAL

Almost every year, there are 40 metric tons of organic waste worldwide (Kumar et al., 2013). Properly managing these large amounts of waste and maximizing its recycling to avoid negative impacts on human health and the environment is one of the biggest challenges facing the world (Zamotaev et al., 2018). The goal of slaughter houses is to improve the recovery of edible meat parts for human consumption, but slaughter operations generate large amounts of protein- and mineral-rich waste (Mondini et al., 2008; Steffen & Kirsten, 1989). Blood is one of the most common byproducts in slaughter houses. Fresh blood contains up to 18% protein (Ockerman & Hansen, 2000) and is usually dried, processed through various thermochemical processes, transformed into blood meal, and used in animal feed (Aniebo et al., 2009). Blood meal contains up to 80% crude protein and contains on average 13% N, 1.0% P, 0.6% K, 0.2% Fe, 160 mg kg-1. and 100 mg kg-1 K (Kikafunda & Sserumaga, 2005). Yunta et al. (2013) reported that the ability of blood meal to provide a nitrogen source and retain iron bound to porphyrin compounds makes it a good candidate for use as an iron fertilizer in organic agriculture. Ciavatta et al. (1997) reported that blood meal incubated in soil for one year increased total nitrogen, iron, and mineral availability and wettability of soil organic matter. In a comparative study of straw and cotton residues with meat and bone meal and blood meal, it was found that animal residues caused more carbon dioxide release and nitrogen mineralization, while crop residues caused the fixation of nitrogen (Cayuela et al., 2009). In a comparative study testing six conventional nitrogen fertilizers and blood meal, Hirzel et al. (2018) reported that conventional fertilizers affected soil pH and the availability of N, P, K, Ca, Mg, and S. In comparison, blood meal has little effect on soil physical and chemical properties. Najjari and Ghasemi (2018) showed that the 5% and 10% (w/w) addition of powdered blood meal to sawdust vermicompost significantly accelerated its maturation and stabilization, and enriched the compost in N, P, K, Fe, Zn, and Mn (Kinley et al., 2010). Although the high fertilization potential of blood meal is well established, knowledge of its impact on soil quality (SQ) and biological activity remains limited. SQ is the ability of soil to function within an ecosystem to support primary production, maintain environmental quality, and promote animal and plant health (Doran & Parkin, 1994). Organic changes affect the soil environment by positively affecting the physical, chemical, and biological properties of soil, thus influencing soil quality, and the extent of these changes can be evaluated by the soil quality index (Qiu et al., 2019). The SQ can be calculated from a minimum data set (MDS) of relevant indicators (Andrews & Carroll, 2001; Muñoz-Rojas et al., 2016) and can be selected using principal component analysis (PCA) (Andrews et al., 2002). Industrial slaughter houses in Sanandaj (Kurdistan region of Iran) produce approximately one ton of blood meal per month, which is mainly recycled locally as organic fertilizer. In blood meal, nutrients are mainly in organic forms and there is still little information on the main mechanisms and timing of release of different nutrients. Table 5.4 shows diverse nutrient sources from animals/manure and their nutrient status.

TABLE 5.4
Diverse Nutrient Sources from Animals/Manure and Their Nutrient Status

	Percentage Content		
Manure	**Nitrogen (N)**	**Phosphoric Acid (P_2O_5)**	**Potash (K_2O)**
Coir pith	1.20	1.20	1.20
Blood meal	10–12	1.2	1.0
Press mud	1–1.5	4–5	2–7
Bone meal			
1) Raw bone meal	3–4	20–25	–
2) Steamed bone meal	1–2	25–30	–
Fish meal	4–10	3.9	0.3–1.5
Animal refuse	0.3–0.4	0.1–0.2	0.1–0.3
Cattle dung, fresh	0.4–0.5	0.3–0.4	0.3–0.4
Horse dung, fresh	0.5–0.5	0.4–0.6	0.3–1.0
Poultry manure, fresh	1.0–1.8	1.4–1.8	0.8–0.9
Sewage sludge, dry	2.0–3.5	1.0–5.0	0.2–0.5
Sewage sludge, activate dry	4.0–7.0	2.1–4.2	0.5–0.7
Cattle urine	0.9–1.2	Trace	0.5–1.0
Horse urine	1.2–1.5	Trace	1.3–1.5
Human urine	0.6–1.0	0.1–0.2	0.2–0.3
Sheep urine	1.5–1.7	Trace	1.8–2.0
Ash, coal	0.73	0.45	0.53
Ash, household	0.5–1.9	1.6–4.2	2.3–12.0
Ash, wood	0.1–0.2	0.8–5.9	1.5–36.0

Source: Hand book of Manures and Fertilizers 1964.

5.12 MEAT AND BONE MEAL

Meat and bone meal is a byproduct of the slaughter houses. It contains about 8% nitrogen (N), 5% phosphorus (P), 1% potassium (K), and 10% calcium (Ca) (Mondini et al., 2008, and Cascarosa et al., 2012) which makes it a valuable source of nutrients for plant production. Bone meal fertilizer is made from steamed animal bones and then ground into a fine powder or granules for use on plants or crops. Bones are rich in nutrients and minerals that are beneficial to plants and help make them healthier and stronger. There are many reasons why bone broth compost is a nutrient-rich soil amendment. Rich in calcium, it is known to build strong bones and is also beneficial for plants. Bone meal fertilizer is rich in this essential mineral. It helps keep plant cell walls strong and healthy and promotes vigorous growth (Brod et al., 2012; Chen et al., 2011). Phosphorus is another key nutrient found in bone meal fertilizers. This mineral plays an important role in photosynthesis; the production of roots, flowers, and seeds; the transfer of energy in plants; and other important functions. Phosphorus is the "P" in NPK that we often see in fertilizers. This increases fruit and seed production, improves flowering, and increases resistance to pests and diseases. Meat and bone meal fertilizer breaks down slowly, making it an ideal long-term fertilizer that only needs to be applied once a year. As the seasons go by, bone meal will continue to nourish your plants. If you grow organic plants, meat and bone meal fertilizer is a good choice. In addition to being completely free of chemicals, the natural decomposition process provides nutrients to soil microorganisms. It contains micronutrients such as magnesium, zinc, and iron that improve plant health and promote microbial growth in the soil. This improves the overall structure and quality of the soil, which in turn improves plant quality and production. Plus, it won't burn your plants if you accidentally apply too much.

5.13 WOOD ASH

Wood ash, an efficient source of minerals, is produced through partial combustion of wood comprising organic and inorganic compounds i.e., mixtures of carbonates, oxides, hydroxides, silicates, and low N contents owing to its volatilization during combustion process, and it is an efficient soil conditioner (Sadia et al., 2024). It is not only alkaline (due to presence of carbonates and oxides of Ca and K) but also contains enough soluble macronutrients such as phosphorus, calcium, magnesium, and potassium. Wood ash application in central Ontario hardwood forest soils led to significant increases in soil pH and essential nutrients like calcium and magnesium (Suzanne et al., 2015). Soil microbial diversity increased, but composition remained stable. Overall, wood ash proved effective in soil improvement without posing risks to soil communities. Nutrient composition of wood ash is largely dependent upon the tree species used, whether it is originated from bark or wood, as well as the efficiency of the combustion process. Wood ash is a soil amendment that has been shown in field experiments to have the ability to enhance soil qualities by modulating pH, in water holding capacity, and increasing microbial activity. Overall, wood ash proved effective in soil improvement without posing risks to soil communities and is treated as organic fertilizer in agriculture.

5.14 FISH MANURE

Natural agribusiness emphasizes supplement reusing through natural inputs to preserve soil plant ripeness frameworks, adjust nourishment generation, and ensure the environment (Ahuja, Dauksas, et al., 2020). A few nations, counting China, the United States, Canada, and Norway, support the fishery industry financially. Norway was the second-largest exporter of angle and angle items in 2018, trailing as it were China. Huge sums of cod, herring, mackerel, and other whitefish and little pelagic species are caught by the Norwegian armada (Ahuja, Dauksas, et al., 2020; FAO, 2020). An average angling vessel within the open ocean and close to the coast can create up to 3,000 tons of angle squander. Globally, the fishing industry creates a critical sum of angle squander. "Angle squander" can incorporate "entirety angle (dead or harmed angle), angle trimmings, and particular tissues such as heads, guts, tails and blades, skins, scales, and bones, among other things" (Ahuja, Dauksas, et al., 2020). In Norway, angle squander is commonly alluded to as "crude materials" or "leftover crude materials." In Norway, the pelagic segment employments 100% of angle squander for angle feast, and there's no angle oil generation or auxiliary squander streams that can be utilized for fertilizer (Richardsen et al., 2019). Between 2012 and 2018, there was a 40–60% increase within the utilize of angle squander for white angle. In 2018, the unutilized 40% squander from white angle totaled around 120,000 tons (Ahuja, Dauksas, et al., 2020). More than half of this angle squander is made up of bone-rich heads and spines, which are an important source of unutilized nitrogen, phosphorus, and calcium. So distant, a noteworthy sum of angle squander from white angle handling has been utilized as bolster for hide creatures. Agriculturists who had hide animals prior to January 2018 can keep them until February 2025; after that date, hide cultivating in Norway will be denied (Ahuja, Dauksas, et al., 2020). In Norway, the shellfish ("skalldyr" in Norwegian) division is little, and the utilization of leftover crude materials in this segment is lower than in white angle (Ahuja, Boer, et al., 2020). This sector's utilization rate in 2018 was 36% (Richardsen et al., 2019). Shrimp shells are presently more broadly accessible (Ween & Sunde, 2019). In case Norway is fruitful in negotiating new snow-crab quantities within the Barents Sea, this may altogether increase the accessible volumes of crab shells. Fish squander composting is an ancient strategy, but it is right now not utilized in the inland fisheries division. Maharashtra incorporates a noteworthy part in inland fisheries. Numerous coastal states are presently moving into the angle squander composting commerce since they have sufficient generation information from mariculture. However, within the state of Maharashtra, where there's inexhaustible angle generation within the shape of seawater and inland fisheries, no expansive industry or angle ranchers lock in fish squander composting business. Fresh water angle squander is an amazing source of supplements. Arranging of this squander is an ancient issue for angle dealers, from expansive commercial nourishment processors to small sport-fishing operations. This will be finished through composting, which is the same handle that domestic cultivators utilize to make their claim soil enhancer. Angle squander composting could be a generally unused, commonsense, and environmentally neighborly strategy of arranging undesirable angle parts and their foul scent, changing angle squander into a profitable item with attractive esteem on soil. Though angle is accessible all year, a huge sum of angle squander as well as non-edible angle related squander is created amid the post-monsoon season in both town markets and family kitchens. The transfer of this squander fabric could be a major issue. Since angle waste contains a high concentration of proteins, it cannot be put away for more than 24 hours. This moreover addresses the issue of pollution and squander disposal. Fish waste composting includes a long history. Composting could be an awesome way to recycle natural squander that can be utilized in present day horticulture. The application of all sorts of natural squander items into compost incorporates a high financial environmental esteem and is considered naturally neighborly (Castaldi et al., 2004; Deguchi et al., 2009). It empowers the collection and application of valuable organic fertilizers, humus substrate (Ranalli et al., 2001). Few tests on marine squanders have been conducted in India (Jayvardhan, 2020; Rebecca et al., 2014; Rebecca et al., 2015). For the primary time in Marathwada, natural fertilizers made from angle squander were made. There hasn't been much research in Maharashtra on composting angle squander to form fertilizer for natural agriculture. When compared to other sorts of ground or dried angle, excrement is an amazing fertilizer with various applications. As a result, plants are safer to bollworms and nematodes, and natural products are way better (in terms of dry matter or coloration). Crops such as lettuce and onions can benefit from angle squander. Amid the wet season, angle feces cause illness and transmit a foul smell. The chemical composition of angle feces was decided utilizing fresh dung tests from 12 commercial rainbow trout ranches in Ontario, Canada. On the other hand, contains more Mn and Cd than lower levels of As, Se, Co, and Ni found in numerous other animals fertilizers. Concurring to this consider, new angle excrement is chemically like other animals fertilizers and can be utilized as a fertilizer (Naylor et al., 1999) (Table 5.4).

5.15 POULTRY DROPPINGS

Poultry droppings are a source of available nutrients, especially nitrogen (N), phosphorus (P), and potassium (K) (Agbede & Oyewumi, 2022) and are known to increase crop yields (Ayyoub et al., 2023). For example, the results of Rasool et al. (2023) revealed that the use of poultry manure enhanced the physical properties of the soil, increased available P, and improved soil organic matter resulting in higher maize yield. A positive effect of poultry manure application is evident in the yield increase of wheat (Azad et al., 2023), as well as horticultural crops such as cabbage and pepper. Poultry manure is organic and environmentally friendly, and its nutritional effect lasts for several years after application, the nutrients being released slowly over time (Ferreira et al., 2022), which prevents them from leaching into the ground water. The poultry litter contains numerous profitable large-scale and microelements, which after pre-processing (due to the odor and nearness of pathogens) can be utilized in farming (Dauda et al., 2008; Montoro et al., 2019; Ippersiel et al., 2012; Oenema et al., 2007; Turan, 2008; Tur-Cardona et al., 2018; Xia et al., 2017; Zarebska et al., 2015). Adding poultry manure to soils not only helps to overcome the disposal issue but also increases the physical, chemical, and biological prolificity of soils (McGrath et al., 2009).

5.16 FOOD WASTES

Every year, 1.3 billion mg of food is wasted worldwide, which represents approximately one third of the edible products produced worldwide (Gustafsson et al., 2013). Food waste includes household and restaurant food waste, treated waste streams and crop residues. These residues contain carbohydrates (starch, cellulose, and hemicellulose), proteins, lignin, fats, and large amounts of water (Esra et al., 2014; Kiran et al., 2014). Food loss and waste prevention, although much debated in theory, is already being implemented by 20% of the world's 50 largest food companies. Losses occur mainly during cultivation and consumption, with approximately 30% at the level of the food supply chain (Principato et al., 2019). Many organizations (such as the United Nations Sustainable Development Goals) call for a 50% reduction in food waste by 2030, a goal that requires the support of policy makers (Principato et al., 2019). Food waste, a useful resource of valuable compounds, is burned or buried with other combustible municipal waste (Esra et al., 2014; Kiran et al., 2014). The effect of orange residue as an organic fertilizer on the growth of durum wheat was evaluated by field studies. When the dose of organic fertilizers is chosen appropriately, yields are similar to those of chemical fertilizers (Tuttobene et al., 2009). Food waste can be used to produce biofertilizers through anaerobic digestion, aerobic composting, and chemical hydrolysis; various agricultural wastes (such as wheat straw) can be dumped directly into the soil. Coffee waste can be used and recycled through proper management (i.e. composting, anaerobic digestion, and pyrolysis) to provide valuable products for agriculture and improve soil structure and fertility (Stylianou et al., 2021). Fertilizers made from coffee residues mixed with ash from biomass burning were tested on four plant species. With the exception of watercress, no toxic effects were observed on the germination of the test plants (Ciesielczuk et al., 2018). The use of organic fertilizers mixed with as much organic waste and minerals as possible will produce the desired results, as long as the composition of the mixture is suitable for the needs of the plants. It has been observed that some fertilizer mixtures can impair the root growth of some plants and stimulate the growth of others (Ciesielczuk et al., 2018). High-temperature aerobic fermentation of food waste can be a suitable method for the rapid production of organic fertilizers (Jiang et al., 2015). The nutrient-rich agricultural dig estate produced can be processed through aerobic digestion to remove excess water and concentrate nutrients, i.e. reverse osmosis (Tampio et al., 2016). Potassium-rich banana waste combined with sewage sludge containing high amounts of phosphorus has been used to produce bio char in thermal processes such as slow pyrolysis and thermal plasma treatment. Both methods are ideal for producing bio char that contains plant-available nutrients. The only aspect of concern is the presence of arsenic, which must be removed before the process, since its presence prevents the use of bio char as a fertilizer for edible plants. When animal products are processed, only 40–60% of the materials are used in food production. Some of the skin and fat is removed, and a small amount of protein is also removed, mostly in the form of blood, meat, and bones. They can be used as ingredients in the production of animal feed. These ends can also serve as a basis for the production of thermoplastic and thermoset materials, as well as coagulants and flocculants used in wastewater treatment (Mekonnen et al., 2016).

Bones are a valuable and concentrated source of phosphorus (Wyciszkiewicz et al., 2017). Of course, replacing some of the phosphoric rock with bone will create new technical problems, such as the solubility of hydrated hydroxyapatite in phosphoric acid, which requires the consumption of more sulfuric acid. In the case of bones, certain acids are necessary to dissolve the organic matrix of the bone. Phosphate fertilizers can be obtained by pyrolysis of slaughter residues, or by the participation of other biomass, such as meat, wood, and maize residues (Zwetsloot et al., 2016). Fish meat and offal, especially with the addition of filler manure, constitute valuable fertilizers rich in nitrogen, phosphorus, and calcium (Radziemska et al., 2018). After hydrolysis, residual keratin (including feathers) forms a mixture of amino acids which, after digestion with sulfuric acid, is a source of nitrogen that can be applied directly to plants as a foliar fertilizer (Chojnacka et al., 2011). Likewise, wool releases valuable nutrients and absorbs moisture even when introduced into the soil untreated. By hydrolysis with hot water, wool can be composted without prior treatment. On the other hand, ash from power plants, especially those that use biomass as feedstock, can provide a renewable source of potassium. The ash resulting from the burning of animal waste is rich in phosphorus and calcium and can be used as an agricultural fertilizer. The mixture of hydrates of animal remains and ash forms a complete set of nutrients for plants. In many cases, microorganisms can convert various biological wastes. It is estimated that 3.7×10^9 mg of agricultural waste and 1.3×10^9 mg of food waste are produced worldwide each year. These large quantities can be processed through incineration, anaerobic fermentation, or composting to produce heat, electricity, or fertilizer. Biotechnological processes make it possible to convert these wastes and use them as a source of nutrients. The idea is based on the primary hydrolysis of organic waste (food processing waste) into monomers such as glucose and amino acids to obtain substrates for the production of various chemical products of industrial interest. Lactic acid, succinic acid, fatty acids, and nutritional supplements. Various organic wastes can be converted into various value-added chemicals. This is the concept of bio economy, which includes a multifunctional and integrated approach to waste recycling, and an optimal sequence of operations aimed to make different products (chemicals polyhydroxyalkanoates, sorbents, polyols, polyurethane foam, carotenoids, phenolic antioxidants) and bioenergy (biodiesel, bio-oil, biogas). The food wastes have significance in organic agriculture in boosting soil and plant health (Zabaniotou & Kamaterou, 2019). The valorization of food waste as the resource for bio-based products and energy in waste bio refineries is of great importance.

When preparing new fertilizers, it is important to study their effects on soil quality and crop productivity in real systems (fields) (Sigurnjak et al., 2016). When developing new composting technologies through waste recycling, it is important to maintain a balance between nutritional, economic, and environmental assessments. Crop productivity and soil quality, as well as NPK use efficiency, should be considered and compared to commercial reference products. Fertilizer performance can be studied through soil chemical analysis, plant response to fertilizer tests, and phosphorus partitioning: phosphorus release patterns (Vaneeckhaute et al., 2016). Large-scale fertilizer production facilities are an obstacle to the implementation of technologies that use renewable raw material bases. Therefore, large quantities of raw materials of reproducible and uniform quality are needed. For example, an ammonia plant uses air to produce 1,500 to 1,800 tons per day. The phosphoric acid plant processes 60 tons of raw materials per hour. This requires the supply of large quantities of standardized raw materials. Farmers prefer to use chemical fertilizers instead of organic fertilizers because the application of the latter requires more labor. Therefore, some additional factors are needed to persuade farmers to use fertilizers produced from renewable resources, such as national policies, subsidies, collection, processing, storage, distribution, and construction of sanitation infrastructure (Wang et al., 2018). Wang et al. (2018) studied the factors that influence the promotion of organic fertilizers as an alternative to chemical fertilizers. Many factors were looked at, including the benefits, risks, and environmental aspects of the prospects. The study found that farmer association membership, support, and farm size influence the choice of fertilizer type. It is important to assess whether fertilizers containing valuable nutrients are marketable. Tur-Cardona et al. (2018) conducted this study in different European countries and found that the mentioned substitution can be done if farmers' preferences are taken into account, i.e. the price (which should be 65% lower than the price of fertilizers). Other properties should include high bioavailability of nutrients, organic carbon content, and sterility of organic fertilizers. If these criteria are met, there is an opportunity to replace chemical fertilizers with organic fertilizers (Tur-Cardona et al., 2018). The fertilizer industry is not under pressure to improve its products. Everything that happens in the fertilizer industry is about process, not product innovation, and the fact that all the fertilizer produced is sold. Ammonia technology was implemented at BASF in 1913 (Haber-Bosch process). This technology has not changed in 100 years. Technological changes have been developed that include increased gas efficiency and decreased processing energy consumption. However, no major changes have been made. A similar situation occurs in the nitrogen fertilizer industry and the phosphate fertilizer industry. The first superphosphate plant was opened in Liverpool in 1846. In the 1920s a continuous method for obtaining phosphoric acid was developed. In 1942, the United Kingdom received a patent for treating phosphate rock with sulfuric acid. Wet phosphoric acid technology has not changed since then. So far, there has been little pressure to change the phosphorus industry. Fertilizers are produced in very large facilities, so a large and repeatable raw material base is essential. Currently, we cannot rely on the fertilizer industry to solve the problem of waste exploitation. The fertilizer industry is not particularly interested in renewable resources because there is no economic justification. However, the fertilizer industry faces some environmental issues that raise some concerns about technological change, such as carbon dioxide emissions in ammonia production. Nitrogen fertilizer plants must pay a carbon dioxide emission quota tax. Although some carbon dioxide is used in the production of urea, some carbon dioxide is released from urea into the soil. On the other hand, the phosphorus industry faces two main problems. The first consumes non-renewable natural resources, and the second is phosphogypsum, the production of which is twice the consumption of raw materials. Another growing problem is the storage of phosphogypsum, for example. There are rock formations in Florida. The concept of the circular economy is based on the reuse, recycling, reuse and development of natural cycles. Although this concept is widely debated scientifically and politically, it has been little used in practice. When developing biofertilizer technologies, the following aspects are important: the impact on the environment must be minimized, resources must be used in a renewable way, taking into account their scarcity, and the technology must guarantee profitability and economic benefits for industrial companies. Reducing natural resources and protecting the environment should be a priority, while maintaining labor requirements for economic efficiency. The fertilizer industry underestimates the benefits of these techniques. For example, the use of feathers containing 11% nitrogen can be a source not only of the macronutrient N, but if the process is properly regulated, it can also be a source of amino acids, which are ion chelators, micronutrients, and bio stimulants. At the same time, this technology produces no emissions and requires no energy. In the manufacture of phosphate fertilizers, bone meal can be converted to hydroxyapatite, which is then decomposed in the same manner as in the wet phosphoric acid process. Thus, raw materials of biological origin can be used to manufacture ultra-modern products. The obligation to dispose of waste falls on agro-food producers, that is to say, the processing plants that generate bio-waste. Large food processing plants offer commercial groceries. Waste from the slaughterhouse is still present and must be recycled. Therefore, plants that produce fertilizers from bio-waste must be close to each other so that there is no need for transport. There is also the issue of disinfection. Biological waste in landfills can cause decomposition, which can lead to emissions. Another barrier to the use of renewable raw materials in the production of fertilizers is the volatility of the raw material. Technology should take this into account. Phosphorus recovery from wastewater is no longer interesting because the wastewater is no longer rich in phosphate. Just a few years ago, this was an interesting trend. However, phosphates have been removed from laundry detergents and are no longer an attractive source of phosphorus. In addition to struvite, dicalcium phosphate and tricalcium phosphate can also be excreted. Demonstration techniques are used in the Netherlands and Japan. Urine can also be recycled if separated into sewer systems (Karak et al., 2015). Dishwasher wastewater also contains phosphates because they are found in detergents and tablets. Struvite crystallization is no longer an interesting trend because the wastewater no longer contains as much phosphorus as it used to. An interesting trend is the mobilization of phosphorus in the soil through the use of phosphorus solubilizing microbes (Bashir et al., 2017; Wei et al., 2018). In addition to nitrogen, phosphorus (P) is an important nutrient for plants. Phosphorus deficiency often occurs in soils with low pH or high concentrations of aluminum and iron. It has been discovered that approximately 30% to 40% of the world's cultivated land is deficient in phosphorus. Studies have found severe phosphorus deficiency in tropical and subtropical regions. The phosphorus deficiency rate in India is 49.3%. In addition, phosphate fertilizers have proven to be expensive as the raw materials are imported from other countries. The raw material is raw phosphorus, but the production of phosphate fertilizers requires high quality raw phosphorus. According to the Food and Agriculture Organization, India has an estimated 260 million tons of available phosphate deposits, but only a small fraction of these (about 5.27 metric tons) meets the specifications of the fertilizer industry due to the low phosphorus content (low quality). Therefore, only 10% of the naturally available raw phosphate in India is suitable for phosphate fertilizer production. Recently, the Ministry of Chemical Fertilizers reported that India has become the "Atmanipal" in the field of phosphate fertilizers. The Government of India (GOI) has taken up a business plan to produce phosphate fertilizers from natural phosphate ore. Low-quality RP can be used as a source of phosphorus fertilizer in phosphorus-deficient soils because it is an abundant and cheap source of phosphorus. However, the use of RP is found to be limited due to its inherent low reactivity, which limits the release of sufficient amounts of P into the soil solution. This is perhaps the best phosphate fertilizer for acid soils and direct application is recommended. Research has focused on the use of FYM and fertilizers to increase RP solubility. Studies have found that applying RP with organic fertilizers, compost, or phosphorus-solubilizing organisms increases the dissolution of RP in the soil, thereby increasing phosphorus utilization by plants.

Element P is a building block of DNA and RNA molecules, as well as ADP and ATP and is responsible for transferring energy and facilitating biochemical processes in plants. Plants (Yagodin, 1984) secrete organic acids (such as malic acid and oxalic acid) through their roots to dissolve naturally occurring phosphate in the soil (called phosphate) into water-soluble forms. $(H_2PO_4)^-$, $(HPO_4)^{-2}$, and $(PO_4)^{-3}$ are absorbed by plants sequentially. A variety of microorganism present in the soil organic matter release organic acids that can dissolve soil phosphates. Single Super Phosphate (SSP) that contains P in water soluble form was produced (Bhattacharya & Jain, 2000) by the scientists of Rothamsted Experimental Station (England) in the year 1840 by reacting rock phosphate and sulfuric acid, with the idea of providing readily available P to the plants. This is followed by production of more complex phosphatic fertilizers that contain water soluble P. Increased agricultural production world over is partly due to the introduction of chemical fertilizers that contain NPK. Unfortunately 60% to 70% of the P applied to the soils, in the water soluble forms is unavailable to the plant (Brady, 1984) as applied phosphorous is fixed by Fe, Al, Mn ions in acidic soils and by Ca, Mg ions in alkaline soils into complexes that plants cannot take up. Phosphorous availability to the plants is maximum in the narrow soil pH range between 5.5 and 7. Excessive and indiscriminate application of chemical fertilizers shows adverse impact on the soils in that soil micro flora and fauna (which impart natural properties to the soils) are destroyed thereby resulting in decreased agricultural production after years of application. The phosphate rocks having high content of P soluble in 2% citric acid are considered for direct application in acidic soils. Phosphate rocks of sedimentary origin, that has PO_4^{3-} partly replaced by CO_3^{2-} isomorphically (carbonate appetites), show (Narayanasamy & Biswas, 1998) high content of P soluble in 2% citric acid. On the other hand phosphate rocks of igneous origin show less solubility of P in 2% citric acid and hence they are not considered for direct application in acidic soils. Organic fertilizers are generally based on their nutrient content (such as N, P, K, etc). Microorganisms that break down organic matter produce many beneficial compounds essential for plant growth, such as gibberellins, auxins, vitamins, fulvic, and humic acids. Therefore, chemical fertilizers and minerals cannot replace chemical fertilizers in agriculture. Peat or lignite can partially replace organic fertilizers or compost, as they also contain humic acid. Soil contains high levels of organic matter in addition to improving aeration by protecting the soil; it also improves the water holding capacity of the soil.

5.17 PHOSPHATE RICH ORGANIC MANURE

It has been observed that farmyard manure (FYM) is rich in fine particles of high quality of natural phosphate (34% P_2O_5) and has better efficiency in agronomy (Sekhar & Aery, 2001; Sekhar et al., 2002). Phosphorus dissolves slowly due to organic acids released by plant roots, soil-colonized microorganisms called phosphate solubilizers. Nitrogen (N) and phosphorus (P) are the primary constituents of plants and animals life playing active roles in many metabolic processes and phenology of crops and vegetables (Mohammed et al., 2013). P has functions of a structural nature in macromolecules such as nucleic acids (DNA/RNA) and of energy transformation as adenosine tri phosphate in metabolic pathways of biosynthesis and degradation. Type of soil, crops, water, fertilizer, management practices, and climatic conditions of an area are important factors to be considered when attempting to formulate sound P-fertilizer source and application for adequate crops yield response (Amanullah & Shah, 2010). Composting is a biological process in which microorganisms like bacteria, fungi, and other organisms convert organic materials, such as leaves, manure, sludge, paper, grass clippings, and food wastes into a soil like material called compost or humus. During composting microbes utilize carbon of organic material as a source of energy and for synthesis of new microbial cells. Carbon serves as both building block and an energy source for microbes. The use of composts offers several potential benefits including manure handling, enhanced soil fertility and reduced environmental risk, enlarged air spaces in soil, improved soil permeability for air and water circulation, improved soil texture, help to retain soil moisture, facilitation of the mechanical treatment of heavy clay soil, added nutrients to the soil, stimulation of biological activities, encouraged vigorous plant rooting system, helping to bind nutrients and preventing them from being leached out of the soil (Sarwar et al., 2007, 2008). Organic materials have beneficial effects on fertility and physical properties of soil (Hoitink & Grebus, 1994). The physical properties of soil play an important role in influencing plant growth thereby contributing to efficient crop production (Zheljazkov & Warman, 2004). Mulching soil surface with different organic materials improves soil biological activities, retains soil moisture for a longer time, and helps to control weeds (Samia et al., 2013). Farmyard manure (FYM) contains on average 0.5% N, 0.2% P_2O_5, and 0.5% K_2O (Brannon & Sommers, 1985). The application of organic matter to soil improves soil conditions by reducing dependence on synthetic fertilizers and helping microorganisms to produce sugars (Guar, 1994; Sharif et al., 2003).

Wheat (*Triticum aestivum* L.) is an important winter crop throughout Pakistan. The total wheat area is 9.046 million hectares and the grain production is 24.033 million tons, while the total wheat area in Khyber Pakhtunkhwa is 769 thousand hectares and the grain production is 1.205 million tons. The total cultivated area of mung bean (*Vigna radiata*) in Pakistan is 2197 ha with a yield of 157 tons and 716 kg ha^{-1}, while the total cultivated area in Khyber Pakhtunkhwa is 7 hectares with a yield of 4 tons and 620 kg ha^{-1}. Worldwide, agricultural soils are deficient in phosphorus (P) after nitrogen (Vance et al., 2003), which impairs various plant physiological and biochemical processes (Wu et al., 2005). It usually represents about 0.2% of plant biomass based on dry weight (Schachtman et al., 1998). Phosphorus is a fixed soil nutrient and requires the use of chemical phosphate fertilizers as an available source for optimal crop production. However, the nutrient use efficiency of chemical fertilizers does not exceed 20%, forcing poor farmers to add twice the optimal use rate of phosphate fertilizers. Adding more phosphorus to the soil makes agricultural production more resource-intensive, thereby increasing the cost of crop production per person (Aziz et al., 2006). In addition, high-quality raw phosphate, the most important source of phosphate fertilizers, is expected to run out by 2050 (Vance et al., 2017). These situations require an economical and environmentally friendly way to increase P utilization while minimizing P loss, and compost-enriched phosphorus solubilizing microorganisms (PSM) and RP can be a good solution. Rock phosphate (RP) is an important natural source of phosphorus and is used as a feedstock for the production of chemical phosphate fertilizers (Reddy & Venkataraman, 2002). Direct application of RP has been shown to work well in acidic soils where the lower pH helps dissolve the RP and increases the forms of phosphorus available to the plant. However, this method is not suitable for alkaline/calcareous soils due to the high pH (Caravaca et al., 2004). Applying compost-enriched RP to an alkaline/or calcareous soil can be a smart way to dissolve the RP. In addition, adding organic fertilizers can also improve the physical, chemical, and biological properties of soil (Adhami et al., 2014; Lim et al., 2015). This approach also has a positive impact on crops, as organic fertilizers contain large amounts of macro- and micronutrients (Kalaivanan & Omar Hattable, 2016). In general, most organic fertilizers are produced from biodegradable solid waste using compost and/or vermicomposting (Lim et al., 2016; Wu et al., 2014). In addition, some rhizobacteria have also been used to improve soil P pools, such as dissolved P and ACC deaminase activity. Ultimately, these approaches also help increase phosphorus availability to crops and reduce ethylene-induced stress (Glick et al., 2017). 1998). Ethylene is a gaseous plant hormone that is produced under biotic and abiotic stress (Arshad & Frankenberger, 2002). Its higher concentrations ($\geq 25\ \mu g\ L{-1}$) inhibit the growth of roots and nodules of leguminous plants (Arshad & Frankenberger, 2002; Mattoo & Suttle, 1991). Inoculation of rhizobacteria with ACC deaminase activity significantly improved the growth parameters of maize and mung bean (Davari et al., 2007, 2012; Shaharoona et al., 2006, 2007), root proliferation (Dey et al., 2004), and soil P transformation (Chen et al., 2006). Therefore, the use of rhizobacteria with the two previous activities could be a good strategy to reduce ethylene production during nodulation and improve the utilization of phosphorus, especially phosphorus in reverse phase. Most of the organic phosphorus is mineralized, and some organic acids are produced during the composting process. These organic acids reduce the pH of the microclimate, thereby increasing the solubility of RP and increasing the bioavailability of P. On the other hand, the combined use of rhizobacteria with P solubilizing activity and ACC deaminase with RP-EC can reduce environmental problems caused, reduce the excessive use of fertilizers, and increase the efficiency in the use of nutrients. In the literature, there are few reports on the combined use of RP, compost, and PSM (Biswas & Narayanasamy, 2006; Ditta et al., 2015, 2018; Ditta & Khalid, 2016; Saleem et al., 2013).

5.18 EGGSHELLS AS FERTILIZERS

An eggshell is the hard, outer covering of the egg. The eggshell has been attractive as an alternative liming source in agricultural production. Eggshells present healthy, balanced Ca due to their trace amounts of other minerals and are probably the best natural source of Ca. Eggshells contain up to 95% Ca carbonate and various macro- and micronutrients such as magnesium, potassium, iron, and phosphorus. Therefore, eggshells are also good for plant growth, stimulate root development, and can be reused to make fertilizers. Application of eggshell powder on several crops (such as cowpea, red chili, tomato, etc) showed better growth and yield (Kurniastuti, 2018; Radha & Karthikeyan, 2019). It will bring out the huge potential to produce and use eggshell powder as an alternative Ca source for organic agriculture production.

5.19 COCO PEAT

Coco peat is a natural byproduct of coconut processing, obtained from the outer husk of coconuts that is usually discarded as waste. It consists of ground-up coconut husk fibers, which are then processed into a usable growing medium. It holds various soil and plant health promoting properties. (a) Water holding properties: Coco peat retains water well, making it ideal for plant growth. (b) Excellent drainage properties: It allows excess water to drain, preventing root rot. (c) Creates perfect aeration: **It** provides good airflow to plant roots. (d) Organic Materials: It adds nutrients to the soil.

Coco peat provides sufficient anchorage or support to the plant, serves as reservoir for nutrients and water, allows oxygen diffusion to the roots, and permits gaseous exchange between the roots and atmosphere outside the root substrate (Argo, 1998; Abad et al., 2002; Bunt, 1988).

5.20 TEA WASTE/RESIDUE AS ORGANIC FERTILIZER

Tea is one of the most common and widely consumed beverages, taking the second position just after drinking water. The wastes derived from tea factories and tea shops is called tea waste or tea residue. Tea waste acts as a natural fertilizer, enriching the soil with essential nutrients. Potent chemicals in tea waste are sugars, antioxidants, cellulose, amino acid, cellulose, macro- and micronutrients. Being rich in minerals, organic material, and bioactive compounds tea waste is a potential fertilizer (Debnath et al., 2021; Hayat et al., 2022). Gurav and Sinalkar (2013) applied tea waste with cow dung to prepare compost and found increased plant growth after applying this compost. In addition, tea compost increased aeration, nutrient contents, and humification. Tea compost, when used as a nutrient source, improved soil fertility and crop growth. A study by Taha et al. (2018) showed that the application of tea compost meaningfully increased N_2 fixing bacterial and fungal populations and boosted N, P, and K uptake by plants. Tea compost performance was significantly better in terms of yields for pepper navel, orange, tomato, and cucumber (Pane et al., 2016).

REFERENCES

Abad, M., Noguera, P., Puchades, R., Maquieira, R., & Noguera, V. (2002). Physico-chemical and chemical properties of some coconut dusts for use as a peat substitute for containerized ornamental plants. *Bioresource Technology*, *82*, 241–245.

Adhami, E., Hosseini, S., & Owliaie, H. (2014). Forms of phosphorus of vermicompost produced from leaf compost and sheep dung enriched with rock phosphate. *International Journal of Recycling of Organic Waste in Agriculture*, *3*, 1–6.

Agbede, T. M., & Oyewumi, A. (2022). Benefits of biochar, poultry manure and biochar–Poultry manure for improvement of soil properties and sweet potato productivity in degraded tropical agricultural soils. *Resources, Environment and Sustainability*, *7*, 100051.

Ahemad, M., & Khan, M. S. (2012). Productivity of greengram in tebuconazole-stressed soil, by using a tolerant and plant growth-promoting *Bradyrhizobium sp.* MRM6 strain. *Acta Physiologiae Plantarum*, *34*, 245–254.

Ahemad, M., & Kibret, M. (2014). Mechanisms and applications of plant growth promoting rhizobacteria: Current perspective. *Journal of King Saud University—Science*, *26*, 1–20.

Ahmad, M., Zahir, A., Zahir, H., Naeem, A., & Asghar, M. (2011). Inducing salt tolerance in mung bean through coinoculation with rhizobia and plant-growth-promoting rhizobacteria containing 1-aminocyclopropane-1-carboxylate deaminase. *Canadian Journal of Microbiology*, *57*(7).

Ahuja, I., Boer, A. d., & Løes, A. K. (2020). Forsøk med gjødsel fra havet (Research with fertilizer from the sea). *Økologisk landbruk*, *39*, 26–29.

Ahuja, I., Dauksas, E., Remme, J. F., Richardsen, R., & Løes, A. K. (2020). Fish and fish waste-based fertilisers in organic farming—With status in Norway: A review. *Waste Management*, *115*, 95–112.

Alloway, B. J. (2009). Soil factors associated with zinc deficiency in crops and humans. *Environmental Geochemistry and Health*, *31*(5), 537–548.

Amadi, A. N., Nwankwoala, H. O., Eze, C. J., Alkali, Y. B., & Waziri, S. H. (2012). A review of waste management techniques in parts of Niger Delta, Nigeria. *Center for Human Settlements and Urban Development Journal*, *3*(1), 98–108.

Amanullah, A. L. K., & Shah, P. (2010). Timing and rate of nitrogen application influence profitability of maize planted at low and high densities in Northwest Pakistan. *Agronomy Journal*, *102*(2), 575–579.

Andrews, S. S., & Carroll, C. R. (2001). Designing a soil quality assessment tool for sustainable agroecosystem management. *Ecological Applications*, *11*, 1573–1585.

Andrews, S. S., Karlen, D. L., & Mitchell, J. P. (2002). A comparison of soil quality indexing methods for vegetable production systems in Northern California. *Agriculture, Ecosystems & Environment*, *90*, 25–45.

Aniebo, A. O., Wekhe, S. N., & Okoli, I. C. (2009). Abattoir blood waste generation in Rivers State and its environmental implications in the Niger Delta. *Toxicological & Environmental Chemistry*, *91*, 619–625.

Anitha, S., Padma Devi, S. N., & Kumari, S. K. (2015). Isolation and identification of zinc solubilizing fungal isolates from agricultural fields. *The Indian Journal of Agricultural Sciences*, *85*(12), 1638–1642.

Argo, W. R. (1998). Root medium physical properties. *HortTechnology*, *8*, 481–485.

Arora, N. K., Maya, V., Prakash, J., & Mishra, J. (2016). Regulation of biopesticides: Global concerns and policies. In Naveen Kumar Arora, Samina Mehnaz, Raffaella Balestrini (Eds.). *Bio formulations: For sustainable agriculture* (pp. 283–299). Springer.

Arshad, M., & Frankenberger Jr, W. T. (2002). Ethylene: Agricultural sources and applications. *Springer Science & Business Media*, *90*(3), 424.

Asad, S. A., Farooq, M., Afzal, A., & West, H. (2019). Integrated phytobial heavy metal remediation strategies for a sustainable clean environment—A review. *Chemosphere*, *217*, 925–941.

Asogwa, E. U., & Dongo, L. N. (2009). Problems associated with pesticide usage and application in Nigerian cocoa production: A review. *African Journal of Agricultural Research*, *4*(8), 675–683.

Ayeni, L., & Adetunji, M. (2010). Integrated application of poultry manure and mineral fertilizer on soil chemical properties, nutrient uptake, yield and growth components of maize. *Nature and Science*, *8*(1), 60–87.

Ayyoub, M. Z., Paracha, P., Abdullah, A., Shehzadi, A., Faizan, M., Mehmood, S., Mehmood, M., Waqas, A., Shehzad, A., & Waqas, M. (2023). Assessing the effectiveness of farmyard manure, poultry manure and nitrogen application for wheat productivity under various tillage systems. *Journal of Positive School Psychology*, *7*(1), 17–30.

Azad, A. K., Ahmed, T., Eaton, T. E. J., & Hossain, M. (2023). Organic amendments with poultry manure and cow dung influence the yield and status of nutrient uptake in wheat (Triticum aestivum). *American Journal of Plant Sciences*, *13*(7), 994–1005.

Aziz, A., Trotel-Aziz, P., Dhuicq, L., Jeandet, P., Couderchet, M., & Vernet, G. (2006). Chitosan oligomers and copper sulfate induce grapevine defense reactions and resistance to gray mold and downy mildew. *Phytopathology*, *96*(11), 1188–1194.

Babalola, O. O., & Glick, B. R. (2010). The use of microbial inoculants in African agriculture: Current practice and future prospects. *Journal of Food, Agriculture & Environment*, *10*(3&4), 540–549.

Babalola, O. O., & Glick, B. R. (2012). The use of microbial inoculants in African agriculture: Current practice and future prospects. *Journal of Food, Agriculture & Environment*, *10*(3&4), 540–549.

Babu, S., Prasanna., A. R., Bidyarani, N., Nain, L., & Shivay, Y. S. (2015). Synergistic action of PGP agents and *Rhizobium spp.* for improved plant growth, nutrient mobilization and yields in different leguminous crops. *Biocatalysis and Agricultural Biotechnology*, *4*(4), 456–464.

Bashir, Z. (2021). Characterization of potassium solubilizing bacteria from Rhizospheric soils of Cherry (*Prunusavium*). *Research Journal of Agricultural Sciences*, *12*(2), 494–496.

Bashir, Z., Hamid, B., Zargar, M. Y., & Kundoo, A. A. (2021). Morphological and biochemical characterization of potassium solubilizing bacteria from Rhizospheric Soils of Pinus (*Pinus wallichiana*). *Journal of Research and Development*, *21*, 58–63.

Bashir, Z., Zargar, M. Y., Baba, Z. A., & Mohiddin, F. A. (2018a). Solubilization of potassium containing mineral by bacteria from Kashmir iris (*Iris kashmiriana*) rhizosphere. *The Pharma Innovation Journal*, *17*(4), 747–750.

Bashir, Z., Zargar, M. Y., Baba, Z. A., Mohiddin, F. A., Peer, F. A., Mir, S. A., & Najar, G. R. (2018b). Isolation and characterization of phosphorus solubilizing bacteria (PSB) from rhizospheric soils of apple (*Malus Domestica*). *Journal of Pharmacognosy and Phytochemistry*, *7*(2), 2997–3000.

Bashir, Z., Zargar, M. Y., Bhat, T. A., Kousar, S., Baba, Z. A., & Mohiddin, F. A. (2018c). Solubilization of phosphorus containing mineral by bacteria from rhizospheric region of walnut (*Juglans regia*). *International Journal of Current Microbiology and Applied Sciences*, *7*(8), 2391–2398.

Bashir, Z., Zargar, M. Y., Mohiddin, F. A., Baba, Z. A., & Hamid, B. (2019a). Isolation and biochemical characterization of phosphorus solubilizing bacteria from rhizospheric region of apricot and peach. *Journal of Research and Development University of Kashmir*, *19*, 0972–5407.

Bashir, Z., Zargar, M. Y., Mohiddin, F. A., Kousar, S., Husain, M., & Rasool, F. (2017). Phosphorus solubilizing microorganisms: Mechanism and diversity. *International Journal of Communication Systems*, *5*(5), 666–673.

Bashir, Z., Zargar, M. Y., & Vishwakarma, D. K. (2019b). Potassium solubilizing microorganisms for sustainable agriculture. *Applied Agricultural Practices for Mitigating Climate Change*, *2*, 17–28.

Batista, B. D., Lacava, P. T., Ferrari, A., Teixeira-Silva, N. S., Bonatelli, M. L., Tsui, S., Mondin, M., Kitajima, E. W., Pereira, J. O., Azevedo, J. L., & Quecine, M. C. (2018). Screening of tropically derived, multi-trait plant growth- promoting rhizobacteria and evaluation of corn and soybean colonization ability. *Microbiological Research*, *206*, 33–42.

Ben, L., & Kamilova, F. (2009). Plant-growth-promoting rhizobacteria. *Annual Review of Microbiology*, 63, 541–556.

Beneduzi, A., Ambrosini, A., & Passaglia, L. M. (2012). Plant growth-promoting rhizobacteria (PGPR): Their potential as antagonists and biocontrol agents. *Genetics and Molecular Biology*, *35*(4), 1044–1051.

Berger, B., Patz, S., Ruppel, S., Dietel, K., Faetke, S., Junge, H., & Becker, M. (2018). Successful formulation and application of plant growth-promoting Kosakonia radicincitans in maize cultivation. *BioMed Research International*, *2018*, 6439481.

Berger, B., Wiesner, M., Brock, A. K., Schreiner, M., & Ruppel, S. (2015). Radicincitans, a beneficial bacteria that promotes radish growth under field conditions. *Agronomy for Sustainable Development*, *35*, 1521–1528.

Bhattacharya, P., & Jain, R. K. (2000). Fertilizer News. *Cambridge University Press*, *25*(10), 45.

Bhattacharyya, P. N., & Jha, D. K. (2012). Plant growth-promoting rhizobacteria (PGPR): Emergence in agriculture. *World Journal of Microbiology and Biotechnology*, *28*, 1327–1350.

Billah, M., Khan, M., Bano, A., Nisa, S., Hussain, A., Muhammad, K., Munir, A., & Khan, N. (2020). Rock phosphateenriched compost in combination with rhizobacteria; A cost-effective source for better soil health and wheat (*Triticum aestivum*) productivity. *Agronomy*, *10*(1390), 1–21.

Bist, V., Niranjan, A., Ranjan, M., Lehri, A., Seem, K., & Srivastava, S. (2020). Silicon-solubilizing media and its implication for characterization of bacteria to mitigate biotic stress. *Frontiers in Plant Science*, *11*, 28.

Biswas, D. R., & Narayanasamy, G. (2006). Rock phosphate enriched compost: An approach to improve low-grade Indian rock phosphate. *Bioresource Technology*, *97*(18), 2243–2251.

Bouajila, K., & Sanaa, M. (2011). Effects of organic amendments on soil physico-chemical and biological properties. *Journal of Materials and Environmental Science*, *2*, 485–490.

Brady, N. C. (1984). *Nature and properties of soils*. Collier Macmillan.

Brannon, C. A., & Sommers, L. E. (1985). Preparation and characterization of model humic polymers containing organic phosphorus. *Soil Biology and Biochemistry*, *17*(2), 213–219.

Brod, E., Haraldsen, T. K., & Breland, T. A. (2012). Fertilization effects of organic waste resources and bottom wood ash: Results from a pot experiment. *Agricultural and Food Science*, *21*, 332–347.

Bunt, A. C. (1988). *Media and mixes for container-grown plants* (1st ed.). Springer.

Caravaca, F., Lax, A., & Albaladejo, J. (2004). Aggregate stability and carbon characteristics of particle-size fractions in cultivated and forested soils of semiarid Spain. *Soil and Tillage Research*, *78*(1), 83–90.

Cascarosa, E., Gloria, G., & Arauzo, J. (2012). Thermochemical processing of meat and bone meal: A review. *Renew and Sustainable Energy Reviews Journal*, *16*, 942–957.

Castaldi, P., Garau, G., & Melis, P. (2004). Influence of compost from sea weeds on heavy metal dynamics in the soil-plant system. *Fresenius Environmental Bulletin*, *13*(11b), 1322–1328.

Cayuela, M. L., Sinicco, T., & Mondini, C. (2009). Mineralization dynamics and biochemical properties during initial decomposition of plant and animal residues in soil. *Applied Soil Ecology*, *41*, 118–127.

Chen, H., Yang, Z. K., Yip, D., Morris, R. H., & Lebreux, S. J. (2019). One-time nitrogen fertilization shifts switchgrass soil microbiomes within a context of larger spatial and temporal variation. *Plos One*, *14*(6), e0211310.

Chen, L., Kivelä, J., Helenius, J., & Kangas, A. (2011). Meat bone meal as fertiliser for barley and oat. *Agricultural and Food Science*, *20*(3), 235–244.

Chen, Y. P., Rekha, P. D., Arun, A. B., Shen, F. T., Lai, W. A., & Young, C. C. (2006). Phosphate solubilizing bacteria from subtropical soil and their tricalcium phosphate solubilizing abilities. *Applied Soil Ecology*, *34*(1), 33–41.

Chojnacka, K., Kowalski, Z., Kulczycka, J., Dmytryk, A., Górecki, H., Ligas, B., & Gramza, M. (2011). Carbon footprint of fertilizer technologies. *Journal of Environmental Management, 231*, 962–967.

Ciavatta, C., Govi, M., Sitti, L., & Gessa, C. (1997). Influence of blood meal organic fertilizer on soil organic matter: A laboratory study. *Journal of Plant Nutrition, 20*(11), 1573–1591.

Ciesielczuk, T., Rosik-Dulewska, C., Poluszyńska, J., Miłek, D., Szewczyk, A., & Sławińska, I. (2018). Acute toxicity of experimental fertilizers made of spent coffee grounds. *Waste and Biomass Valorization, 9*, 2157–2164.

Das, P. P., Singh, K. R., Nagpure, G., Mansoori, A., Singh, R. P., Ghazi, I. A., & Singh, J. (2022). Plant-soil-microbes: A tripartite interaction for nutrient acquisition and better plant growth for sustainable agricultural practices. *Environmental Research, 214*, 113821.

Dauda, S. N., Ajayi, F. A., & Ndor, E. (2008). Growth and yield of water melon (*Citrullus lanatus*) as affected by poultry manure application. *Journal of Agriculture and Social Sciences, 4*(3), 121–124.

Davari, M. R., Sharma, S. N., & Mirzakhani, M. (2007). Effect of cropping systems and crop residue incorporation on production and properties of soil in an organic agroecosystem. *Biological Agriculture & Horticulture, 28*(3), 206–222.

Davari, M. R., Sharma, S. N., & Mirzakhani, M. (2012). Residual influence of organic materials, crop residues, and biofertilizers on performance of succeeding mung bean in an organic rice-based cropping system. *International Journal of Recycling of Organic Waste in Agriculture, 1*, 1–9.

Davarinrjad, G., Haghnia, G., & Lakzian, A. (2004). Effect of enrichment manure and compost on wheat yield. *Agricultural Industrial Science Journal, 18*, 75–83.

Dawwam, G. E., Elbeltagy, A., Emara, H. M., Abbas, I. H., & Hassan, M. M. (2013). Beneficial effect of plant growth promoting bacteria isolated from the roots of potato plant. *Annals of Agricultural Research, 58*, 195–201.

Debnath, B., Haldar, D., & Purkait, M. K. (2021). Potential and sustainable utilization of tea waste: A review on present status and future trends. *Journal of Environmental Chemical Engineering, 9*(5), 106179.

Deguchi, S., Kawamoto, H., Tanaka, O., Fushimi, S., & Uozumi, S. (2009). Compost application increases the soil temperature on bare Andosol in a cool climate region, *Soil Science and Plant Nutrition, 55*(6), 778–782.

Devi, K., Seth, N., & Kothamasi, D. (2007). Hydrogen cyanide-producing rhizobacteria kill subterranean termite *odontotermes obesus* (Rambur) by cyanide poisoning under in vitro conditions. *Current Microbiology Article, 54*, 74–78.

Dey, R. K. K. P., Pal, K. K., Bhatt, D. M., & Chauhan, S. M. (2004). Growth promotion and yield enhancement of peanut (Arachis hypogaea L.) by application of plant growth-promoting rhizobacteria. *Microbiological Research, 159*(4), 371–394.

Dhaked, B. S., Triveni, S., Reddy, R. S., & Padmaja, G. (2017). Isolation and screening of potassium and zinc solubilizing bacteria from different rhizosphere soil. *International Journal of Current Microbiology and Applied Sciences, 6*(8), 1271–1281.

Dharma, A. K. (1996). *Organic farming for sustainable agriculture* (pp. 115–130). Agro Botanical Publishers (India).

Ditta, A., Arshad, M., Zahir, Z. A., & Jamil, A. (2015). Comparative efficacy of rock phosphate enriched organic fertilizer vs. mineral phosphatic fertilizer for nodulation, growth and yield of lentil. *International Journal of Agriculture and Biology, 17*, 589–595.

Ditta, A., Imtiaz, M., Mehmood, S., Rizwan, M. S., Mubeen, F., Aziz, O., & Tu, S. (2018). Rock phosphate-enriched organic fertilizer with phosphate-solubilizing microorganisms improves nodulation, growth, and yield of legumes. *Communications in Soil Science and Plant Analysis, 49*(21), 2715–2725.

Ditta, A., & Khalid, A. (2016). Bio-organo-phos: A sustainable approach for managing phosphorus deficiency in agricultural soils. *Organic Fertilizers-From Basic Concepts to Applied Outcomes*, *30*, 109–136.

Dittmar, H., Drach, M., Vosskamp, R., Trenkel, M. E., & Gutser, R. (2000). *Fertilizers, 2. types*. Ullmann's Encyclopedia of Industrial Chemistry.

Doran, J. W., & Parkin, T. B. (1994). Defining and assessing soil quality. *Define Soil Quality A Sustainable Environment: Proceedings of A Symposium Minneapolis*, *35*, 3–21.

Edwards, S., & Hailu, A. (2011). How to make compost and use. In L. L. Ching, S. Edwards & H. S. Nadia (Eds.). *Climate Change and Food Systems Resilience in Sub-Sarahan Africa* (pp. 379–436). FAO.

Egamberdieva, D., Kucharova, Z., Davranov, K., Berg, G., Makarova, N., Aza-rova, T., Chebotar, V., Tikhonovich, I., Kamilova, F., Validov, S., & Lugtenberg, B. (2010). Bacteria able to control foot and root rot and to promote growth of cucumber in salinated soils. *Biology and Fertility of Soils*, *47*, 197–205.

El-Shemy, S., & Aly, B. (1997). Report on agriculture wastes in Egypt. In A. El-Galaa (Ed.), *Organic agriculture* (p. 308).

Esra, U., Antoine, K., Trzcinski, P., Wun, J. N., & Liu, Y. (2014). Bioconversion of food waste to energy: A review. *Fuel*, *134*, 389–399.

FAO. (2020). *World fertilizer trends and outlook to 2022* (p. 28). Food and Agriculture Organization of United Nations.

Fasim, F., Ahmed, N., Parsons, R., & Gadd, G. M. (2002). Solubilization of zinc salts by a bacterium isolated from the air environment of a tannery. *FEMS Microbiology Letters*, *213*(1), 1–6.

Ferreira, P. A. A., Ceretta, C. A., Lourenzi, C. R., De Conti, L., Marchezan, C., Girotto, E., & Tiecher, T. L. (2022). Long-term effects of animal manures on nutrient recovery and soil quality in acidtypic hapludalf under no-till conditions. *Agronomy*, *12*(2), 243.

Gerard, F., Mayer, K., Hodson, M., & Ranger, J. (2008). Modelling the biogeochemical cycle of silicon in soils: Application to a temperate forest ecosystem. *Geochimica et Cosmochimica Acta*, *72*, 741–758.

Ghaly, A. E., & Alhattab, M. (2013). Drying poultry manure for pollution potential reduction and production of organic fertiliser. *American Journal of Environmental Sciences*, *9*(2), 88–102.

Glick, B. R., Cheng, Z., Czarny, J., & Duan, J. (2017). Promotion of plant growth by ACC deaminase-producing soil bacteria. In *New perspectives and approaches in plant growth-promoting rhizobacteria research* (pp. 329–339). Springer Nature.

Glick, B. R., Penrose, D. M., & Li, J. (1998). A model for the lowering of plant ethylene concentrations by plant growth-promoting bacteria. *Journal of Theoretical Biology*, *190*(1), 63–68.

González-Chang, M., Wratten, S. D., Lefort, M. C., & Boyer, S. (2016). Food webs and biological control: A review of molecular tools used to reveal trophic interactions in agricultural systems. *Food Webs*, *9*, 4–11.

Guar, A. C. (1994). Bulky organic manures and crop residues. In H. L. S. Tandon (Ed.). *Fertilizers, Organic Manures, Recyclable Wastes and bio-fertilizers* (pp. 36–51). Fertilizers Development and Consultation Organization.

Guene, O. (Ed.). (2002). Integrated traditional composting within domestic solid waste management. In *Proceedings of International Symposium on Environmental Pollution Control and Waste Management, Tunisia*.

Gurav, M., & Sinalkar, S. (2013). Preparation of organic compost using waste tea powder. In *National Conference on Biodiversity: Status and Challenges in Conservation* (pp. 97–99). FAVEO.

Gustafsson, J., Cederberg, C., Sonesson, U., & Emanuelsson, A. (2013). *The methodology of the FAO study: Global food losses and food waste-extent, causes and prevention*. FAO.

Hamid, B., & Bashir, Z. (2019a). Potassium solubilization microorganisms: An Alternative technology to chemical fertilizers. *Journal of Research and Development University of Kashmir*, *19*, 0972–5407.

Haque, M. M., & Biswas, J. C. (2021). Emission factors and global warming potential as influenced by fertilizer management for the cultivation of rice under varied growing seasons. *Environmental Research*, *197*, 111156.

Hayat, T., Adeel, S., Batool, F., Amin, N., Ahmad, T., & Ozomay, M. (2022). Waste black tea leaves (*Camelia sinensis*) as a sustainable source of tannin natural colorant for bio-treated silk dyeing. *Environmental Science and Pollution Research*, *29*(16), 24035–24048.

Hirzel, J., Donnay, D., & Fernández, C. (2018). Evolution of nutrients and soil chemical properties of seven organic fertilizers in two contrasting soils under controlled conditions. *Chilean Journal of Agricultural & Animal Sciences*, *34*, 77–88.

Hoitink, H. A., & Grebus, M. E. (1994). Status of biological control of plant diseases with composts. *Compost Science & Utilization*, *2*(2), 6–12.

Igiehon, N., & Babalola, O. (2018). Rhizosphere microbiome modulators: Contributions of nitrogen fixing bacteria towards sustainable agriculture, *International Journal of Environmental Research and Public Health*, *15*(4), 574.

Ippersiel, D., Mondor, M., Lamarche, F., Tremblay, F., Dubreuil, J., & Masse, L. (2012). Nitrogen potential recovery and concentration of ammonia from swine manure using electrodialysis coupled with air stripping. *Journal of Environmental Management*, *95*, S165–S169.

Jacob, S., & Banerjee, R. (2017). Nutrient enrichment of organic manure through biotechnological means. *Waste and Biomass Valorization*, *8*, 645–657.

Jayvardhan, V. B. (2020). Devising of organic fertilizer from fish and crab wastes: Waste to best technology. *International Journal of Fisheries and Aquatic Studies*, *8*(2), 1–5.

Jiang, Y., Ju, M., Li, W., Ren, Q., Liu, L., Chen, Y., & Liu, Y. (2015). Rapid production of organic fertilizer by dynamic high-temperature aerobic fermentation (DHAF) of food waste. *Bioresource Technology*, *197*, 7–14.

Kalaivanan, D., & Omar Hattab, K. (2016). Recycling of sugarcane industries byproducts for preparation of enriched pressmud compost and its influence on growth and yield of rice (*Oryza sativa L.*). *International Journal of Recycling of Organic Waste in Agriculture*, *5*, 263–272.

Kalpana, B., & Dinesh, K. M. (2020). Zinc solubilizing bacteria (*Bacillus megaterium*) with multifarious plant growth promoting activities alleviates growth in *Capsicum annuum* L., *3 Biotech*, *10*, 36. https://doi.org/10.1007/s13205-019-2033-9

Kannan, P., Saravanan, A., Krishna Kumar, S., & Natarajan, S. (2001). Biological properties of soil as influenced by different organic manures. *Research Journal of Agriculture and Biological Sciences*, *1*(2), 181–183.

Karak, T., Sonar, I., Nath, J. R., Paul, R. K., Das, S., Boruah, R. K., & Das, K. (2015). Struvite for composting of agricultural wastes with termite mound: Utilizing the unutilized. *Bioresource Technology*, *187*, 49–59.

Kaur, T., Rani, R., & Manhas, R. K. (2019). Bio control and plant growth promoting potential of phylogenetically new Streptomyces sp. MR14 of rhizospheric origin. *AMB Express*, *9*(125).

Kikafunda, J. K., & Sserumaga, P. (2005). Production and use of a shelf-stable bovine blood powder for food fortification as a food-based strategy to combat iron deficiency anemia in subsaharan Africa. *African Journal of Food, Agriculture, Nutrition and Development*, *5*.

Kinley, B., Rieck, J., Dawson, P., & Jiang, X. (2010). Analysis of Salmonella and enterococci isolated from rendered animal products. *Canadian Journal of Microbiology*, *56*, 65–73.

Kiran, E. U., Trzcinski, A. P., Ng, W. J., & Liu, Y. (2014). Bioconversion of food waste to energy: A review. *Fuel*, *134*, 389–399.

Kumar, D. S., Kumar, P. S., Kumar, V. U., & Anbuganapathi, G. (2013). Impact of bio fertilizers on growth and reproductive performance of *Eisenia fetida* (Savigny 1926) during flower waste vermicomposting process. *Annual Research & Review in Biology*, *3*(4), 574–583.

Kurniastuti, T. (2018). Effect of rice husk ash and eggshell on the growth and yield of red chili (Capsicum annuum L.). *Journal Academic Research and Science*, *3*(1), 46–52.

Lim, P. N., Wu, T. Y., Clarke, C., & Nik Daud, N. N. (2015). A potential bioconversion of empty fruit bunches into organic fertilizer using *Eudrilus eugeniae. International Journal of Environmental Science and Technology*, *12*, 2533–2544.

Lim, S. L., Lee, L. H., & Wu, T. Y. (2016). Sustainability of using composting and vermicomposting technologies for organic solid waste biotransformation: Recent overview, greenhouse gases emissions and economic analysis. *Journal of Cleaner Production*, *111*, 262–278.

Liu, D., Lian, B., & Dong, H. (2012). Isolation of *Paenibacillus sp.* and assessment of its potential for enhancing mineral weathering. *Geomicrobiology Journal*, *29*(5), 413–421.

Long, H. H., Schmidt, D. D., & Baldwin, I. T. (2008). Native bacterial endophytes promote host growth in a species specific manner. *Phytohormone Manipulations Do Not Result in Common Growth Responses*, *3*(7), e2702.

López-Mosquera, M. E., Fernández-Lema, E., Villares, R., Corral, R., Alonso, B., & Blanco, C. (2011). Composting fish waste and seaweed to produce a fertilizer for use in organic agriculture. *Procedia Environmental Sciences*, *9*, 113–117.

Lucy, M., Reed, E., & Glick, B. R. (2004). Applications of free living plant growth promoting rhizobacteria. *Antonie van Leeuwenhoek Article*, *86*, 1–25.

MacFeely, S. (2020). Measuring the sustainable development goal indicators: An unprecedented statistical challenge. *Journal of Official Statistics*, *36*(2), 361–378.

Manoj, S., & Krishna, C. K. (2020). Understanding the molecular mechanisms for the enhanced phytoremediation of heavy metals through plant growth promoting rhizobacteria: A review. *Journal of Environmental Management*, *254*, 109779.

Mattoo, A. K., & Suttle, J. C. (1991). *The plant hormone ethylene* (p. 337). CRC Press.

McGrath, S., Maguire, R. O., Tacy, B. F., & Kike, J. H. (2009). Improving soil nutrition with poultry litter application in low input forage systems. *Agronomy Journal*, *102*, 48–54.

Meena, V. S., Meena, S. K., & Verma, J. P. (2017). Plant beneficial rhizospheric microorganism (PBRM) strategies to improve nutrients use efficiency: A review. *Ecological Engineering*, *107*, 8–32.

Mekonnen, M. M., Lutter, S., & Martinez, A. (2016). Anthropogenic nitrogen and phosphorus emissions and related grey water footprints caused by EU-27′ s crop production and consumption. *Water*, *8*(1), 30.

Mohammed, R., Olorukooba, M. M., Akinyaju, M. M., & Kambai, E. A. (2013). Evaluation of different concentrations and frequency of foliar application of moringa extract on growth & yield of onion, *Allium cepa Lam. Agrosearch*, *13*(3), 196–205.

Mondini, C., Cayuela, M. L., Sinicco, T., Sánchez-Monedero, M. A., Bertolone, E., & Bardi, L. (2008). Soil application of meat and bone meal: Short-term effects on mineralization dynamics and soil biochemical and microbiological properties. *Soil Biology and Biochemistry*, *40*, 462–474.

Montoro, S. B., Lucas Jr, J., Santos, D. F. L., & Costa, M. S. S. M. (2019). Anaerobic co-digestion of sweet potato and dairy cattle manure: A technical and economic evaluation for energy and biofertilizer production. *Journal of Cleaner Production*, *226*, 1082–1091.

Muñoz-Rojas, M., Erickson, T. E., Dixon, K. W., & Merritt, D. J. (2016). Soil quality indicators to assess functionality of restored soils in degraded semiarid ecosystems. *Restoration Ecology*, *24*, S43–S52.

Najjari, F., & Ghasemi, S. (2018). Changes in chemical properties of sawdust and blood powder mixture during vermicomposting and the effects on the growth and chemical composition of cucumber. *Scientia Horticulturae (Amsterdam)*, *232*, 250–255.

Narayanasamy, G., & Biswas, D. R. (1998). Phosphate rocks of India—Potentialities and constraints. *Fertiliser News*, *43*(10), 21–28, 31–32.

Naureen, Z., Aqeel, M., Hassan, M. N., Gilani, S. A., Bouqellah, N., Mabood, F., Hussain, J., & Hafeez, F. Y. (2015). Isolation and screening of silicate bacteria from various habitats for biological control of phytopathogenic fungi. *American Journal of Plant Sciences*, *6*, 2850–2859.

Naylor, S. J., Moccia, R. D., & Durant, G. M. (1999). The chemical composition of settleable solid fish waste (manure) from commercial rainbow trout farms in Ontario, Canada. *North American Journal of Aquaculture*, *61*(1), 21–26.

Niassy, S., Diarra, K., Niang, Y., Niang, S., & Pfeifer, H. R. (2010). Effect of organic fertilizers on the susceptibility of tomato Lycopersicon esculentum: Solanaceae to Helicoverpa armigera Lepidoptera: Noctuidae in the Niayes area Senegal. *Research Journal of Agriculture and Biological Sciences*, *6*(6), 708–712.

Nnedinma, U. (2016). Approaches, drivers and motivators of health and safety self-regulation in the Nigerian construction industry: A scoping study. *Architectural Engineering and Design Management*, *12*(6).

Ockerman, H., & Hansen, C. L. (2000). *Animal by-product processing and utilization, Lancaster*. Technomic Pub. Co. Inc, c2000. https://doi.org/10.1201/9781482293920

Oenema, O., Oudendag, D., & Velthof, G. L. (2007). Nutrient losses from manure management in the European Union. *Livestock Science*, *112*(3), 261–272.

Pane, C., Palese, A. M., Spaccini, R., Piccolo, A., Celano, G., & Zaccardelli, M. (2016). Enhancing sustainability of a processing tomato cultivation system by using bioactive compost teas. *Scientia Horticulturae*, *202*, 117–124.

Penhallegon, R. (2003). *Nitrogen-phosphorus-potassium values of organic fertilizers*. Oregon State University, Extension Service.

Principato, L., Ruini, L., Guidi, M., & Secondi, L. (2019). Adopting the circular economy approach on food loss and waste: The case of Italian pasta production. *Resources, Conservation and Recycling*, *144*, 82–89.

Qiu, X., Peng, D., & Wang, H. (2019). Minimum data set for evaluation of stand density effects on soil quality in Larix principis-rupprechtii plantations in North China. *Ecological Indicators*, *103*, 236–247. https://doi.org/10.1016/j.ecolind.2019.04.010

Qurashi, A. W., & Sabri, A. N. (2012). Bacterial exopolysaccharide and biofilm formation stimulate chickpea growth and soil aggregation under salt stress. *Brazilian Journal of Microbiology*, *43*, 1183–1191.

Radha, T., & Karthikeyan, G. (2019). Hen eggshell waste as fertilizer for the growth of phaseolus vulgaris (Cowpea seeds). *Research Journal of Life Sciences, Bioinformatics, Pharmaceutical and Chemical Sciences*, *51*(1), 398–406.

Radziemska, M., Vaverkova, M. D., Adamcova, D., Brtnicky, M., & Mazur, Z. (2018). Valorization of fish waste compost as a fertilizer for agricultural use. *Waste and Biomass Valorizationiz*, *10*, 2537–2545.

Ramyasmruthi, S., Pallavi, O., Pallavi, S., Tilak, K., & Srividya, S. (2012). Chitinolytic and secondary metabolite producing Pseudomonas fluorescens isolated from Solanaceae rhizosphere effective against broad spectrum fungal phytopathogens. *Asian Journal of Plant Sciences*, *2*(1), 16–24.

Ranalli, G., Botturea, G., Taddei, P., Garavni, M., Marchetti, R., & Sorlini, G. (2001). Composting of solid and sludge residues from agricultural and food industries: Bioindicators of monitoring and compost maturity. *Journal of Environmental Science Health*, *36*, 415–436.

Rani, N., Kaur, R., & Kaur, S. (2020). Zinc solubilizing bacteria to augment soil fertility—A comprehensive review. *International Journal of Agricultural Sciences and Veterinary Medicine*, *8*(1), 38–44.

Rasool, A., Ghani., A., Naway, R., Ahmad, R., Shahyad, R., Rebi, R. B., & Zhou, A. J. (2023). Effects of poultry manure on the growth, physiology, yield, and yield-related traits of maize varieties. *ACS Omega*, *29*, 25766–25779.

Rebecca, J. L., Anbuselvi, S., Medok, P., & Sarkar, D. (2014). Effect of marine waste on seed germination. *Journal of Chemical and Pharmaceutical Research*, *6*(4), 581–584.

Rebecca, J. L., Anbuselvi, S., Sharmila, S., Prathiba, M., & Sarkar, D. (2015). Effect of marine waste on plant growth. *Der Pharmacia Lettre*, *7*(10), 299–301.

Reddy, M. S., & Venkataraman, C. (2002). Inventory of aerosol and sulphur dioxide emissions from India: I-Fossil fuel combustion. *Atmospheric Environment*, *36*(4), 677–697.

Richardsen, R., Mhyre, M., Nystøyl, R., Strandheim, G., & Marthinussen, A. (2019). Analyse marint restråstoff 2018—Tilgang og anvendelse av marint restråstoff 2018 (Analysis of marine raw material 2018—Access and use of marine raw material 2018), Rep. No. 2019:00475. SINTEF Ocean AS.

Sadia, B. B., Ali, S., Bernas, S., Moudrý, J., Konvalina, P., Mushtaq, Z., & Theoneste, Y. M. (2024). Wood ash application for crop production, amelioration of soil acidity and contaminated environments. *Chemosphere*, *357*, 141865.

Saleem, M. M., Muhammad Arshad, M. A., & Muhammad Yaseen, M. Y. (2013). Effectiveness of various approaches to use rock phosphate as a potential source of plant available P for sustainable wheat production. *International Journal of Agriculture and Biology*, *15*(2), 223–230.

Samia, K., Abdelkarim, C., Yassin, H., & Teruo, H. (2013). The effect of long-term soil irrigation by wastewater on organic matter, polycyclic aromatic hydrocarbons, and heavy metals evolution: Case study of Zaouit Sousse (Tunisia). *Arabian Journal of Geosciences*, *6*, 4337–4346.

Santana, E. B., Marques, E. L. S., & Dias, J. C. T. (2016). Effects of phosphate-solubilizing bacteria, native microorganisms and rock dust on *Jatropha curcas* L. growth. *Genetics and Molecular Research*, *15*, 4.

Sarwar, G., Hussain, N., Schmeisky, H., Suhammad, S., Ibrahim, M., & Ahmad, S. (2008). Efficiency of various organic residues for enhancing rice-wheat production under normal soil conditions. *Pakistan Journal of Botany*, *40*(5), 2107–2113.

Sarwar, G., Nisa, M., Shahzad, M. A., & Hassan, Z. (2007). Effect of replacement of varying level of concentrate with urea molasses treated wheat straw ensiled with manure on ruminal characteristics, in situ digestion kinetics and nitrogen metabolism in Niliravi Buffalo Bulls. In *Proceedings of International Nutrition Conference*. General Animal Nutrition Section.

Sattar, A., Naveed, M., Ali, M., Zahir, Z. A., Nadeem, S. M., Yaseen, M., Meena, V. S., Farooq, M., Singh, R., Rahman, M., & Meena, H. N. (2019). Perspectives of potassium solubilizing microbes in sustainable food production system: A review. *Applied Soil Ecology*, *133*, 146–159.

Schachtman, D. P., Reid, R. J., & Ayling, S. M. (1998). Phosphorus uptake by plants: From soil to cell. *Plant Physiology*, *116*(2), 447–453.

Sekhar, D. M. R., & Aery, N. C. (2001). Phosphate rock with farmyard manure as P fertilizer in neutral and weakly alkaline soils. *Current Science*, *80*(9), 1113–1115.

Sekhar, D. M. R., Aery, N. C., & Gupta, D. K. (2002). *Indian Chemical Engineer*, *44*(3).

Shaharoona, B., Arshad, M., & Zahir, Z. A. (2006). Effect of plant growth promoting rhizobacteria containing ACC-deaminase on maize (*Zea mays* L.) growth under axenic conditions and on nodulation in mung bean (*Vigna radiata* L.). *Letters in Applied Microbiology*, *42*(2), 155–159.

Shaharoona, B., Jamro, G. M., Zahir, Z. A., Arshad, M., & Memon, K. S. (2007). Effectiveness of various *Pseudomonas spp.* and *Burkholderia caryophylli* containing ACC-Deaminase for improving growth and yield of wheat (Triticum aestivum I.). *Journal of Microbiology and Biotechnology*, *17*(8), 1300.

Sharif, M., Khattak, R. A., & Sarir, M. S. (2003). Residual effect of humic acid and chemical fertilizers on maize yield and nutrient accumulation. *Sarhad Journal of Agriculture*, *19*(4), 543–550.

Sharma., R. Z. S., Trivedi, M. H., & Gobi, T. A. (2013). Phosphate solubilizing microbes: Sustainable approach for managing phosphorus deficiency in agricultural soils, *Springer Plus*, *2*, 587.

Sigurnjak, I., Michels, E., Crappé, S., Buysens, S., Tack, F. M. G., & Meers, E. (2016). Utilization of derivatives from nutrient recovery processes as alternatives for fossil-based mineral fertilizers in commercial greenhouse production of *Lactuca sativa L. Scientia Horticulturae*, *198*, 267–276.

Singh, S. J., & Gupta, V. K. (2018). Soil microbial biomass: A key soil driver in management of ecosystem functioning. *Science of the Total Environment*, *634*, 497–500.

Singh, V. K., Dwivedi, B. S., Rathore, S. S., Mishra, R. P., Satyanarayana, T., & Majumdar, K. (2021). Timing potassium applications to synchronize with plant demand. *Improving Potassium Recommendations for Agricultural Crops*, 363–384.

Soumare, A., Djibril, S. A. R. R., & Diédhiou, A. G. (2023). Potassium sources, microorganisms and plant nutrition: Challenges and future research directions. *Pedosphere*, *33*(1), 105–115.

Sousa, L. B., de Stamford, N. P., Oliveira, W., da, S., Silva, E. V. N. da, Martins, M., dos S., Santos, C. E., & de, R. S. (2018). Evolution of nutrient availability in maturation phase of composting using proportions of different residues inoculated with *Beijerinckia indica. Acta Scientiarum Agronomy*, *1*(7), 40.

Steffen, R., & Kirsten, I. (1989). *Water and waste-water management in the red meat industry* (pp. 36). WRC Report.

Stylianou, M., Panagiotou, C. F., Andreou, E., Frixou, F., Christou, A., & Papanastasiou, P. (2021). Assessing the influence of biochars on the hydraulic properties of a loamy sand soil. *Biomass Conversion and Biorefinery*, *11*, 315–323.

Suzanne, B., Bélanger, N., & Guillemette, T. (2015). Wood ash and N fertilization in the Canadian boreal forest: Soil properties and response of jack pine and black spruce. *Forest Ecology and Management*, *348*, 1–14.

Taha, S. S., Seoudi, O. A., Abdelaliem, Y. F., Tolba, M. S., & El-Sayed, S. S. F. (2018). Influence of bio-spent mushroom compost tea and potassium humate as sustainable partial alternate source to mineral-N influence of bio-spent mushroom compost tea and potassium humate as a sustainable partial alternate source to mineral-N fertigation. *Egyptian Journal of Basic and Applied Sciences*, *33*(1), 103–122.

Tampio, E., Marttinen, S., & Rintala, J. (2016). Liquid fertilizer products from anaerobic digestion of food waste: Mass, nutrient and energy balance of four digestate liquid treatment systems. *Journal of Cleaner Production*, *125*, 22–32.

Teja, G., & Bhosle, S. (2013). Effect of metals on a siderophore producing bacterial isolate and its implications on microbial assisted bioremediation of metal contaminated soils. *Chemosphere*, *93*(9), 1835–1843.

Timmusk, S., Behers, L., Muthoni, J., Muraya, A., & Aronsson, A. C. (2017). Perspectives and challenges of microbial application for crop improvement. *Frontiers in Plant Science*, *8*, 49.

Turan, Z. M. (2008). Responses of potato (Solanum tuberosum L.) to green manure cover crops and nitrogen fertilization rates. *American Journal of Potato Research*, *85*, 150–158.

Tur-Cardona, J., Bonnichsen, O., Speelman, S., Verspecht, A., Carpentier, L., Debruyne, L., & Buysse, J. (2018). Farmers' reasons to accept bio-based fertilizers: A choice experiment in seven different European countries. *Journal of Cleaner Production*, *197*, 406–416.

Tuttobene, R., Avola, G., Gresta, F., & Abbate, V. (2009). Industrial orange waste as organic fertilizer in durum wheat. *Agronomy for Sustainable Development*, *29*, 557–563.
Vance, C., Gaston, L., & Beasley, J. (2017). Bermudagrass establishment on infertile soil: Growth and phosphorus losses with poultry litter ash and triple superphosphate. *Communications in Soil Science and Plant Analysis*, *52*(8), 886–895.
Vance, C. P., Uhde-Stone, C., & Allan, D. L. (2003). Phosphorus acquisition and use: Critical adaptations by plants for securing a nonrenewable resource. *New Phytologist*, *157*(3), 423–447.
Vandana, N., & Manu, S. (2021). Isolation screening and molecular characterization of zinc solubilizing bacteria and their effect on the growth of wheat (*Triticum aestivum*). *Molecular Biology and Biotechnology*, *29*(2), 85–97.
Vaneeckhaute, C., Janda, J., Vanrolleghem, P. A., Filip, M. G., & Meers, E. (2016). Phosphorus use efficiency of bio-based fertilizers: Bioavailability and fractionation. *Pedosphere*, *26*(3), 310–325.
Vasanthi, N., Saleena, L. M., & Raj, S. A. (2018). Silica solubilization potential of certain bacterial species in the presence of different silicate minerals. *Silicon*, *10*, 267–275.
Wang, T., Shi, F., Zhang, Q., Qian, X., & Hashimoto, S. (2018). Exploring material stock efficiency of municipal water and sewage infrastructures in China. *Journal of Cleaner Production*, *181*, 498–507.
Ween, O., & Sunde, J. (2019). Bioconversion of crustacean processing discards to added value compounds. In L. K. Akslen-Hoel & B. Egilsson (Eds.). *International perspectives on regional research and practice* (p. 17). Orkana Forlag.
Wei, Y., Zhao, Y., Shi, M., Cao, Z., Lu, Q., Yang, T., & Wei, Z. (2018). Effect of organic acids production and bacterial community on the possible mechanism of phosphorus solubilization during composting with enriched phosphate-solubilizing bacteria inoculation. *Bioresource Technology*, *247*, 190–199.
Wu, H., Li, L., Du, J., Yuan, Y., Cheng, X., & Ling, H. Q. (2005). Molecular and biochemical characterization of the Fe (III) chelate reductase gene family in *Arabidopsis thaliana*. *Plant and Cell Physiology*, *46*(9), 1505–1514.
Wu, T. Y., Lim, S. L., Lim, P. N., & Shak, K. P. Y. (2014). Biotransformation of biodegradable solid wastes into organic fertilizers using composting or/and vermicomposting. *Chemical Engineering Transactions*, *39*, 1579–1584.
Wyciszkiewicz, M., Saeid, A., Samoraj, M., & Chojnacka, K. (2017). Solid-state solubilization of bones by B. megaterium in spent mushroom substrate as a medium for a phosphate enriched substrate. *Journal of Chemical Technology & Biotechnology*, *92*(6), 1397–1405.
Xia, L., Lam, S. K., Yan, X., & Chen, D. (2017). How does recycling of livestock manure in agroecosystems affect crop productivity, reactive nitrogen losses, and soil carbon balance? *Environmental Science & Technology*, *51*(13), 7450–7457.
Yagodin, B. A. (1984). *Agricultural Chemistry*. Mir Publishers.
Yahaya, S. M., Mahmud, A. A., Abdullahi, M., & Haruna, A. (2023). Recent advances in the chemistry of nitrogen, phosphorus and potassium as fertilizers in soil: A review. *Pedosphere*, *33*(3), 385–406.
Yunta, F., Di Foggia, M., Bellido-Díaz, V., Morales-Calderón, M., Tessarin, P., López-Rayo, S., Tinti, A., Kovács, K., Klencsár, Z., Fodor, F., & Rombolà, A. D. (2013). Blood meal-based compound. Good choice as iron fertilizer for organic farming. *Journal of Agricultural and Food Chemistry*, *61*(17), 3995–4003. https://doi.org/10.1021/jf305563b
Zabaniotou, A., & Kamaterou, P. (2019). Food waste valorization advocating Circular Bioeconomy-A critical review of potentialities and perspectives of spent coffee grounds bio refinery. *Journal of Cleaner Production*, *211*, 1553–1566.
Zake, J., Tenywa, J., & Kabi, F. (2005). Enhancement of manure utilization in smallholder cattle management systems for crop production in Central Uganda. *African Crop Science Conference Proceedings*, 1067–1071.

Zamotaev, I. V., Ivanov, I. V., Mikheev, P. V., & Belobrov, V. P. (2018). Assessment of the state of soils and vegetation in areas of landfills and municipal solid waste sites (A review). *Eurasian Soil Science*, *51*, 827–842.

Zarebska, A., Amor, Á. C., Ciurkot, K., Karring, H., Thygesen, O., Andersen, T. P., & Norddahl, B. (2015). Fouling mitigation in membrane distillation processes during ammonia stripping from pig manure. *Journal of Membrane Science*, *484*, 119–132.

Zhang, Y. S., Lu, X., Liu, B. Y., Wu, D. T., Fu, G., Zhao, Y. T., & Sun, P. L. (2021). Spatial relationships between ecosystem services and socioecological drivers across a large-scale region: A case study in the Yellow River Basin. *Science of The Total Environment*, *766*, 142480.

Zheljazkov, V. D., & Warman, P. R. (2004). Phytoavailability and fractionation of copper, manganese, and zinc in soil following application of two composts to four crops. *Environmental Pollution*, *131*(2), 187–195.

Zwetsloot, M. J., Lehmann, J., Bauerle, T., Vanek, S., Hestrin, R., & Nigussie, A. (2016). Phosphorus availability from bone char in a P-fixing soil influenced by root-mycorrhizae-biochar interactions. *Plant and Soil*, *408*, 95–105.

6 Soil Health

Role of Microorganisms in Soil Health

Zaffar Bashir, Shugufta Majeed, Asma Jan, Burhan Hamid, Parvaze Ahmad Wani, Rehana Rasol, Mehrun Nisa, and Bilal Ahmad Dar

6.1 INTRODUCTION

Worldwide food production needs to be increased in order to feed the world's growing population and at the same time, the reliance on inorganic fertilizers and pesticides should be diminished. To accomplish this goal, the various beneficial associations between plants and soil microorganisms should be explored. The soil microbial community are bacteria, actinomycetes, viruses, fungi, and protozoa. They have an important soil function that has fulfilled several useful tasks in the soil system. Microorganisms support biological nitrogen fixation of different biological transformations that support the accumulation and utilization of key nutrients, support root and shoot growth processes, control disease, and improve soil quality. They play an essential role in decomposing organic matter; cycling nutrients like N_2, P, K, Zn, C, and Si; and fertilizing the soil. Besides, they improve plant growth on various physiological parameters of plants by a number of mechanisms. The mechanism involved in growth promotion includes plant growth regulators, production of different metabolites, and conversion of atmospheric nitrogen into ammonia in direct and indirect ways. In addition, soil microbes offer resistance against diseases (Riaz et al., 2021).

Soil health is defined as the capacity of soil to function as a vital living system, by recognizing that it contains biological elements that are key to ecosystem function within land use boundaries. In the context of agriculture, it may refer to its capability to sustain productivity. Microbial community possess the ability to give an integrated measure of soil health, an aspect that is impossible with physical/chemical measures or higher organisms. Microbial community are the key players in the cycling of N_2, phosphorus, potassium, zinc, and silicon, and the decomposition of organic residues like cellulose, hemicellulose, starch, pectin, and lignin. Production of extra-cellular polysaccharides and other cellular debris by microorganisms help in maintaining soil structure as well as soil health. Various microbial species like *Pseudomonas* spp., *Bacillus* spp., *Serratia* spp., *Micrococcus* spp., *Azotobacter* spp., *Azospirillum* spp., *Rhizobium* spp., Nitrobacteria, Nitrosamines (Bashir et al., 2017, 2018a, 2018b, 2019), *Aspergillus* spp., *Penicillium* spp., *Fusarium* spp., *Mucor*, *Alternaria* spp., Nostoc, Anabaena, Azolla, and protozoa play essential roles in maintaining soil

DOI: 10.1201/9781003491460-6

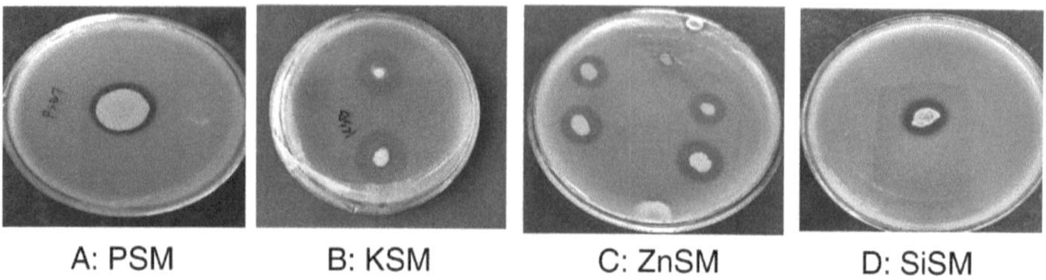

FIGURE 6.1 Plant growth promoting and soil health improving bacteria on their specific medium.

health. Fluctuations in microbial community can lead to evident changes in the soil's physical and chemical properties, thereby providing an early sign of soil improvement or an early warning of soil degradation (Fozia et al., 2015).

Microorganisms also affect the physical properties of the soil by production of extra-cellular polysaccharides and other cellular debris by microorganisms help in maintaining soil structure as well as soil health. Thereby, they also affect water holding capacity, infiltration rate, crusting, erodibility, and susceptibility to compaction (Elliott et al., 1996). Figure 6.1 shows how microbial community increases soil health. Changes in microbial populations or activity can precede detectable changes in the soil's physical and chemical properties, thereby providing an early sign of soil improvement or an early warning of soil degradation (Pankhurst et al., 1995).

6.2 DIVERSE BACTERIAL COMMUNITIES AND THEIR ROLE IN SOIL HEALTH

Single-celled, tiny organisms are called bacteria. Bacteria come in millions of diversities and inhabit every imaginable type of environment on the planet. They inhabit the earth's crust, the sea, the soil, and even hydrothermal vents. It has even been documented that certain bacteria may survive in radioactive waste. Numerous bacteria exist on human and animal bodies, both on the skin and in the mouth, airways, digestive, reproductive, and urinary tracts, all without endangering them. These microorganisms are known as resident flora or the microbiome. The indigenous bacterial flora can occasionally act as pathogens and cause disease under specific conditions. The most common type of microorganism found inside plants is called endophytes. They are able to colonize several plant components, including the leaves, roots, stems, and flowers. Certain bacteria develop mutualistic connections with plants in which the bacteria supply vital nutrients and promote plant growth in exchange for the plant's habitat, nutrients, and other resources. For example, the bacteria *Rhizobium* produce nodules in the roots of legumes, where they fix atmospheric nitrogen—a necessary element for plant the growth and development of plants (Patil et al., 2023).

The complex network of relationships among the soil, its resident microorganisms, and the plants that grow there is known as the "soil-microbe-plant system."

This system aids in the growth and development of plants and is essential to the cycling of nutrients like phosphorus, nitrogen, carbon, potassium, sulfur, zinc, etc. Numerous activities, including the cycling of macro- and micronutrients, the inhibition of disease, and the promotion of plant growth, are aided by microorganisms (Yadav et al., 2021).

One of the most significant microbial groups engaged in the cycling of nutrients is N_2-fixing microorganisms like blue green algae (BGA) and bacteria. These bacteria are special because they can change air nitrogen into ammonium, a form that plants can use. Another significant microbial community engaged in the cycling of nutrients is nitrifying bacteria (*Nitrosomonas* and *Nitrobacter*), which transform ammonium into nitrite and subsequently nitrate. Nitrate is easily leached from soil and is an essential source of nitrogen for plant growth. Thus, the amount of nitrogen available for plant growth can be greatly influenced by the activity of nitrifying bacteria. Phosphorus (P) and potassium (K) solubilizing bacteria are the group of microbial strains which plays an essential role in cycling of macronutrients (phosphorus and potassium) (Bashir et al., 2017, 2018a). These microbes are capable of converting phosphorus/potassium from the insoluble forms found in the soil to the soluble forms that plants can utilize. Although phosphorus/potassium is often a limiting component in soils, it is necessary for plant growth and development. Because they replenish the soil with nutrients like carbon, nitrogen, and phosphorus, bacteria and other decomposer microorganisms are essential to the decomposition of organic matter in the soil. Bacteria are necessary to preserve the integrity and structure of the soil, which is necessary for plant roots to grow and absorb nutrients (Bashir et al., 2018b; Boltenstern et al., 2015).

In soil ecosystems, a number of important microbial communities like *Bacillus* spp., *Pseudomonas* spp., *Micrococcus* spp., *Rhizobium* spp., *Azotobacter* spp., and *Serratia* spp. are involved in the cycling of nutrients (Bashir et al., 2018a). Maintaining soil fertility and supplying nutrients for plant growth is mostly dependent on the activities of decomposer microbes, nitrifying bacteria, phosphorus-solubilizing bacteria, potassium solubilizing bacteria, and nitrogen-fixing bacteria (Bashir et al., 2017, 2018a, 2018b, 2019; Sanchez et al., 2017). Plant growth-promoting rhizobacteria (PGPR) are free-living beneficial bacteria that provide health benefits to crop plants. This group of bacteria includes several different genera, including *Azospirillum*, *Pseudomonas*, *Micrococcus*, *Serratia*, *Rhizobium*, *Azospirillum*, *Azotobacter*, *Bacillus*, and others (Bashan et al., 2014; Bashir et al., 2018b).

By promoting biotic and abiotic stress tolerance and assisting with host plant nutrition, bacteria that live in the rhizosphere can serve as a source of nutrients (biofertilizers) to boost plant growth and development. A class of bacteria called bacterial biofertilizers assists in fixing various soil nutrients required for plant growth. They have the ability to fix nitrogen; solubilize phosphate, potassium, and other micronutrients; and release organic molecules like HCN, lytic enzymes to control plant diseases or substances (IAA) that promote plant growth. *Azotobacter*, *Azospirillum*, *Rhizobium*, and *Bacillus* are a few of the most widely used bacterial biofertilizers that have been used worldwide (Ortiz & Sansinenea, 2022).

Numerous plant-associated bacteria have a variety of beneficial effects on the physiology and health of plants. Recently, three well researched biological disease prevention and plant health promotion mechanisms facilitated by plant-associated bacteria were examined. The complexity and multifactorial nature of biological control, however, has been masked by the emphasis on finding and defining particular processes. Numerous distinct bacteria and bacterial metabolites have been recognized as significant contributors to the biological management of plant diseases. Bacterial phytohormones control plant growth and development, and under some conditions, they may be essential to a plant's capacity for survival and adaptation. These phytohormones include cytokinins, gibberellins, auxin, and ethylene. Several of these phytohormones are also produced by plant growth-promoting rhizobacteria (PGPR). Many PGPR also create a variety of VOCs, or volatile organic compounds. Some of these VOCs have the ability to stimulate plant growth directly, while others work via changing how the phytohormones are expressed. Additionally, plants may use the phytohormone-degrading activity found in many soil bacteria to maintain maximally effective levels of different phytohormones (Orozco-Mosqueda et al., 2023).

Bacterial endophytes enter plants similarly to phytopathogens and can be employed as biological agents without endangering them. Endophytic bacteria colonize a wide variety of plant species, and it is commonly asserted that Endophytic bacteria are found in all plant species. Because the bacteria found there and those found inside the roots have a great deal in common, the rhizosphere and phyllosphere are assumed to be the main locations of Endophytic colonization. By means of creative or innovative synthesis of bacterial chemicals and unique metabolites, they are able to regulate plant diseases. Furthermore, a number of researchers have found that Endophytic bacteria use IAA, gibberellins, and cytokinins to promote plant development in addition to carrying distinct antibiotic genes (Rajendra et al., 2023). Phosphorus solubilizing microorganisms (PSMs), potassium solubilizing microorganisms (KSMs), zinc solubilizing microorganisms (ZnSMs), and silicate solubilizing microorganisms (SiSMs) are the best example of plant growth promoting microbial communities (Bashir et al., 2018, 2019a, 2019b, 2021; Bashir, 2021). Figure 6.1 shows these microbial strains on specific media (qualitative assay). Tables 6.1 and 6.2 show some plant growth promoting micobes (bacteria) and soil health improving microorganisms.

TABLE 6.1
Plant Growth Promoting Micobes (Bacteria) and Soil Health Improving Microorganisms

PSM	Reference	KSM	Reference
Bacillus amyloliquefaciens	Zhang et al., 2023	*Burkholderia paludis*	Imam et al., 2023
Serratia marcescens	Gong et al., 2022	*Bacillus* sp.	Zhou et al., 2022
P. libanensis	Kour et al., 2020	*Burkholderia* sp.	Giang et al., 2020
Bacillus spp., *Pseudomonas* spp.	Bashir et al., 2018a	*Pseudomonas* sp.	Sarikhani et al., 2018

TABLE 6.2
Plant Growth Promoting and Soil Health Improving Microorganisms

ZnSM	Reference	SiSM	Reference
Bacillus cereus	Vandana & Manu, 2021	*Bacillus tequilensis* SKSSB09	Karuganti et al., 2023
Bacillus tropicus	Vandana & Manu, 2021	*Pseudomonas frederiksbergensi*	Jayvee et al., 2022
Bacillus megaterium	Kalpana et al., 2020	*Pseudomonas plecoglossicida*	Jayvee et al., 2022
Bacillus aryabhattai	Vidyashree et al., 2018	*Rhizobium* sp. CCNWYC119	Chandrakala et al., 2019
Pseudomonas taiwenensis	Vidyashree et al., 2018	*Enterobacter ludwigii GAK2*	Lee et al., 2019

6.3 FUNGI AND THEIR ROLE IN SOIL HEALTH

The eukaryotic organisms known as fungi are a source of food, antibiotics, organic acids, alcohol, enzymes, amino acids, and growth-promoting secretions. Microorganisms like mold, yeast, and mushrooms are among them. Fungi are excellent decomposers of organic waste and target cellulose, lignin, gums, and other organic complexes the most quickly. Together, fungi and other organisms play a fundamental part in a variety of physiological processes, including mineral and water uptake, chemical alterations, stomatal movement, and the biosynthesis of substances known as bio stimulants, auxins, lignin, and ethylene to increase plants' capacity to recognize and adapt to environmental stresses like drought, salinity, heat, cold, and significant metals.

Fungi are highly significant microorganisms that have the potential to be used in a variety of industries, including agriculture. Fungi can interact with plants in a beneficial or detrimental way. Valuable fungi play a significant part in the sustainability of agriculture. These fungi come in a variety of forms, including symbiotic and Endophytic varieties. They are an alternative to chemical pesticides in the management of plant diseases. Although they might belong to several groups, the useful fungi are often mycorrhiza, Endophytic, entomopathogenic, mushroom, and dark septate fungi. The value of fungus in several domains led to their introduction into the industry to create biopesticides and biofertilizers, which may be used in place of synthetic chemicals and further reduce the amount of chemical residue in the environment. Fungi first break down the organic residues so that various microorganisms can start degrading and converting them into usable form. Through the formation of hyphae network, 80 to 90% of all plants establish symbiotic relationships with mycorrhizal fungi. Improved plant health and development, water uptake, nutrient availability, stress tolerance, and biocontrol are all possible benefits of using fungi in both agriculture and horticulture. The process of mycoremediation is another important biological instrument in the sustainable agriculture ecosystem (Thakur, 2020). Arbuscular mycorrhizal fungi (AMF) facilitating a series of intricate communication processes between the plants and fungus that result in increased photosynthetic rate and other gas exchange-related features, mycorrhizal fungi help host plants develop enthusiastically under adverse situations (Birhane et al., 2012).

Arbuscular mycorrhizal fungi (AMF) improve nutrient availability and transmission, hence improving plant nutrition. Arbuscular mycorrhizal fungi enhance soil quality by affecting its composition and texture, which benefits plant health (Zou et al., 2016). AMF-mediated growth promotion safeguards plants from fungus infections in addition to enhancing their ability to absorb water and nutrients from the adjacent soil (Jung et al., 2012). AMF are essential endosymbionts that contribute significantly to plant productivity and ecological health (Gianinazzi et al., 2010). AMF inoculation has an impact on growth-related processes including as stomatal conductance, leaf water potential, relative water content (RWC), PSII efficiency, and CO_2 assimilation (Chandrasekaram et al., 2019). AMF alters the physiological state of the organs and tissues above ground, which helps increase tolerance to water stress (Barzanal et al., 2012).

AMF is a blend of naturally occurring compounds that is used as a biofertilizer or bioagents to increase soil fertility or soil health. Both plant growth and soil health is improved by the usage of these fertilizers (Sadhana, 2014). Over the past two decades, numerous research studies on AMF have revealed their numerous advantages for agricultural productivity and soil health. Because mycorrhizal treatment can successfully reduce the quantitative usage of chemical fertilizer input, particularly of phosphorus, it is widely assumed that AMF could be considered as a replacement for inorganic fertilizers in the future (Ortas, 2012). The soil-dwelling fungi significantly boost plant nutrition uptake and resistance to various abiotic stresses (Sun et al., 2018).

The term "PGPF" refers to the plant growth-promoting fungi that live in the rhizosphere of crop plants and exhibit characteristics that promote the sustainability of the plants. They are biotic inducers that contribute to the sustainability of agriculture and serve essential purposes. PGPF including *Trichoderma* spp., *Gliocladium* virens, *Penicillium digitatum*, *Aspergillus flavus*, *Actinomucor elegans*, Arbuscular mycorrhizal fungi, etc. have proven their ecofriendly nature to improve the production of crops by improving the growth of the shoots and roots of crop plants, the germination of seeds, the production of chlorophyll for photosynthesis, and the abundant production of crops. The mineralization of the major and minor elements necessary to support plant development and production is the potential mode of action for PGPF.

As a broad and multipurpose class of soil microorganisms, fungi are well known for their exceptional ability to adapt to a wide range of unfavorable abiotic environments. It's also critical to note that, while growing on solid substrates, fungi demonstrate a strong resistance to high osmotic pressure and low water activity values while maintaining a high level of metabolic activity. There are 1.5 million different types of fungi in soil, with mycorrhizae being the most well-studied beneficial fungus-plant relationship. These fungi can either live freely in the bulk soil or as endophytes colonizing plant tissues. Many fungi, the most effective of which are *Aspergillus*, *Penicillium*, and *Trichoderma*, are capable of solubilizing insoluble phosphates or facilitating P-acquisition by plants. As a result, they play a significant role in the production of commercial microbial product (Vassileva et al., 2022).

Over the past ten years, a growing body of research has demonstrated the important role endophytes play in controlling plant development, communication, and stress tolerance. Endophytic fungi have been thoroughly studied and are a significant help to host plants in maintaining growth and development in a variety of climatic circumstances. Fungal endophytes are bio stimulants that increase crop protection and productivity in agriculture. This will give farmers environmentally friendly methods to improve yield and quality without endangering plant health or disrupting the ecosystem.

6.4 CYANOBACTERIA OR BLUE GREEN ALGAE AND THEIR POTENTIAL TO IMPROVE SOIL HEALTH

Cyanobacteria, which are also called blue green algae, are one of the most important groups of prokaryotes (no cell organelles are present) that have ability to bear extreme environmental conditions and are present in almost all terrestrial and aquatic habitats (Kumar & Singh, 2017; Tamaru et al., 2005). These are present in extreme environmental surroundings such as polar and alpine regions. These are found as pelagic, benthic, and symbiotic communities in marine and freshwater ecologies (Elliot, 2012; Scott & Marcarelli, 2012). Eutrophication from industrial sources has led to the formation of blooms of pelagic freshwater cyanobacteria, many of which are awfully destructive for the environment (Nowruzi & Porzani, 2021; Rajabpour et al., 2019). Cyanobacteria play a significant role in maintaining soil health and promoting plant growth, the soil and plant health promoting properties of BGA. They are known to fix nitrogen, degrade organic waste, and remediate heavy metals, agrochemicals, and other pollutants. Cyanobacteria are also pioneers in biocrust formation, promoting soil stability, nutrient availability, and water retention, leading to the development of other communities. Diverse cyanobacterial species like *Nostoc*, *Aulosira*, *Calothrix* sp., *Tolypothrix* sp., Anabaena, and *Scytonema* sp. are the most efficient N_2-fixing cyanobacteria present in rice growing areas (Prasad & Prasad, 2001). Numerous secretary substances including auxins, gibberellins, cytokinins, ethylene, and abscisic acid can increase growth and output in plants such as *Daucus carota*, *Santalum album*, and *Oryzae sativa* (Manickavelu et al., 2006). About 20–25 kg/ha of atmospheric nitrogen is fixed up by some species of cyanobacteria belonging to Aphanizomenonaceae and Nostocaceae.

Agrochemicals that can cause eutrophication in aquatic ecosystems could be replaced by nitrogen-fixing cyanobacteria (Godlewska et al., 2019). To improve soil health and crop productivity, BGA are applied as bioagents or biofertilizers, and these bioagents release bioactive substances such as IAA, polysaccharides, amino acids, vitamins and siderophores; they can stimulate plant growth and help to fight against various abiotic and biotic stresses (Řezanka et al., 2018; Singh et al., 2016). The most common seed genera for plant growth-promotion are filamentous cyanobacteria which include *Nostoc*, *Cylindrospermum*, *Synechococcus*, *Aulosira Gloeotrichia*, *Spirulina*, *Oscillatoria*, and *Anabaena* (Khushwaha & Banerjee, 2015; Osman et al., 2010; Verma et al., 2016).

The genus *Nostoc* can form symbiotic associations with gymnosperms and angiosperms, thus allowing cyanobacteria to fix atmospheric N_2 enzymatically into ammonia using nitrogenase enzyme (Sivalingam & Batiri, 2020). Many cyanobacterial species such as *N. muscorum*, *A. fertilissima*, and *T. tenuis* release extracellular bioactive substances, which produce similar responses as that of phytohormones such as IAA, jasmonic acid, or abscisic acids (Singh, 2014). These phytohormones induce regulation of seed germination, stimulation of root growth, branching, fruit ripening, and inflorescence (Bareke, 2018). These phytohormones also synthesize enzymes, pigments, and metabolites; induce or suppress the expression of genes; and promote resistance of plants against environmental factors (Singh, 2014). Cyanobacteria and their extracts are the source of both biofertilizers (Chittora et al., 2020) and bioactive compounds for disease control in plants (Gonçalves, 2021).

6.4.1 Benefits of BGA as Biofertilizer

- BGA play a vital role in agriculture by fixing atmospheric N_2, providing macronutrient (N_2) to plants.
- These prokaryotic strains contribute to soil fertility or soil health by releasing P, K, and micronutrients like Zn, thereby enriching the overall nutrient content of the soil.
- BGA have positive influence on soil structure, water retention, aeration, and drainage, which collectively increases soil health and thus promote plant growth and development.
- Stress tolerance and resistance of cyanobacteria contribute to increasing plant resilience against adverse situations, including water stress, pests, and diseases.
- Because of biological control properties, BGA act as natural bioagents to suppress soil pathogens, bolstering plant disease resistance.
- Their cost-effective production and application present an economical alternative to agrochemicals.
- Environmentally sustainable, BGA offer a means to reduce reliance on synthetic fertilizers, contributing to more ecologically sound agricultural practices. By raising soil health and minimizing environmental impact, the use of BGA or cyanobacteria promotes sustainable agriculture.
- Stimulate plant growth
- Solubilizing macro- and micronutrients
- As a source of organic matter
- Mineralizing simpler organics
- Plant protection from pathogens
- Forming soil aggregates, providing O_2 to subsurface
- Enhancing resilience against stressors

6.5 ALGAE A SUSTAINABLE WAY TO IMPROVE SOIL FERTILITY AND PLANT GROWTH

Algae are defined as eukaryotic (nucleus bearing) photosynthetic organisms but lack the specialized multicellular reproductive structures of plants. These organisms are agriculturally important as they are used as biofertilizers or bioagents and soil stabilizers. The size of algae may range from single cells of one micrometer to large seaweeds that may grow up to 50 meters (Vymazal, 1995). These are diverse in nature;

these are present almost in every part of earth, that is in soils, ice, snow fields, hot springs, hot and cold deserts (South & Whittick, 1987). These possess chlorophyll-a, carbohydrates, and proteins similar to that of higher plants. In aquatic ecosystems these play an important role as they occur as a base of the food chain. These also produce O_2 which is essential for the metabolism of the consumer organisms (Lee et al., 1989). Algae are used to improve the characteristics of the soil such as carbon content, nitrates, texture, and aeration (Hamed, 2007). Marine algae such as large brown and red algae are used as organic fertilizers and are usually richer in potassium content but poorer in nitrogen and phosphorus content (Raouf et al., 2012). Seaweeds are widely used in agriculture as liquid fertilizers (Povolny, 1981). In the United Kingdom and France, calcareous red algae which are also known as maerl are used to reduce soil acidity (Blunden et al., 1981). Algae are an important source of organic matter in soil (Ibraheem, 2007). The organic matter resulting from death and decay of soil may get mixed in the soil, and mucilage acts as a binding agent for soil texture; therefore the humus content of soil is increased and after some years it becomes more habitable for other plants (Marathe & Chaudhari, 1975).

In the U.S.S.R., it was found that filamentous forms of cyanophyceae, mostly *Oscillatoria*, *Schizothrix*, and *Plectonema*, played an important role in soil formation (Gayel & Shtina, 1974). When seaweed manure is applied to fruits, vegetables, and flower crops, there is an increase in crop yield, improved vigor, increased uptake of soil nutrients, increased resistance to some pests and diseases, longer shelf life of fruits, improved seed germination, and increased defense against abiotic stresses (Khan et al., 2009). Seaweed extracts contain 1.3–1.6 D-glucane, oligosaccharide which is also known as laminarine, which acts as a stimulator or precursor for polyamine synthesis. It also contains a group of growth regulators such as auxins, cytokinins, gibberellins, and betaines that stimulate cell division and provide defense mechanisms (Patier et al., 1993). The components of seaweeds such as macro- and microelement nutrients, amino acids, vitamins, auxins, cytokinins, and abscisic acid (ABA) affect the cellular metabolism in treated plants which leads to increased crop growth and crop yield (Khan et al., 2009). Seaweed products also show plant growth stimulating activities. The seaweed formulations are used as bio stimulants in crop production. Bio stimulants are also known as "metabolic enhancers" and are defined as materials other than fertilizers, which promote growth of plants when applied in small quantities (Khan et al., 2009). The extracts of seaweeds are active at low concentrations that is diluted as 1:1000 or more (Crouch & Van Staden, 1993). Algal products have useful effects on root growth and development. These affect the root development by improving lateral root formation (Atzmon & Van Staden, 1994; Vernieri et al., 2005) and increasing total volume of the root system (Mancuso et al., 2006). The application of extract from *Ascophyllum nodosum* resulted in the increase in root length (up to 32%) in *Arabidopsis thaliana* (Rayorath et al., 2008), and the seaweed concentrate prepared from Ecklonia maxima increased the root growth of tomato plants (Crouch & Van Staden, 1992). The application of seaweed extracts increases the seed germination and seedling vigor in soils (Rayorath et al., 2008). Algae, usually blue green algae, play an important role in nitrogen fixation in many agricultural soils. BGA among algae serve as the cheapest source of biofertilizer in rice fields (Ladha & Reddy, 2003). Giant Kelp (an algal species) is used as a fertilizer in combination with other fertilizers, which then provide plants a high amount of nutrients they require. Kelp fertilizers are present in solid, soluble powder and liquid forms. It provides all the major nutrients including potassium which is an important plant nutrient. This practice is usually performed in the Pacific Coast of North America (Prakash & Nikhil, 2014). Thus algae play an important role in agriculture, soil fertility, biological nitrogen fixation, P-availability and P-uptake, and waste water treatment (Raouf et al., 2012).

6.6 ACTINOMYCETES BENEFACTION ROLE IN SOIL AND PLANT HEALTH

Gram-positive, aerobic, spore-forming bacteria belong to the actinomycetales order and are known as actinomycetes. Their capacity to grow both aerially and on a substrate sets them apart. These are the most common organisms that form thread-like filaments in the soil, and they are responsible for the characteristic "earthy" smell of freshly turned, healthy soil. Certain non-leguminous plants are associated with actinomycetes, which fix nitrogen (N) and make it available to the host and other nearby plants. They assist in moving vegetative matter around in the soil, which is why they are important to the environment's soil nutrient cycling (Kapoor et al., 2021). Saprophytes, which make up the bulk of actinomycetes, produce the enzyme required to decompose complex polysaccharides from plants, such as lignin, pectin, starch, cellulose, and hemicellulose. Actinomycetes are also being investigated and used to develop remedies for a variety of bacterial infections, illnesses, and terminal diseases. Because of their varied biochemical characteristics and soil-derived origin, actinomycetes are also utilized to improve the health of the soil and plants' tolerance to a variety of biotic and abiotic stresses (Sharma et al., 2014). The rhizosphere or hot spot of plants is home to actinomycetes. Through the synthesis of antifungal enzymes that break down the fungal cell wall or the synthesis of antifungal chemicals, they can suppress the growth of fungal infections and protect roots (Silva et al., 2022). Some actinomycetes strains were isolated from the rhizosphere zone of four different plant species: rosemary, acacia, strawberry, and olive. Actinomycetes have a wide range of dwelling places and can be isolated from terrestrial soil, rhizospheres of plants, and marine sediments. Actinomycetes produce several bioactive secondary metabolites with antibacterial, antifungal, and antiviral properties (Elshafie et al., 2023).

The ability of actinomycetes to resist the effects of antibiotics, siderophores, lytic enzymes, volatile organic compounds, space and nutrition limits, and induction of host resistance are their main antagonistic strategies for regulating phytopathogenic fungi. By producing phytohormones, solubilizing minerals, and fixing nitrogen from the atmosphere, actinomycetes also support the growth and development of plants. Actinomycetes have been the subject of several published research; however, due to their influence on agriculture and capacity to resist phytopathogenic fungi, only a small number of these studies have focused on actinomycetes that have been isolated from various settings and used as biocontrol agents. Actinomycetes invade plant roots and produce a variety of bioactive compounds in the rhizospheric environment, which promotes plant growth. Actinomycetes, which are vital components of the rhizosphere, encourage plant growth in its surroundings. They break down and either directly or indirectly encourage the growth of the plant species (Javed et al., 2020).

By generating phytohormones, soil actinomycetes influence plant growth and development, hence inducing plant health. Actinoplanes, Micromonospora, and Nocardiopsis are examples of endophytic actinobacteria that are major producers of phytohormones such as indole-3-acetic acid and indole-3-pyruvic acid, which are primarily involved in regulating gene expression and coordinating cell growth. Along with the growth of rhizobia supplied with the soil nutrients, soil actinomycetes, particularly Streptomyces, typically improve the capacity of plants to absorb nutrients. The variety of cultivable endophytic actinomycetes associated with plants, particularly those used medicinally, are identified and described. Numerous studies have documented the enhanced endophytic actinomycetes' colonization and dispersion in plants. Different genera of endophytic actinomycetes are found in plants, including the prominent genus Streptomyces, as well as Micromonospora, Microbispora, Nocardiopsis, Rhodococcus, and Nocardia. In addition, a large number of newly identified strains fall into these taxa. Endophytic actinomycetes appear to aid in the growth of plants and exhibit advantageous characteristics that can be used to maintain plants, provide nutrients, and provide defense and protection.

This community of microorganisms are the source of enzymes, immunosuppressive drugs, anticancer medicines, and antibiotics, etc. These metabolites have also been shown to have neurogenic, antibacterial, antifungal, anticancer, antihelmintic, and antioxidant qualities (Rastimesina et al., 2022; Shengping et al., 2021).

6.7 VIRUSES AND SOIL HEALTH

Viruses, with an estimated abundance of 10^{31} on Earth, are an important component of soil ecosystems. As obligate parasites, ubiquitous and genetically diverse entities that entirely depend on hosts for reproduction and survival, viruses have been linked to microbial community diversity and metabolic activities in soil. Emerging evidence indicates that soil viruses influence a broad spectrum of processes that sustain soil biodiversity, biogeochemical cycling, fertility, and plant health. Research on soil viruses is in its early stages. Even observational assessments of viral ecology such as abundance, diversity, distribution, life strategies, ecological relevance, and functions, are only just beginning to be revealed (Breitbart & Rohwer, 2005). Viruses play an important role in horizontal gene transfer. Bacteriophages comprise a majority among viruses in aquatic environments (Wommack & Colwell, 2000). The findings that phages comprise the majority among viruses in aquatic environments cannot be applied to soil environments because fungal biomass is often greater than bacterial biomass in soils (Anderson & Domsch, 1980). Viruses which infect *Rhizobium* spp. are known as rhizobiphages. Many studies were conducted to explain rhizobiphages populations, the host range of rhizobiphages on the nodule formation and yield of leguminous plants (Abebe et al., 1992). The viral effects on beneficial bacteria and soil borne plant pathogens have been extensively studied (Keel et al., 2002). Mostly phages are largely present in soils where bacterial hosts multiply, such as rhizobiphages in the fields of legumes (Sharma et al., 2002). Bacterial blight which is caused by *Xanthomonas oryzae* pv. *Oryzae* is a harmful disease caused in rice plants in Asia. It is an air borne disease and many tadpole shaped and filamentous phages of the pathogen were found in paddy fields and diseased rice leaves (Marsh & Wellington, 1994). The term phytobiome is defined as plants, their environment, and their associated communities of organisms. The first beneficial phytobiome virus was the virus-infected fungus that provides thermal tolerance in geothermic soils in Yellowstone National Park (Márquez et al., 2007). Fungi have a variety of interactions with plants, many of which are positive known as endophytes. Many endophyte viruses play an important role in plant-fungal interactions (Bao & Roossinck, 2013). Bacteriophages can be lytic or lysogenic. In plants bacteriophages can be divided into phyllosphere viruses and rhizosphere viruses (Lamichhane & Venturi, 2015). Tobacco Mosaic Virus was the first virus found in late 19th century and infected tobacco plants (Beijerinck, 1898). Mostly the plant viruses are transmitted by insects (Jones, 2018).

6.8 PGPR: SUSTAINABLE WAY TO IMPROVE SOIL HEALTH

Plant growth-promoting rhizobacteria (PGPR) are a class of bacteria that live in the soil region known as the rhizosphere. Rhizosphere refers to the area of soil where a plant root is found. It is an area with strong microbial activity that produces a restricted nutrient pool from which important macro- and micronutrients are extracted. Only bacterial strains that can fulfill at least two criteria, such as vigorous invasion, promotion of plant growth, and biocontrol, are eligible for the designation of PGPR (Arjumend et al., 2022).

Plant growth-promoting rhizobacteria (PGPR) are a category of advantageous soil bacteria that colonize the rhizosphere or root surface of plants and offer the host plants a number of advantages. The potential of PGPR to improve plant growth and general health through a variety of mechanisms has been extensively investigated. Phosphates solubilization, potassium solubilization, zinc solubilization, silicon solubilization, and atmospheric nitrogen fixation is performed by PGPR, increasing the availability of these vital nutrients to plants. They create organic acids and enzymes that aid in the digestion of complicated organic substances and release nutrients that plants can absorb. PGPR can boost nutrient uptake and utilization, resulting in improved plant growth and productivity (Bashir et al., 2019; Zaib et al., 2023).

In studies, it has been found that PGPR increases root length and density, which facilitates nutrient and water uptake, boosts crop output, and maintains the osmotic balance in the host plant by controlling relative water content. Plant roots are colonized by PGPR, which promotes direct or indirect plant development and offers protection from disease and drought stress. Through interactions with plants and their metabolic processes, some PGPR have an impact on plant development (Chieb et al., 2023).

Based on location, PGPR is widely divided into two types: Extracellular PGPR and intracellular PGPR. Intracellular PGPR will live in the nodular structure of root cells, while extracellular PGPR will live in the rhizosphere between the cells of the root cortex. Genera like *Bacillus*, *Aztobacter*, *Serratia*, *Pseudomonas*, *Arthrobacter*, *Micrococcus*, etc. are examples of those that belong to extracellular PGPR. *Allorhizobium*, *Bradyrhizobium*, *Mesorhizobium*, *Rhizobium*, and *Frankia* species are members of the intracellular PGPR and are capable of fixing atmospheric nitrogen (Neetu et al., 2022).

6.8.1 PGPR Can Enhance the Growth of Plant via Various Direct/Indirect Mechanisms

6.8.1.1 Direct Mechanism

Nutrient fixation, phosphate, zinc, silicate and potassium solubilization, nutrient uptake, siderophore generation, and phytohormone production are all direct mechanisms of PGPR. As plants have a tendency to increase the concentration of nutrients, PGPR can directly promote plant growth. By fixing nitrogen in the soil, PGPR such as *Azospirillum*, *Mesorhizobium*, and *Sinorhizobium*, among others, would improve plant growth and development and decrease attacks by various phytopathogens (Ahemad & Kibret, 2014).

6.8.1.2 Indirect Mechanism

The production of hydrolytic enzymes and antibiotics are two examples of indirect mechanisms used by PGPR to prevent or delay the negative effects of phytopathogens and unpleasant or stressful conditions on plants. Plants are exposed to a variety of biotic and abiotic stress situations as a result of the continually changing climate circumstances. The main abiotic stresses that crop plants experience are high wind speeds, droughts, hot temperatures, excessive salinity, floods, etc. These stresses have an adverse effect on plant development and productivity. Several solutions are put forth to lessen the effects of stress brought on by abiotic causes. The use of PGPR to osmolytes like trehalose, proline, etc. has drawn a lot of interest. PGPR such as *Bacillus amyloliquefaciens*, *Bacillus licheniformis*, *Bacillus subtilis*, and *Paenibacillus polymyxa* can address biotic stress-related problems (Ngumbi & Kloepper, 2016).

The use of PGPR is potentially enhanced in organic farming because they are used as best nutrient solubilizers/fixers (Bhat et al., 2023). We can better understand PGPR's role by using the best PGPR strains found in commercial biofertilizers (Riaz et al., 2021).

Plants may more easily resist stress and adjust to shifting environmental conditions with the help of PGPR. They govern the amount of water in the soil, the osmotic balance, and ion homeostasis as well as aid the soil in acquiring nutrients. Plant development is impacted by the modulation of plant hormonal homeostasis via PGPR-mediated external release of hormones, metabolites, and other substances. PGPR are crucial for encouraging plant development and growth. The synthesis of indole-3-acetic acid (IAA) is a fairly typical property of PGPR that enhances the fitness of plants growing under stress situations (Bashandy et al., 2019; Bhat et al., 2023).

The diverse groups of plant-associated microbes known as plant growth-promoting rhizobacteria (PGPR) can influence the host plant's systemic resistance to elicit a defense response, which in turn can lessen the severity or incidence of disease. By boosting systemic resistance against numerous crop-damaging diseases, an amalgamation of different PGPR strains has increased efficacy (Meena et al., 2020).

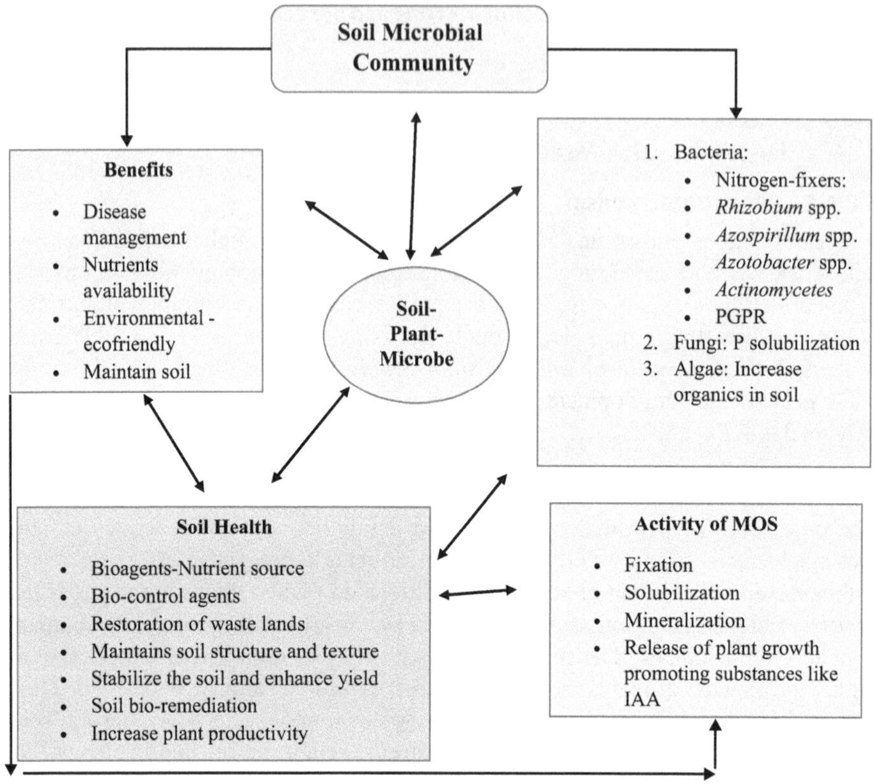

FIGURE 6.2 Role of microbial community in promoting soil health.

PGPR are a broad set of microorganisms that are associated with many different plant species. They provide plants with a number of benefits, including stress relief and growth promotion. The potential use of PGPR as biological control agents, as well as their ability to replace chemical fertilizers, herbicides, and fungicides, has been extensively documented in the literature (Besset-Manzoni et al., 2018). Figure 6.2 shows the role of microbial community in promoting soil health.

6.9 PROTOZOANS AND SOIL HEALTH

Protozoa are single-celled animals that feed primarily on bacteria but also eat other protozoa, soluble organic matter, and sometimes fungi. These are aquatic, larger than bacteria ranging from 5 to 500 μm in diameter (Hoorman, 2011). On the basis of their shape, protozoa are classified into three groups. A. Species of Mastigophora or *flagellates* are smallest (5 to 10 μm) and contain one to four whip-like flagella for locomotion. These feed mainly on bacteria and are most abundant in number. B. Ciliophora or *ciliates* are largest (10 to 80 μm) protozoa and are least in number and move with the help of hair-like cilia. These eat bacteria as well as the other two types of protozoa (Esteban & Fenchel, 2020). C. Sarcodina or *amoebae* are also quite large and move with the help of a temporary foot or "pseudopod." These exist in in the rhizosphere and at the root surface where they graze on bacterial populations (Jahn et al., 1974).

Protozoa are found in abundant quantity near the surface of the soil, particularly in the upper 15 cm (six inches) (Finlay et al., 2000). Species richness of protozoa at the global level is very inaccurately known. The number of known non-fossil protozoan species in the world is estimated as 40,000 (Vickerman, 1992). Protozoa play a significant role in mineralizing nutrients, thus making them accessible for the use by plants and other soil organisms (Chandarana & Amaresan, 2022). Protozoa contain lower concentrations of carbon and nitrogen in their cells than the bacteria they eat. Bacteria eaten by protozoa contain too much nitrogen for the amount of carbon protozoa need. Thus, they release the excess nitrogen in the form of ammonium (NH_4^+). This process usually occurs near the root system of plant. Bacteria and other organisms quickly take up most of the ammonium, but some is also used up by the plant (Hoorman, 2011). Protozoa also play an important role in regulating bacterial and algal populations (Finlay & Esteban, 1998). Single protozoa (*Paramecium*) can consume about 5 million bacteria in one day. These protozoa thus help to maintain ecological balance in soil (Hoorman, 2011). Protozoa are also an important source of food for other soil organisms and help to suppress disease by competing with or feeding on pathogens (Hoorman, 2011). Protozoa have selective feeding habits, thus they influence the bacteria composition and bacteria population in the rhizosphere (Chen et al., 2007).

The capability for nutrient release from bacteria by protozoan grazing is quite well recognized. However, in pot and microcosm experiments in which nutrients were added in addition to plant demand, an increase in plant growth was witnessed in the presence of protozoan as compared to treatments with bacteria only (Bonkowski, 2004). It was seen that protozoan grazing increased the capacity of bacterial strains of producing IAA plant hormone (Bonkowski, 2004). Protozoan predation can be a problem when beneficial bacteria are experimentally inoculated into soil. In moist soil containing active protozoa, the introduced bacteria can be quickly consumed when they enter the larger sites in the soil pore space that can be readily reached by protozoa. In drier soil with a higher proportion of narrow protective pore space available, the added bacteria may have a better chance of survival, but their chances for contact with fine roots (100–500 μm diameter) on which they should establish to be effective will decrease. Protozoa and bacteria form the oldest predator-prey relationship on earth, thus bacteria have therefore developed many diverse approaches to escape protozoan grazing (Matz & Kjelleberg, 2005).

The role of protozoa in the sulphur cycle is less known. All three major groups of protozoa appear to have capacity for stimulating mineralization and nutrient release. The enhancement of phosphorus cycling by protozoan grazers is well known in aquatic ecosystems, where they release and isolate major portions of the total phosphorus (Barsdate et al., 1974). In soil, a similar role was proposed by Coleman et al. (1977). Most of the protozoa assimilate amino acids, and some authors propose that they excrete ammonia (Stout, 1980). The effect of protozoa on the nitrogen cycle is mostly significant for plant growth. Several studies suggest that the plants grown in the presence of protozoa had larger nitrogen contents and dry weights than plants grown in absence of protozoa (Kuikman et al., 1990). (Figure 6.2)

6.10 CONCLUSION

The soil microbial community include bacteria, fungi, actinomycetes, viruses, nematode, and protozoa. They produce plant growth regulators and metabolites that affect the plant growth and development. Soil microorganisms are critical to decompose organic residues and recycling soil nutrients. They provide nutrients to crops and enhance soil health and crop outputs. Soil microbial community play a significant part in controlling plant diseases. Considering the environmental damage associated with use of chemical fertilization, a research priority on optimizing plant-soil microbe nutritional interactions is essential for more sustainable agricultural systems.

REFERENCES

Abebe, H. M., Sadowsky, M. J., Kinkle, B. K., & Schmidt, E. L. (1992). Lysogeny in Bradyrhizobium Japonicum and its effect on soybean nodulation. *Applied and Environmental Microbiology*, *58*(10), 3360–3366.

Ahemad, M., & Kibret, M. (2014). Mechanisms and applications of plant growth promoting rhizobacteria: Current perspective. *Journal of King Saud University-Science*, *26*(1), 1–20.

Anderson, J. P. E., & Domsch, K. H. (1980). Quantities of plant nutrients in the microbial biomass of selected soils. *Soil Science*, *130*(4), 211–216.

Arjumend, T., Sarhan, E. O., & Yildirim, M. U. (2022). Plant-bacterial symbiosis: An ecologically sustainable agriculture production alternative to chemical fertilizers. In *İntechopen License*. Revisiting Plant Bio stimulants Publisher.

Atzmon, N., & Van Staden, J. (1994). The effect of seaweed concentrate on the growth of Pinus pinea seedlings. *New Forests*, *8*, 279–288.

Bao, X., & Roossinck, M. J. (2013). Multiplexed interactions: Viruses of endophytic fungi. *Advances in Virus Research*, *86*, 37–58.

Bareke, T. (2018). Biology of seed development and germination physiology. *Advances in Plants & Agriculture Research*, *8*(4), 336–346.

Barsdate, R. J., Prentki, R. T., & Fenchel, T. (1974). Phosphorus cycle of model ecosystems: Significance for decomposer food chains and effect of bacterial grazers. *Oikos*, 239–251.

Barzanal, G., Aroca, R., Paz, J. A., Chaumont, F., & Martinez-Ballesta, M. A. (2012). Abruscular mycorrhizal symbiosis increases relative apoplastic water flow in roots of the host plant under both well-watered and drought stress conditions. *Annals of Botany*, *109*, 1009–1017.

Bashan, Y., de-Bashan, L., Prabhu, S. R., & Hernandez, J. P. (2014). Advances in plant growth-promoting bacterial inoculants technology: Formulations and practical perspectives (1998–2013). *Plant and Soil*, *378*, 1–33.

Bashandy, S. R., Abd-Alla, M. H., & Bagy, M. M. K. (2019). Biological nitrogen fixation and biofertilizers as ideal potential solutions for sustainable agriculture. *Integrating Green Chemistry and Sustainable Engineering*, *1*(1), 343–396.

Bashir, Z. (2021). Characterization of potassium solubilizing bacteria from Rhizospheric soils of cherry (*Prunusavium*). *Research Journal of Agricultural Sciences*, *12*(2), 494–496.

Bashir, Z., Hamid, B., Zargar, M. Y., & Kundoo, A. A. (2021). Morphological and Biochemical Characterization of potassium solubilizing bacteria from Rhizospheric soils of Pinus (*Pinus wallichiana*). *Journal of Research and Development*, *21*, 58–63.

Bashir, Z., Zargar, M. Y., Baba, Z. A., & Mohiddin, F. A. (2018). Solubilization of potassium containing mineral by bacteria from Kashmir iris (*Iris kashmiriana*) Rhizosphere. *The Pharma Innovation Journal*, *17*(4), 747–750.

Bashir, Z., Zargar, M. Y., Baba, Z. A., Mohiddin, F. A., Peer, F. A., Mir, S. A., & Najar, G. R. (2018a). Isolation and characterization of phosphorus solubilizing bacteria (PSB) from rhizospheric soils of apple (*Malus Domestica*). *Journal of Pharmacognosy and Phytochemistry*, *7*(2), 2997–3000.

Bashir, Z., Zargar, M. Y., Bhat, T. A., Kousar, S., Baba, Z. A., & Mohiddin, F. A. (2018b). Solubilization of phosphorus containing mineral by bacteria from rhizospheric region of Walnut (*Juglans regia*). *International Journal of Current Microbiology and Applied Sciences*, *7*(8), 2391–2398.

Bashir, Z., Zargar, M. Y., Mohiddin, F. A., Baba, Z. A., & Hamid, B. (2019). Isolation and biochemical characterization of phosphorus solubilizing bacteria from rhizospheric region of Apricot and Peach. *Journal of Research and Development University of Kashmir*, *19*, 0972–5407.

Bashir, Z., Zargar, M. Y., Mohiddin, F. A., Kousar, S., Husain, M., & Rasool, F. (2017). Phosphorus solubilizing microorganisms: Mechanism and diversity. *International Journal of Communication Systems*, *5*(5), 666–673.

Bashir, Z., Zargar, M. Y., & Vishwakarma, D. K. (2019b). Potassium solubilizing microorganisms for sustainable agriculture. *Applied Agricultural Practices for Mitigating Climate Change*, 2, 17–28.

Beijerinck, M. W. (1898). Concerning a *contagium vivum fluidum* as cause of the spot disease of tobacco leaves. *Phytopathology Class*, *7*(1), 33–52.

Besset-Manzoni, Y., Rieusset, L., Joly, P., Comte, G., & Prigent-Combaret, C. (2018). Exploiting rhizosphere microbial cooperation for development sustainable agriculture strategies. *Environmental Science and Pollution Research*, *25*(30), 29953–29970.

Bhat, M. A., Mishra, A. K., Jan, S., Bhat, M. A., Kamal, M. A., Rahman, S., & Jann, A. T. (2023). Plant growth promoting rhizobacteria in plant health: *A perspective* study of the underground interaction. *Plants*, *12*(3), 629.

Birhane, E., Sterck, F. J., Fetene, M., Bongers, F., & Kuyper, T. W. (2012). Arbuscular mycorrhizal fungi enhance photosynthesis, water use efficiency, and growth of frankincense seedlings under pulsed water availability conditions. *Oecologia*, *169*, 895–904.

Blunden, G., Barwell, C. J., Fidgen, K. J., & Jewers, K. (1981). A survey of some British marine algae for anti-influenza virus activity. *Environmental Sciences, Biology, Medicine, Corpus,* 84711029.

Boltenstern, S., Keiblinger, K. M., Mooshammer, M., Penuelas, J., Richter, A., Sardans, J., & Wanek, W. (2015). The application of ecological stoichiometry to plant-microbial-soil organic matter transformations. *Ecological Monographs*, *85*(2), 133–155.

Bonkowski, M. (2004). Protozoa and plant growth: The microbial loop in soil revisited. *New Phytologist*, *162*(3), 617–631.

Breitbart, M., & Rohwer, F. (2005). Here a virus, there a virus, everywhere the same virus? *Trends in Microbiology*, *13*(6), 278–284.

Chandarana, K. A., & Amaresan, N. (2022). Soil protists: An untapped microbial resource of agriculture and environmental importance. *Pedosphere*, *32*(1), 184–197.

Chandrasekaram, M., Chanratana, M., Kim, K., Seshadri, S., & Sa, T. (2019). Impact of arbuscular mycorrhizal fungi on photosynthesis, water status, and gas exchange of plants under salt stress—A meta-analysis. *Frontiers in Plant Science*, *10*, 457.

Chen, X., Liu, M., Hu, F., Mao, X., & Li, H. (2007). Contributions of soil micro-fauna (protozoa and nematodes) to rhizosphere ecological functions. *Acta Ecologica Sinica*, *27*(8), 3132–3143.

Chieb, M., & Gachomo, E. W. (2023). The role of Plant growth promoting rhizobacteria in plant drought stress responses. *BMC Plant Biology*, *23*(1), 407.

Chittora, D., Meena, M., Barupal, T., Swapnil, P., & Sharma, K. (2020). Cyanobacteria as a source of biofertilizers for sustainable agriculture. *Biochemistry and Biophysics Reports*, *22*, 100737.

Coleman, D. C., Cole, C. V., Anderson, R. V., Blaha, M., Campion, M. K., Clarholm, M., & Schaefer, B. (1977). An analysis of rhizosphere-saprophage interactions in terrestrial ecosystems. *Ecological Bulletins*, 299–309.

Crouch, I. J., & Van Staden, J. (1992). Effect of seaweed concentrate on the establishment and yield of greenhouse tomato plants. *Journal of Applied Phycology*, *4*, 291–296.

Crouch, I. J., & Van Staden, J. (1993). Evidence for the presence of plant growth regulators in commercial seaweed products. *Plant Growth Regulation*, *13*, 21–29.

Elliott, J. A. (2012). Is the future blue-green? A review of the current model predictions of how climate change could affect pelagic freshwater cyanobacteria. *Water Research*, *46*(5), 1364–1371.

Elliott, L. F., Lynch, J. M., & Papendick, R. I. (1996). The microbial component of soil quality. In G. Stotzky & J. M. Bollag (Eds.). *Soil Biochemistry* (pp. 1–21). Marcel Dekker.

Elshafie, H. S., De Martino, L., Formisano, C., Caputo, L., De Feo, V., & Camele, I. (2023). Chemical identification of secondary metabolites from rhizospheric actinomycetes using LC-MS analysis: In silico antifungal evaluation and growth-promoting effects. *Plants*, *12*(9), 1869.

Esteban, G. F., & Fenchel, T. M. (2020). *Ecology of protozoa*. The Biology of Free-living Phagotrophic Protists: Springer International Publishing.

Finlay, B. J., Black, H. I., Brown, S., Clarke, K. J., Esteban, G. F., Hindle, R. M., & Vickerman, K. (2000). Estimating the growth potential of the soil protozoan community. *Protist*, *151*(1), 69–80.

Finlay, B. J., & Esteban, G. F. (1998). Freshwater protozoa: Biodiversity and ecological function. *Biodiversity & Conservation*, *7*, 1163–1186.

Fozia, S. W., Ahmad, L., Ali, T., & Mushtaq, A. (2015). Role of microorganisms in nutrient mobilization and soil health. *Journal of Pure and Applied Microbiology*, *9*(2), 1401–1410.

Gayel, A. G., & Shtina, E. A. (1974). Algae on the sands of arid regions and their role in soil formation. *Soviet Soil Science*, *6*, 67–75.

Giang, N., Hoai, T., Trang, M., Trung, K., Duong, V., Huyen, P., Diep, N., Nguyen Thanh, T., Khanh, T., Ly, P., & Dang Xuan, T. (2020). Isolation and characterization of potassium solubilizing bacteria in some Vietnamese soil samples. *The International Journal on Emerging Technologies*, *11*(3), 639–643.

Gianinazzi, S., Golotte, A., Binet, M. N., Van Tuinen, D., Redecker, D., & Wipf, D. (2010). Agroecology: The key role of arbuscular mycorrhizas in ecosystem services. *Mycorrhiza*, *20*, 519–530.

Godlewska, K., Michalak, I., Pacyga, P., Baśladyńska, S., & Chojnacka, K. (2019). Potential applications of cyanobacteria: *Spirulina platensis* filtrates and homogenates in agriculture. *World Journal of Microbiology and Biotechnology*, *35*, 1–18.

Gonçalves, A. L. (2021). The use of microalgae and cyanobacteria in the improvement of agricultural practices: A review on their biofertilising, bio stimulating and biopesticide roles. *Applied Sciences*, *11*(2), 871.

Gong, A., Wang, G., & Sun, Y. (2022). Dual activity of *Serratia marcescens* Pt-3 in phosphate-solubilizing and production of antifungal volatiles. *BMC Microbiology*, 22–26.

Hamed, S. M. M. (2007). Studies on nitrogen fixing Cyanobacteria. *Egyptian Journal of Phycology*, 205–210.

Hoorman, J. J. (2011). The role of soil protozoa and nematodes. *Fact sheet: Agriculture and natural resources* (pp. 1–5). The Ohio State University Extension.

Ibraheem, I. (2007). Cyanobacteria as alternative biological conditioners for bioremediation of barren soil. *Egyptian Journal of Phycology*, *8*(1), 99–117.

Imam, T., Justine, S., Gibson, Suman Das Gupta, Mohammad Foysal, Shetu B. Das, Md. Ahasanul Hoque, Guillaume Fournié, & Joerg Henning. (2023). Social and cognitive factors influencing commercial chicken farmers' antimicrobial usage in Bangladesh. *Scientific Reports*, *13*, 572. https://doi.org/10.1038/s41598-022-26859-8

Jahn, T. L., Bovee, E. C., & Griffith, D. L. (1974). Taxonomy and evolution of the Sarcodina: A reclassification. *Taxon*, *23*(4), 483–496.

Javed, Z., Tripathi, G. D., Mishra, M., & Dashora, K. (2020). Actinomycetes–The microbial machinery for the organic-cycling, plant growth, and sustainable soil health. *Biocatalysis and Agricultural Biotechnology*, *31*, 101893.

Jayvee, B., Fraser, J., Jong, B., & Sarah, A. (2022). Atypical ileocaecal valve polyp at endoscopy: A rare case of intestinal lymphangioma in an adult woman. *ANZ of Surgery*, 1–10. https://doi.org/10.1111/ans.17755

Jones, R. A. (2018). Plant and insect viruses in managed and natural environments: Novel and neglected transmission pathways. *Advances in Virus Research*, *101*, 149–187.

Jung, S. C., Martinez-Medina, A., Lopez-RAEZ, J. A., & Pozo, M. J. (2012). Mycorrhiza-induced resistance and priming of plant defenses. *Journal of Chemical Ecology*, *38*, 651–664.

Kalpana, et al. (2020). Applications of data mining in hydrocarbon exploration, constraints on geology and petroleum reservoir. *International Journal of Physics*, 1–15.

Kapoor, R. T., Salvadori, M. R., Rafatullah, M., Siddiqui, M. R., Khan, M. A., & Alshareef, S. A. (2021). Exploration of microbial factories for synthesis of nanoparticles–A sustainable approach for bioremediation of environmental contaminants. *Frontiers in Microbiology*, *12*, 1404.

Karuganti, S., Ganti, S., & Muralitharan, G. (2023). Experimental studies on isolation and characterization of silicate solubilizing *Bacillus tequilensis* SKSSB09. *Journal of Applied Biology & Biotechnology*, *11*(4), 135–140. https://doi.org/10.7324/JABB.2023.11911

Keel, C., Ucurum, Z., Michaux, P., Adrian, M., & Haas, D. (2002). Deleterious impact of a virulent bacteriophage on survival and biocontrol activity of *Pseudomonas fluorescens* strain CHA0 in natural soil. *Molecular Plant-Microbe Interactions*, *15*(6), 567–576.

Khan, W., Rayirath, U. P., Subramanian, S., Jithesh, M. N., Rayorath, P., Hodges, D. M., & Prithiviraj, B. (2009). Seaweed extracts as bio stimulants of plant growth and development. *Journal of plant growth regulation*, *28*, 386–399.

Khushwaha, M., & Banerjee, M. (2015). A novel method of seed germination and growth of three staple crop plants: Effect of low temperature and cyanobacterial culture addition. *Journal of Algal Biomass Utilization*, *6*(1), 26–32.

Kour, D., Rana, K. L., Sheikh, I., Kumar, V., Yadav, A. N., Dhaliwal, H. S., & Saxena, A. K. (2020). Alleviation of drought stress and plant growth promotion by Pseudomonas libanensis EU-LWNA-33, a drought-adaptive phosphorus-solubilizing bacterium. *Proceedings of the National Academy of Sciences, India Section B: Biological Sciences*, *90*(4), 785–795.

Kuikman, P. J., Van Elsas, J. D., Jansen, A. G., Burgers, S. L. G. E., & Van Veen, J. A. (1990). Population dynamics and activity of bacteria and protozoa in relation to their spatial distribution in soil. *Soil Biology and Biochemistry*, *22*(8), 1063–1073.

Kumar, A., & Singh, J. S. (2017). Cyanoremediation: A green-clean tool for decontamination of synthetic pesticides from agro-and aquatic ecosystems. *Agro-Environmental Sustainability: Managing Environmental Pollution*, *2*, 59–83.

Ladha, J. K., & Reddy, P. M. (2003). Nitrogen fixation in rice systems: State of knowledge and future prospects. *Plant and Soil*, *252*, 151–167.

Lamichhane, J. R., & Venturi, V. (2015). Synergisms between microbial pathogens in plant disease complexes: A growing trend. *Frontiers in Plant Science*, *6*, 385.

Lee, K. E., Adhikari, A., Kang, S. M., You, Y. H., Joo, G. J., & Kim, J. H. (2019). Isolation and characterization of the high silicate and phosphate solubilizing novel strain Enterobacter ludwigii GAK2 that promotes growth in rice plants. *Agronomy*, *9*, 144.

Lee, R. F., Valkirs, A. O., & Seligman, P. F. (1989). Importance of microalgae in the biodegradation of tributyltin in estuarine waters. *Environmental Science & Technology*, *23*(12), 1515–1518.

Mancuso, S., Briand, X., Mugnai, S., & Azzarello, E. (2006). Marine bioactive substances (IPA Extract) improve foliar ion uptake and water stress tolerance in potted "Vitisvinifera" plants. *Marine Bioactive Substances (IPA Extract) Improve Foliar Ion Uptake and Water Stress Tolerance in Potted" Vitisvinifera" Plants*, 1000–1006.

Manickavelu, A., Nadarajan, N., Ganesh, S. K., Ramalingam, R., Raguraman, S., & Gnanamalar, R. P. (2006). Organogenesis induction in rice callus by cyanobacterial extracellular product. *African Journal of Biotechnology*, *5*(5), 437–439.

Marathe, K. V., & Chaudhari, P. R. (1975). An example of algae as pioneers in the lithosphere and their role in rock corrosion. *The Journal of Ecology*, 65–69.

Márquez, L. M., Redman, R. S., Rodriguez, R. J., & Roossinck, M. J. (2007). A virus in a fungus in a plant: Three-way symbiosis required for thermal tolerance. *Science*, *315*(5811), 513–515.

Marsh, P., & Wellington, E. M. H. (1994). Phage-host interactions in soil. *FEMS Microbiology Ecology*, *15*(1–2), 99–107.

Matz, C., & Kjelleberg, S. (2005). Off the hook–how bacteria survive protozoan grazing. *Trends in Microbiology*, *13*(7), 302–307.

Meena, M., Swapnil, P., Divyanshu, K., Kumar, S., Harish, Tripathi, Y. N., & Upadhyay, R. S. (2020). PGPR-mediated induction of systemic resistance and physiochemical alterations in plants against the pathogens. *Journal of Basic Microbiology*, *60*(10), 828–861.

Neetu, S., Singh, A., & Dahiya, (2022). Plant growth promoting rhizobacteria bacteria in sustaining plant health under changing climate. In *Microbiome under changing climate* (257–269). Woodhead Publishing.

Ngumbi, E., & Kloepper, J. (2016). Bacterial-mediated drought tolerance: Current and future prospects. *Applied Soil Ecology*, *105*, 109–125.

Nowruzi, B., & Porzani, S. J. (2021). Toxic compounds produced by cyanobacteria belonging to several species of the order Nostocales: A review. *Journal of Applied Toxicology*, *41*(4), 510–548.

Orozco-Mosqueda, M. D. C., Santoyo, G., & Glick, B. R. (2023). Recent advances of plant growth. *Plants*, *12*(3), 606.

Ortas, I. (2012). The effect of mycorrhizal fungal inoculation on plant yield, nutrient uptake and inoculation effectiveness under long-term field conditions. *Field Crops Research*, *125*, 35–48.

Ortiz, A., & Sansinenea, E. (2022). The role of beneficial microorganisms in soil quality and plant health. *Sustainability*, *14*(9), 5358.

Osman, M. E. H., El-Sheekh, M. M., El-Naggar, A. H., & Gheda, S. F. (2010). Effect of two species of cyanobacteria as biofertilizers on some metabolic activities, growth, and yield of pea plant. *Biology and Fertility of Soils*, *46*, 861–875.

Pankhurst, C. E., Hawke, B. G., McDonald, H. J., Kirkby, C. A., Buckerfield, J. C., Michelsen, P., O'Brien, K. A., Gupta, V. V. S. R., & Doube, B. M. (1995). Evaluation of soil biological properties as potential bioindicators of soil health. *Australian Journal of Experimental Agriculture*, *35*, 1015–1028.

Patier, P., Yvin, J. C., Kloareg, B., Liénart, Y., & Rochas, C. (1993). Seaweed liquid fertilizer from Ascophyllum nodosum contains elicitors of plant D-glycanases. *Journal of Applied Phycology*, *5*, 343–349.

Patil, R., Satpute, R., & Nalage, D. (2023). Plant microbiome and their role in plant health. *Microenvironment & Microecology Research*, *5*(1), 2.

Povolny, M. (1981). The effect of the steeping of peat-cellulose flowerpots (Jiffypots) in extracts seaweeds on the quality of tomato seedlings. In *Proceedings of the International Seaweed Symposium*, *8*, 730–733.

Prakash, S., & Nikhil, K. (2014). Algae as a soil conditioner. *International Journal of Engineering Research*, *2*, 68–70.

Prasad, R. C., & Prasad, B. N. (2001). Cyanobacteria as a source biofertilizer for sustainable agriculture in Nepal. *Botanica Orientalis: Journal of Plant Science*, *1*, 127–133.

Rajabpour, N., Nowruzi, B., & Ghobeh, M. (2019). Investigation of the toxicity, antioxidant and antimicrobial activities of some cyanobacterial strains isolated from different habitats. *Acta Biologica Slovenica*, *62*(2), 3–14.

Rajendra, l., Durgadevi, D., Kavitha, R., Archana, T., Harish, S., Sendhilvel, V., & Karthikeyan, G. (2023). Deciphering the role of growth-promoting bacterial endoohytes in harmonizing plant health. In *Microbial Symbionts and Plant Health. Trends and Applications for Changing Climate* (pp. 265–300). Springer Nature Singapora.

Raouf, N., Al-Homaidan, A. A., & Ibraheem, I. B. M. (2012). Agricultural importance of algae. *African Journal of Biotechnology*, *11*(54), 11648–11658.

Rastimesina, I., Postolachi, O., Bogdevich, O., Josan, V. V., Nicolau, E., & Culîghin, E. (2022). Microbiological characteristic of soil for the bioremediation of POPs contaminated sites. In *Ecological and Environmental Chemistry* (pp. 178–178).

Rayorath, P., Jithesh, M. N., Farid, A., Khan, W., Palanisamy, R., Hankins, S. D., & Prithiviraj, B. (2008). Rapid bioassays to evaluate the plant growth promoting activity of Ascophyllum nodosum (L.) Le Jol. Using a model plant, Arabidopsis thaliana (L.) Heynh. *Journal of Applied Phycology*, *20*, 423–429.

Řezanka, T., Palyzová, A., & Sigler, K. (2018). Isolation and identification of siderophores produced by cyanobacteria. *Folia Microbiologica*, *63*, 569–579.

Riaz, U., Murtaza, G., Anum, W., Samreen, T., Sarfraz, M., & Nazir, M. Z. (2021). Plant Growth-Promoting Rhizobacteria (PGPR) as biofertilizers and biopesticides. In *Miocrobiota and biofertilizers: A sustainable continuum for plant and soil health* (pp. 181–196). Springer.

Sadhana, B. (2014). Arbuscular mycorrhizal fungi (AMF) AS A Biofertilizers-a review. *International Journal of Current Microbiology and Applied Sciences*, *3*(4), 384–400.

Sanchez, O. J., Ospina, D. A., & Montoya, S. (2017). Compost supplementation with nutrients and microorganisms in composting process. *Waste Managemententement*, *69*, 136–153.

Sarikhani, M. R., Oustan, S., Ebrahimi, M., & Aliasgharzad, N. (2018). Isolation and identification of potassium-releasing bacteria in soil and assessment of their ability to release potassium for plants. *European Journal of Soil Science*, *69*(6), 1078–1086.

Scott, J. T., & Marcarelli, A. M. (2012). Cyanobacteria in freshwater benthic environments. In *Ecology of Cyanobacteria II: Their diversity in space and time* (pp. 271–289). Springer Netherlands.

Sharma, M., Dangi, P., & Choudhary, M. (2014). Actinomycetes source, identification, and their application. *International Journal of Current Microbiology and Applied Sciences*, *3*(2), 801–832.

Sharma, R. S., Mohmmed, A., & Babu, C. R. (2002). Diversity among rhizobiophages from rhizospheres of legumes inhabiting three ecogeographical regions of India. *Soil Biology and Biochemistry*, *34*(7), 965–973.

Shengping, X. U. E., & Xiaohuan, W. A. N. G. (2021). Inoculation of soil with cadmium-Resistant actinomycetes flora reduces cadmium accumulation in rice (*Oryza Sativa* L.). *International Journal of Environmental Pollution and Remediation*, *9*, 1–11.

Silva, G. D. C., Kitano, I. T., Ribeiro, I. A. D. F., & Lacava, P. T. (2022). The potential use of actinomycetes as microbial inoculants and biopesticides in agriculture. *Frontiers in Soil Science*, *2*.

Singh, N. K., Dhar, D. W., & Tabassum, R. (2016). Role of cyanobacteria in crop protection. *Proceedings of the National Academy of Sciences, India Section B: Biological Sciences*, *86*, 1–8.

Singh, S. (2014). A review on possible elicitor molecules of cyanobacteria: Their role in improving plant growth and providing tolerance against biotic or abiotic stress. *Journal of Applied Microbiology*, *117*(5), 1221–1244.

Sivalingam, K. M., & Batiri, A. (2020). Isolation and identification of some cyanobacteria and their plant growth promoting effect on wheat (Triticum aestivum L.), Ethiopia. *Journal of Science and Inclusive Development*, 17–37.

South, G. R., & Whittick, A. (1987). *Introduction to Phycology*. Blackwell Scientific Publications.

Stout, J. D. (1980). The role of protozoa in nutrient cycling and energy flow. In *Advances in Microbial Ecology* (pp. 1–50). Springer US.

Sun, Z., Song, J., Xin, X., Xie, X., & Zhao, B. (2018). Arbuscular mycorrhizal fungal proteins are involved in arbuscular formation and responses to abiotic stresses during AM symbiosis. *Frontiers in Microbiology*, *5*, 9–19.

Tamaru, Y., Takani, Y., Yoshida, T., & Sakamoto, T. (2005). Crucial role of extracellular polysaccharides in desiccation and freezing tolerance in the terrestrial cyanobacterium Nostoc commune. *Applied and Environmental Microbiology*, *71*(11), 7327–7333.

Vandana, N., & Manu, S. (2021). Isolation screening and molecular characterization of zinc solubilizing bacteria and their effect on the growth of wheat (*Triticum aestivum*). *Molecular Biology and Biotechnology*, *29*(2), 85–97.

Vassileva, M., Mendes, G. D. O., Deriu, M. A., Benedetto, G. D., Flor-Peregrin, E., Mocali, S., & Vassilev, N. (2022). Fungi, P-Solubilization, and plant nutrition. *Microorganisms*, *10*(9), 1716.

Verma, S., Adak, A., Prasanna, R., Dhar, S., Choudhary, H., Nain, L., & Shivay, Y. S. (2016). Microbial priming elicits improved plant growth promotion and nutrient uptake in pea. *Israel Journal of Plant Sciences*, *63*(3), 191–207.

Vernieri, P., Borghesi, E., Ferrante, A., & Magnani, G. (2005). Application of bio stimulants in floating system for improving rocket quality. *Journal of Food Agriculture and Environment*, *3*(3/4), 86.

Vickerman, K. (1992). The diversity and ecological significance of Protozoa. *Biodiversity & Conservation*, *1*, 334–341.

Vidyashree, D. N., Muthuraju, N., Panneerselvam, P., & Mitra, D. (2018). Organic acids production by zinc solubilizing bacterial isolates. *International Journal of Current Microbiology and Applied Sciences*, *7*(10), 626–633.

Vymazal, J. (1995). *Algae and element cycling in wetlands* (p. 64). Lewis Publishers Boca Rotan.

Wommack, K. E., & Colwell, R. R. (2000). Virioplankton: Viruses in aquatic ecosystems. *Microbiology and Molecular Biology Reviews*, *64*(1), 69–114.

Yadav, A. N., Kour, D., Kaur, T., Devi, R., Yadav, A., Dikilitas, M., & Saxena, A. K. (2021). Biodiversity, and biotechnological contribution of beneficial soil microbiomes for nutrient cycling, plant growth improvement and nutrient uptake. *Biocatalysis and Agricultural Biotechnology*, *33*, 102009.

Zaib, M., Sandhu, M., Mehmood Zar, U. G., Rehman, A., Musharraf, M. H., Zubair, M., & Raza, I. (2023). A brief review on plant growth promoting rhizobacteria (PGPR): A key role in plant growth. *International Research Journal of Engineering and Technology*, 5(7).

Zhang, C., Chen, H., Dai, Y., Chen, Y., Tian, Y., & Huo, Z. (2023). Isolation and screening of phosphorus solubilizing bacteria from saline alkali soil and their potential for Pb pollution remediation. *Frontiers in Bioengineering and Biotechnology*, *11*, 1134310.

Zhou, Z., Chang, N., Lv, Y., Jiang, H., Yao, C., Wan, X., Li, Y., & Zhang, X. (2022). K-solubilizing bacteria (Bacillus) promote theanine synthesis in tea roots (*Camellia sinensis*) by activating CsTSI activity. *Tree Physiology*, *42*(8), 1613–1627.

Zou, Y. N., Srivastava, A. K., & Wu, Q. S. (2016). Glomalin: A potential soil conditioner for perennial fruits. *International Journal of Agriculture and Biology*, *18*, 293–297.

7 Silicate Solubilizers

Role of Silicate Solubilizing Microorganisms in Organic Agriculture

Rizwana Ali, Zaffar Bashir, Ufaq Fayaz, and Mehrun Nisa

7.1 INTRODUCTION

The current state of food security is regarded as one of society's most severe problems. Currently, there are about 8 billion people on the planet, and it is predicted that number will increase to close to 10 billion in the next five decades (Glick, 2014). Due to the increasing demand on the food supply worldwide, land degradation, unfavorable impacts on agriculture biodiversity and the environment, and a 70% decline in crop output are all results of the current burden on agriculture (Veatch-Blohm, 2007). Significant yield losses in important cereal species including wheat, maize, rice, and barley have been observed due to change in weather, which increases the occurrence and severity of various non-living environmental pressures, notably drought and elevated temperatures (Carmen & Roberto, 2011; Lobell & Field, 2007). Encouraging connection between plant roots and advantageous soil microbes is an advanced approach in farming which leads to improve health and productivity of crops (Lugtenberg et al., 2002). Silicon (Si), being the most prevalent and the eighth most plentiful element in the universe, has received scant attention in crop production until recently (Heckman et al., 2012). It has been recognized as a beneficial nutrient and in 2013, the Association of American Plant Food Control Officials categorized it as an "almost indispensable" component due to high comprehension of silicon's impact on enhancing growth and safeguarding plants (Heckman, 2013).

In comparison to other species of plant, the build-up of silicon is often greater in ferns and monocots (Farooq & Dietz, 2015; Hodson et al., 2005). In an unfavorable environment, it's seen that Si supplementation improves plant fitness (Sathe et al., 2021). In numerous plant-pathogen systems, increased resistance to a number of diseases under conditions of Si supplementation has been documented (Hawerroth et al., 2019; Liang et al., 2015; Bhat et al., 2022). According to separate studies by Etesami and Beattie (2017) and Etesami and Jeong (2018), Si has the ability to relieve various abiotic challenges, including salinity, toxicity by metals, nutritional imbalance, and drought. Si application may lead to stress resistance through changing the homeostasis of phytohormones, according to an increasing number of researches (Etesami &

DOI: 10.1201/9781003491460-7

Jeong, 2018). Additionally, it helps in the absorption of phosphorous, magnesium, potassium, iron, copper, and zinc (Chen et al., 2000). Si plays positive roles in a variety of other species also, including cattle, horses, and poultry (Farooq & Dietz, 2015). It has been discovered that cows given such Si rich fodder generate more milk of good quality (Radkowski et al., 2017). Si application has the potential to cure a number of illnesses in people, including alopecia, Alzheimer's, and cardiovascular disease (Martin, 2013). Likewise, silicon also has positive impact on neural health, and the condition of the hair, nails, and skin can be also improved by Si (Martin, 2013).

The typical types of silicon found in soil are insoluble silicates, including iron, sodium, potassium, magnesium, and aluminium silicates (Bist et al., 2020). Orthosilicic acid includes that type of Si which is usually preferred by plants and is absorbed together with water (Rodrignes & Datnoff, 2005). In vascular bundles, sclerenchyma, epidermal cells, and florescence brackets of cereals, silica gel that has accumulated as silicate is deposited (Lanning et al., 1958). $Si(OH)_4$ has a maximum solubility of 2 mM in solution, and in soil solutions, its concentration generally falls between 0.1 and 0.6 millimolar. Various mechanisms within soil environments, including the weathering of soil minerals, release from the soil matrix, and irrigation water, can lead to the formation of monosilicic acid ($Si(OH)_4$) (Klotzbucher et al., 2014). Solubilizing insoluble forms of silicates is a critical function of silicate solubilizing microbes, and this process has a crucial role in boosting both plant defense systems and soil fertility (Vasanthi et al., 2012; Cruz et al., 2022).

While Si was traditionally called an essential element, emerging research has shed light on its importance in various plant physiological processes. Technical challenges in developing plants in soil which is free from Si have made it tough to evaluate the necessity of silicon for plants accurately (Guntzer et al., 2012). Moreover, Si is naturally abundant in the Earth's crust and is non-toxic to plants, animals, and microorganisms. Unlike some synthetic fertilizers, silicon fertilizer does not contribute to soil or water pollution, making it environmentally safe (Zhu & Gong, 2014). Silicon fertilizer can be a boon for agroindustries. Silicon plays a crucial role in strengthening cell walls, thus making them more resilient to diseases and environmental stresses like drought and heat. Additionally, it enhances the soil's water retention capacity, ensuring that plants have access to moisture when needed (Meena et al., 2014a). Use of slags, such as calcium silicate slages, fly ash, and sodium silicate, as sources of silicon in crops has proved to offer various benefits, including acquisition of nutrient and control of disease (Datnoff et al., 1997; Ishiguro, 2001; Parab et al., 2015; Peera et al., 2016). Sodium and potassium silicates are water-soluble and have been utilized effectively as foliar fertilizers in agriculture. Their solubility allows for easy application as foliar sprays, which can be absorbed by the plant leaves (Liang et al., 2015). Debates are there related to the effectiveness of foliar application for silicon uptake in plants, but there's limited evidence supporting significant silicon uptake through the leaf surface. Fertilizers of silicon consist of: (i) Si-rich industrial by-products or slags, such as calcium silicate slag and fly ash which are by-products of industrial processes like steel manufacturing and power generation. These can be rich sources of silicon but may also contain trace metals that can potentially contaminate soils if applied in large quantities over. (ii) Fertilizers which are biologic in origin. This category includes natural materials like bentonite, diatomaceous earth, and micas, which naturally contain Si. However, their silicon content may not always be readily available to plants due to low solubility or slow release rates. As a result, high application rates may be necessary to achieve desired silicon amount in soil, which can increase its costs and subsequent imbalances in other soil nutrients. (iii) Potassium silicate which can be quite expensive compared to other fertilizers (Gascho, 2001). One of the main challenges with Si fertilizers is their limited availability and relatively high cost. Unlike conventional fertilizers like nitrogen, phosphorus, and potassium, which are widely produced and readily accessible to farmers, Si fertilizers may be less abundant in the market (Meena et al., 2014). Recycling mineral, mining wastes from metal, rocks of silicon, and demolition wastes, as well as crop residues which are rich in Si to produce low-cost silicate fertilizers presents a promising solution to the challenges caused by availability and affordability of these fertilizers. Different physical and chemical interventions can be used to enhance the availability of nutrients from recycled materials when used as fertilizers. Si fertilizers can offer a range of benefits that contribute to increased productivity by employing various mechanisms such as increasing resistance against stresses, adjusting pH, and nutrient acquisition contained in the silicate fertilizers.

The amount of silicon removed from the soil is about 224 million tons per year (Tubana et al., 2016). This depletion of silicon from soil due to crop uptake can have significant implications for crop production, especially in regions where mono-cropping of silicon high-accumulator species such as rice is prevalent. While the application of Si fertilizers can help alleviate silicon deficiency in such soils and improve crop productivity, it's essential to consider potential environmental concerns, sustainability implications, and increased production costs associated with their use. In contrast, exploring the potential of silicon solubilizing bacteria (SSB) offers a promising alternative that aligns with sustainability, environmental friendliness, and economic viability for enhancing silicon availability to crop plants and improving agricultural production.

Leveraging bacteria for weathering of soil silicates presents a novel and promising strategy to increase bioavailable Si in soil solutions. Bacteria have a promising role in accelerating the solubilization of silicates through the process of weathering, thereby increasing the availability of reactive Si in the plant-soil system (Raturi et al., 2021). Plants and their associated microflora play a significant role in influencing silicate weathering through various mechanisms, including the secretion of ligands for chelation, changes in pH, and alteration of soil physical properties (Cornelis et al., 2011). SSB such as *Burkholderiaeburnea*, *Bacillusglobisporous*, *Bacillus mucilaginosus*, *Bacillus tequilensis*, and *Enterobacter ludwigii* have the ability to solubilize silicate minerals, thereby releasing silicon into the soil solution and making it more available for plants (Kang et al., 2017; Lee et al., 2019; Sheng et al., 2008; Sukumar et al., 2023). The potential for using SSB to reduce root rot by *Rigidoporos microporus* is an interesting concept that aligns with the multifunctional benefits of these bacteria in agriculture (Shabbir et al., 2020).

7.2 SILICON: ITS BIOAVAILABILITY AND IMPORTANCE

After O_2, Si ranks second in its abundance and is an important element in Earth's crust (Ali et al., 2020; Shahzad et al., 2021). In soil the concentration of silicon ranges from 50 to 400 g per kg of soil. Different tissues of plants have varying concentrations of Si because before deposition and storage in these tissues, Si undergoes various chemical modifications. Varying amount in these tissues is also due to different absorption mechanisms (Epstein, 1994; Ma et al., 2006a, b; Mandlik et al., 2020). Various forms of insoluble silicates are calcium silicate, sodium silicates, magnesium silicate, iron silicate, and aluminium silicates (Bist et al., 2020). Although various Si forms such as quartz, amorphous silica, and vermiculite are present, other forms like plagioclase and orthoclase have been also reported (El-Ramady et al., 2022). These forms are inert and less soluble (Raturi et al., 2021). In soil liquid form, silicon is ergonomically important which includes monosilicic acid (H_4SiO_4) as one of the major forms and at pH less than 9 its amount can vary between 3 and 17 mg of Si per liter (Tubaña & Heckman, 2015). If pH increases, monosilicic acid separates into silicate ions, because it's an uncharged molecule (H_4SiO_4) between pH 2 and 9, and at pH>9 it presents as ionized form ($H_3SiO^{4-}/H_2SiO_4^{2-}$). ($H_4SiO_4$) is the most dissolved form of silicon. In soil, silica minerals occur as primary silicates and secondary silicates. Pedogenic processes are important for the distribution of primary and secondary silicates in soil; for example, secondary silicates are present in clay and primary silicate in sand and silt. Therefore, pH is a crucial factor which determines whether Si will be present as monomer or in dissociated form (Epstein, 1994).

The beneficial impact of soluble forms of silicon on the development of plants has been showed by several studies. In addition to the positive impact of soluble Si yield, Si can mitigate the effect of various stresses in several species of plant (Epstein & Bloom, 2006), including rice. Si enhances plant growth by mitigating the adverse impacts of both living organisms and non-living factors during unfavorable conditions (Ma & Yamaji, 2006). Silicon can offer resistance to infections caused by both fungal and bacterial pathogens such as rust, sheath blight, gray leaf spot, and rice blast (Fauteux et al., 2005). Silicon is also inhibitory for the growth of various pests like stem borer, white backed plant hopper, and brown plant hopper (Savant et al., 1997). In wheat, protection against damage caused by rabbits (wild) is also provided by Si (Cotterill et al., 2007). Si has an emerging role in monocotyledonous in stress relief, delaying of wilting and providing protection against biotic and abiotic stresses including toxicity by metals, salinity, drought, and nutritional imbalance (Etesami & Beattie, 2017; Etesami & Jeong, 2018). The presence of Si below the cuticle has several roles in plants; for instance, it increases physical strength, acts as a physical barrier for phytopathogens, and decreases transpiration rate. Furthermore, soluble silicon can elevate the resistance in plants against phytopathogens by interacting with the signaling molecule and by enhancing the production of plant defense substances. Figure 7.1 shows the importance of silicon in enhancing plant growth.

7.3 THE FUNCTION OF BACTERIA THAT SOLUBILIZE SILICATE IN THE LIBERATION OF SILICON

Bashan and Holguin (1998) introduced two significant terms, namely "biocontrol plant growth promoting bacteria" (Biocontrol-PGPB) and "plant growth promoting bacteria" (PGPB). As biocontrol agents, PGPB can inhibit the growth of plant pathogens by secreting inhibitory compounds like antibiotics, HCN, and siderophores.

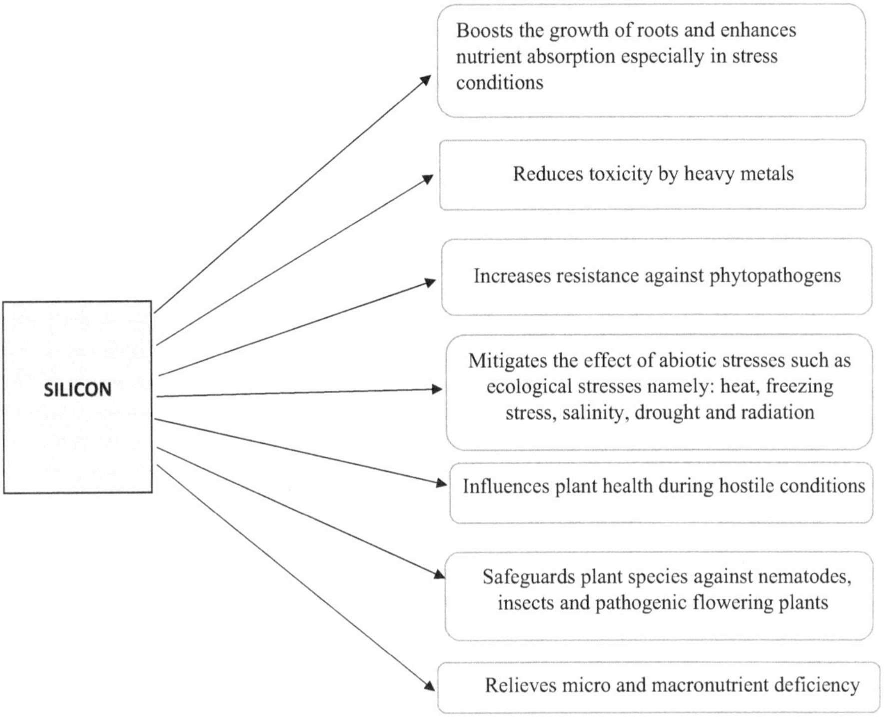

FIGURE 7.1 Silicon against various stresses.

Plant growth promoting bacteria use both direct and indirect ways to increase plant growth. Indirect methods of plant growth promotion involve the production of inhibitory substances such as HCN, siderophores, and antibiotics against phytopathogens as noted by Wang et al. (2021b). Directly PGPB can promote the growth of plants through the solubilization of insoluble minerals like phosphorous, zinc, and potassium silicates (Bashir et al., 2017; Bashir et al., 2018; Bashir et al., 2018a; Bashir et al., 2018b; Bashir et al., 2019; Bashir et al., 2019b; Bashir et al., 2021a; Bashir, 2021b; Hamid & Bashir, 2019a). One of the potential candidates among PGPB is silicon solubilizing bacteria (SSB). These bacteria are efficient in solubilization of insoluble silicate minerals that are present in soil and are less available for plant utilization. So SSB employs different mechanisms for solubilization of these silicate minerals and makes them available for plant nutrition. SSB are found in different environments like water, marine sediments, and soil, but compared to other bacteria they are present in lower numbers which tells us about their uniqueness in nature (Naureen et al., 2015; Vasanthi et al., 2018).

Silicon is the second plentiful element in the Earth's crust (Bist et al., 2020), and biogeochemical cycle is responsible for its distribution in different parts of ecosystems. It plays important roles such as reduction of metal toxicity, increased development in plants, formation of soil, and attributes in several chemical reactions (Gerard et al., 2008).

The breakdown of rocks in the Earth's crust facilitated by microbial activity is a vital component of the nutrient cycle, crucial for maintaining the sustainability of the biosphere. Bio weathering is a geochemical process that uses living organisms for the disintegration and deterioration of minerals and rocks (Burford et al., 2003; Gadd, 2007). It includes the basic step in the separation of minerals from rocks (Burford etal., 2003; Li et al., 2006). In soil, insoluble silicates can be transformed into a usable form by the action of beneficial microorganism, weathering of rocks, and by the roots of plants. In this regard, SSB are known as potential candidates, which have the power of solubilization of insoluble silicates and make these soluble forms ready for plant nutrition. It is known that bacteria can enhance mineral weathering from rocks, but the exact mechanism by which these bacteria solubilize minerals from weathered feldspar surfaces is unknown (Lee & Fein, 2000; Wightman & Fein, 2004).

Silicate solubilization by SSB involves many steps, and the first step involves cations (K^+, Na^+ and Ca^+) exchange for proton at the surface of minerals. After this step, there is separation of different Si forms from the framework (Aagaard & Helgeson, 1982; Helgeson et al., 1984; Wollast, 1967). Various studies have reported that bacteria are the efficient solubilizers of silicate minerals and the different ways of Si solubilization involve solubilization by acids (organic and inorganic), ligands, extracellular polysaccharides, and alkali (nucleophilic attack). Of all the possible mechanisms lysis by acids is a frequently occurring mechanism for weathering of Si minerals (Jongmans et al., 1997).

7.3.1 Scientific Protocol for Isolation of Silicate Solubilizing Microbes from Environmental Samples

The isolation of silicate solubilizing microbe involves a specific medium for isolation. The silicate medium composition involves: glucose (10g/L), insoluble magnesium trisilicate [$Mg_2O_8Si_3$] (2.5g/L), agar (20g/L) as a solidifying reagent, and 1000 ml of distilled H_2O (Lee et al., 2019) is prepared and autoclaved at 121°C for 15 minutes at 15 PSI for sterilization. Under aseptic conditions, the silicate medium is allowed to cool and transferred to sterile petri plates. All plates were spot inoculated with fresh cultures of bacterial isolate or any microbial strain (24 hours) and incubated at 37°C for 24 hours. After incubation, the formation of a clear zone around the bacterial colonies indicates the silicate solubilizing activity of microbial strain. Figure 7.2 Shows the zone of clearance on silicate solubilizing medium after incubation which

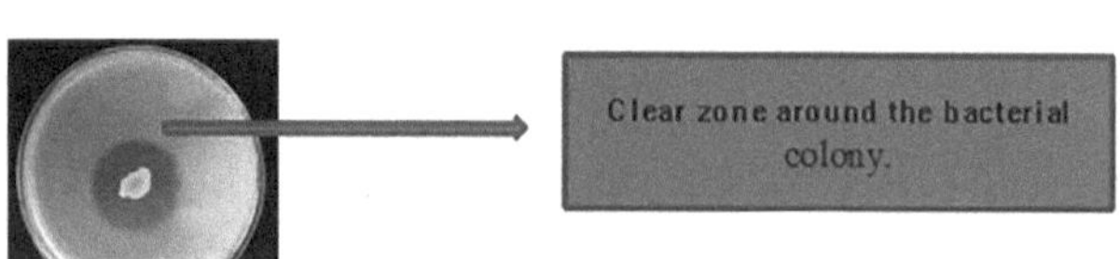

FIGURE 7.2 Silicate solubilization assay on silicate solubilizing medium by bacterium.

indicates the silicate solubilizing activity of particular isolate. The solubilization index was calculated by using the formula:

$$\text{Silicate solubilization index (SSI)} = \frac{\underset{\text{(mm)}}{\text{Colony diameter}} + \underset{\text{(mm)}}{\text{Halo zone diameter}}}{\text{Colony diameter (mm)}}$$

7.4 UPTAKE MECHANISM OF SILICON

Different tissues of a plant have varying concentration of Si, and the reason for this is difference in the ways by which a plant absorbs Si from the soil and the mechanism of transporting Si in plants (Epstein, 1994; Jarvis, 1987). The first transport system for Si was discovered in diatoms; after that, diverse life forms were found to have this transport system (Hildebrand et al., 1997). Because of the absence of their homolog (s) in plants, there are not much evidences about their occurrence in plants (Deshmukh & Belanger, 2016). Specific trans membrane proteins are used by plants for the uptake of Si (Marron et al., 2016). These transporters do not require energy for the passage of silicic acid across the plasma membrane; hence, these are called passive transporters. There are some examples of crops in which there is energy dependent transport of Si like banana, barley, maize, cucumber, and rice (Nikolic et al., 2007). Studies done by Ma et al. (2006a, b) proved the presence of genes on plasma membrane which are required for Si loading in xylem. In cucumber plants, Mitani and Ma (2005) proved that the solution (external) contains higher amounts of Si than the xylem sap, which means that loading of silicon in its xylem is entirely passive. Through an induced mutagenesis approach Ma et al. (2006a, b) discovered Lsi1 (low Si 1), the first transporter of Si in a rice plant. It has been also confirmed that the Si accumulators and intermediators use both passive and active mechanisms for Si uptake (Liang et al., 2006). Chromosome 2 has a gene for transporter (Lsi1) and this gene has four introns and five exons (Ma et al., 2004, 2006a, b). It belongs to nodulin-26 like intrinsic proteins (NIPs), a subfamily of aquaporins (Kumar et al., 2020; Singh et al., 2020). Similar in function, another transporter, namely Lsi2, was discovered by Ma et al. (2007a) which is used for efflux of Si. Both of them differ in their presence across exodermis and the endodermis cells (Ma et al., 2007a). Studies have shown the location of Lsi2 on the plasma membrane. Same as Lsi1, Lsi2 genes are present in the roots, but there are lower numbers of Lsi2 transporters in the root tip than in the maturation zone (Ma et al., 2007b). Once Si is absorbed by the roots it's first transported into the stele, and from stele it then move towards the shoot by transpirational pull through the xylem. Concentration of silicon (monosilicic acid) in xylem is high (Casey et al., 2003; Mitani et al., 2005). The maximum portion of Si absorbed by roots is translocated to the shoots, and inside shoots polymerization reaction transforms it into amorphous silica (Ma & Yamaji, 2006). Once (SiO_2-nH_2O) is formed it's condensed in the walls of cells, and in the cells of plant roots (Ma & Yamaji, 2006). Homology of Lsi1 such as Lsi6 has been found in rice plant which is also used in xylem unloading. The dorsal side of the xylem parenchyma shows the presence of Lsi6, and it's known to cause uniform distribution of silica in the shoot. Lsi6 also uptakes Si in the root tips but as compared to Lsi1 its absorption efficiency is lower (Yamaji et al., 2008). Among several types of Si transporter such as Lsi1 and Lsi2, Lsi6 unlike root tip is also found in other tissues like leaf blades. Knockout experiment of Lsi6 does not have negative impact on Si uptake through roots, but it affects the concentration of Si in the leaf blades and sheaths. Similarly in maize and barley other transporters namely ZmLsi1 and HvLsi1 have been identified to do influx of silicon (Chiba et al., 2009; Mitani et al., 2009). Both HvLsi1 and ZmLsi1 have the same function as OsLsi1, but the level of expression and localization within the specific cell is entirely different from OsLsi1. For example, HvLsi1 and ZmLsi1 are present in cortex cells, epidermal, and hypodermal cells (Chiba et al., 2009; Mitani et al., 2009). It has been found that there is a strong relation between the expression of HvLsi2 and uptake of Si (Mitani et al., 2009), which indicates that HvLsi2 is a determining factor for Si uptake in barley. Si do not affect the expression level of HvLsi1 (Chiba et al., 2009) whereas it down regulates OsLsi1 expression (Ma et al., 2006a, b). Recent research have showed that in rice panicle OsLsi3 is responsible for uptake of silicon (Yamaji et al., 2015). Despite all the Si transporters mentioned in this chapter, many transporters are yet to be discovered (Mandlik et al., 2020). For example, transporters for xylem loading are not yet known.

7.5 SILICON AS FERTILIZER

Unsustainable farming practices which cause depletion of nutrients from soil are the main cause of low productivity. Various studies relate this decline in productivity with low amounts of nitrogen supply from the soil (Cassman et al., 1995). Another reason contributing to this decline is the low availability of Si for plant nutrition. In agroindustries silicon holds its significance not only for being a nutritionally important element but also for enhancing capability of plants to withstand abiotic stresses (Liang et al., 2005). Application of silicon fertilizers in agricultural fields has a beneficial impact on plants such as it enhances physiochemical properties of soil, protects plants against phytopathogens, improves cell wall strength, and increases the amount of soluble nutrients in soil. Table 7.1 displays the type and effect of Si fertilizers under different stresses on various crops. Among all the crops, rice is known to accumulate Si in the highest amount (Meharg & Meharg, 2015). Also, the impact of silicon fertilizers has been intensively studied for rice cultivation. To meet the demand of population, continuous cultivation of crops especially rice has decreased the amount Si in paddy soils and thus deteriorates the yield and health of soil (Sasaki et al., 2014). Many reports have shown that addition of silicon in soil enhances the capability of both plant and microbes to uptake nutrients from soil. For example, Neu et al. (2017) have shown that 10.4-15.6 μg/kg concentration of Si can increase the rate at which fertilizers are utilized, while concentration higher than 20.8 μg/kg has adverse effects on plant growth. Growth of reeds is also affected by Si (Schaller et al., 2012). Dicotyledonous plants, fruit crops, and some monocotyledonous plants can accumulate the higher concentration of silicon in their tissues (Devanna et al., 2021). When Si fertilizers are applied several factors including fertility of soil, crop nutrient requirements and moisture content of soil are considered (Zargar et al., 2019). Positive impact of Si fertilizers includes enhanced productivity, protection against phytopathogens and tolerance towards abiotic stresses (Borawska-Jarmułowicz et al.,

TABLE 7.1
Displaying the Type and Effect of Si Fertilizers under Different Stresses on Various Crops

Kind of Si Fertilizer Used	Stress	Crop	Results on Plants	References
K_2SiO_3 (Potassium silicate)	Salt stress	Cucumber (*Cucumis sativus*)	Reduces the oxidative damage that salt stress causes to plant tissue, improves metabolic and physiological functions	Zhu et al. (2004)
H_4SiO_4 (Monosilicic acid)	Drought stress	Papaya (*Carica papaya* L)	Enhances quality and thickness of the fruit in relatively dry and well-watered environments	Barberena (2014)
H_4SiO_4 (Monosilicic acid)	Drought stress	Banana (*Musa*)	Improves the productivity of irrigation water, enhances fruit quantity and productivity	Ako et al. (2011)
K_2SiO_3, $CaSiO_3$ (Potassium silicate and Calcium silicate)	Drought stress	Rice (*Oryza sativa*)	Enhances nutrient absorption, improves biomass yield	Tayade et al. (2022)
Na_2SiO_3 (Sodium silicate)	Salinity stress	Cucumber (*Cucumis sativus L.*)	Increases plant growth and tolerance to salinity	Roe et al. (1995)
K_2SiO_3 (Potassium silicate)	Thermal	Banana (*Musa*)	Enhances growth and development of plants	Takrim et al. (2021)
$CaSiO_3$, K_2SiO_3 (Calcium silicate and Potassium silicate)	Effect on embroygenesis	Dry date palm (*Phoenix Dactylifer*)	Delays the impact of disease and ceases its development during the growth phase	Chigwaya et al. (2016)
SiO2.nH2O	Drought stress	Different crops	Role in enhancing water retention within plant tissues	Nedukha et al. (2022)
(Sodium Silicate)	Wound stress	Wound stress	Heightened enzyme activity of antioxidants in reaction to wounding stress	Khan et al. (2022)
Na_2SiO_3 (Sodium silicate)	Salinity	Various crops	Production of phytohormones	Sabir et al. (2022)
(SiO_2–nH_2O)	Cell wall strength		Used in xylem loading	Ma and Yamaji (2006)
Na_2SiO_3 (Sodium silicate)	Herbivores		Increases the attraction of herbivore-induced plant volatile mixtures	Liu et al. (2017)
Si-mediated	Leaf folder		For the sensitive varieties, it provides tolerance to leaf folders	Han et al. (2015)
Liquid silicon fertilizer	Light stress	Red-lettuce (*Lactuca sativa*)	Under the influence of LED lighting, red-leaf lettuce output biomass improves	Semenova et al. (2021)
Silicon and iron fertilizer	Arsenic stress	Water spinach (*Ipomoea aquatica*)	Improves plant growth and lowers arsenic uptake of plants to lower the risk of consumption to human health	Yu et al. (2016)
OSiF (Organosilicon fertilizer)	Heavy metal toxicity	Wheat (*Triticum aestivum*)		Huang et al. (2019)

2022; Rao et al., 2019; Thakral et al., 2021). As per recent findings silicon not only plays a role in nitrogen metabolism but it's also involved in utilization of phosphorous (Liao et al., 2020; Neu et al., 2017; Saudy & Mubarak, 2015). Therefore, use of silicon fertilizer for Si poor soil is highly recommended for the overall performance of plants. Further, iron and steel manufacturing are producing a huge amount of slag which draws our attention towards its safe use in agriculture. Using slag as a by-product of iron and steel industries is gaining attention as a potential soil amendment. Slag can improve soil fertility by providing essential nutrients such as calcium, magnesium, and silicon. Wastes from these industries such as steel slag and blast furnace slag are considered as good silica fertilizers (Anggria & Masunaga, 2021). Properties of slag depend upon various factors such as manufacturing, cooling, and valorization processes. Slag as fertilizers has not only reduced its negative consequences on environment but can also enhance productivity (Das et al., 2019; Gwon et al., 2018). Slag based fertilizers typically contain high levels of Ca, Si, P, Mg, Fe, Mn, and Fe, however they may have low concentration of nitrogen, potassium, and phosphorous. It is therefore recommended to use SSF in combination with organic fertilizer that has good amounts of N, P, and K. Rice husks, a by-product of rice milling, are rich in silica (SiO_2) and constitute a significant source of Si. It's estimated that worldwide around 80 million tons of rice husks or hulls are produced annually. This production amounts to approximately 3 million tons of silicon, primarily as opaline silica (Chaudhary et al., 2002). Converting opaline silica-rich rice husks into biochar under controlled conditions, such as partial oxygen, offers an innovative approach to utilizing this nutrient-rich material as a soil (Ebe et al., 2019). Rice husk biochar (RHB) offers multiple benefits in agriculture, including its potential to decrease insect infestation and increase Si levels in plants like brinjal (Bakhat et al., 2021). Similarly combined use of silicon rich-rice husk ash and rice husk biochar can inhibit the assemblage of arsenic in grains (Leksungnoen et al., 2019). Si obtained from RHB and RHA tends to be more soluble compared to Si from silicate minerals. This increased solubility means that Si from RHB and RHA can be easily available for plant absorption, especially under conditions where the soil pH is favorable for Si dissolution. Monosilicic acid, a form of soluble silicon, is crucial for reducing toxicity by cadmium in rice plants. When monosilicic acid is absorbed by rice plants, it is taken up and transported through the apoplastic pathway, which is the space between plant cell walls (Bakhat et al., 2021) (Table 7.1).

7.6 FACTORS INFLUENCING UPTAKE OF SILICON

Si plays a crucial role in soil dynamics and plant nutrition. Its availability can indeed be influenced by various factors such as pH, temperature, type of soil, weathering of minerals, and soil microorganisms. In soil silicon dissolution in soil depends on various conditions such as redox potential and moisture content (Savant et al., 1996). The pH of soil has a significant role in presence and availability of monosilicic acid (H_4SiO_4) in soil because adsorption or desorption of H_4SiO_4 onto soil particles is dependent on pH which is around 9. However, when the pH goes above or below this range, desorption of monosilicic acid can occur. At higher pH levels, such as around pH 9 and pH 11, monosilicic acid (H_4SiO_4) can undergo dissociation into its respective forms: disilicic acid (H_2SiO_3) and orthosilicic acid (H_2SiO_4), respectively. This dissociation occurs due to the alkaline conditions present in the soil (Tubaña & Heckman, 2015). Soil temperature also influences silicates solubility in soil and subsequently affects the availability of soluble silicon in soil. In temperate regions where soil temperatures are lower, the solubility of soil silicates tends to be higher compared to tropical regions with warmer temperatures. This higher solubility means that more silicon is available in a form that plants can uptake (Savant et al., 1996). In soil, microbes are the critical component of silicon cycling and can significantly affect the availability of Si in soil. Weathering by microbes is a geochemical process wherein living organisms contribute to the disintegration and deterioration of rocks (Burford et al., 2003; Gadd, 2007). The release of essential nutrients like phosphorus (P), potassium (K), and Si through microbial activity plays a crucial role in supporting plant development and overall ecosystem productivity (Sharma et al., 2023). The depolymerization and solubilization of naturally occurring Si by bacteria is an important process in biogeochemical cycle. Bacteria can release organic ligands or chelating compounds that bind to divalent cations present in silicate minerals. This binding weakens the mineral structure, facilitating its disintegration and liberation of silicon ions into the soil solution. Acid solubilization (organic, inorganic acids) can directly attack the mineral surfaces, promoting their dissolution and the release of silicon (Jongmans et al., 1997).

7.7 SILICON-ASSISTED RELIEF FROM ABIOTIC STRESS

7.7.1 Silicon in Alleviating Heavy Metal Toxicity

Both silicon and rhizobacteria can mitigate the adverse impact of heavy metal contamination on plants cultivated in polluted soils (Etesami & Jeong, 2018). Metal toxicity has negative impacts not only on plants but also on the entire ecosystem in which they reside. When plants are subjected to elevated levels of heavy metal concentrations in polluted environments, they often exhibit a range of adverse effects, including reduced growth and development, altered physiological processes, and high accumulation of metals within their tissues (Pang et al., 2023). Binding of these toxic metals to protein sulfhydryl (SH) groups is a key mechanism underlying the toxicity of these metals in plants and other organisms. Heavy metals like cadmium (Cd), lead (Pb), mercury (Hg), among others, exhibit a strong attraction to sulfhydryl groups found in proteins, enzymes, and various biomolecules. This binding can disrupt the structure and function of these biomolecules, leading to a range of physiological and molecular disturbances, ultimately affecting homeostasis (Riyazuddin et al., 2021). Plant growth-promoting rhizobacteria (PGPR) have demonstrated the capacity to mitigate heavy metal toxicity through diverse mechanisms. One significant mechanism by which PGPR reduce heavy metal toxicity is through metal chelation, particularly by siderophore-producing PGPR (Etesami & Jeong, 2018). Production of 1-aminocyclopropane-1-carboxylat deaminase is a fascinating mechanism utilized by PGPR to mitigate the adverse effects of heavy metal stress. The enzyme ACC deaminase, which is produced by specific strains of plant growth-promoting rhizobacteria (PGPR), plays a pivotal role in plant stress tolerance. This enzyme catalyses the cleavage of 1-aminocyclopropane-1-carboxylate (ACC) into α-ketobutyrate and ammonia. This effectively reduces the levels of ACC available for ethylene synthesis within plant tissues (Yavuz et al., 2023). Citrate and oxalate are organic acids which play a crucial role in mitigating heavy metal toxicity in plants, including Cd, by elevating the response of antioxidant enzyme (Etesami & Jeong, 2018). Bio surfactants released by plant growth-promoting rhizobacteria are amphiphilic molecules that can reduce surface tension and can also enhance the solubilization and mobilization of hydrophobic compounds, including heavy metals, in soil (Sangwan et al., 2022). EPS are complex biopolymers composed of polysaccharides, mucopolysaccharides, humic substances, proteins, and other organic components; they contribute to metal detoxification and remediation through various mechanisms which include: binding and sequestration, synthesis of metal complexes and enhanced metal immobilization (Etesami & Jeong, 2018).

Similar to plant growth-promoting rhizobacteria (PGPR), silicon has been found to employ mechanisms that mitigate the adverse effects of heavy metal stress on plants (Etesami & Jeong, 2018). The role of Si in alleviating toxicity is multifaceted and involves both internal and external mechanisms. While Si can enhance metal tolerance within plants by various means such as stimulating antioxidant defense systems and forming complexes with metals inside plant tissues, its effects on the soil environment are equally significant (Liang et al., 2005). As per recent research, Si can enhance the release of various chelating agents including flavonoids and phenolic from plant roots. These compounds have a high affinity for metal ions and can form stable complexes with them, effectively immobilizing metals in the rhizosphere and reducing their toxicity to plants. For example, acids like malate and aconitate are known to chelate metals, making them less available for absorption by roots (Dodampe Mudiyanselage, 2024). The precipitation of Si with heavy metals in the rhizosphere can change the structure and properties of plant root cell walls, including a reduction in cell wall porosity which can then release the heavy metal (Rana et al., 2023). Another mechanism through which Si reduces metal toxicity is by dilution effect. In soil Si can enhance the flow of xylem sap, the nutrient-rich fluid that moves through a plant's vascular system. This enhanced flow can effectively dilute the concentration of heavy metals absorbed by the plant roots, reducing their toxicity (Liang et al., 2005).

7.7.2 Silicon vs Arsenic Stress

Using silicon to mitigate arsenic toxicity in contaminated sites is a promising approach. Silicon deposition in plant tissues, particularly as silica bodies, can reinforce the walls of cells and strengthen plant structures. This enhancement of plant defense mechanisms makes rice plants more resilient to arsenic-induced stress (Mawia et al., 2021). Arsenic exposure can impair photosynthetic processes in plants, leading to reduced growth and productivity. Silicon supplementation likely alleviates this effect by promoting the efficient functioning of chloroplasts and photosynthetic machinery. Studies have also shown that silicon supplementation likely influences the genes which have a role in absorption and detoxification of arsenic in plants like rice. By regulating the activity of these genes, silicon can restrict the absorption of arsenic and regulate its distribution in plant tissues. Marmiroli et al. (2014) proved that the implementation of Si has beneficial impact on germination of seeds and inhibited the accumulation of arsenic in root and shoot tissues of tomato plants.

7.7.3 Silicon vs Lead Stress

Pb is one of the most concerning heavy metal pollutants due to its widespread presence in the environment and its profound detrimental impacts on both plant vitality and human well-being. Its toxicity affects not only terrestrial ecosystems but also aquatic environments, as lead can leach into water reserves and contaminate them. Lead pollution poses significant ecological concerns, as it can disrupt ecosystems and harm biodiversity. The investigation conducted by Liu et al. (2015) examined the influence of nano conjugated Si and conventional Si on the absorption and transportation of Pb in various rice cultivars. Their findings indicated that Si, whether in the form of Nano conjugated silicon or commonly occurring silicon, significantly reduced the absorption of lead in rice plants. A decrease in lead translocation from roots to shoots and subsequently from shoots to grains has been observed following the application of both Nano conjugated silicon and conventional silicon. Research by Bhat et al. (2019) provides further evidence of positive impact of Si supplementation in mitigating Pb toxicity, in cotton plants. Their experiment demonstrated that the addition of silicon significantly can reduce electrolyte leakage, Pb uptake, hydrogen peroxide (H_2O_2), and levels of malondialdehyde (MDA) in cotton plants following treatment with a combination of silicon and lead. An increase in the activity of antioxidant enzymes in both root and leaf tissues of cotton plants has also been noted in the presence of Si.

7.7.4 Silicon vs Cadmium Stress

Cadmium is an extremely hazardous environmental pollutant prevalent in developed regions. Even though it's necessary in small amounts, excessive accumulation in plants over their lifespan can pose health risks. Hu et al. (2023) proved that Si can counteract the negative effects of cadmium, correlating with reduced cadmium absorption and its translocation from roots to shoots. Silicon supplementation resulted in elevated cadmium accumulation within root tissues, concurrently mitigating its translocation to aerial parts, with a reduction observed by as much as 30%. Hou et al. (2023) similarly showed that silicon, when present with cadmium, improved overall growth parameters compared to cadmium-only treatments. This suggests silicon supplementation could be a beneficial strategy for mitigating cadmium-induced plant health problems.

7.7.5 Silicon vs Chromium Stress

In recent times, environmental contamination with chromium (Cr) due to industrial activities and human actions has emerged as a significant concern. The global release of chromium surpasses that of other metals like Cd, Hg, and Pb (Gautam et al., 2016). Cr is found in multiple forms, but the most harmful to organisms is Cr (VI) due to its ability to readily traverse cell membranes, enter the cytoplasm, and disrupt biochemical activities (Bakshi & Panigrahi, 2022). Cr induced negative effects on plans include poor germination of seeds, damage by (ROS), impaired nutrient absorption, interference with processes of photosynthesis, and alterations in nitrogen metabolism (Bhat et al., 2023). Increase in Cr contamination to enhance crop production and reduce health risks is associated with consuming crops that contain chromium in low levels (there are less health risks for those crops which contain too less Cr concentration). Tripathi et al. (2015) observed that Si nutrition could play a crucial role in mitigating the toxic effects of Cr in wheat plants. As Si is known to enhance plant resilience to various stressors, including heavy metals like chromium, so providing silicon supplements to wheat plants may reduce the negative effects of chromium toxicity, which leads to improved plant growth and health. The findings from Ali et al. (2015) further support the possible function of Si in remedying adverse effects of chromium in crops. Their study suggests that the addition of silicon had noticeable positive effects on various physiological parameters of plants, including growth, net carbon assimilation rate, intracellular carbon dioxide levels, leaf stomatal aperture, and water loss rate through transpiration. The ameliorative effects of Si on metal phytotoxicity extend beyond its actions within the plant itself. Another important aspect is the effect of silicon on the availability of metals in the soil. Silicon-treated soils often exhibit reduced bioavailability of metals like Cr, which means that plants grown in such soils have less absorption of toxic metals (Liang et al., 2005).

7.7.6 Silicon in Alleviating Salinity Stress

Application of Si is crucial for enhancing plant resistance to environmental stresses, especially to salinity. Salinity, a significant abiotic stressor for crop plants, has led to the degradation of arable area by at least 20%. However, silicon employs different ways to mitigate salt stress. These ways include maintaining plant water balance, increasing mineral uptake, reducing ion toxicity and enhancing mineral uptake assimilation, enhancing the production solutes and phytohormones, alleviating oxidative stress, and influencing gene expression.

One fundamental mechanism through which silicon enhances plants' tolerance to salt stress involves the mitigation of sodium ion (Na^+) uptake and accumulation. This reduction in Na^+ uptake is facilitated by several means:

- Decreasing the permeability of root cell membranes: Silicon helps in fortifying the membranes of root cells, making them less permeable to sodium ions.
- Enhancing the architecture and resilience of root cell structures: By enhancing the structural integrity of root cells, silicon contributes to better resistance against the harmful effects of salt stress (Liu et al., 2019).
- Facilitating the activation of the root plasma membrane H^+-ATPase: This enzyme is important for pumping sodium ions out of the cell, thereby reducing their accumulation within the plant.
- Assisting in the removal of Na^+ ions from the cell: Silicon assists in the efficient export of sodium ions from the cell, further reducing their harmful effects.

Moreover, silicon application has been shown to address nutritional imbalances induced by salt stress. Studies have reported increased levels of essential minerals such as calcium, magnesium, potassium, and phosphorus in plants treated with silicon under the conditions of salt. Moreover, the external supplementation of silicon improves exchange of gases in diverse species of plant experiencing stress by salinity (Etesami & Jeong, 2018). The accumulation of silicon as silica gel within root cell walls serves as a crucial mechanism in alleviating the detrimental effects of salt stress on plants. This mechanism offers attachment points for salts, inhibiting their excessive movement to the aerial parts of the plant due to which plant growth is enhanced. Studies, such as those by Adil et al. (2023), have demonstrated the beneficial impact of Si application on plant health due to salt stress. For instance, in barley roots, the introduction of silicon resulted in a notable rise in the activity levels of several antioxidant enzymes, such as peroxidase, superoxide dismutase, catalase, and glutathione reductase, as documented by Liang et al. (2003). Similarly, Amerian et al. (2024) observed similar effects in cucumber plants subjected to salt stress. The application of silicon enhances the expression level of the aquaporin and augments water absorption in cucumber plants experiencing salt stress. Silicon has the added capability of triggering the genes linked to salt stress responses in tomato plants experiencing salt stress. These genes include leDREB-3, leDREB-1, and leDREB-2 and are involved in responses related to salt stress (Dhiman et al., 2021). Silicon-treated salt-stressed plants exhibit higher efficiency in antioxidant defense enzymes, including superoxide dismutase (leSOD), peroxidase (leAPX), and catalase (leCAT). These enzymes play crucial roles in neutralizing (ROS) generated by salt stress (Zargar et al., 2019).

7.7.7 Silicon in Alleviating Drought Stress

Drought is a significant challenge for agriculture worldwide. When plants experience drought stress, their growth and productivity are severely impacted. Drought conditions occur when there is insufficient water available for plant uptake to fulfill the demand of the plant. In plants this water deficit leads to various physiological changes, such as stomatal closure, reduced photosynthesis, decreased cell expansion, and altered metabolism (Kumar et al., 2023). The consequences of drought stress can be profound, resulting in yield losses, poor crop quality, and even crop failure. Agricultural crops vary in their tolerance to drought, with some being more resilient than others. However, even drought-tolerant crops can suffer reduced yields under prolonged or severe drought conditions. Several mechanisms by which Si application increases the ability of plants to withstand in drought stress are discussed next.

- Stomatal Regulation and Transpiration: Drought stress typically leads to stomatal closure and reduced transpiration rates, which can hinder plant growth. Si application may help mitigate this effect by up-regulating aquaporin genes (such as PIP) and preventing the inhibition of aquaporin activity induced by reactive oxygen species (ROS).
- Osmo-regulation: Application of silicon under drought stress conditions can bolster osmoregulation by augmenting the assembly of amino acids and other solutes in the sap of xylem, consequently sustaining osmotic

equilibrium. It may also activate the translocation of potassium ions (K^+) to the xylem sap via genes like SKOR, which helps maintain water balance in the plant.

- Deposition of Silicon in Cell Walls: In the realm of plant physiology, silicon deposition within cell walls emerges as a pivotal mechanism in reducing drought tolerance. This process includes the accumulation of silicon dioxide (SiO2) within cellular structures, which reduces transpiration rates and an augmentation of internal water retention capacities. Nedukha (2022) showed how the hydrophilic properties of $SiO_2.nH_2O$ play a crucial role in enhancing water retention within plant tissues, thereby protecting them against the adverse effects of drought stress.
- Osmotic Adjustment: Osmotic adjustment plays a vital role in maintaining moisture within plant cells even when water availability is low. Silicon application supports this process, contributing to drought tolerance (He et al., 2024).
- Modification in Root Morphology: The alteration of root structure facilitated by silicon enhances plants' capacity to absorb water, as reported by Wang et al. (2021a). This could lead to improved drought tolerance by enhancing water absorption from the soil.
- Enhanced Structural Stability: Silicon binding with cell wall hemicellulose improves structural stability, which can be particularly beneficial under water shortage conditions. This enhances the ability of plants to tolerate the stress by drought (Ganie & Ahammed, 2021).
- Improvement in Seed Germination and Seedling Protection: The external administration of silicon enhances seed germination and biochemical functions, and shields seedlings from oxidative stress by strengthening antioxidant defense mechanisms.
- Increased Calcium Concentration: The provision of silicon leads to an elevation in calcium levels, crucial for upholding permeability of membrane and stability during periods of drought stress (Islam et al., 2024).

7.7.8 Silicon in Alleviating Radiation Stress

Research conducted by Tripathi et al. (2017) has showed the positive impacts of exogenous Si application in alleviating the effects of UV-radiations on various crops, such as wheat and maize. UV-B radiation can induce oxidative stress in plants, resulting in damage to the components of cells like DNA, lipids, and proteins (Bokor et al., 2021; Tripathi et al., 2017). Tripathi et al. (2017) noted a considerable improvement in the resilience of wheat seedlings to UV-B stress following the administration of silicon nanoparticles. This discovery suggests that silicon nanoparticles may offer superior efficacy in reducing the detrimental effects of UV-B radiation compared to conventional forms of silicon application.

7.7.9 Silicon in Alleviating Wound Stress

Wounding stress occurs when plants experience physical injury, which can be caused by various factors such as wind, herbivore attack, mechanical damage, or human activities like pruning. When plant tissues are wounded, it can lead to cell damage and the disruption of cellular membranes. This damage exposes the plant to oxidative stress, where reactive oxygen species accumulate, leading to further cellular damage. Silicon-treated wounded rice plants exhibit significantly higher antioxidant activities of catalase and peroxidase compared to control plants (Islam et al., 2023). Catalase and peroxidase, pivotal antioxidant enzymes responsible for clearing (ROS) and safeguarding cells against harm caused by reactive oxygen species, play crucial roles. The heightened activities of these enzymes observed in silicon-treated plants imply that silicon bolsters the antioxidant defense mechanism, thereby aiding in the reduction of this stress triggered by injuries. Research conducted by Khan et al. (2022) illustrates that plants treated with silicon manifest heightened activity of antioxidant enzymes in reaction to wounding stress.

7.7.10 Silicon in Alleviating Thermal Stress

Temperature is undeniably a crucial environmental factor that significantly impacts development of plants. Plants are highly sensitive to changes in temperature, and variations can affect various physiological processes, including photosynthesis, respiration, flowering, and fruiting (Raza et al., 2019). Stress, whether it's from cold or heat, can disrupt the delicate balance between reactive oxygen species and antioxidant activity in plants. Silicon, while not classified as a vital nutrient for plant development, has been found to confer several benefits, including enhanced tolerance to abiotic stresses like heat (Epstein, 1994). For example, leakage of electrolytes, which is a common indicator of cellular membrane damage, can occur under stress conditions such as high temperatures. The fact that supplementation of Si can lower electrolyte leakage in rice suggests its involvement in maintaining the thermal stability of membrane lipid (Bishnoi et al., 2023). The findings cited from Malhotra and Kapoor (2019) indicate that the incorporation of silicon into the vessels of xylem is pivotal in mitigating vessel compression, especially under elevated transpiration rates triggered by high temperature. This finding is supported by earlier studies of Liang et al. Examples of the beneficial impact of silicon supplementation on plants under high-temperature conditions include reduction in transpiration loss, increased antioxidant activity, and enhanced enzyme concentrations. Increasing the concentration of ribulose bisphosphatase carboxylase enzyme in tissues of leaves could potentially enhance the efficiency of carbon dioxide metabolism in plants, particularly in stressful environments where CO_2 availability might be limited. The study by Hossain and Islam (2021) suggests that silicon treatment may enhance the antioxidant defense system in Scarlet sage (Salvia splendens) under high-temperature conditions by increasing the activities of superoxide dismutase (SOD), ascorbate peroxidase (APX), and guaiacol peroxidase (GPX).

Cold stress and limited light exposure can induce an excess production of reactive oxidative species within plant cells. These reactive species which encompass superoxide radicals, hydrogen peroxide, and hydroxyl radicals, are notably reactive entities capable of inflicting oxidative harm upon diverse cellular constituents like lipid within membrane, proteins, and nucleic acids (Checa & Aran, 2020). Studies like the one by Souri et al. (2021) demonstrate the potential of Si supplementation to reduce the detrimental effects of low temperature on plants. Silicon has been observed to mitigate oxidative damage induced by chilling or frost conditions through the regulation of the antioxidant defense system (Ma et al., 2006a, b, 2007a, 2007b). Silicon deposition in the cuticle forms a protective layer that can act as a barrier against various environmental stresses, including low temperatures and frost. This reinforced cuticular layer helps to reduce water loss and provides mechanical support to plant tissues (Siam et al., 2022). Extreme weather conditions, cold temperatures, typhoons, and limited sunlight lead to lodging and decreased rice yield. The application of silicon fertilizer reduces such harm by strengthening rigidity and mechanical resilience through the silicification of shoots, thus preventing lodging (Kovács et al., 2022). When subjected to silicon treatment, plant leaves, including those of rice, cucumber, and maize, cultivated hydroponically in low temperatures (0-4 °C), demonstrated enhanced resilience against cold-induced wilting and showed improved root growth for nutrient uptake. The alleviation of freezing damage facilitated by silicon is probably attributable to increased antioxidant defense activities, leading to decreased oxidative damage to membranes and improved water retention in leaf tissues (Bashir et al., 2023).

7.7.11 Silicon Mediated Alleviation of Biotic Stress

Silicon is renowned for its protective effects against a diverse group of fungal and bacterial pathogens, facilitating the development of monocotyledonous plants by imparting rigidness in stems. Examination of Si-treated plants utilizing chemical analysis and energy dispersive X-ray microanalysis uncovered Si presence in awns, improving mechanical stability (Seal et al., 2018). Two principal mechanisms are there for Si-amplified resistance to diseases and pests. Primarily, Si could accumulate below the cuticle or on the upper parts of the leaf, constituting a physical barrier that impedes pathogen infiltration, thus curtailing infections (Singh, 2012). Secondarily, Si may serve as a signalling agent to instigate the synthesis of defense molecules, encompassing escalated synthesis of phenolic compounds, liginin, and phytoalexins (Souri et al., 2021).

The supplementation of silicon augmented cucumber plants' resilience against powdery mildew fungus. Furthermore, it heightened the photosynthetic processes by amplifying the number of chlorophyll molecules and RuBP carboxylase activity, while also activating defense mechanisms that reinforced resistance (Singla et al., 2020). Silicon also prompted the accumulation of lignin, phenolic compounds, and phytoalexins, thus amplifying plant defense mechanisms (Gulzar et al., 2021). Notably, callose, and methylaconitate production were discerned in the epidermal cells of plants treated with silicon fertilizer. The positive impacts of silicon against pathogens can be seen in wheat, mitigating infections like powdery mildew and eyespot (Islam et al., 2023). Silicon also confers resistance to rice blast, fusarium wilt, and leaf spot in rice plant (Rachappanavar et al., 2024). During attacks by pathogens, damaged tissues generate stress signals including jasmonic acid, salicylic acid, and ethylene, in addition to resistance against microbial compounds. Consequently, silicic acid assumes a beneficial role in both local and systemic resistance, amplifying the secretion of stress hormones without itself functioning as a secondary messenger (Guin et al., 2024).

Silicon also mitigates crop damage from animals, particularly insects. Grasses treated with silicon fertilizer are less susceptible to grazing by animals. The efficacy of silicon in mitigating attacks from various species, including insect borers, rice leaf hoppers, brown leaf hoppers, and aphids, has been substantiated (Guntzer et al., 2012).

7.8 CONCLUSION

The positive impact of Si supplementation in various crops, as documented in recent decades, has led to an increased use of silicon-based fertilizers. However, incorporating silicon into the fertilizer routine of various crop systems continues to prompt several apprehensions. One of the key priorities involves ensuring the efficient dissolution of silicon fertilizer to enable its accessibility to plants, determining the optimal dosage for different crops, addressing issues of sustainability, and assessing the long-term environmental effects. This conversation also delves into silicon solubilizing bacteria (SSB), their possible uses, and existing technological progressions. Furthermore, it investigates the present comprehension of silicon absorption, deposition, and the ensuing advantages for plants. pH plays a critical role in agricultural soils, impacting both the solubilization of silicate and the formation of a variety of silicon compounds. Multiple research findings have also indicated the significance of substances like Indole-3-Acetic Acid (IAA) and organic acids such as malic and oxalic acids, which are generated by silicon solubilizing bacteria, in the process of silicate solubilization.

REFERENCES

Aagaard, P., & Helgeson, H. C. (1982). Thermodynamic and kinetic constraints on reaction rates among minerals and aqueous solutions; I, Theoretical considerations. *American Journal of Science*, *282*, 810–818.

Adil, M., Shah, A. N., Khan, A. N., Younas, T., Mehmood, M. S., Mahmood, A., Muhammad, R., Asghar, R., & Javed, M. S. (2023). Amelioration of harmful effects of soil salinity on plants through silicon application: A review. *Pakistan Journal of Botany*, *55*(1), 9–18.

Ako Andrew, A. (2011). Hydrological study on groundwater in the Banana Plain and Mount Cameroon area-Cameroon volcanic line (Ph.D thesis). CVL.

Ali, N., Rethore, E., Yvin, J. C., & Hosseini, S. A. (2020). The regulatory role of silicon in mitigating plant nutritional stresses. *Plants*, *9*(12), 1779.

Ali, S., Chaudhary, A., Rizwan, M., Anwar, H. T., Adrees, M., Farid, M., & Anjum, S. A. (2015). Alleviation of chromium toxicity by glycinebetaine is related to elevated antioxidant enzymes and suppressed chromium uptake and oxidative stress in wheat (Triticum aestivum L.). *Environmental Science and Pollution Research*, *22*, 10669–10678.

Amerian, M., Palangi, A., Gohari, G., & Ntatsi, G. (2024). Enhancing salinity tolerance in cucumber through Selenium biofortification and grafting. *BMC Plant Biology*, *24*(1), 24.

Anggria, L., & Masunaga, T. (2021). A method for production of pure silica as fertilizer from industrial waste material. *IOP Conference Series: Earth and Environmental Science*, *648*, 012213.

Bakhat, H. F., Bibi, N., Fahad, S., Hammad, H. M., Abbas, S., Shah, G. M., Zakir, A., Murtaza, B., & Ashraf, M. R. (2021). Rice husk bio-char improves brinjal growth, decreases insect infestation by enhancing silicon uptake. *Silicon*, *13*, 3351–3360.

Bakshi, A., & Panigrahi, A. K. (2022). Chromium contamination in soil and its bioremediation: An overview. In *Advances in bioremediation and phytoremediation for sustainable soil management: Principles, monitoring and remediation* (pp. 229–248). Springer Nature.

Barberena, O. A. M. (2014). *Effect of soil applied potassium silicate on papaya (Carica papaya L.) Plant growth, development, yields, physiology and, postharvest fruit quality*. University of Florida.

Bashan, Y., & Holguin, G. (1998). Proposal for the division of plant growth-promoting rhizobacteria into two classifications: Biocontrol-PGPB (plant growth-promoting bacteria) and PGPB. *Soil Biology and Biochemistry*, *30*, 1225–1228.

Bashir, S., & John, R. (2023). Alleviation of chilling stress by supplementation of brassinosteroid and silicon in Solanum lycopersicum L. *Plant and Soil*, *486*(1), 165–181.

Bashir, Z. (2021b). Characterization of potassium solubilizing bacteria from rhizospheric soils of cherry (*Prunusavium*). *Research Journal of Agricultural Sciences*, *12*(2), 494–496.

Bashir, Z., Hamid, B., Zargar, M. Y., & Kundoo, A. A. (2021a). Morphological and biochemical characterization of potassium solubilizing bacteria from Rhizospheric soils of Pinus (*Pinus wallichiana*). *Journal of Research and Development*, *21*, 58–63.

Bashir, Z., Zargar, M. Y., Baba, Z. A., & Mohiddin, F. A. (2018). Solubilization of potassium containing mineral by bacteria from Kashmir iris (*Iris kashmiriana*) rhizosphere. *The Pharma Innovation Journal*, *17*(4), 747–750.

Bashir, Z., Zargar, M. Y., Baba, Z. A., Mohiddin, F. A., Peer, F. A., Mir, S. A., & Najar, G. R. (2018a). Isolation and characterization of phosphorus solubilizing bacteria (PSB) from rhizospheric soils of apple (*Malus Domestica*). *Journal of Pharmacognosy and Phytochemistry*, *7*(2), 2997–3000.

Bashir, Z., Zargar, M. Y., Bhat, T. A., Kousar, S., Baba, Z. A., & Mohiddin, F. A. (2018b). Solubilization of phosphorus containing mineral by bacteria from Rhizospheric region of Walnut (*Juglans regia*). *International Journal of Current Microbiology and Applied Sciences*, *7*(8), 2391–2398.

Bashir, Z., Zargar, M. Y., Mohiddin, F. A., Baba, Z. A., & Hamid, B. (2019). Isolation and biochemical characterization of phosphorus solubilizing bacteria from Rhizospheric region of Apricot and Peach. *Journal of Research and Development University of Kashmir*, *19*, 972–5407.

Bashir, Z., Zargar, M. Y., Mohiddin, F. A., Kousar, S., Husain, M., & Rasool, F. (2017). Phosphorus solubilizing microorganisms: Mechanism and diversity. *International Journal of Communication Systems*, *5*(5), 666–673.

Bashir, Z., Zargar, M. Y., & Vishwakarma, D. K. (2019b). Potassium solubilizing microorganisms for sustainable agriculture. *Applied Agricultural Practices for Mitigating Climate Change*, *2*, 17–28.

Bhat, J. A., Shivaraj, S. M., Singh, P., Navadagi, D. B., Tripathi, D. K., Dash, P. K., & Deshmukh, R. (2019). Role of silicon in mitigation of heavy metal stresses in crop plants. *Plants*, *8*(3), 71.

Bhat, K. A., Batool, A., Mansoor, M., Manzoor, M., Bashir, Z., Nazir, M., & Zargar, S. M. (2022). Silicon- and nanosilicon-mediated disease resistance in crop plants. In Hassan Etesami, Abdullah H. Al Saeedi, Hassan El-Ramady, Masayuki Fujita. Mohammad Pessarakli, Mohammad Anwar Hossain (Eds.). *Silicon and nano-silicon in environmental stress management and crop quality improvement* (pp. 193–205). Academic press is an inprint of Elsevier, London (UK).

Bhat, M. A., Mishra, A. K., Jan, S., Shah, S. A., Asgher, M., Rahman, S., & Jan, A. T. (2023). Chromium and plant health: A prospective study of toxicity to remediation and restoration of the environment. *South African Journal of Botany*, *162*, 505–518.

Bishnoi, A., Jangir, P., Shekhawat, P. K., Ram, H., & Soni, P (2023). Silicon supplementation as a promising approach to induce thermotolerance in plants: Current understanding and future perspectives. *Journal of Soil Science and Plant Nutrition*, *23*(1), 34–55.

Bist, V., Niranjan, A., Ranjan, M., Lehri, A., Seem, K., & Srivastava, S (2020). Silicon-solubilizing media and its implication for characterization of bacteria to mitigate biotic stress. *Frontiers in Plant Science, 11*, 28.

Bokor, B., Santos, C. S., Kostoláni, D., Machado, J., da Silva, M. N., Carvalho, S. M., & Vasconcelos, M. W. (2021). Mitigation of climate change and environmental hazards in plants: Potential role of the beneficial metalloid silicon. *Journal of Hazardous Materials, 416*, 126193.

Borawska-Jarmułowicz, B., Mastalerczuk, G., Janicka, M., & Wróbel, B. (2022). Effect of silicon-containing fertilizers on the nutritional value of grass–legume mixtures on temporary grasslands. *Agriculture, 12*, 145.

Burford, E. P., Fomina, M., & Gadd, G. M. (2003). Fungal involvement in bio weathering and biotransformation of rocks and minerals. *Mineralogical Magazine, 67*, 1127–1155. https://doi.org/10.1180/0026461036760154

Carmen, B., & Roberto, D. (2011). Soil bacteria support and protect plants against abiotic stresses. In A. Shan (Ed.). *abiotic stress in plants mechanisms and adaptations* (pp. 143–170). InTech.

Casey, W. H., Kinrade, S. D., & Knight, C. T. (2003). Aqueous silicate complexes in wheat, Triticum aestivum L. *Plant, Cell & Environment, 27*, 51–54.

Cassman, K. G., De Datta, S. K., Oik, D. C., Alcantara, J., Samson, M., Descalsota, J., & Dizon, M. (1995). Yield decline and the nitrogen economy of long-term experiments on continuous, irrigated rice systems in the tropics. In R. Lal & B. A. Stewart (Eds.). *Soil Management Experimental Basis for Sustainability and Environmental Quality* (pp. 181–222). CRC Publisher.

Chaudhary, D. S., Jollands, M. C., & Cser, F. (2002). Understanding rice hull ash as fillers in polymers: A review. *Silicon Chemistry, 1*, 281–289.

Checa, J., & Aran, J. M. (2020). Reactive oxygen species: Drivers of physiological and pathological processes. *Journal of Inflammation Research*, 1057–1073.

Chen, J., Caldwell, R. D., Robinson, C. A., & Stein, K. R. (2000). Silicon: The estranged medium element. In *Institute of Food and Agricultural Sciences* (p. 341). University of Florida.

Chiba, Y., Yamaji, N., & Mitani, N. (2009). HvLsi1 is a silicon influx transporter in barley. *The Plant Journal, 57*, 810–818.

Chigwaya, K. (2016). *Moisture loss studies in nectarine (Prunus persica var. necterina)*. (Doctoral diss.), Stellenbosch University.

Cornelis, J. T., Delvaux, B., Georg, R. B., Lucas, Y., Ranger, J., & Opfergelt, S. (2011). Tracing the origin of dissolved silicon transferred from various soil-plant systems towards rivers. *Biogeography Science, 8*, 89–112.

Cotterill, J. V., Watkins, R. W., Brennon, C. B., & Cowan, D. P. (2007). Boosting silica levels in wheat leaves reduces grazing by rabbits. *Pest Management Science, 63*, 247–253. https://doi.org/10.1002/ps.1302

Cruz, J. A., Tubana, B. S., Fultz, L. M., Dalen, M. S., & Ham, J. H. (2022). Identification and profiling of silicate-solubilizing bacteria for plant growth-promoting traits and rhizosphere competence. *Rhizosphere, 23*, 10056.

Das, S., Gwon, H. S., Khan, M. I., Van Nostrand, J. D., Alam, M. A., & Kim, P. J. (2019). Taxonomic and functional responses of soil microbial communities to slag-based fertilizer amendment in rice cropping systems. *Environment International, 127*, 531–539.

Datnoff, L. E., Deren, C. W., & Snyder, G. H. (1997). Silicon fertilization for disease management of rice in Florida. *Crop Protection, 16*(6), 525–531.

Deshmukh, R., & Belanger, R. R. (2016). Molecular evolution of aquaporins and silicon influx in plants. *Functional Ecology, 30*, 1277–1285.

Devanna, B. N., Mandlik, R., Raturi, G., Sudhakaran, S. S., Sharma, Y., Sharma, S., Rana, N., Bansal, R., Barvkar, V., & Tripathi, D. K. (2021). Versatile role of silicon in cereals: Health benefits, uptake mechanism, and evolution. *Plant Physiology and Biochemistry, 165*, 173–186.

Dhiman, P., Rajora, N., Bhardwaj, S., Sudhakaran, S. S., Kumar, A., Raturi, G., & Deshmukh, R. (2021). Fascinating role of silicon to combat salinity stress in plants: An updated overview. *Plant Physiology and Biochemistry, 162*, 110–123.

Dodampe Mudiyanselage, S. N. M. D. (2024). *Potential use of reclaimed struvite as a phosphorus fertilizer source and effects of low molecular weight organic acids (LMWOAs) in the rhizosphere on its performance in three contrasting soils* (Doctoral diss.).

Ebe, S., Ohike, T., Okanami, M., & Ano, T. (2019). Components of rice husk biochar in promoting the growth, sporulation and iturin A production of *Bacillus* sp. strain IA. *Zeitschrift für Naturforschung C, 74*, 211–217.

El-Ramady, H., Verma, K. K., Rajput, V. D., Minkina, T., Elbehery, F., Elbasiony, H., Elsakhawy, T., Omara, A. E.-D., & Amer, M. (2022). Sources of silicon and nano-silicon in soils and plants. In Hassan Etesami, Abdullah H. Al Saeedi, Hassan El-Ramady, Masayuki Fujita. Mohammad Pessarakli, Mohammad Anwar Hossain (Eds.). *Silicon and nano-silicon in environmental stress management and crop quality improvement* (pp. 1–15). Academic press is an inprint of elsevier, London (UK).

Epstein, E. (1994). The anomaly of silicon in plant biology. *Proceedings of the National Academy of Sciences of the United States of America, 91*, 11–17.

Epstein, E., & Bloom, A. J. (2006). *Mineral nutrition of plants: Principles and perspectives* (2nd ed.). John Wiley and Sons.

Etesami, H., & Beattie, G. A. (2017). Plant-microbe interactions in adaptation of agricultural crops to abiotic stress conditions. In *Probiotics and plant health* (pp. 163–200). Springer.

Etesami, H., & Jeong, B. R. (2018). Silicon (Si): Review and future prospects on the action mechanisms in alleviating biotic and abiotic stresses in plants. *Ecotoxicology and Environmental Safety, 147*, 881–896.

Farooq, M. A., & Dietz, K. J. (2015). Silicon as versatile player in plant and human biology: Overlooked and poorly understood. *Frontiers in Plant Science, 6*, 994.

Fauteux, F., Rémus-Borel, W., Menzies, J., & Belanger, R. (2005). Silicon and plant disease resistance against pathogenic fungi. *FEMS Microbiology Letters, 249*, 1–6.

Gadd, G. M. (2007). Geomycology: Biogeochemical transformations of rocks, minerals, metals and radionuclides by fungi, bio weathering and bioremediation. *Mycological Research, 111*, 3–49. https://doi.org/10.1016/j.mycres.2006.12.001

Ganie, S. A., & Ahammed, G. J. (2021). Dynamics of cell wall structure and related genomic resources for drought tolerance in rice. *Plant Cell Reports, 40*, 437–459.

Gascho, G. J. (2001). Silicon sources for agriculture. In L. E. Datnoff, G. H. Snyder, & G. H. Korndorfer (Eds.), *Silicon in agriculture* (pp. 197–207). Elsevier.

Gautam, M., Pandey, D., Agrawal, S. B., & Agrawal, M. (2016). Metals from mining and metallurgical industries and their toxicological impacts on plants. In *Plant responses to xenobiotics* (pp. 231–272). Springer.

Gerard, F., Mayer, K., Hodson, M., & Ranger, J. (2008). Modelling the biogeochemical cycle of silicon in soils: Application to a temperate forest ecosystem. *Geochimica et Cosmochimica Acta, 72*, 741–758.

Glick, B. R. (2014). Bacteria with ACC deaminase can promote plant growth and help to feed the world. *Microbiological Research, 169*, 30–39.

Guin, A., Korav, S., Banik, B., & Unjia, D. (2024). Current knowledge on the role of salicylic acid for stress tolerance on field crops. *International Journal of Environment and Climate Change, 14*(4), 447–457.

Gulzar, N., Kamili, A. N., & Shah, M. A. (2021). Silicon, the multifunctional element in reducing biotic and abiotic stress in plants. *International Research Journal of Plant Science*, 12 (4), 1–9, DOI: http:/dx.doi.org/10.14303/irjps.2021.20

Guntzer, F., Keller, C., & Meunier, J. D. (2012). Benefits of plant silicon for crops: A review. *Agronomy for Sustainable Development, 32*, 201–213.

Gwon, H. S., Khan, M. I., Alam, M. A., Das, S., & Kim, P. J. (2018). Environmental risk assessment of steel-making slags and the potential use of LD slag in mitigating methane emissions and the grain arsenic level in rice (*Oryza sativa* L). *Journal of Hazardous Materials, 353*, 236–243.

Hamid, B., & Bashir, Z. (2019a). Potassium solubilization microorganisms: An Alternative technology to chemical fertilizers. *Journal of Research and Development University of Kashmir, 19*, 0972–5407.

Han, Y., Lei, W., Wen, L., & Hou, M. (2015). Silicon-mediated resistance in a susceptible rice variety to the rice leaf folder, *Cnaphalocrocis medinalis* Guenée (Lepidoptera: Pyralidae). *PLoS ONE, 10*, e0120557.

Hawerroth, C., Araujo, L., Bermúdez-Cardona, M. B., Silveira, P. R., Wordell Filho, J. A., & Rodrigues, F. A. (2019). Silicon-mediated maize resistance to macrospora leaf spot. *Tropical Plant Pathology, 44*, 192–196.

He, J., Ng, K., Qin, L., Shen, Y., Rahardjo, H., Wang, C. L., & Ghosh, S. (2024). Photosynthetic gas exchange, plant water relations and osmotic adjustment of three tropical perennials during drought stress and re-watering. *Plos One, 19*(2), e0298908.

Heckman, J. (2012). Silicon and soil fertility. *The Soil Profile, 20*, 1–12.

Heckman, J. (2013). Silicon: A beneficial substance. *Better Crops with Plant Food, 97*(4), 14–16.

Helgeson, H. C., Murphy, W. M., & Aagaard, P. (1984). Thermodynamic and kinetic constraints on reaction rates among minerals and aqueous solutions. II. Rate constants, effective surface area, and the hydrolysis of feldspar. *Geochimica et Cosmochimica Acta, 48*, 2405–2432.

Hildebrand, M., Volcani, B. E., Gassmann, W., & Schroeder, J. I. (1997). A gene family of silicon transporters. *Nature, 385*, 688–689.

Hodson, M., White, P., Mead, A., & Broadley, M. (2005). Phylogenetic variation in the silicon composition of plants. *Annals of Botany, 96*, 1027–1046.

Hossain, A., & Islam, T. (2021). Silicon and selenium transporters in plants under abiotic stresses. In *Metal and nutrient transporters in abiotic stress* (87–116). Academic Press.

Hou, L., Ji, S., Zhang, Y., Wu, X., Zhang, L., & Liu, P. (2023). The mechanism of silicon on alleviating cadmium toxicity in plants: A review. *Frontiers in Plant Science, 14*, 1141138.

Hu, Y., Zhou, X., Shi, A., Yu, Y., Rensing, C., Zhang, T., & Yang, W. (2023). Exogenous silicon promotes cadmium (Cd) accumulation in Sedum alfredii Hance by enhancing Cd uptake and alleviating Cd toxicity. *Frontiers in Plant Science, 14*, 1134370.

Huang, H., Rizwan, M., Li, M., Song, F., Zhou, S., He, X., Ding, R., Dai, Z., Yuan, Y., & Cao, M. (2019). Comparative efficacy of organic and inorganic silicon fertilizers on antioxidant response, Cd/Pb accumulation and health risk assessment in wheat (*Triticum aestivum* L.). *Environmental Pollution, 255*, 113146.

Ishiguro, K. (2001). Review of research in Japan on the roles of silicon in conferring resistance against rice blast. *Studies in Plant Science, 8*, 277–291.

Islam, M. S., Islam, M. R., Hasan, M. K., Hafeez, A. G., Chowdhury, M. K., Pramanik, M. H., & El Sabagh, A. (2024). Salinity stress in maize: Consequences, tolerance mechanisms, and management strategies. *OBM Genetics 2024, 8*(2), doi:10.21926/obm.genet.2402232

Islam, W., Tauqeer, A., Waheed, A., Ali, H., & Zeng, F. (2023). Silicon in plants mitigates damage against pathogens and insect pests. In Renato de Mello Prado (Ed.). *Benefits of silicon in the nutrition of plants* (pp. 347–378). Springer.

Jarvis, S. C. (1987). The uptake and transport of silicon by perennial ryegrass and wheat. *Plant Soil, 97*, 429–437.

Jongmans, A., Van Breemen, N., Lundstrom, U., Van Hees, P., Finlay, R., Srinivasan, M., Unestam, T., Giesler, R., Melkerud, P.-A., & Olsson, M. (1997). Rock-eating fungi. *Nature, 389*, 682–683.

Kang, S. M., Waqas, M., Shahzad, R., You, Y. H., Asaf, S., Khan, M. A., & Lee, I. J. (2017). Isolation and characterization of a novel silicate-solubilizing bacterial strain *Burkholderiaeburnea* CS4-2 that promotes growth of japonica rice (*Oryza sativa* L. cv. Dongjin). *Soil Science and Plant Nutrition*, *63*(3), 233–241.

Khan, A., Khan, V., Pandey, K., Sopory, S. K., & Sanan-Mishra, N. (2022). Thermo-priming mediated cellular networks for abiotic stress management in plants. *Frontiers in Plant Science*, *13*, 866409.

Klotzbucher, T., Marxen, A., Vetterlein, D., Schneiker, J., Türke, M., Sinh, N. V., Manh, N. H., Chien, H. V. (2014). Plant-Available silicon in paddy soils as a key factor for sustainable rice production in Southeast Asia. *Basic and Applied Ecology*, *16*(8), 665–673.

Kovács, S., Kutasy, E., & Csajbók, J. (2022). The multiple role of silicon nutrition in alleviating environmental stresses in sustainable crop production. *Plants*, *11*(9), 1223.

Kumar, A., Sengar, R. S., Pathak, R. K., & Singh, A. K. (2023). Integrated approaches to develop drought-tolerant rice: Demand of era for global food security. *Journal of Plant Growth Regulation*, *42*(1), 96–120.

Kumar, N., Kumawat, S., Khatri, P., Singla, P., Tandon, G., Bhatt, V., Shinde, S., Patil, G., Sonah, H., & Deshmukh, R. (2020). Understanding aquaporin transport system in highly stress-tolerant and medicinal plant species Jujube (*Ziziphus jujuba* Mill.). *Journal of Biotechnology*, *324*, 103.

Lanning, F. C., Ponnaiya, B. W. X., & Crumpton, C. F. (1958). The chemical nature of silica in plants. *Plant Physiology*, *33*, 339–343.

Lee, J. U., & Fein, J. B. (2000). Experimental study of the effects of Bacillus subtilis on gibbsite dissolution rates under near-neutral pH and nutrient-poor conditions. *Chemical Geology*, *166*, 193–202.

Lee, K. E., Adhikari, A., Kang, S. M., You, Y. H., Joo, G. J., Kim, J. H., & Lee, I. J. (2019). Isolation and characterization of the high silicate and phosphate solubilizing novel strain *Enterobacter ludwigii GAK2* that promotes growth in rice plants. *Agronomy*, *9*(3), 144.

Leksungnoen, P., Wisawapipat, W., Ketrot, D., Aramrak, S., Nookabkaew, S., Rangkadilok, N., & Satayavivad, J. (2019). Biochar and ash derived from silicon-rich rice husk decrease inorganic arsenic species in rice grain. *Science of the Total Environment*, *684*, 360–370.

Li, F. C., Li, S., Yang, Y. Z., & Cheng, L. J. (2006). Advances in the study of weathering products of primary silicate minerals, exemplified by mica and feldspar. *Acta Petrol Mineral*, *25*, 440–448.

Liang, Y., Chen, Q., Liu, Q., Zhang, W., & Ding, R. (2003). Exogenous silicon (Si) increases antioxidant enzyme activity and reduces lipid peroxidation in roots of salt-stressed barley (*Hordeum vulgare* L.). *Journal of Plant Physiology*, 1157–1164.

Liang, Y., Nikolic, M., Belanger, R., Gong, H., & Song, A. (2015). Silicon sources for agriculture. *Silicon in Agriculture* (pp. 225–232). Springer.

Liang, Y., Sun, W., Si, J., & Romheld, V. (2005). Effects of foliar-and root-applied silicon on the enhancement of induced resistance to powdery mildew in *Cucumis sativus*. *Plant Pathology*, *54*, 678–685.

Liang, Y. C., Hua, H. X., Zhu, Y. G., Zhang, J., Cheng, C. M., & Romheld, V. (2006). Importance of plant species and external silicon concentration to active silicon uptake and transport. *New Phytologist*, *172*(1), 63–72.

Liao, M., Fang, Z. P., & Liang, Y. Q. (2020). Effects of supplying silicon nutrient on utilization rate of nitrogen and phosphorus nutrients by rice and its soil ecological mechanism in a hybrid rice double-cropping system. *Journal of Zhejiang University-Science B (Biomedicine & Biotechnology)*, *21*(6), 474–484.

Liu, B., Soundararajan, P., & Manivannan, A. (2019). Mechanisms of silicon-mediated amelioration of salt stress in plants. *Plants*, *8*(9), 307.

Liu, J., Cai, H., Mei, C., & Wang, M. (2015). Effects of nano-silicon and common silicon on lead uptake and translocation in two rice cultivars. *Frontiers of Environmental Science & Engineering*, *9*, 905–911.

Liu, J., Zhu, J., Zhang, P., Han, L., Reynolds, O. L., Zeng, R., Wu, J., Shao, Y., You, M., & Gurr, G. M. (2017). Silicon supplementation alters the composition of herbivore induced plant volatiles and enhances attraction of parasitoids to infested rice plants. *Frontiers in Plant Science*, *8*, 1265.

Lobell, D. B., & Field, C. B. (2007). Global scale climate-crop yield relationships and the impacts of recent warming. *Environmental Research Letters*, *2*, 014002.

Lugtenberg, B. J. J., Chin-A-Woeng, T. F. C., & Bloemberg, G. V. (2002). Microbe-plant interactions: Principles and mechanisms. *Antonie Van Leeuwenhoek*, *81*, 373–383.

Ma, J. F., Mitani, M., & Nagao, S. (2004). Characterization of Si uptake system and molecular mapping of Si transporter gene in rice. *Plant Physiology*, *136*, 3284–3289.

Ma, J. F., Tamai, K., & Yamaji, N. (2006a). A silicon transporter in rice. *Nature*, *440*, 688–691.

Ma, J. F., Tamai, K., Yamaji, N., Mitani, N., Konishi, S., & Katsuhara, M. (2006b). Silicon uptake and accumulation in higher plants. *Trends Plant Science*, *11*, 392–397.

Ma, J. F., & Yamaji, N. (2006). Silicon uptake and accumulation in higher plants. *Trends Plant Science*, *11*, 392–397.

Ma, J. F., Yamaji, N., & Mitani, N. (2007a). An efflux transporter of silicon in rice. *Nature*, *448*, 209–211.

Ma, J. F., Yamaji, N., Mitani, N., Tamai, K., Konishi, S., & Fujiwara, T. (2007b). An efflux transporter of silicon in rice. *Nature*, *448*, 209–212.

Malhotra, C., & Kapoor, R. T. (2019). Silicon: A sustainable tool in abiotic stress tolerance in plants. In *Plant abiotic stress tolerance: Agronomic, molecular and biotechnological approaches* (pp. 333–356). Springer.

Mandlik, R., Thakral, V., Raturi, G., Shinde, S., Nikolić, M., Tripathi, D. K., Sonah, H., & Deshmukh, R. (2020). Significance of silicon uptake, transport, and deposition in plants. *Journal of Experimental Botany*, *71*(21), 6703–6718.

Marmiroli, M., Pigoni, V., Savo-Sardaro, M. L., & Marmiroli, N. (2014). The effect of silicon on the uptake and translocation of arsenic in tomato (Solanum lycopersicum L.). *Environmental and Experimental Botany*, *99*, 9–17.

Marron, A. O., Sarah, R., Wheeler, G. L., Goldstein, R. E., King, N., & Not, F. (2016). The evolution of Si transport in eukaryotes. *Molecular Biology and Evolution*, *33*, 3226–3248. https://doi.org/10.1093/molbev/msw209

Martin, K. R. (2013). Silicon: The health benefits of a metalloid. *Metal Ions in Life Sciences*, *13*, 451–473.

Mawia, A. M., Hui, S., Zhou, L., Li, H., Tabassum, J., Lai, C., & Hu, P. (2021). Inorganic arsenic toxicity and alleviation strategies in rice. *Journal of Hazardous Materials*, *408*, 124751.

Meena, V. D., Dotaniya, M. L., Coumar, V., Rajendiran, S., Ajay, Kundu, S., & Subba Rao, A. (2014). A case for silicon fertilization to improve crop yields in tropical soils. In *Proceedings of the National Academy of Sciences, India Section B: Biological Sciences*, *84*(3), 505–518.

Meena, V. D., Dotaniya, M. L., Coumar, V., Rajendiran, S., Kundu, S., & Rao, A. S. (2014a). A case for silicon fertilization to improve crop yields in tropical soils. In *Proceedings of the National Academy of Sciences, India Section B: Biological Sciences*, *84*, 505–518.

Meharg, C., & Meharg, A. A. (2015). Silicon, the silver bullet for mitigating biotic and abiotic stress, and improving grain quality, in rice? *Environmental and Experimental Botany*, *120*, 8–17.

Mitani, N., & Ma, J. F. (2005). Uptake system of silicon in different plant species. *Journal of Experimental Botany*, *56*, 1255–1261.

Mitani, N., Ma, J. F., & Iwashita, T. (2005). Identification of silicon form in the xylem of rice (Oryza sativa L.). *Plant and Cell Physiology*, *46*, 279–283.

Mitani, N., Yamaji, N., & Ma, J. F. (2009). Identification of maize silicon influx transporters. *Plant and Cell Physiology*, *50*, 5–12.

Naureen, Z., Aqeel, M., Hassan, M. N., Gilani, S. A., Bouqellah, N., Mabood, F., Hussain, J., & Hafeez, F. Y. (2015). Isolation and screening of silicate bacteria from various habitats for biological control of phytopathogenic fungi. *American Journal of Plant Sciences*, *6*, 2850–2859.

Nedukha, O. (2022). Tolerance of plant cell wall to environment. In *Advances in Plant Defense Mechanisms*. IntechOpen.

Neu, S., Schaller, J., & Dudel, G. E. (2017). Silicon availability modifies nutrient use efficiency and content, C: N: P stoichiometry, and productivity of winter wheat (*Triticum aestivum* L.). *Scientific Reports*, *7*, 40829.

Nikolic, M., Nikolic, N., Liang, Y. C., Kirkby, E. A., & Römheld, V. (2007). Germanium-68 as an adequate tracer for Si transport in plants. Characterization of Si uptake in different crop species. *Plant Physiology*, *143*, 495–503. https//doi.org/10.1104/pp.106.090845

Pang, Z., Yin, W., Wang, Y., Zeng, W., Peng, H., & Liang, Y. (2023). Silicon-phosphorus pathway mitigates heavy metal stress by buffering rhizosphere acidification. *Science of the Total Environment*, *904*, 166887.

Parab, N., Sinha, S., & Mishra, S. (2015). Coal fly ash amendment in acidic field: Effect on soil microbial activity and onion yield. *Applied Soil Ecology*, *96*, 211–216.

Peera, S. P. G., Balasubramaniam, P., Reddy, M. R., & Chandramani, P. (2016). Effect of silicate solubilizing bacteria and fly ash on the incidence of stem borer (*Scirpophagaincertulas*) (*Lepidoptera*: Pyralidae) in rice. *International Journal of Plant Protection*, *9*(1), 183–186.

Rachappanavar, V., Kumar, M., Negi, N., Chowdhury, S., Kapoor, M., Singh, S., & Yadav, A. N. (2024). Silicon derived benefits to combat biotic and abiotic stresses in fruit crops: Current research and future challenges. *Plant Physiology and Biochemistry*, 108680.

Radkowski, A., Sosin-Bzducha, E., & Radkowska, I. (2017). Effects of silicon foliar fertilization of meadow plants on the nutritional value of silage fed to dairy cows. *Journal of Elementology*, *22*, 1311–1322.

Rana, S., Zia Ur Rehman, M., Umair, M., Ayub, M. A., & Arif, M. (2023). Silicon-mediated alleviation of heavy metal stress in plants. In Sangeeta Pandey, Durgesh Kumar Tripathi, Vijay Pratap Singh, Shivesh Sharma, Devendra Kumar Chauhan (Eds.). *Beneficial chemical elements of plants: Recent developments and future prospects* (pp. 256–276). John Wiley & Sons Ltd.

Rao, B., Gao, L., Dai, H., Hong, Z., & Xie, H. (2019). An efficient and sustainable approach for preparing silicon fertilizer by using crystalline silica from ore. *JOM*, *71*, 3915–3922.

Raturi, G., Sharma, Y., Rana, V., Thakral, V., Myaka, B., Salvi, P., Singh, M., Dhar, H., & Deshmukh, R. (2021). Exploration of silicate solubilizing bacteria for sustainable agriculture and silicon biogeochemical cycle. *Plant Physiology and Biochemistry*, *166*, 827–838.

Raza, A., Razzaq, A., Mehmood, S. S., Zou, X., Zhang, X., Lv, Y., & Xu, J. (2019). Impact of climate change on crops adaptation and strategies to tackle its outcome: A review. *Plants*, *8*(2), 34.

Riyazuddin, R., Nisha, N., Ejaz, B., Khan, M. I. R., Kumar, M., Ramteke, P. W., & Gupta, R. (2021). A comprehensive review on the heavy metal toxicity and sequestration in plants. *Biomolecules*, *12*(1), 43.

Rodrignes, F. A., & Datnoff, L. E. (2005). Silicon and rice disease management. *Fitopatologia Brasileira*, *30*, 457–469.

Roe, N. E., & Stoffella, P. J. (1995). Biosolids-yard trimming composts increases bell pepper (*Capsicum annuum* L.) and Cucumber (Cucumis sativus L.) Yields. *HortScience*, *30*(4), 892–893.

Sabir, F., Noreen, S., Malik, Z., Kamran, M., Riaz, M., Dawood, M., & Ali, M. (2022). Silicon improves salinity tolerance in crop plants: Insights into photosynthesis, defense system, and production of phytohormones. In *Silicon and Nano-Silicon in Environmental Stress Management and Crop Quality Improvement* (pp. 91–103). Academic Press.

Sangwan, S., Kaur, H., Sharma, P., Sindhu, M., & Wati, L. (2022). Routing microbial biosurfactants to agriculture for revitalization of soil and plant growth. In *New and future developments in microbial biotechnology and bioengineering* (pp. 313–338). Elsevier.

Sasaki, S. M., Kakuda, K., Sasaki, Y., & Ando, H. (2014). Effects of slag silicate fertilizer on silicon content of rice plants grown in paddy fields on the Shounai Plain, Yamagata, Japan. *Journal of Soil Science and Plant Nutrition*, *60*, 708–721.

Sathe, A., Kumar, A., Mandlik, R., Raturi, G., Yadav, H., Kumar, N., Shivaraj, S., Jaswal, R., Kapoor, R., Gupta, S., Sharma, T., & Sonah, H. (2021). Role of silicon in elevating resistance against sheath blight and blast diseases in rice (*Oryza sativa* L.). *Plant Physiology Biochemisty*, *166*, 128–139.

Saudy, H. S., & Mubarak, M. (2015). Mitigating the detrimental impacts of nitrogen deficit and fenoxaprop-*p*-ethyl herbicide on wheat using silicon. *Communications in Soil Science and Plant Analysis*, *46*(7), 897–907. https://doi.org/10.1080/00103624.2015.1011753

Savant, N. K., Snyder, G. H., & Datnoff, L. E. (1996). Silicon management and sustainable rice production. *Advances in Agronomy*, *58*, 151–199.

Schaller, J., Brackhage, C., & Gessner, M. O. (2012). Silicon supply modifies C: N: P stoichiometry and growth of *Phragmites australis*. *Plant Biology*, *14*(2), 392–396.

Seal, P., Das, P., & Biswas, A. K. (2018). Versatile potentiality of silicon in mitigation of biotic and abiotic stresses in plants: A review. *American Journal of Plant Sciences*, *9*(7), 1433–1454.

Semenova, N. A., Smirnov, A. A., Grishin, A. A., Pishchalnikov, R. Y., Chesalin, D. D., Gudkov, S. V., & Chilingaryan, N. O. (2021). The effect of plant growth compensation by adding silicon-containing fertilizer under light stress conditions. *Plants*, *10*, 1287.

Shabbir, I., Abd-Samad, M. Y., Othman, R., Wong, M. Y., Sulaiman, Z., Jaafar, N. M., & Bukhari, S. A. H. (2020). Silicate solubilizing bacteria UPMSSB7, a potential biocontrol agent against white root rot disease pathogen of rubber tree. *Journal of Rubber Research*, *23*(3): 227–235.

Shahzad, S., Ali, S., Ahmad, R., Ercisli, S., & Anjum, M. A. (2021). Foliar application of silicon enhances growth, flower yield, quality and postharvest life of tuberose (*Polianthes tuberosa* L.) under saline conditions by improving antioxidant defense mechanism. *Silicon*, 1–8.

Sharma, B., Kumawat, K. C., Tiwari, S., Kumar, A., Dar, R. A., Singh, U., & Cardinale, M. (2023). Silicon and plant nutrition—Dynamics, mechanisms of transport and role of silicon solubilizer microbiomes in sustainable agriculture. *Pedosphere*, *33*(4), 534–555.

Sheng, X. F., Zhao, F., He, L. Y., Qiu, G., & Chen, L. (2008). Isolation and characterization of silicate mineral solubilizing *Bacillus globisporus* Q12 from the surfaces of weathered feldspar. *Canadian Journal of Microbiology*, *54*(12), 1064–1068.

Siam, H. S., Mr, A. E. M., Taalab, S. A. M. A., Ageeb, G. W., & Abd El Kader, M. G. (2022). Ameliorative Effect of Silicate in Soil and Plant. A Review. *Current Science International*, *11*(1).

Singh, S., Bhatt, V., Kumar, V., Kumawat, S., Khatri, P., Singla, P., Shivaraj, S., Nadaf, A., Deshmukh, R., Sharma, T., & Sonah, H. (2020). Evolutionary understanding of aquaporin transport system in the basal eudicot model species *Aquilegia coerulea*. *Plants*, *9*(6), 799.

Singh, V. (2012). *Studies on plant-aphid interactions: A novel role for trehalose metabolism in Arabidopsis defense against green peach aphid*. University of North Texas.

Singla, P. (2020). Plant synthetic biology: A paradigm shift targeting stress mitigation, reduction of ecological footprints and sustainable transformation in agriculture. *Plant Stress Biology: Strategies and Trends*, 435–489.

Souri, Z., Khanna, K., Karimi, N., & Ahmad, P. (2021). Silicon and plants: Current knowledge and future prospects. *Journal of Plant Growth Regulation*, *40*, 906–925.

Sukumar, K., Swarnabala, G., & Gangatharan, M. (2023). Experimental studies on isolation and characterization of silicate solubilizing *Bacillus tequilensis* SKSSB09. *Journal of Applied Biology and Biotechnology*, *11*(4), 135–140. https://doi.org/10.7324/JABB.2023.119115

Takrim, M. A., Zaharah, S. S., Ismail, M. R., Kasim, S., & Ahmad, K. (2021). Effectiveness of K2siO3 on growth and physio-biochemical changes of banana seedlings grown under tropical climate as influenced by application frequency. In *IOP Conference Series: Earth and Environmental Science, IOP Publishing*, *948*, 012011.

Tayade, R., Ghimire, A., Khan, W., Lay, L., Attipoe, J. Q., & Kim, Y. (2022). Silicon as a smart fertilizer for sustainability and crop improvement. *Biomolecules*, *12*, 1027.

Thakral, V., Bhat, J. A., Kumar, N., Myaka, B., Sudhakaran, S., Patil, G., Sonah, H., Shivaraj, S. M., & Deshmukh, R. (2021). Role of silicon under contrasting biotic and abiotic stress conditions provides benefits for climate smart cropping. *Environmental and Experimental Botany*, *189*, 104545.

Tripathi, D. K., Singh, S., Singh, V. P., Prasad, S. M., Dubey, N. K., & Chauhan, D. K. (2017). Silicon nanoparticles more effectively alleviated UV-B stress than silicon in wheat (Triticum aestivum) seedlings. *Plant Physiology and Biochemistry*, *110*, 70–81.

Tripathi, D. K., Singh, V. P., Prasad, S. M., Chauhan, D. K., Dubey, N. K., & Rai, A. K. (2015). Silicon-mediated alleviation of Cr (VI) toxicity in wheat seedlings as evidenced by chlorophyll florescence, laser induced breakdown spectroscopy and anatomical changes. *Ecotoxicology and Environmental Safety*, *113*, 133–144.

Tubaña, B. S., Babu, T., & Datnoff, L. E., (2016). A review of silicon in soils and plants and its role in US agriculture: History and future perspectives. *Soil Science*, *181*, 393–411.

Tubaña, B. S., & Heckman, J. R. (2015). Silicon in soils and plants. *Silicon and Plant Diseases*, 7–51.

Vasanthi, N., Saleena, L. M., & Raj, S. A. (2012). Silicon in day today life. *World Applied Sciences Journal*, *17*, 1425–1440.

Vasanthi, N., Saleena, L. M., & Raj, S. A. (2018). Silica solubilization potential of certain bacterial species in the presence of different silicate minerals. *Silicon*, *10*, 267–275.

Veatch-Blohm, M. E. (2007). Principles of plant genetics and breeding. *Crop Science*, *47*, 1763–1763.

Wang, D., Hou, L., Zhang, L., & Liu, P. (2021a). The mechanisms of silicon on maintaining water balance under water deficit stress. *Physiologia Plantarum*, *173*(3), 1253–1262.

Wang, H., Liu, R., You, M. P., Barbetti, M. J., & Chen, Y. (2021b). Pathogen biocontrol using plant growth-promoting bacteria (PGPR): Role of bacterial diversity. *Microorganisms*, *9*(9), 1988. https://doi.org/10.3390/microorganisms9091988

Wightman, P. G., & Fein, J. B. (2004). The effect of bacterial cell wall adsorption on mineral solubilities. *Chemical Geology*, *212*, 247–254.

Wollast, R. (1967). Kinetics of the alteration of K-feldspar in buffered solutions at low temperature. *Geochimica et Cosmochimica Acta*, *31*, 635–648.

Yamaji, N., Mitatni, N., & Ma, J. F. (2008). A transporter regulating silicon distribution in rice shoots. *Plant Cell*, *20*, 1381–1389.

Yamaji, N., Sakurai, G., Mitani-Ueno, N., & Ma, J. F. (2015). Orchestration of three transporters and distinct vascular structures in node for intervascular transfer of silicon in rice. *Proceedings of the National Academy of Sciences of the United States of America*, *112*, 11401–11406.

Yavuz, D., Baştaş, K. K., Seymen, M., Yavuz, N., Kurtar, E. S., Süheri, S., & Kıymacı, G. (2023). Role of ACC deaminase-producing rhizobacteria in alleviation of water stress in watermelon. *Scientia Horticulturae*, *321*, 112288.

Yu, T.-H., Peng, Y.-Y., Lin, C.-X., Qin, J.-H., & Li, H.-S. (2016). Application of iron and silicon fertilizers reduces arsenic accumulation by two *Ipomoea aquatica* varities. *Journal of Integrative Agriculture*, *15*, 2613–2619.

Zargar, S. M., Mahajan, R., Bhat, J. A., Nazir, M., & Deshmukh, R. (2019). Role of silicon in plant stress tolerance: Opportunities to achieve a sustainable cropping system. *3 Biotech*, *9*(3), 73.

Zhu, Y., & Gong, H. (2014). Beneficial effects of silicon on salt and drought tolerance in plants. *Agronomy for Sustainable Development*, *34*, 455–472.

Zhu, Z., Wei, G., Li, J., Qian, Q., & Yu, J. (2004). Silicon alleviates salt stress and increases antioxidant enzymes activity in leaves of salt-stressed cucumber (*Cucumis sativus* L.). *Plant Science*, *167*, 527–533.

8 Microbial Zinc (Zn) Solubilization

Role in Soil and Plant Health

Zaffar Bashir, Burhan Hamid, Saima Shafi, Farahanaz Rasool, Irfan Ahmad, Bilal Ahmad Dar, and Mehrun Nisa

8.1 INTRODUCTION

Micronutrients such as copper, boron, manganese, iron, molybdenum and zinc are vital elements that plants require in small quantities for their metabolic processes. Micronutrients contribute to cell protection and gene upregulation. They have an active participation in activating or inhibiting enzymes. Some micronutrients alter the osmotic potential of the plant cell. Micronutrients are involved in primary and secondary metabolite synthesis. A micronutrient like iron (Fe) is essential for chlorophyll synthesis and electron transport (ET) in photosynthetic mechanism; boron is required for the formation of the cell wall and pollen tube growth; silicon (Si) enhances plant resistance to pests and diseases; cobalt (Co) is involved in nitrogen fixation by leguminous plants; selenium (Se) acts as an antioxidant; and sodium (Na) is required by some plants for osmotic balance (Frassinetti et al., 2006). Zn is one of the fundamental trace elements required for plants. It is also an important metal constituent compulsory for all the enzymatic activities of plants (Rani et al., 2020). The Zn deficiency would lead to physiological strain due to which different enzymes are not able to perform their function properly (Alloway, 2009). Amongst all micronutrients, Zn is needed by the majority of proteins. Around 2800 proteins prerequisite zinc for their structural as well as biological processes (Ghosh et al., 2014). In both prokaryotic and eukaryotic organisms, zinc plays an essential part in their nutrition and physiology. It upholds structural as well as functional veracity of cell membrane and facilitates protein synthesis and genes expression. In eukaryotic cells zinc-binding proteins make up nearly 10% of the proteomes, and around 36% of these proteins are involved in gene expression (Claudia et al., 2006). Zinc also helps plants to overcome the environmental stress conditions and protect the cells from the detrimental effects of stress (Cakmak, 2008). Plants need zinc in the ionic form (divalent cation -Zn^{+2}) to perform many metabolic processes like fertilization, photosynthesis, auxin biosynthesis and resistance to many pathogens (Goteti et al., 2013). It plays a vital role in the progression of a plant as it is an essential component of enzymes that drive the metabolic reactions. It is also a component of enzyme carbonic anhydrase which

DOI: 10.1201/9781003491460-8

helps in carbon dioxide fixation during photosynthesis. It shows a potential effect on basic plant life processes like photosynthesis, chlorophyll synthesis, pollen synthesis, nitrogen metabolism, protein quality, protection against oxidative damage and helping plants to resist biotic and abiotic stresses (Hafeez et al., 2013; Potarzycki & Grzebisz, 2009).

Several parts around the world like India, China, Australia, Pakistan, Afghanistan, Turkey, Brazil, Iran and USA have reported the deficiency of zinc and it is influencing the soil crop-system and thereby decreasing the productivity of crops (Alloway, 2009; Cakmak, 2008). There are multiple reasons for Zn deficiency in soil which include reduced recycling of agricultural residues, excess fertilization, less organic matter and intensive cropping pattern (Hafeez et al., 2013). Approximately 50% of the agricultural land around the world is deficient of zinc. Several countries demonstrated low zinc levels in maize (staple food) when cultivated on zinc deficient soil (Noulas et al., 2018). It has been recorded that wheat produced on calcareous soils and rice cultivated on flooded paddy soils shows zinc deficiency (Mori et al., 2016). It is one of the most critical nutrients and its deficiency results in hidden hunger among women and pre-school children. According to a study, deficiency of micronutrients like iodine, zinc, selenium and iron affects around 2 billion individuals globally (White & Broadley, 2009; WHO, 2012). Zinc deficiency is omnipresent worldwide and around 50% of soils in India are zinc deficient (Singh et al., 2005). Agricultural soils of Kashmir valley have also been reported as zinc deficient (Wani et al., 2013). The zinc deficient soils affect the crops which negatively affect human as well as animal health when they consume these zinc deficient crops. In plants deficiency of zinc leads to delayed nitrogen metabolism, photosynthesis, growth of flowers and fruits, and carbohydrate synthesis (Vandana & Manu, 2021). Nearly one third of the world population is facing the deficit of zinc as it causes many nutritional and health problems in human beings (Hotz & Brown, 2004).

To prevail over Zn deficiency, various conventional techniques and chemical fertilizers are being used. Most of the time Zn is available in insoluble form, so it is not readily accessible to plants. The bioavailability of zinc to plants is fulfilled through the exogenic application of zinc based chemical fertilizers (like zinc sulphate, $ZnSO_4$) (Rani et al., 2020). However, only up to 5% of zinc (Zn) is used by the crop and the remaining 95% is reverted back into different insoluble forms of zinc like zinc carbonate (in calcium rich soil), zinc phosphate (phosphate rich soil), zinc hydroxide (high pH soil) etc (Kabata-Pendias, 2000; Kumar et al., 2017). Chemical fertilizers are not environmentally friendly and are expensive. They make crops liable to diseases and also impact soil health. Over use of fertilizers is hazardous to plants as well as animals as they cause water and soil pollution. High content of zinc is present in soils but in the form of minerals like zincite (ZnO), smithsonite ($ZnCO_3$), hopeite ($Zn_3(PO_4)_2.4H_2O$), sphalerite (ZnS) and wellemite (Zn_2SiO_4) (Lindsay, 1979). Therefore, it is most important to formulate approaches and use different technologies which can improve the soluble zinc content by solubilizing insoluble zinc reserves present in the soil. Several biogeochemical reactions occur in the soil between minerals and microbes, which results in unlocking the trapped nutrient metal ions. There are different mechanisms by which solubilization occurs which include excretion of metabolites like organic acids like malic acid, lactic acid, formic acid, propionic acid, glycolic acid, acetic acid, succinate and fumaric acid, manufacture of chelating agents or proton extrusion (Sayer & Gadd, 1997).

Conventional breeding techniques and use of chemical fertilizers have many disadvantages. They are overpriced, sluggish and arduous with high contamination degree, truncated shelf life and low farmer recognition. They make crops vulnerable to pathogens, decrease the soil fertility in the long run and are not eco-friendly as they cause environmental pollution such as soil and water pollution (Rani et al., 2020). To prevail over these issues, the scientific community is fluctuating its attention towards the manufacture of biofertilizers in which zinc solubilizing microorganisms (ZnSMs) are used to increase the concentration of soluble Zn in the soil and making Zn available to plants. Biofertilizers comprising of living microbial strains are eco-friendly and maintain soil fertility in the extended period. Microbes play an essential role in the solubilization of inorganic zinc components to improve the bioavailability of zinc to crops. There are several potential zinc solubilizing strains which belong to *Bacillus* spp., *Azotobacter* sp., *Micrococcus* sp., *Pseudomonas* sp., *Acinetobacter* sp., *Gluconacetobacter* sp. etc (Azadeh et al., 2012). The application of these peculiar strains through the intermediation of different formulations, compounds, biofertilizers and most commonly rhizobacteria in soils is inexpensive, sustainable and eco-friendly for zinc availability, either through solubilizing locked or fixed zinc or making available Zn from a remote area of the rhizosphere. Zinc solubilization is a direct mechanism of PGPR that plays an important role in plant growth and yield (Glick, 1995). The Zn solubility in soil greatly depends upon moisture and soil pH. The amount and nature of organic acids produced by different soil microbes mainly depends upon carbon source, pH and buffering capacity (Mattey, 1992).

ZnSMs are a boon for sustainable agriculture. The only possible way to increase crop productivity as well as quality of food without creating environmental issues or concern is the use of zinc solubilizing microbial strains (Manasa et al., 2019; Kalpana & Dinesh, 2020). They secrete various compounds like organic acids which include malic acid, oxalic acid and so on that separate Zn cations and decrease the pH in nearby soils (Alexander, 1997). Solubilization of zinc involves the production of low molecular weight siderophores (Saravanan et al., 2011) and protons, oxidoreductive systems on cell membranes and chelating ligands (Chang et al., 2005; Wakatsuki, 1995). It has been testified that many beneficial bacterial strains cause zinc solubilization on laboratory scale like *Pseudomonas striata, Bacillus* spp., *Azotobacter* spp., *P. fluorescens, Burkholderia cenocepacia* (Pawar et al., 2015), *Pseudomonas aeruginosa* (Fasim et al., 2002), *Bacillus thuringiensis, S. marcescens, Serratia liquefaciens* (Abaid-Ullah et al., 2015), *Gluconacetobacter diazotrophicus* (Saravanan et al., 2007), *B. megaterium* (Kalpana & Dinesh, 2020) and *Bacillus cereus* (Vandana & Manu, 2021). To increase the zinc content in crops, chemical fertilizers are widely used, but it is expensive and non-sustainable (Bahrani et al., 2010; Bulut, 2013; Cakmak et al., 2010; Hafeez et al., 2013; White & Broadley, 2011). An important strategy to raise the content of zinc in major crops is the bio fortification approach, and it would be an economically viable and socially acceptable tool as well (Saran et al., 2019). The most important concept is to understand the zinc mechanism so that it will give us an insight in increasing crop productivity and mitigate any accumulation of harmful elements (Upadhyay & Srivastava, 2014).

The world population is continuously rising and the requirement of food to feed the people is also increasing. To counter this, farmers apply various agrochemicals like chemical fertilizers in large amounts to increase the crop production per unit area. Indiscriminate application of chemical fertilizers causes various unacceptable changes in structure of soil and plant-animal health which becomes the future threat to the ecosystem. Therefore, to address such issues the microbial world is the only source to retain soil and plant health. Use of microbial strains in organic agriculture is an eco-friendly approach. Zinc based bioagents are the only source to retain soil health and maintain plant growth.

8.2 AVAILABILITY OF ZINC IN SOIL

Rocks naturally contain zinc. The parent rock composition of the soil determines how much zinc is contained in that soil. The parent rock's weathering and the soil's geochemical makeup determine the zinc content. The crust of the Earth has a zinc content of 78 mg/kg, which varies depending on the parent rock. For example, the Zn content in sedimentary rocks ranges from 80 to 120 mg/kg, in magmatic rocks it is 40–120 mg/kg, in sandstones it is 15–30 mg/kg, and in dolomites and limestones it is 10–25 mg/kg (Rani et al., 2020). In polluted soils Zn composition depends on soil location and sources of pollution. The zinc composition in polluted soils depends upon the soil location and pollution (Kabata-Pendias & Pendias, 2001). The existence of Zn in soil is found as hopeite [$Zn_3(PO_4)_2 4H_2O$], zincite (ZnO), smithsonite ($ZnCO_3$), zinkosite ($ZnSO_4$), sphalerite (ZnS) and franklinite ($ZnFe_2O_4$); however, accessibility of Zn from these varying sources relies on contrasting factors. The naturally occurring sources of zinc are due to physical and chemical weathering of parent rock (Alloway, 1995; Malik & Khan, 2022) and atmospheric indulgence of zinc to soils (e.g., dust, huge forest fires and volcanic eruptions) (Friedland, 1990; International Zinc Association, 2011). Table 8.1 shows a list of zinc mineral ores.

The active Zn in soil occurs either in divalent (Zn^{2+}) or in complex forms like Zn $(OH)^{3-}$, $ZnOH^{+}$ etc. There are five different pools in which zinc exists within the soil like exchangeable, organically bound, water soluble, chelated and absorbed (Rani et al., 2020). The Zn availability to plants is determined by total Zn content in soil. Moreover, the quantity of Zn in soil depends on calcareous elevated pH soils, crumbled soil, sandy soil and acid deposited soil. There are many other factors also

TABLE 8.1
List of Zinc Mineral Ores

Mineral Zn	Absorbed Zn	Complexed Zn	References
Franklinite ($ZnFe_2O_4$)			
Hopeite [$Zn_3(PO_4)_2.4H_2O$]	Zn-$MgCO_3$	Manure and Organic residues	Rani et al. (2020)
Smithsonite ($ZnCO_3$)	Zn-FeO		
Sphalerite (ZnS)	Zn-$CaCO_3$		
Zincite (ZnO)	Zn-MnO		
Wellemite (Zn_2SiO_4)			

which include calcium carbonate and organic matter content, elevated concentrations of cations and anions and phosphate soluble forms (Alloway, 2009). Zn is strongly adsorbed on $CaCO_3$, $MgCO_3$, FeO or MnO. The critical value of zinc in soil varies from 0.6 to 1.2 ppm; less than 0.6 ppm means the soil is zinc deficient and above 1.2 ppm indicates zinc toxicity (Vandana & Manu, 2021) (Table 8.1). Surface assimilation of replaceable Zn cations in soils is given as:

$$Zn^{2+} + \text{M-soil} \qquad \text{Zn-soil} + M^{2+}$$

8.3 ROLE OF ZINC IN PLANTS

Zinc plays a dynamic role in physiological and metabolic functions of a plant. Plants require around 30–100 mg/kg zinc. It influences the activities of carbonic anhydrase and ribulose 1, 5-bisphosphate carboxylase which are very essential for the process of photosynthesis. The deficiency of Zn causes 50–70% reduction in photosynthesis (Rani et al., 2020). It plays a critical role in RNA, DNA synthesis and regulations. It prevents peroxidation of proteins and lipids due to ROS and also protects ribosomal RNA against ribonuclease attack. It also controls the membrane integrity by interacting with sulfydryl groups and phospholipids (Welch & Graham, 2004). The gene expression regulation and maintenance required for tolerance of environmental stresses in plants are Zn dependent. It is essential for the regulation, pollen formation and normal fruit development (Hafeez et al., 2013). Its deficit results in the development of abnormalities in plants which become visible as the deficiency symptoms such as chlorosis, stunted growth, smaller leaves, very short internodes and leaf resetting. In plants Zn unavailability leads to hampered photosynthesis, metabolism of nitrogen, carbohydrate synthesis and delayed growth of flowers and fruits (Vandana & Manu, 2021). Crops may be more susceptible to injury or infection due to zinc deficiency as the plants do not gain their full production potential (Gandhi & Muralidharan, 2016; Hafeez et al., 2013).

Zinc is an essential element for the growth of plants, humans and animals. It plays a dynamic role in crop nutrition since it is necessary for the smooth execution of several enzymatic reactions, metabolic processes and oxidation-reduction reactions. Zinc plays a significant part in the synthesis of RNA and DNA and is a component of numerous enzymes that contain zinc and are essential for the growth and differentiation of cells (Rani et al., 2020). It also interacts with membrane proteins and phospholipids to maintain the integrity of the cell membrane. Lack of zinc promotes membrane permeability, which may lead to changes in metabolism (Fariduddin et al., 2022).

Because of a lack of information and technology, malnutrition affects over half of the world's population, particularly in underdeveloped nations. This nutrient is extremely important to human health. Utilizing soil microorganisms that can mobilize unavailable zinc, boost zinc assimilation and stimulate plant growth, thereby making it indirectly available to humans, is one way to combat zinc deficiency (Rana et al., 2012). Zinc deficiency affects several crops worldwide, including fruits, vegetables and cotton. The most significant cereal crops, including wheat, rice and maize, are zinc deficient in many nations (Hafeez et al., 2013). Zinc deficiency can be related to various factors affecting total Zn content as shown in Figure 8.1.

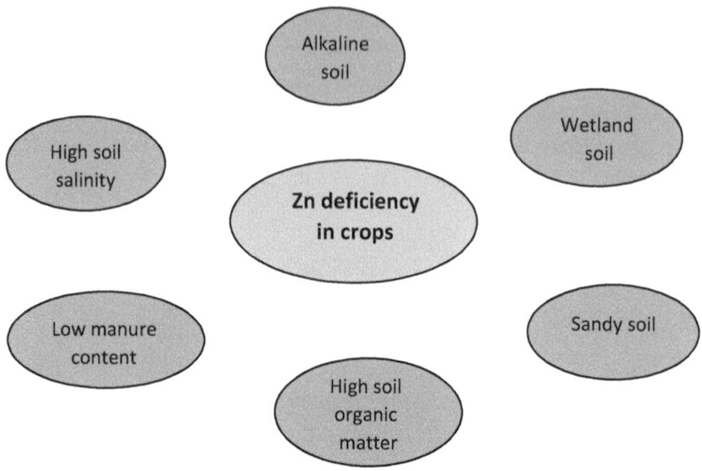

FIGURE 8.1 Factors affecting availability of zinc in soil.

8.4 ZN UPTAKE MECHANISM IN PLANTS

In poorly reducing neutral or alkaline soils Zn is immobile, while it is easily mobile into plants in oxidizing acid soils. The transporter systems like mass flow, diffusion and root extension carry zinc as a divalent cation (Zn^{2+}) or complexed with ligand in the direction of roots (Suzuki et al., 2008; Yoneyama et al., 2015). Maximum zinc absorption occurs through an active transport system where it is transported to shoots from the root. Basal node retains the small quantity of Zn which facilitates zinc distribution in plants. Both xylem and phloem tissues are responsible for Zn transport (Rani et al., 2020). Inside phloem Zn is coupled with nicotinamine which is predominant ligand in rice phloem sap and in xylem zinc combines with a soluble protein called phytate and translocation occurs through apoplast and symplast pathway by the transpiration (Sadeghzadeh, 2013; Yang et al., 2021).

8.5 EFFICACY OF ZINC FERTILIZERS IN SOIL

Nowadays, the common method to solve zinc deficiency is the use of fertilizers, whose selectivity would rely on economic compatibility, convenience, cost, acceptability of the environment and mode of implementation. Commonly Zn fertilizers are divided into three types, namely inorganic, natural organic and synthetic chelates complexes. There are several inorganic zinc sources like zinc oxide, zinc chloride, zinc phosphate, zinc carbonate and zinc sulphate. The cheapest, easily available and most effective fertilizer used is zinc sulphate heptahydrate ($ZnSO_4.7H_2O$). Zinc oxide contains 80% Zn and has very low water solubility while zinc phosphate is cheaper than zinc oxide and very useful for seed coating (Mortvedt & Gilkkes, 1993). Zn percentage in Zn fertilizer is given in Figure 8.2.

There are many Zn organic fertilizers like Zn-lignosulfonate, Zn-phenolate and Zn-EDTA. Zn-EDTA is more efficiently transported to plants as compared to zinc

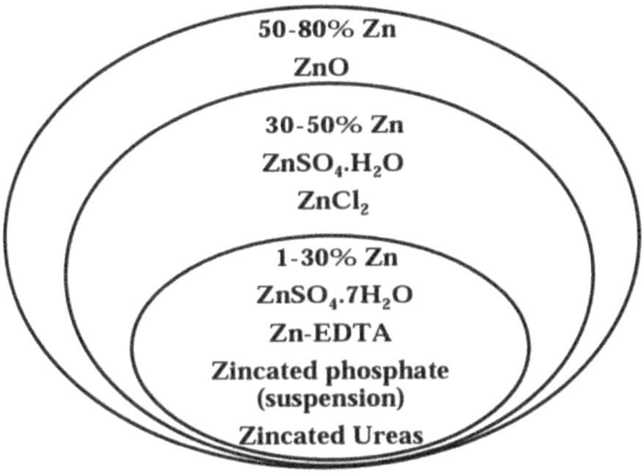

FIGURE 8.2 Zinc percentage in zinc fertilizer.

sulphate and also facilitates the phosphorus uptake. Many fertilizers like zincated super, boronated super and zincated urea, when combined with micronutrients, have effectively increased the soil fertility and micronutrient deficiency in plants. The activity of some complexes such as gluconates and lignosulfonates which are the by-products of the paper and food industry has shown great results (López-Rayo et al., 2016).

8.6 ZINC SOLUBILIZING MICROORGANISMS

Microorganisms can be employed to solubilize insoluble forms of zinc and make plants healthy with strong yield and output in order to address zinc shortage (Fasim et al., 2002). Producing the agricultural commodity would be less expensive if Zn-solubilizing microbial strains were chosen and injected, either alone or in combination with inexpensive insoluble Zn compounds. It is well known that a variety of microorganisms are crucial to the solubilization of potassium, phosphorus, iron and zinc in plants. Most active rhizospheric fungus in solubilizing insoluble zinc compounds include *Aspergillus niger*, *A. nomius* and *A. oryzae*, which were isolated from a zinc mining site in Thailand (Anitha et al., 2015; Dhaked et al., 2017; Shaikh & Saraf, 2017). Fungi *B. caledonica* converts insoluble zinc phosphate into soluble zinc through the process of acidolysis, complex lysis and metal accumulation (Fomina et al., 2004).

A. niger has been reported as solubilizer of insoluble ZnO, Zn_3 $(PO_4)_2$ and Ca_3 $(PO_4)_2$ to soluble form through the production of citric, malic and oxalic acid (White et al., 1997). Furthermore, *A. terreus* (ZSF-9) isolated from Tirupur District, India also reported the solubilization of ZnO, $ZnCO_3$, Zn_3 $(PO_4)_2$ via decrease in pH through the production of gluconic acid (Anitha et al., 2015). *Trichoderma harzianum Rifai* 1295-22 (T-22) converts insoluble Zn present in the soil into soluble form by releasing Zn^{2+} ion via oxidative dissolution process.

The *Bacillus* sp. has shown the property of zinc solubilization and it is suggested to use it alone or in combination with cheaper insoluble forms of Zn like

ZnO, $ZnCO_3$ and ZnS and it could be the best alternative to costly $ZnSO_4$ (Mahdi et al., 2010a, 2010b). Many bacteria have been reported to solubilize insoluble zinc and can be employed as biofertilizers such as *Bacillus subtilis*, *Bacillus thuringiensis*, *Thiobacillus thioxidans*, *Pseudomonas striata*, *Pseudomonas fluoregnence*, *B. cenocepacia*, *Cynobacteria*, *Acinetobacter*, *S. marcescens*, *S. liquefaciens*, *Bacillus megaterium* and *Bacillus cerius* (Abaid-Ullah et al., 2015; Altomare et al., 1999; Bapiri et al., 2012; Fasim et al., 2002; Hussain et al., 2015; Kalpana & Dinesh, 2020; Vandana & Manu, 2021). Some of the effective microbes with their zinc solubilizing mechanism are listed in Table 8.1.

8.7 MECHANISM OF ZINC SOLUBILIZING BACTERIA

The zinc in the soil is mainly present in an insoluble form and is unavailable to plants (Barber, 1995). The important feature of PGPR is the solubilization of different metal salts as they become available to plant uptake. Bacterial comparative and functional genomics research has unlocked new ways for approaching these underlying mechanisms at the molecular and biochemical level. To check the mechanisms of Zn solubilization by zinc solubilizing bacteria (ZnSB) various studies have been performed. There are different mechanisms of solubilization such as release of organic acids, acidification and chelation and through exchange reactions (Chung et al., 2005; Hafeez et al., 2005). The mechanism of mobilization of zinc likely involves siderophore formation (Burd et al., 2000; Saravanan et al., 2011; Tariq et al., 2007; Wani et al., 2007) and production of gluconate or the derivatives of gluconic acid, e.g., 2-keto-gluconic acid (Fasim et al., 2002), 5-keto-gluconic acid (Saravanan et al., 2007a, 2007b) and many other organic acids (Di Simine et al., 1998; Kalpana & Dinesh, 2020; Tariq et al., 2007; Wani et al., 2007; Vandana & Manu, 2021). It has been reported that soil bacteria such as *Bacillus* sp., *Azotobacter* spp. and *Pseudomonas* sp. solubilize inorganic Zn through the production of organic acids like oxalic acid, glycolic acid, lactic acid, formic acid, propionic acid, acetic acid, succinic acid and fumaric acid (Vazquez et al., 2000). Acidification is the most preferred mechanism of zinc solubilization by microorganisms.

During an experiment performed on pot culture, it was reported that inoculants of combined *Gluconacetobacter diazotrophicus*, a gram-negative bacteria, and insoluble zinc compound correcting the Zn deficient in maize seedling (Sarathambal et al., 2010). It was also found that *G. diazotrophicus* is a potential insoluble Zn solubilizer especially zinc carbonate, zinc oxide and zinc phosphate (Saravanan et al., 2007). *Bacillus* sp. (AZ6) when grown on Bunt and Revira medium containing ZnO forms halo area and a comparable halo area by *Acinetobacter* sp. (AGM3) was reported (Goteti et al., 2013; Hussain et al., 2015). The three strains, *B. aryabhattai*, *Pseudomonas taiwenensis* and *Bacillus* sp. were investigated for organic acid production. It has been found that 11 organic acids were produced in different zinc sources like ZnO, ZnCO3 and Zn-phosophate. The major organic acids produced by these three isolates were lactic acid, malic acid, malonic acid, citric acid and succinic acid. During Zn solubilization, *B. aryabhattai* produced almost all organic acids. Therefore, these reports clearly indicate that secretions of organic acids by different bacteria will vary depending on Zn substrate (Vidyashree et al., 2018) (Table 8.2).

TABLE 8.2
Effective Bacterial Strains with Zinc Solubilizing Mechanism

Name	Insoluble Zn Form	Solubilization Mechanism	Location	References
P. fluorescens	$Zn_3(PO_4)_2$	Production of organic acid like malic acid	UK	Di Simine et al. (1998)
P. aeruginosa	ZnO, $Zn_3(PO_4)_2$	Production of gluconic acid and keto-gluconic acid	Scotland, UK	Fasim et al. (2002)
Pseudomonas sp. and *Bacillus* sp.	ZnO, ZnS and $ZnCO_3$	Production of organic acids like oxalic acid	Tamil Nadu, India	Saravanan et al. (2004)
Acinetobacter calcoaceticus	ZnO, $ZnCO_3$	Organic acid production like succinic acid	Tami Nadu, India	Goteti et al. (2013)
Bacillus sp.	ZnO, $ZnCO_3$	Production of organic acids like propionic acid	Faisalabad, Pakistan	Azhar et al. (2015)
B. cereus	ZnO, $Zn_3(PO_4)_2$	Organic acid production	Kerala, India	Kumar et al. (2017)
Bacillus aryabhattai	ZnO, $ZnCO_3$	Production of malic acid, keto-D-gluterate and gluconic acid	Odisha, India	Vidyashree et al. (2018)
Pseudomonas taiwenensis	ZnO, $ZnCO_3$ and $Zn_3(PO_4)_2$	Production of keto-D-gluterate, citric acid, propionic acid and oxalic acid	Odisha, India	Vidyashree et al. (2018)
Bacillus megaterium	ZnO, $ZnCO_3$	Production of organic acids	India	Kalpana and Dinesh (2020)
Bacillus cereus & *Bacillus tropicus*	ZnO	Production of organic acids	Haryana, India	Vandana and Manu (2021)

8.8 HOTSPOT (RHIZOSPHERE): A HEART FOR ZINC SOLUBILIZING BACTERIA FOR DIRECT TALK WITH HOST PLANT

The small area of soil known as the rhizosphere, or hotspot, is directly impacted by the root xudates or secretions and the related soil bacteria called the root microbiome. It is a hive of biological activity where plant roots exchange nutrients and other chemicals with the soil surrounding it and microbes (Hartmann et al., 2008). The rhizosphere plays a significant role in the cycling of nutrients such as phosphorus (P), zinc (Zn), nitrogen (N_2), potassium (K), manganese (Mn), silicon (Si), magnesium (Mg) and sulphur (S)(Bashir et al., 2017; Bashir et al., 2018a; Bashir et al., 2018b; Bashir et al., 2019; Bashir et al., 2018; Bashir et al., 2019b; Bashir et al., 2021a; Bashir, 2021b; Hamid & Bashir, 2019a). It also functions as a site for disease suppression, where PGPR release compounds such as lytic enzymes, antibiotics and HCN (Singh et al., 2004).

Plant metabolic happenings affect the rhizosphere by releasing root exudates (Ahkami et al., 2017) which are either attractive or repulsive (Ortíz-Castro et al., 2009) and determine the microorganisms diversity of hotspot (Singh & Mukerji, 2006). Moreover root exudates, the cellular excretion in the form of several chemicals, viz. chemical compounds, antipathogenic metabolites, growth regulators like IAA, and nutrient mobilization, siderophore production, N_2 fixation, P solubilization are the other unique properties that belong to both plants and microbes (Chauhan et al., 2022; Upadhyay & Chauhan, 2022). The primary tactic used by coexisting soil microorganisms in the hotspot zone is root colonization; this region's bacterial population is commonly referred to as rhizobacteria (Schmidt et al., 2018). Numerous substances generated by both plants and bacteria facilitate the contact between the microbe and the root, as demonstrated by the root colonization zone, which displays the mutual interaction between rhizobacteria and the root. However, by altering the structural characteristics of the soil and related microbial populations as well as encouraging root surface colonization, root exudates serve as communication signals that start a biological and physiological connection between the soil flora and the roots. Rhizobacteria, which are bacteria that can dissolve zinc, are well known for their capacity to do so (Shakeel et al., 2015). Numerous approaches of PGPR are directly attributed to the proliferation of plant growth (Chauhan et al., 2022). The secretion of metabolites by rhizospheric bacteria such as organic acids (malic acid, oxalic acid and succinic acid), enzymes and siderophores play an important part in Zn solubilization; other secretions like antibiotics, hydrolytic enzymes and antifungal compounds improve plant health by inhibiting the growth of pathogens in the hotspot region. The solubilized form (Zn^{2+}) can be easily taken up by plant roots and translocated into several parts including edible parts or grains as an additional microbial-assisted biofortification step and further utilized for various metabolic activities (Ambreetha & Balachandar, 2019).

8.9 EFFECT OF ZINC SOLUBILIZING BACTERIA (ZNSB) FOR PLANT GROWTH PROMOTION

Zinc is a vital micronutrient for the development of plants. Zn is prerequisite for numerous functions such as plant metabolism, reproduction, membrane integrity and resistance to biotic and abiotic stresses (Gondal et al., 2021). It has a vital role in crop nutrition for performing various metabolic processes, enzymatic reactions and redox reactions. It also plays an important role in DNA and RNA synthesis and is a constituent of many enzymes which have an important role in cellular growth and differentiation (Ashok Kumar et al., 2019). Due to lack of knowledge and technology in developing countries, Zn deficiency causes malnutrition to 50% of the population. To overcome this problem, there is an application of soil microbial community that can solubilize insoluble forms of zinc, increase the bioavailability of zinc in soil and promote plant growth and development (Rana et al., 2012). The combined use of biofertilizers and chemical fertilizers not only improved the crop productivity but also increased the nutrient efficacy in crops. There are many bacterial strains which can be used as biofertilizers and biocontrol agents such as *Bacillus*, *Micrococcus*, *Pseudomonas*, *Enterobacter*, *Azotobacter*, *Rhizobium* and *Burkholderia* and have also been shown to promote plant growth (Ashok Kumar et al., 2019). Zn solubilizing PGPR inoculants are used as biofertilizers to increase soil fertility and plant health. They have also been proved to increase the growth and survival rate of plants, maximize crops yield, control the use of chemical fertilizers and also reduce malnutrition rates (Hafeez et al., 2001). The effectiveness of zinc solubilizing bacteria (ZnSB) has an optimistic impact on various growth parameters like length of the root, root dry matter, root area, volume of the root, shoot dry matter and panicle emergence index (Manasa et al., 2019; Vandana & Manu, 2021). PGPR work with the symbiotic association of microbes with plants. Different zinc solubilizing bacterial (ZnSB) strains have shown the growth promotion activities in different crops (Table 8.3).

8.10 SCIENTIFIC WAY FOR ISOLATION OF ZINC SOLUBILIZERS

A number of zinc solubilizing bacterial strains are accessible as biofertilizers and are gaining traction as an alternative to synthetic agrochemicals and sustainable management in agricultural practices. By adding living microorganisms (bacteria, fungi and actinomycetes) to plant surfaces or soil, biofertilizers encourage plant growth and yield by colonizing the rhizosphere or plant roots. Through a number of processes, including biological nitrogen fixation, the synthesis of growth regulators, and the solubilization of insoluble elements like zinc, and silicate, biofertilizers improve nutrient availability to the root zone, which in turn stimulates plant growth and development (Ferguson et al., 2019). The activity and survivability of biofertilizers within the host plant is drastically affected by both biotic and abiotic factors including degree of interaction of PGPR with soil microbiota, extreme temperature, oxygen level in soil, presence of heavy metals, acidity and alkalinity of soil and deficiency

TABLE 8.3
Inoculation Effect of Various Zn Solubilizing Strains on Crops

Zn Solubilizers	Crops	Enhanced Parameters	References
Zinc solubilizing promoters	Rice	Elevated zinc concentration in yield, total plant biomass, root dimension area, plant area index, bulk of the roots as well as shoot heft or weight	Tariq et al. (2007)
Rhizobium spp. RL9	Lentil	Amendments in grain protein, nodule number, dry matter and seed yield	Wani et al. (2008)
Zn solubilizing PGPR	Wheat	Boosted root area, weight, volume, length and shoot weight	Kutman et al. (2010)
Pseudomonas spp. P17 and *bacillus* spp. B40	Maize	Heightened uptake of Zn, K, N and Mn as well as the overall dry mass	Goteti et al. (2013)
Bacillus aryabhattai strains MDSR7 and MDSR11	Soybean and wheat	Boosted plant height, shoot dry weight, root dry weight and zinc absorption in seeds	Ramesh et al. (2014)
Burkholderia and *Acinetobacter*	Rice	Improved total zinc uptake, yield, number of grains and mean dry matter	Vaid et al. (2014)
P. fragi, *P. agglomerans* and *P. dispersa*	Wheat	Enhanced Zn content and plant growth promotion	Kamran et al. (2017)
Bacillus spp.	Soybean and wheat	Optimized yield, growth and zinc biofortification	Khande et al. (2017)
Pseudomonas sp. and *Bacillus* sp.	Rice	Improved Zn conc. and plant physiological growth	Manasa et al. (2019)
Bacillus megaterium	Pepper	Adjusted yield, growth and zinc biofortification	Kalpana and Dinesh (2020)
Bacillus cereus	Wheat	Elevated zinc content and encouragement of plant growth	Vandana and Manu (2021)
Priestia spp.	kale	Kale production and quality	Kordatzaki et al. (2022)
P. aryabhattai KEX505 and *P. megaterium* KAH109	Green soybean	8.79% and 26.96% increase in the dry weight of green soybean plants	Vandana and Manu (2021)

of essential nutrients (Basu et al., 2021). Therefore, a specific climate is needed for inoculants to successfully adapt in agricultural fields. The method by which zinc solubilizing microbes are isolated involves zinc solubilizing agar (ZnSA) containing glucose 10 g, ammonium sulphate 1 g, potassium chloride 0.2 g, dipotassium hydrogen phosphate 0.1 g, magnesium sulphate heptahydrate 0.2 g, zinc oxide 1 g as insoluble source of zinc, agar powder 15 g, and distilled water 1000 ml. It will be prepared and autoclaved at 121°C for 15 minutes at 15 PSI for sterilization. Under aseptic conditions, zinc solubilizing agar will be poured into sterile petri plates and allowed to solidify. After solidification, the plates will be spot inoculated with a loopful of freshly grown bacterial culture (24 hours) and incubated at 28±2°C for 3–5 days. After incubation, the formation of a clear zone around colonies (Figure 8.2) will indicate the ability of bacterial isolates to solubilize insoluble zinc (Saravanan et al., 2007).

8.11 PHASES FOR ISOLATION AND CHARACTERIZATION OF ZNSM BIOAGENTS/BIOFERTILIZER

The following are phases for isolation and characterization of ZnSM bioagents/biofertilizer given as

- Assembly of rhizospheric soil.
- Serial dilution.
- Streaking/inoculating on specific medium.
- Presence of Zone of Clearance around colony after incubation at 37°C indicates ZnSM activity (qualitative assay).
- Liquid assay for Zn dissolution.
- Re-inoculation for pure culture.
- Morphological assessment and colony characterization.
- Biochemical assessment (IMVIC etc) and Molecular Characterization (16SrRNA).
- Experimentation on modal plants.
- Poly house trial followed by field experimentation later on tested for quality control and then finally commercial biofertilizers/bioagent formulated (Figure 8.3).

8.12 ZINC SOLUBILIZING MICROBES: A POTENTIAL APPROACH FOR BIOREMEDIATION

Attributable to heavy metal pollution brought on by the industrial revolution and the use of naturally occurring resources for metal mining and industry, the ecosystem has suffered severe harm. Scientists are faced with the challenge of detoxifying living creatures from high quantities of heavy metals, which can have hazardous effects. The dangerous impacts of heavy metals continue to plague the planet despite efforts to reduce environmental contamination. A novel technique like bioremediation may be able to lessen the negative impacts of heavy metal pollution and help contain the catastrophe (Golla, 2021). Heavy metal contamination has been cleaned up using

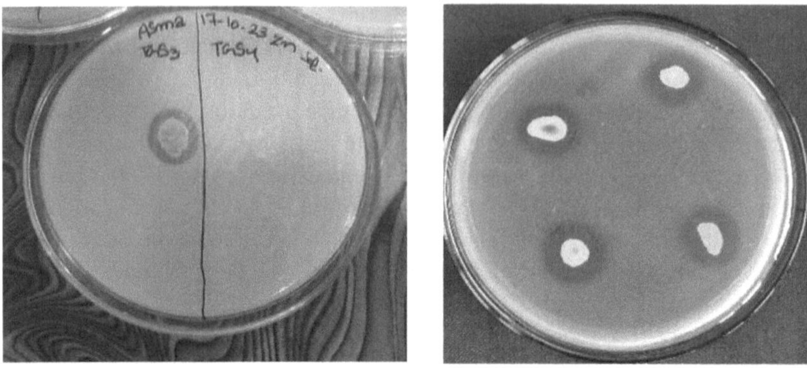

FIGURE 8.3 Qualitative assay of zinc solubilization by bacterial strains on ZnSA.

physico-chemical procedures such as membrane filtration, ion exchange, redox, electrochemical processes and precipitation. These techniques do not, however, have the capacity to permanently detoxify heavy metals, nor are they cost-effective or free of toxic by-products. They work well in areas that are heavily polluted and big but have low concentrations of heavy metals (Singh et al., 2020). Researchers are currently working on new technologies that prioritize total heavy metal removal, with microbial remediation offering one of the best prospects for decontamination. Heavy metals can be tolerated at high concentrations by soil microorganisms, and some of them are necessary for their metabolic processes as micronutrients. Bioremediation can be carried out in situ by introducing the biological source to contaminated land, or ex situ by moving contaminated areas to be treated (Shilev et al., 2020). Two utmost important processes are used by a majority of microorganisms: either heavy metals are oxidized or are reduced to less harmful forms, or they are immobilized and/or sequestered. Both aerobic and anaerobic conditions can be used for bioremediation, although aerobic settings are faster and more effective than anaerobic ones.

Utilizing biological variety, bioremediation is a comprehensive method that transforms harmful contaminants into safe ones. To detoxify heavy metals, plants, fungi, bacteria, actinomycetes and algae are used as biological agents. There are two approaches: ex situ, in which the polluted site is moved to another location for processing, and in situ, in which the procedure takes place at the contaminated site itself. The idea behind it all is that microbes use hazardous metals as a food supply. Research and investigations are needed in the multidisciplinary subject of microbial bioremediation. The procedure entails either introducing the biological agent to the polluted area or encouraging naturally occurring species to cope with pollutants (Albdaiwi et al., 2019). By removing hazardous chemicals, oil spills and heavy metals from changed locations using bioremediation, a safer way to manage environmental degradation, harmful chemicals and metals are eliminated. The biological response to environmental maltreatment is known as bioremediation. In the bioremediation process, plant growth enhancers (PGPR) are essential for removing organic pollutants, fungicides and PTE from contaminated soils (Afridi et al., 2019). Certain photosynthesis-regulated plants (PGPR) have the ability to bind hazardous metals from soil to their cell surfaces or incorporate them into their metabolism after being captured. As a model for the complete removal of metals from soil, several microorganisms have the ability to convert metal ions to Hg^{2+} and Ag^{+}. Extracellular enzymes produced by rhizobacteria include glutathione-S-transferase, reductases, lacases, peroxidases and Cytochome P450. These enzymes are crucial in the breakdown of polycyclic aromatic hydrocarbons, which are compounds found in crude oil and are known to be harmful and carcinogenic. Meliani's assessment of the use of PGPR in soil decontamination concentrates on the bioremediation techniques used by the genus *Pseudomonas*, including siderophore synthesis, bio surfactants, biofilms and the solubilization of hazardous metals (Zhou et al., 2020).

Zinc Solubilizing bacteria improve soil quality and plant growth in stressful environments. They keep ions from building up and water shortages prevent by giving root cells suitable solutes. In the rhizosphere, some bacteria produce exopolysaccharides (EPS), which improve ion and water uptake and are crucial for stabilizing soil structure. In response to ionic stress, PGPR express ion antiporters to preserve the cytoplasmic water balance. They generate antioxidant and stress-mitigating enzymes to get rid of reactive oxygen species (ROS) that are created when abiotic stressors occur. Under water stress, the physiology of plants is stabilized by these microorganisms' phytohormones (Jian et al., 2019). Beneficial soil bacteria employ processes in PTE bioremediation to change and adsorb harmful substances from the soil. Among greengram plants, *P. aeruginosa* markedly increased both shoot as well as root nutrient content including total nitrogen uptake, phosphorus uptake and seed output at all tested dosages of tebuconazole, a fungicide frequently used in irrigation. However, high concentrations can accumulate in soils and plants, thus becoming potentially toxic as well as damaging to plant tissues. *P. protegens* were found to have a positive correlation with the plant physiology including shoot length, plant biomass and germination percentage rate in a 2016 study by Bensidhoum on barley growth restoration under the influence of hazardous metal contamination. In PTE-contaminated soil, *A. feacalis* as well as *P. aeruginosa* were shown to have considerable potential as strong bioremediant in wheat development during heavy metal contamination. *Azospirillum lipoferum* strain 5 has been shown by Muratova et al. (2005) to improve wheat root systems in soil contaminated with crude oil. In a greenhouse experiment, Gomaa et al. (2012) investigated the ability of *Azospirillum lipoferum* and *Rhizobium leguminosarum* to stimulate wheat growth at varying zinc and copper concentrations. In hydrocarbon and toxic metal-contaminated soils, a *Pseudomonas* sp. strain induced root and shoot elongation in rice and improved its ultimate yield. Table 8.4 records the potential of some zinc solubilizing microbes that are used in bioremediation.

8.13 ZNSB AS A STRESS ALLEVIATOR

"Stresses" are terms used to describe external conditions that negatively impact a plant's ability to grow and develop. Plants are supposedly considered exceptional for their ability to sense and react to stress signals, such as those caused by extreme temperatures, salt, drought and pathogen. This knack of plants to survive a variety of biotic as well as abiotic stimuli depends on their ability to quickly adapt and adjust their innate tolerance mechanisms. However, under highly stressed settings, plants

TABLE 8.4
Potential of Some Zinc Solubilizing Bacteria for Bioremediation

Bacterial Genera	Soil-Related Problem	Targeted Crops	Reference
Bacillus	Fungal toxicity and soil-borne pathogens	Rice	Shen et al. (2019)
		Groundnut	Ahmad et al. (2019)
	Chromium metal intoxication	Durum wheat	Mazhar et al. (2020)
Pseudomonas	Arsenic metal intoxication	Rice	Xiao et al. (2020)
	Cadmium, lead and zinc metal intoxication	Spinach	Shilev et al. (2020)
Citrobacteria and *Enterobacteria*	Cadmium, lead and nickel metal intoxication	Common wheat	Ajmal et al. (2022)
Pantoea and *Klebsiella*	Cadmium and zinc metal intoxication	Napier grass	Sumranwanich et al. (2022)
Agrobacteria and *Sinorhizobia*	Copper and zinc metal intoxication	Hop clover	Jian et al. (2019)
Bacillus and *Azotobacter*	Metal intoxication	Garden pea	Singh et al. (2019)
Anabaena Nostoc	Fungicidal toxicity and soil-borne pathogens	Rice	Zhou et al. (2020)
Serratia spp.	Augmented plant physiological parameters and activity of antioxidant enzymes	Maize	Anzuay et al. (2017)
Lysinibacillus spp.	Elevated levels of ascorbic acid, proline, total phenol content and chlorophyll content (both a and b)	Maize	Jinal et al. (2021)
B. seminalis and *Burkholderia vietnamiensis*	Stimulated the breakdown of potassium and phosphate and the synthesis of indole acetic acid (IAA)	Rice	Bhakat et al. (2021)
Serratia sp.	Increased overall chlorophyll content, shoot length and root length	Maize	Kour et al. (2019)

are unable to withstand these unfavorable conditions; hence, exogenous stress-alleviating and environmentally friendly techniques are needed. PGPR, especially ZnSB, have a variety of functions, which makes them popular for reducing abiotic pressures brought on by climate change. Augmenting ZSB in the plant rhizosphere may be a useful substitute strategy to boost host plant growth and production since ZSB can confer resistance on plants against specific abiotic stresses. *Arthrobacter protophormiae* and *Dietzia natronolimnaea* were found to improve salt tolerance by Barnawal et al. (2017) whereas *Bacillus subtilis* modified the levels of phytohormone in wheat plants to boost drought resistance. Under their study, Jha and Subramanian (2018) observed that rice plants treated with *Bacillus pumilus* and *Pseudomonas pseudoalcaligenes* expressed more genes linked to stress and had greater levels of osmoprotectants pursuant to salinity-stressed circumstances.

Methylobacterium sp. inoculated potato plants showed enhanced resistance to salt and several fungal diseases, as well as an increase in the number of auxiliary roots as well as leaves and rosette diameter. Wheat plants infected with ZnSB showed increased commotion of many antioxidant enzymes, including catalase, peroxidase and superoxide dismutase, which improved the plants' tolerance to salt stress. In low-nutrient and drought-prone conditions, petunia plants grew more rapidly when exposed to *Pseudomonas fluorescens* and *P. poae*. *Bacillus subtilis* and *pseudomonas fluorescens* increased the build-up of proline, which is an active osmolyte, which counteracted the effects of salinity. In the same vein, *Pseudomonas* sp. shielded tomatoes against salt stress, according to Orozco-Mosqueda et al. (2019). According to El-Esawi et al. (2019) *Azospirillum lipoferum* modulated chickpea's stress induced gene expression, antioxidant enzyme production and osmolytes synthesis, therefore mitigating the adverse effects of salt stress. In addition to abiotic elements, microorganism's viz. bacteria, viruses, fungi, viroids and others can expose plants to a variety of biotic agents, which can lead to biotic stress and eventually lower agricultural output. ZnSB demonstrate a variety of direct and indirect strategies, including the creation of antioxidants, hormonal substances, cell wall-degrading enzymes and secondary metabolites, to decrease illnesses caused by different pathogens.

Antibiotics, which are antagonistic low-molecular-weight chemical compounds produced by many ZnSB, are very efficient in preventing the growth of phytopathogens. *Bacillus* species, for instance, are capable of producing antibiotics such as circulin, colistin and polymyxin. According to Ali et al. (2020) *Pseudomonas fluorescens*, *P. putida*, *P. aeruginosa* and *P. stutzeri* generated rhamnolipids, violacein, phenazine, pyoverdines and pyochelin which considerably were quite effective against several fungal phyto-pathogens. Fernandez et al. (2020) research revealed that *Pseudomonas fluorescens* generated two distinct S-type bacteriocins: tailocins, which resemble phage tails, and colicin. Chitinase, protease, β-1, 3-glucanase, cellulase and other lytic enzymes are synthesized in response to phytopathogens, and certain ZnSB have inhibitory effects on these enzymes. Numerous investigations have demonstrated that lytic enzymes produced by rhizospheric bacteria degrade cell walls of plant root diseases, such as *Fusarium oxysporum* and *Rhizoctonia solani*, causing cell death (Table 8.4). The anti-maize rot pathogen characteristics of *Bacillus cereus*, *B. megaterium* and other *Bacillus* sp. were documented by Cheng et al. (2019). Furthermore, for promoting plant growth, some non-pathogenic organisms in soil

TABLE 8.5
Discrepancy of ZnSB in Extenuating the Bearing of Biotic as Well as Abiotic Stresses in Various Crops

Stresses	Zinc Solubilizing Strain	Plant	Mode of Action	Reference
Salt stress	*B. amyloliquefaciens*	*T. aestivum* L.	Elevated calcium and potassium absorption and translocation	Singh et al. (2021)
	P. pseudoalcaligenes and *B. pumilu*	*O. sativa*	Enhanced levels of antioxidant enzymes, carotenoids and chlorophyll	Jha and Mohamed (2023)
	P. fragi and *P. agglomerans*	*P. vulgaris*	Elevated levels of osmoprotectants, carotenoid and chlorophyll, as well as enhanced activity of antioxidant enzymes	Gupta et al. (2023)
	Bacillus spp.	*Triticum aestivum L.*	Elevated Zn concentration in grains and shoots, as well as increased plant growth metrics	Ain et al. (2020)
Famine	*Bacillus* spp.	*Zea mays*	Enhanced biochemical and physiological characteristics along with decreased activity of antioxidant enzymes	Azeem et al. (2022)
	Azotobacter	*Zea mays*	Elevated development of plants	Shirinbayan et al. (2019)
Temperature	*Stenotrophomonas*	*Zea mays*	Elevated levels of chlorophyll, auxins and carbohydrates as well as increased resistance to heat stress	Rezaee Niko et al. (2018)
	L. sphaericus and *L. fusiformis*		Increased levels of total phenols and ACC deaminase, as well as antioxidant defense system. Increased cell viability, lignin content and osmolytes accumulation	Pantoja-Guerra et al. (2023)
Disease	*Bacillus* sp. and *B. cereus*	*O. sativa*	Enhanced the yield while inhibiting the growth of Fusarium moniliforme and Pyricularia oryzae	Rais et al. (2017)
	T. lixii	*Solanum lycopersicum*	Lessened severity of early blight and Fusarium wilt	Sachdev et al. (2022)
	B. pumilus	*Oryza sativa*	Decreased brown spot illness and slowed the growth of fungi	Weinand et al. (2023)

have the ability to induce systemic resistance, or ISR. Because of PGPR, a defense mechanism against pathogen infections, many plants acquire systemic resistance (both ISR and SAR, or systemic acquired resistance) to a variety of biotic stressors. *Vigna radiata* growing in disease-prone conditions was given systemic resistance by *P. putida* and *B. subtilis* bacterial strains. By up-regulating the expression of pathogenesis related (PR) genes and augmenting the activity of β-1,3 glucanase, peroxidase, phenylalanine ammonia lyase, chitinase and polyphenol oxidase in the leaves, *B. amyloliquefaciens* increased susceptibility of tomato plant to yellow leaf curl virus disease. To put it briefly, soil rhizobacteria have the capacity to sustainably lessen the effects of a variety of environmental restrictions, both biotic and abiotic.

8.14 CONCLUSION

Zinc (Zn) is considered as one of the vital micronutrients that play an incredible part in various metabolic activities of plants like photosynthesis, DNA and RNA synthesis, pollen synthesis, development of root system and shoot system. It protects plant species from biotic and abiotic stresses. Its presence in available form (Zn^{2+}) is highly needed by the plants. Zinc deficiency was reported worldwide especially in our country India. As per published work, Jammu and Kashmir soil is also deficient in Zn. The paucity of zinc in the soil is abridged by applications of zinc-based supplements like fertilizer, but soon after the fertilization maximum amount of zinc gets fixed or locked in the soil. On other hand synthetic fertilizers have a damaging impact on plant as well as soil health. Plant growth promoting microbes are probed by the research community to overcome the zinc deficit in soils and eventually in plants through countless mechanisms. The role of several zinc solubilizing microbes has been sightseen and testified in the literature; however, finding new potential ZnSMs and using them in biofertilizers or microbial formulations or bioagents is the need of the hour as they can help to enhance the microbial ecology in the desired regions. Future studies utilizing metabolomics, proteomics and functional genomics may create ZnSMs formulations for long-term nutritional deficits in biofertilizers. Organic substrates are frequently broken down by microbes, excluding inorganic contaminants in the process. It is worthwhile to think about using microbial enzymes and microorganisms in the bioremediation of radioactive waste. However, through algal blooms, bacteria can also be a part of pollution. Various temperatures cause various chemicals to degrade, and different environments favor different microbial survival. To find the ideal temperature for deterioration and strike a balance between ambient temperature and microbe survival, field study is required. It's critical to monitor how bacteria affect the ecosystem after bioremediation since some species may mutate and cause contamination. Regulators must keep an eye out for any hazards that can arise from microorganisms in certain settings.

REFERENCES

Abaid-Ullah, H. S., Munis, M. F. H., Fahad, S., & Yang, X. (2015). Phytoremediation of heavy metals assisted by plant growth promoting (PGP) bacteria: A review. *Environmental and Experimental Botany*, *117*, 28–40.

Afridi, M. S., Amna, Sumaira, Mahmood, T., Salam, A., Mukhtar, T., Mehmood, S., Ali, J., Khatoon, Z., & Bibi, M. (2019). Induction of tolerance to salinity in wheat genotypes by plant growth promoting endophytes: Involvement of ACC deaminase and antioxidant enzymes. *Plant Physiology and Biochemistry*, *139*, 569–577.

Ahkami, A. H., Allen, White, R., Handakumbura, P. P., & Jansson, C. (2017). Rhizosphere engineering: Enhancing sustainable plant ecosystem productivity. *Rhizosphere*, *3*, 233–243.

Ahmad, A. G. M., Attia, A. Z. G., Mohamed, M. S., & Elsayed, H. E. (2019). Fermentation, formulation and evaluation of PGPR *Bacillus subtilis* isolate as a bioagent for reducing occurrence of peanut soil-borne diseases. *Journal of Integrative Agriculture*, *18*, 2080–2092.

Ain, N. U., Naveed, M., Hussain, A., Mumtaz, M. Z., Rafique, M., Bashir, M. A., Alamri, S., & Siddiqui, M. H. (2020). Impact of coating of urea with *bacillus*-augmented zinc oxide on wheat grown under salinity stress, *Plants*, *9*, 1375.

Ajmal, A. W., Yasmin, H., Hassan, M. N., Khan, N., Jan, B. L., & Mumtaz, S. (2022). Heavy Metal–Resistant Plant Growth–Promoting *Citrobacter werkmanii* Strain WWN1 and *Enterobacter cloacae* Strain JWM6 Enhance Wheat (*Triticum aestivum* L.) Growth by Modulating Physiological Attributes and Some Key Antioxidants Under Multi-Metal Stress. *Frontiers in Microbiology*, *13*, 500.

Albdaiwi, R. N., Khyami-Horani, H., Ayad, J. Y., Alananbeh, K. M., & Al-Sayaydeh, R. (2019). Isolation and characterization of Halotolerant Plant Growth Promoting Rhizobacteria from Durum Wheat (*Triticum turgidum* subsp. durum) Cultivated in Saline Areas of the Dead Sea Region. *Frontiers in Microbiology*, *10*, 1639.

Alexander, M. (1997). *Introduction to soil microbiology*. Wiley.

Ali, S., Hameed, S., Shahid, M., Iqbal, M., Lazarovits, G., & Imran, A. (2020). Functional characterization of potential PGPR exhibiting broad-spectrum antifungal activity. *Microbiological Research*, *232*, 126389.

Alloway, B. J. (1995). Heavy metals in soils (2nd Edn.). Blackie Academic & Professional.

Alloway, B. J. (2009). Soil factors associated with zinc deficiency in crops and humans. *Environmental Geochemistry and Health*, *31*(5), 537–548.

Altomare, C., Norvell, W. A., Björkman, T., & Harman, G. E. (1999). Solubilization of phosphates and micronutrients by the plant-growth-promoting and biocontrol fungus *Trichoderma harzianum Rifai* 1295–22. *Applied and Environmental Microbiology*, *65*(7), 2926–2933.

Ambreetha, S., & Balachandar, D. (2019). Rhizobacteria-mediated root architectural improvement: A hidden potential for agricultural sustainability. In A. Kumar & V. Meena (Eds.). *Plant Growth Promoting Rhizobacteria for Agricultural Sustainability* (pp. 111–128). Springer.

Anitha, S., Devi, P. S. N., & Kumari, S. K. (2015). Isolation and identification of zinc solubilizing fungal isolates from agricultural fields, *The Indian Journal of Agricultural Sciences*, *85*(12), 1638–1642.

Anzuay, M. S., Ciancio, M. G. R., Ludueña, L. M., Angelini, J. G., Barros, G., Pastor, N., & Taurian, T. (2017). Growth promotion of peanut (Arachis hypogaea L.) and maize (*Zea mays* L.) plants by single and mixed cultures of efficient phosphate solubilizing bacteria that are tolerant to abiotic stress and pesticides. *Microbiological Research*, *199*, 98–109.

Azadeh, B., Asgharzadeh, A., Mujallali, H., Khavazi, K., & Pazira, E. (2012). Evaluation of Zinc solubilization potential by different strains of Fluorescent Pseudomonads. *Journal of Applied Sciences and Environmental Management, 16*(3), 295–298.

Azeem, M., Haider, M. Z., Javed, S., Saleem, M. H., & Alatawi, A. (2022). Drought stress amelioration in maize (*Zea mays* L.) by inoculation of *Bacillus* spp. strains under sterile soil conditions. *Agriculture*, *12*, 50.

Azhar, H., Zahir, M. A., Zahir, Z., & Asghar, M. (2015). Prospects of zinc solubilizing bacteria for enhancing growth promotion of wheat. *Pakistan Journal of Agricultural Sciences*, *52*(4), 915–922.

Bahrani, A., Pourreza, J., & Joo, M. H. (2010). Response of winter wheat to co-inoculation with *Azotobacter* and arbuscular mycorrhizal fungi (amf) under different sources of nitrogen fertilizer. *American-Eurasian Journal of Sustainable Agriculture*, *8*, 95–103.

Barber, S. A. (1995). *Soil nutrient bioavailability* (2nd Edn.). Wiley.

Barnawal, D., Bharti, N., Pandey, S. S., Pandey, A., Chanotiya, C. S., & Kalra, A. (2017). Plant growth-promoting rhizobacteria enhance wheat salt and drought stress tolerance by altering endogenous phytohormone levels and TaCTR1/TaDREB2 expression. *Physiologia Plantarum*, *161*, 502–514.

Bashir, Z. (2021b). Characterization of Potassium Solubilizing Bacteria from Rhizospheric Soils of Cherry (*Prunusavium*). *Research Journal of Agricultural Sciences*, *12*(2), 494–496.

Bashir, Z., Hamid, B., Zargar, M. Y., & Kundoo, A. A. (2021a). Morphological and Biochemical Characterization of Potassium Solubilizing Bacteria from Rhizospheric Soils of Pinus (*Pinus wallichiana*). *Journal of Research and Development*, *21*, 58–63.

Bashir, Z., Zargar, M. Y., Baba, Z. A., & Mohiddin, F. A. (2018). Solubilization of potassium containing mineral by bacteria from Kashmir iris (*Iris kashmiriana*) rhizosphere. *The Pharma Innovation Journal*, *17*(4), 747–750.

Bashir, Z., Zargar, M. Y., Baba, Z. A., Mohiddin, F. A., Peer, F. A., Mir, S. A., & Najar, G. R. (2018a). Isolation and characterization of phosphorus solubilizing bacteria (PSB) from rhizospheric soils of apple (*Malus Domestica*). *Journal of Pharmacognosy and Phytochemistry*, *7*(2), 2997–3000.

Bashir, Z., Zargar, M. Y., Bhat, T. A., Kousar, S., Baba, Z. A., & Mohiddin, F. A. (2018b). Solubilization of Phosphorus Containing Mineral by Bacteria from Rhizospheric Region of Walnut (*Juglans regia*). *International Journal of Current Microbiology and Applied Sciences*, *7*(8), 2391–2398.

Bashir, Z., Zargar, M. Y., Mohiddin, F. A., Baba, Z. A., & Hamid, B. (2019). Isolation and Biochemical Characterization of phosphorus solubilizing bacteria from Rhizospheric region of Apricot and Peach. *Journal of Research and Development University of Kashmir*, *19*, 0972–5407.

Bashir, Z., Zargar, M. Y., Mohiddin, F. A., Kousar, S., Husain, M., & Rasool, F. (2017). Phosphorus solubilizing microorganisms: Mechanism and diversity. *International Journal of Communication Systems*, *5*(5), 666–673.

Bashir, Z., Zargar, M. Y., & Vishwakarma, D. K. (2019b). Potassium solubilizing microorganisms for sustainable agriculture. *Applied Agricultural Practices for Mitigating Climate Change*, *2*, 17–28.

Basu, A., Prasad, P., Das, S. N., Kalam, S., Sayyed, R. Z., Reddy, M. S., & El Enshasy, H. (2021). Plant growth promoting rhizobacteria (PGPR) as green bioinoculants: Recent developments, constraints, and prospects. *Sustainability*, *13*(3), 1140.

Bensidhoum, L., Nabti, E., Tabli, N., Kupferschmied, P., Weiss, A., Rothballer, M., Schmid, M., Keel, C., & Hartmann, A. (2016). Heavy metal tolerant *Pseudomonas protegens* isolates from agricultural well water in northeastern Algeria with plant growth promoting, insecticidal and antifungal activities. *European Journal of Soil Biology*, *75*, 38–46.

Bhakat, K., Chakraborty, A., & Islam, E. (2021). Characterization of zinc solubilization potential of arsenic tolerant Burkholderia spp. isolated from rice rhizospheric soil. *World Journal of Microbiology and Biotechnology*, *37*, 39.

Bulut, S. (2013). Evaluation of yield and quality parameters of phosphorous solubilizing and N-fixing bacteria inoculated in wheat (*Triticum aestivum* L.), *Turkish Journal of Agriculture and Forestry*, *37*, 545–554.

Burd, G. I., Dixon, D. G., & Glick, B. R. (2000). Plant growth promoting bacteria that decrease heavy metal toxicity in plants. *Canadian Journal of Microbiology*, *46*, 237–245.

Cakmak, I. (2008). Enrichment of cereal grains with zinc: Agronomic or genetic Biofortification. *Journal of Plant and Soil*, *302*(1), 1–17.

Cakmak, I., Pfeiffer, W. H., & McClafferty, B. (2010). Biofortification of durum wheat with zinc and iron. *Cereal Chemistry*, *87*, 10–20.

Chang, H. B., Lin, C. W., & Huang, H. J. (2005). Zinc-induced cell death in rice (*Oryza sativa* L.) roots. *Plant Growth Regulation*, *46*, 261–266.

Chauhan, P. K., Upadhyay, S. K., Tripathi, M., Singh, R., Krishna, D., Singh, S. K., & Dwivedi, P. (2022). Understanding the salinity stress on plant and developing sustainable management strategies mediated salt-tolerant plant growth-promoting rhizobacteria and CRISPR/Cas9. *Biotechnology & Genetic Engineering Reviews*, *725*, 213–1958.

Cheng, X., Ji, X., Ge, Y., Li, J., Qi, W., & Qiao, K. (2019). Characterization of antagonistic *Bacillus methylotrophicus* isolated from rhizosphere and its biocontrol effects on maize stalk rot. *Phytopathology*, *109*, 571–581.

Chung, H., Park, M., Madhaiyan, M., Seshadri, S., Song, J., & Cho, H. (2005). Isolation and characterization of phosphate solubilizing bacteria from the rhizosphere of crop plants of Korea. *Soil Biology and Biochemistry*, *37*, 1970–1974.

Claudia, A., Banci, L., Bertini, I., & Rosato, A. (2006). Zinc through the three domains of life. *Journal Proteome Research*, *5*(11), 3173–3178.

Dhaked, B. S., Triveni, S., Reddy, R. S., & Padmaja, G. (2017). Isolation and Screening of Potassium and Zinc Solubilizing Bacteria from Different Rhizosphere Soil. *International Journal of Current Microbiology and Applied Sciences*, *6*(8), 1271–1281.

Di Simine, C. D., Sayer, J. A., & Gadd, G. M. (1998). Solubilization of zinc phosphate by a strain of *Pseudomonas fluorescens* isolated from forest soil. *Biol Ferti Soils*, *28*(1), 87–94.

El-Esawi, M. A., Al-Ghamdi, A. A., Ali, H. M., & Alayafi, A. A. (2019). *Azospirillum lipoferum* FK1 confers improved salt tolerance in chickpea (*Cicer arietinum* L.) by modulating osmolytes, antioxidant machinery and stress-related genes expression, *Environmental and Experimental Botany*, *159*, 55–65.

Fariduddin, Q., Saleem, M., Khan, T. A., & Hayat, S. (2022). Zinc as a versatile element in plants: An overview on its uptake, translocation, assimilatory roles, and deficiency and toxicity symptoms. In S. T. Khan & A. Malik (Eds.). *Microbial Biofertilizers and Micronutrient Availability*. Springer.

Fasim, F., Ahmed, N., Parsons, R., & Gadd, G. M. (2002). Solubilization of zinc salts by a bacterium isolated from the air environment of a tannery, *FEMS Microbiology Letters*, *213*(1), 1–6.

Ferguson, B. J., Mens, C., Hastwell, A. H., Zhang, M., Su, H., Jones, C. H., & Gresshoff, P. M. (2019). Legume nodulation: The host controls the party. *Plant, Cell & Environment*, *42*(1), 41–51.

Fernandez, M., Godino, A., Príncipe, A., Ramírez, V. L., Quesada, J. M., Rigo, V., Espinosa-Urgel, M., Morales, G. M., & Fischer, S. (2020). Characterization of the bacteriocins and the PrtR regulator in a plant-associated *Pseudomonas* strain. *Journal of Biotechnology*. *307*, 182–192.

Fomina, M., Alexander, I. J., Hillier, S., & Gadd, G. M. (2004). Zinc phosphate and pyromorphite solubilization by soil plant-symbiotic fungi. *Geomicrobiology Journal*, *21*(5), 351–366.

Frassinetti, S., Bronzetti, G. L., Caltavuturo, L., Cini, M., & Della Croce, C. (2006). The role of zinc in life: A review. *Journal of Environmental Pathology, Toxicology*, *25*(3), 597–610.

Friedland, A. J. (1990). The movement of metals through soils and ecosystems. In A. J. Shaw (Ed.). *Heavy metal tolerance in plants: Evolutionary aspects* (pp. 7–19). CRC.

Gandhi, A., & Muralidharan, G. (2016). Assessment of zinc solubilizing potentiality of *Acinetobacter* sp. Isolated from rice rhizosphere. *European Journal of Soil Biology*, *76*, 1–8.

Ghosh, P. K., Hazra, K. K., Kumar, N., Nadarajan, N., & Venkatesh, M. S. (2014). *Resource Conservation Technology in Pulses, 'Zinc in crops and human health—An overview* (pp. 213–223). Scientific Publishers.

Glick, B. R. (1995). The enhancement of plant growth by free-living bacteria. *Canadian Journal of Microbiology, 41*(2), 109–117.

Golla, B. (2021). Agricultural production system in arid and semi-arid regions. *The Journal of Agricultural Science Food Technology, 7*, 234–244.

Gomaa, A. M., Al-Fassi, F. A., Al-Kenawy, Z., & Al-Gharbawi, H. T. (2012). Role of *Azospirillum* and *Rhizobium* in bio-remediating Cd and Zn polluted soil cultivated with wheat plant. *Australian Journal of Basic and Applied Sciences, 6*, 550–556.

Gondal, A. H., Zafar, A., Zainab, D., Toor, M. D., Sohail, S., & Ameen, S. (2021). A detailed review study of Zn involvement in animal, plant and human nutrition. *Indian Journal of Pure & Applied Biosciences, 9*(2), 262–271.

Goteti, P. K., Daniel, L., Emmanuel, A., Desai, S., & Hassan, M. (2013). Prospective zinc solubilising bacteria for enhanced nutrient uptake and growth promotion in maize (*Zea mays* L.). *International Journal of Microbiology, 20333*, 1.

Gupta, S., Pandey, S., Kotra, V., & Kumar, A. (2023). Assessing the role of ACC deaminase-producing bacteria in alleviating salinity stress and enhancing zinc uptake in plants by altering the root architecture of French bean (*Phaseolus vulgaris*) plants. *Planta, 258*, 3.

Hafeez, F. Y., Abaid-Ullah, M., & Hassan, M. N. (2013). Plant growth-promoting rhizobacteria as zinc mobilizers: A promising approach for cereals biofortification. In Dinesh K. Maheshwari, Meenu Saraf, Abhinav Aeron (Eds.). *Bacteria in agrobiology: Crop productivity* (pp. 217–235). Springer.

Hafeez, F. Y., Hameed, S., Ahmad, T., & Malik, K. A. (2001). Competition between effective and less effective strains of *Bradyrhizobium* spp. For nodulation on *Vigna radiata. Biology and Fertility of Soils, 33*, 382–386.

Hafeez, F. Y., Naeem, F. I., Naeem, R., Zaidi, A. H., & Malik, K. A. (2005). Symbiotic effectiveness and bacteriocin production by Rhizobium *leguminosarum* bv. Viciae isolated from agriculture soils in Faisalabad. *Environmental and Experimental Botany, 54*, 142–147.

Hamid, B., & Bashir, Z. (2019a). Potassium solubilization microorganisms: An Alternative technology to chemical fertilizers. *Journal of Research and Development University of Kashmir*, 19, 0972–5407.

Hartmann, A., Rothballer, M., Schmid, M., & Hiltner, L. (2008). A pioneer in rhizosphere microbial ecology and soil bacteriology research. *Plant Soil, 312*, 7–14.

Hotz, C., & Brown, K. H. (2004). Assessment of the risk of zinc deficiency in populations and options for its control. *Food and Nutrition Bulletin, 25*, 94–204.

Hussain, A., Arshad, M., Zahir, Z. A., & Asghar, M. (2015). Prospects of zinc solubilizing bacteria for enhancing growth of maize, Pak. *The Journal of Agricultural Science, 52*(4), 915–922.

International Zinc Association (IZA). (2011). http://www.zinc.org/sustainability

Jha, Y., & Mohamed, H. I. (2023). Inoculation with *Lysinibacillus fusiformis* strain YJ4 and *Lysinibacillus sphaericus* strain YJ5 alleviates the effects of cold stress in maize plants. *Gesunde Pflanz, 75*, 77–95.

Jha, Y., & Subramanian, R. B. (2018). From interaction to gene induction: An eco-friendly mechanism of PGPR-mediated stress management in the plant. In *Plant microbiome: Stress response* (pp. 217–232). Springer.

Jian, L., Bai, X., Zhang, H., Song, X., & Li, Z. (2019). Promotion of growth and metal accumulation of alfalfa by coinoculation with *Sinorhizobium* and *Agrobacterium* under copper and zinc stress. *Peer J, 7*, e6875.

Jinal, H. N., Gopi, K., Kumar, K., & Amaresan, N. (2021). Effect of zinc-resistant Lysinibacillus species inoculation on growth, physiological properties, and zinc uptake in maize (*Zea mays* L.). *Environmental Science and Pollution Research, 28*, 6540–6548.

Kabata-Pendias, A. (2000). *Trace elements in soils and plants*. CRC Press.
Kabata-Pendias, A., & Pendias, H. (2001). *Trace elements in soils and plants* (3rd ed.). CRC Press.
Kalpana, B., & Dinesh, K. M. (2020). Zinc solubilizing bacteria (*Bacillus megaterium*) with multifarious plant growth promoting activities alleviates growth in *Capsicum annuum* L. *3 Biotech*, *10*(2), 36.
Kamran, S., Shahid, I., Baig, D. N., Rizwan, M., Malik, K. A., & Mehnaz, S. (2017). Contribution of zinc solubilizing bacteria in growth promotion and zinc content of wheat. *Front Microbiology*, *8*, 2593.
Khande, R., Sushil, K. S., Ramesh, A., & Mahaveer, P. S. (2017). Zinc solubilizing *Bacillus* strains that modulate growth, yield and zinc biofortification of soybean and wheat. *Rhizosphere*, *4*, 126–138.
Kordatzaki, G., Katsenios, N., Giannoglou, M., Andreou, V., Chanioti, S., Katsaros, G., Savvas, D., & Efthimiadou, A. (2022). Effect of foliar and soil application of plant growth promoting bacteria on kale production and quality characteristics. *Scientia Horticulturae*, *301*, 111094.
Kour, R., Jain, D., Bhojiya, A. A., Sukhwal, A., Sanadhya, S, Saheewala, H., Jat, G., Singh, A., & Mohanty, S. R. (2019). Zinc biosorption, biochemical and molecular characterization of plant growth-promoting zinc-tolerant bacteria. *Biotech*, *9*, 421.
Kumar, A. S., Meenakumari, K. S., & Anith, K. N. (2017). Screening for Zn solubilisation potential of soil bacteria from Zn deficient soils of Kerala. *Journal of Tropical Agriculture*, *54*(2), 194.
Kutman, U. B., Yildiz, B., Ozturk, L., & Cakmak, I. (2010). Biofortification of durum wheat with zinc through the soil and foliar applications of nitrogen. *Cereal Chemistry*, *87*(1), 1–9.
Lindsay, W. L. (1979). *Chemical equilibrium in soils*. Wiley Interscience.
López-Rayo, S., Nadal, P., & Lucena, J. J. (2016). Novel chelating agents for iron, manganese, zinc and copper mixed fertilisation in high pH soil-less cultures. *Journal of the Science of Food and Agriculture*, *96*(4), 1111–1120.
Mahdi, S. S., Dar, S. A., Ahmad, S., & Hassan, G. I. (2010b). Zinc availability: A major issue in agriculture. *Research Journal of Agricultural Sciences*, *3*(3), 78–79.
Mahdi, S. S., Hassan, G. I., Samoon, S. A., Rather, H. A., Dar, S. A., & Zehra, B. (2010a). Biofertilizers in organic agriculture. *Journal of Phytopathology*, *2*(10), 42–54.
Malik, A., & Khan, S. T. (Eds.). (2022). *Microbial biofertilizers and micronutrient availability: The role of Zn in agriculture and human health*. Springer.
Manasa, S. G., Mahadevaswamy Nagaraj, M., Ramesh, Y., & Gundappagol, R. C. (2019). Isolation and characterization of zinc solubilizing bacteria from rhizosphere soils of paddy grown in Tungabhadra Command Area. *International Journal of Current Microbiology and Applied Sciences*, *8*(10), 215–222.
Mattey, M. (1992). The production of organic acids. *Critical Reviews in Biotechnology*, *12*, 87–132.
Mazhar, R., Ilyas, N., Arshad, M., Khalid, A., & Hussain, M. (2020). Isolation of Heavy Metal-Tolerant PGPR strains and amelioration of chromium effect in Wheat in combination with Biochar. *Iranian Journal of Science and Technology, Transaction A, Science*, *44*, 1–12.
Mori, A., Kirk, G. J., Lee, J. S., Morete, M. J., Nanda, A. K., Johnson-Beebout, S. E., & Wissuwa, M. (2016). Rice genotype differences in tolerance of zinc-deficient soils: Evidence for the importance of root-induced changes in the rhizosphere. *Frontiers in Plant Science*, *6*, 1160.
Mortvedt, J. J., & Gilkkes, R. J. (1993). Zinc fertilizers. In A. D. Robson (Ed.), *Zinc in soils and plants*. Kluwer Academic Publishers.
Muratova, A. Y., Turkovskaya, O. V., Antonyuk, L. P., Makarov, O. E., Pozdnyakova, L. I., & Ignatov, V. V. (2005). Oil-oxidizing potential of associative rhizobacteria of the genus *Azospirillum*, *Microbiology*, *74*, 210–215.

Noulas, C., Tziouvalekas, M., & Karyotis, T. (2018). Zinc in soils, water and food crops, *Journal of Trace Elements in Medicine and Biology*, *49*, 252–260.

Orozco-Mosqueda, M. d. C., Duan, J., DiBernardo, M., Zetter, E., Campos-García, J., Glick, B. R., & Santoyo, G. (2019). The production of ACC deaminase and trehalose by the plant growth promoting bacterium *Pseudomonas* sp. UW4 synergistically protect tomato plants against salt stress. *Frontiers in Microbiology*, *10*, 1392.

Ortíz-Castro, R., Contreras-Cornejo, H. A., Macías-Rodríguez, L., & López-Bucio, J. (2009). The role of microbial signals in plant growth and development. *Plant Signaling & Behavior*, *4*, 701–712.

Pantoja-Guerra, M., Burkett-Cadena, M., Cadena, J., Dunlap, C. A., & Ramírez, C. A. (2023). *Lysinibacillus* spp.: An IAA-producing endospore forming-bacteria that promotes plant growth. *Antonie van Leeuwenhoek*, *116*, 615–630.

Pawar, A., Ismail, S., Mundhe, S., & Patil, V. D. (2015). Solubilization of insoluble zinc compounds by different microbial isolates in vitro condition. *International Journal of Tropical Agriculture*, *33*, 865–869.

Potarzycki, W., & Grzebisz, W. (2009). Effect of zinc foliar application on grain yield of maize and its yielding components. *Plant Soil Environment*, *55*(12), 519–527.

Rais, A., Jabeen, Z., Shair, F., Hafeez, F. Y., & Hassan, M. N. (2017). *Bacillus* spp., a bio-control agent enhances the activity of antioxidant defense enzymes in rice against *Pyricularia oryzae*. *PLoS ONE*, *12*, e0187412.

Ramesh, A., Sharma, S. K., Sharma, M. P., Yadav, N., & Joshi, O. P. (2014). Inoculation of zinc solubilizing *Bacillus aryabhattai* strains for improved growth, mobilization, and biofortification of zinc in soybean and wheat cultivated in Vertisols of Central India. *Applied Soil Ecology*, *73*, 87–96.

Rana, A., Joshi, M., Prasanna, R., Shivay, Y. S., & Nain, L. (2012). Biofortification of wheat through inoculation of plant growth promoting rhizobacteria and *cyanobacteria*. *European Journal of Soil Biology*, *50*, 118–126.

Rani, N., Kaur, R., & Kaur, S. (2020). Zinc solubilizing bacteria to augment soil fertility—A comprehensive review. *International Journal of Agricultural Sciences and Veterinary Medicine*, *8*(1), 38–44.

Rezaee Niko, B., Enayatizamir, N., & Norozi Masir, M. (2018). Effect of zinc solubilizing growth promoter bacterium on plant growth under laboratory conditions. *Agricultural Engineering*, *41*, 113–132.

Sachdev, S., Bauddh, K., & Singh, R. P. (2022). Native rhizospheric microbes mediated management of biotic stress and growth promotion of tomato. *Sustainability*, *15*, 593.

Sadeghzadeh, B. (2013). A review of zinc nutrition and plant breeding. *Journal of Soil Science and Plant Nutrition*, *13*(4), 905–927.

Saran, N. S., Gomathy, M., Manikandan, K., & Suresh, K. (2019). Isolation and characterization of zinc solubilizing bacteria from soils of Thoothukudi district. *International Journal of Microbiology Research*, *11*(6), 1620–1623.

Sarathambal, C., Thangaraju, M., Paulraj, C., & Gomathy, M. (2010). Assessing the Zinc solubilization ability of *Gluconacetobacter diazotrophicus* in maize rhizosphere using labelled 65 Zn compounds. *Indian Journal of Microbiology*, *50*(1), 103–109.

Saravanan, V. S., Kumar, M. R., & Sa, T. M. (2011). Microbial zinc solubilization and their role on plants. In D. K. Maheshwari (Ed.). *Bacteria in agrobiology: Plant nutrient management* (pp. 47–63). Springer.

Saravanan, V. S., Madhaiyan, M., & Thangaraju, M. (2007a). Solubilization of zinc compounds by the diazotrophic, plant growth promoting bacterium *Gluconacetobacter diazotrophicus*. *Chemosphere*, *66*, 1794–1798.

Saravanan, V. S., Osborne, J., Madhaiyan, M., Mathew, L., Chung, J., Ahn, K., & Sa, T. (2007b). Zinc metal solubilization by *Gluconacetobacter diazotrophicus* and induction of pleomorphic cells. *Journal of Microbiology and Biotechnology*, *17*(9), 1477–1482.

Saravanan, V. S., Subramoniam, S. R., & Raj, S. A. (2004). Assessing in vitro solubilization potential of different zinc solubilizing bacterial (ZSB) isolates. *Brazilian Journal of Microbiology*, *35*(1–2), 121–125.

Sayer, A. J., & Gadd, M. G. (1997). Solubilization and transformation of insoluble inorganic metal compounds to insoluble metal oxalates by *Aspergillus niger*. *Mycologicalogy Research*, *101*, 653661.

Schmidt, H., Nunan, N., Höck, A., Eickhorst, T., Kaiser, C., & Woebken, D. (2018). Recognizing patterns: Spatial analysis of observed microbial colonization on root surfaces. *Frontiers in Environmental Science*, *6*, 61–65.

Shaikh, S. S., & Saraf, M. S. (2017). Optimization of growth conditions for Zinc solubilizing plant growth associated with bacteria and fungi. *Journal of Advanced Research*, *2*(1), 1–9.

Shakeel, M., Rais, A., Hassan, M. N., & Hafeez, F. Y. (2015). Root associated *Bacillus* sp. improves growth, yield and zinc translocation for Basmati rice (*Oryza sativa*) varieties. *Front Microbiology*, *6*, 1286.

Shen, F. T., Yen, J. H., Liao, C. S., Chen, W. C., & Chao, Y. T. (2019). Screening of rice endophytic biofertilizers with fungicide tolerance and plant growth-promoting characteristics. *Sustainability*, *11*, 1133.

Shilev, S., Babrikova, I., & Babrikov, T. (2020). Consortium of plant growth-promoting bacteria improves spinach (*Spinacea oleracea* L.) growth under heavy metal stress conditions. *Journal of Chemical Technology & Biotechnology*, *95*, 932–939.

Shirinbayan, S., Khosravi, H., & Malakouti, M. J. (2019). Alleviation of drought stress in maize (*Zea mays*) by inoculation with *Azotobacter* strains isolated from semi-arid regions. *Applied Soil Ecology*, *133*, 138–145.

Singh, B. K., Millard, P., Whiteley, A. S., & Murrell, J. (2004). Unravelling rhizosphere-microbial interactions: Opportunities and limitations. *Trends Microbiology*, *12*, 386–393.

Singh, B. K., Natesan, S. K. A., Singh, B. K., & Usha, K. (2005). Improving zinc efficiency of cereals under zinc deficiency. *Current Science*, *88*(1), 36–44.

Singh, D. P., Singh, V., Gupta, V. K., Shukla, R., Prabha, R., Sarma, B., & Patel, J. S. (2020). Microbial inoculation in rice regulates antioxidative reactions and defense related genes to mitigate drought stress. *Scientific Reports*, *10*, 4818.

Singh, G., & Mukerji, K. G. (2006). Root exudates as determinant of rhizospheric microbial biodiversity. In K. G. Mukerji, C. Manoharachary & J. Singh (Eds.). *Microbial Activity in the Rhizosphere* (pp. 39–53). Springer.

Singh, S., Kumar, V., Sidhu, G. K., Datta, S., Dhanjal, D. S., Koul, B., Janeja, H. S., & Singh, J. (2019). Plant growth promoting rhizobacteria from heavy metal contaminated soil promote growth attributes of *Pisum sativum* L. *Biocatalysis and Agricultural Biotechnology*, *17*, 665–671.

Singh, U. B., Malviya, D., Singh, S., Singh, P., Ghatak, A., Imran, M., Rai, J. P., Singh, R. K., Manna, M. C., & Sharma, A. K. (2021). Salt-tolerant compatible microbial inoculants modulate physio-biochemical responses enhance plant growth, Zn biofortification and yield of wheat grown in saline-sodic soil. *International Journal of Environmental Research and Public Health*, *18*, 9936.

Sumranwanich, T., Leartsiwawinyu, W., Meeinkuirt, W., & Chayapan, P. (2022). Application of plant growth-promoting rhizobacteria (PGPR) associated with energy plant, Pennisetum purpurenum, in cadmium and zinc contaminated soil, *Research Square*, 1–15.

Suzuki, M., Tsukamoto, T., Inoue, H., Watanabe, S., Matsuhashi, S., Takahashi, M., Nakanishi, H., Mori, S., & Nishizawa, N. K. (2008). Deoxymugineic acid increases Zn translocation in Zn-deficient rice plants. *Plant Molecular Biology*, *66*(6), 609–617.

Tariq, M., Hameed, S., Malik, K. A., & Hafeez, F. Y. (2007). Plant root-associated bacteria for zinc mobilization in rice. *Pakistan Journal of Botany*, *39*(1), 245–253.

Upadhyay, A., & Srivastava, S. (2014). Mechanism of zinc resistance in plant growth promoting *Pseudomonas fluorescens* strain. *World Journal of Microbiology and Biotechnology, 30*(8), 2273–2282.

Upadhyay, S. K., & Chauhan, P. K. (2022). Optimization of eco-friendly amendments as sustainable asset for salt-tolerant plant growth promoting bacteria mediated maize (*Zea mays* L.) plant growth, Na uptake reduction and saline soil restoration. *Environmental Research, 211*, 113081.

Vaid, S. K., Kumar, B., Sharma, A., Shukla, A. K., & Srivastava, P. C. (2014). Effect of zinc solubilizing bacteria on growth promotion and zinc nutrition of rice. *Journal of Soil Science and Plant Nutrition, 14*(4), 889–910.

Vandana, N., & Manu, S. (2021). Isolation screening and molecular characterization of zinc solubilizing bacteria and their effect on the growth of wheat (*Triticum aestivum*). *Molecular Biology and Biotechnology, 29*(2), 85–97.

Vazquez, P., Holguin, G., Puente, M. E., Lopez-Cortes, A., & Bashan, Y. (2000). Phosphate-solubilizing microorganisms associated with the rhizosphere of mangroves in a semiarid coastal lagoon. *Biology and Fertility of Soils, 30*(5–6), 460–468.

Vidyashree, D. N., Muthuraju, N., Panneerselvam, P., & Mitra, D. (2018). Organic acids production by zinc solubilizing bacterial isolates. *International Journal of Current Microbiology and Applied Sciences, 7*(10), 626–633.

Wakatsuki, T. (1995). Metal oxidoreduction by microbial cells. *Journal of Industrial Microbiology and Biotechnology, 14*(2), 169–177.

Wani, M. A., Wani, J. A., Bhat, M. A., Kirmani, N. A., Wani, Z. M., Bhat, M. A., & Nazir, S. (2013). Mapping of soil micronutrients in Kashmir agricultural landscape using ordinary kriging and indicator approach. *Journal of the Indian Society of Remote Sensing, 41*, 319–329.

Wani, P. A., Khan, M. S., & Zaidi, A. (2007). Impact of zinc tolerant plant growth promoting rhizobacteria on lentil grown in zinc amended soil. *Agronomy for Sustainable Development, 28*, 445–449.

Wani, P. A., Khan, M. S., & Zaidi, A. (2008). Effect of metal-tolerant plant growth-promoting Rhizobium on the performance of pea grown in metal-amended soil. *Archives of Environmental Contamination and Toxicology, 55*(1), 33–42.

Weinand, T., El-Hasan, A., & Asch, F. (2023). Role of *Bacillus* spp. Plant Growth Promoting Properties in Mitigating Biotic and Abiotic Stresses in Lowland Rice (*Oryza sativa* L.). *Microorganisms, 11*, 2327.

Welch, R. M., & Graham, R. D. (2004). Breeding for micronutrients in staple food crops from a human nutrition perspective. *Journal of Experimental Botany, 55*(396), 353–364.

White, C., Sayer, J. A., & Gadd, G. M. (1997). Microbial solubilization and immobilization of toxic metals: Key biogeochemical processes for treatment of contamination. *FEMS Microbiology Reviews, 20*(3–4), 503–516.

White, P. J., & Broadley, M. R. (2009). Biofortification of crops with seven mineral elements often lacking in human diets—Iron, zinc, copper, calcium, magnesium, selenium and iodine. *New Phytologist, 182*, 49–84.

White, P. J., & Broadley, M. R. (2011). Physiological limits to zinc biofortification of edible crops. *Frontiers in Plant Science, 80*, 1–11.

WHO. (2012). *The world health report*. World Health Organization.

Xiao, A. W., Li, Z., Li, W. C., & Ye, Z. H. (2020). The effect of plant growth-promoting rhizobacteria (PGPR) on arsenic accumulation and the growth of rice plants (*Oryza sativa* L.). *Chemosphere, 242*, 125136.

Yang, Z., Yang, F., Liu, J. L., Wu, H. T., Yang, H., & Shi, Y. (2021). Heavy metal transporters: Functional mechanisms, regulation, and application in phytoremediation. *Science of the Total Environment*, 151099.

Yoneyama, T., Ishikawa, S., & Fujimaki, S. (2015). Route and regulation of zinc, cadmium and iron transport in rice plants (*Oryza sativa* L.) during vegetative growth and grain filling: Metal transporters, metal speciation, grain Cd reduction and Zn and Fe biofortification. *International Journal of Molecular Sciences*, *16*(8), 1619111–1619129.
Zhou, Y., Bao, J., Zhang, D., Li, Y., Li, H., & He, H. (2020). Effect of heterocystous nitrogen-fixing cyanobacteria against rice sheath blight and the underlying mechanism. *Applied Soil Ecology*, *153*, 103580.

9 PGPR

The Role of Beneficial Plant-Microbe Interactions in Enhancing Crop Productivity and Resilience

Asma Jan, Zaffar Bashir, Shagufta Majeed, and Mehrun Nisa

9.1 INTRODUCTION

Agriculture is the primary food source of the world (Fariha et al., 2021). Population expansion forces the agriculture sector to meet a wide range of food needs while taking social, economic, and environmental factors (including labor scarcity, land degradation, biodiversity loss, and water scarcity) into account. There are currently multiple limits on its growth, assuming seasonal unpredictability and a hard habitat. The agriculture industry's progress depends on the discovery of novel, long-lasting techniques (Attri et al., 2023). A nation's economic health depends heavily on its agricultural sector. Among the many businesses that make up a country, the agricultural sector is essential to its survival as well as to its ability to satisfy the demands of an expanding population and its economic endeavors. In order to support more people, the world will have to withstand the growing demand for food (Berger et al., 2018). With 7.41 billion people living on 6.38 billion hectares of land, the world's population is predicted to grow to 10 billion people in the next 50 years. Numerous scientific developments have improved crop output in the agroindustries, resulting in enough food to feed all of these people. This method of increasing agricultural production is not sustainable because it heavily relies on chemical fertilizers and pesticides to reduce the infectivity of plant pathogens, increase production levels, and disrupt the nutrient cycle, all of which are essential for crop productivity but have been linked to health risks and the destruction of biological communities (Gomiero et al., 2011). Reducing the use of synthetic agrochemicals, managing agricultural land well, and utilizing microbial formulations or biofertilizers efficiently are the only ways to guarantee competitive crop productivity, environmental safety, and improved soil and plant health. These metabolites derived from microorganisms are non-toxic, eco-friendly, and have the potential to stimulate plant growth. They can solubilize phosphorus, zinc, and silicate, and fixed nutrients, fix atmospheric

DOI: 10.1201/9781003491460-9

nitrogen from the atmosphere, and act as antifungal agents/biocontrol agents without compromising crop productivity (Ilangumaran & Smith, 2017).

Soil quality, or health, refers to a soil's ability to function within ecosystems and land use boundaries in order to maintain biological productivity and environmental quality while preventing activities like nutrient mining and erosion that harm the health of the soil and plants (Bhat et al., 2017). Chemical fertilizers provide the key macronutrients needed by plants, such as nitrogen, potassium, and phosphorus, but abuse of them can harm the soil's organic matter level, humic content, acidity, erosion, and salinization (Sinha & Tandon, 2020). The acidification increases the toxic ion concentration in the soil and inhibits the ability of crops to absorb phosphorous from soil thus affecting overall growth and development of plants. Chemical fertilizers, on the other hand, promote rapid plant growth, however this results in plants developing too quickly to have adequate root development. Plants cannot produce robust stems or nutrient-dense fruits and vegetables. Given their increased susceptibility to illnesses and pests, their chances of survival are probably reduced in these conditions (Pahalvi et al., 2021).

Soil microbial activity is a crucial aspect of healthy soil. The soil microflora, which is made up of a wide variety of microorganisms such as fungi, bacteria, actinomycetes, algae, and protozoa, is an essential part of the agricultural environment. These microorganisms are crucial to soil health and play a number of important roles, including digesting both organic and inorganic waste and improving production through increased availability of rare minerals and soil fertility and nutrient cycle. The usage of agrochemicals has a detrimental effect on the soil's fertility and microbial diversity (Maurya et al., 2020). It also causes an imbalance in the nutrients in the soil. Excessive nitrogen fertilizer depletes micronutrients and reduces crop yields because it damages topsoil and eventually throws off the balance between the three macronutrients (N, P, and K). Because plants can absorb a large amount of ammonium from nitrogen fertilizers in the form of ammonium nitrate and do not require symbiotic bacteria to do so, this reduces biological nitrogen fixation in the soil. Excessive application of potassium chloride in the field causes plants to suffer from a calcium deficit and loss of soil structure (low oxygen content) (Rehman et al., 2020). Reduced soil oxygen levels prevent roots from respiring, which causes plants to produce toxic compounds. Lack of calcium and deterioration of the soil's structure lower the number of beneficial microorganisms in the soil, which lowers the amount of vital nutrients that plants receive (Shrivastav et al., 2020).

Due to the widespread usage of agrochemicals, crop production and global agriculture productivity have drastically decreased. It is essential to create a replacement given the escalating problems with food security and sustainability in agriculture. In this scenario, bacterially based climate-smart farming methods offer chances to satisfy the global population's ongoing food needs. Because fertilizers, herbicides, and pesticides significantly affect the plant defense system, applying PGPR may reduce the long-term harm that these substances do to the ecosystem.

Microbes and plants have always interacted during the growth and development of plants (Gouda et al., 2018). The majority of the nutrients needed by plants for growth are actually provided by the soil, and microorganisms in the soil compete with plants for these resources. Breaking down organic materials and cycling nutrients are only two of the many roles that microbes play in soil fertility and plant health. In agricultural and natural environments alike, the interactions between plants and bacteria in the soil and on above-ground shoots are crucial (Devi et al., 2022). A zone of soil interaction between plants and bacteria is closely monitored as a "hot spot" due to the high level of microbial activity (Vidal et al., 2021). Plant growth promoting rhizobacteria (PGPR) are different groupings of bacteria found in the limited zone around plant roots. This zone is referred to as the rhizosphere. These diverse free-living soil bacteria, known as rhizobacteria, can actively colonize the root surface of a plant and use direct or indirect methods to support the growth of the plant in response to both biotic and abiotic stressors (Abhilash et al., 2016), fixing atmospheric nitrogen, producing siderophores, synthesizing phytohormones, and solubilizing minerals including phosphorus, potassium, silicates, and zinc are examples of direct methods of promoting plant development (Chauhan et al., 2021). Functionally, PGPR are categorized as biopesticides (production of antagonistic chemicals), rhizoremediators (degrading heavy metals), biofertilizers (raising nutrient availability), and phytostimulators (promotes plant growth through growth regulators) (Riaz et al., 2021). Biofertilizers containing various PGPR bacterial strains are becoming more popular as an alternative to synthetic agrochemicals and sustainable management in agricultural practices. By increasing the availability of nutrients in the root zone through a number of processes, such as nitrogen fixation, the solubilization of mineral nutrients, the mineralization of organic compounds, and the creation of phytohormones, PGPR improves plant nutrition (Kumar et al., 2022). The effectiveness of PGPR-based inoculants is dependent on several factors including the bacteria's ability to colonize in significant quantities following inoculation, their compatibility with other bacteria in the rhizosphere, and the condition of the soil (nutrient availability, soil type, and concentration of toxic metals) (Kumari et al., 2022). According to research on plant inoculation, possible PGPR strains that are effective biofertilizers include symbiotic rhizobacteria (*Bradyrhizobium*, *Rhizobium*, and *Mesorhizobium*) and non-symbiotic rhizobacteria (*Pseudomonas*, *Azospirillum*, *Azotobacter*, *Bacillus*, and *Klebsiella*) (Fahde et al., 2023). Certain PGPR strains have dual roles as biocontrol agents and biofertilizers. For instance, strains of *Burkholderia cepacia* exhibit antifungal action against *Fusarium sp.*, while at the same time, by producing *siderophores*, they facilitate the growth of maize in environments lacking in iron (Guardiola-Márquez et al., 2023). Biofertilizers are formulated products containing living microorganisms (bacteria, fungi, actinomycetes) which colonize the rhizosphere or plant roots and enhance the growth and yield when applied to plant surfaces or soil. The use of PGPR as a biocontrol agent is increasing nowadays in the world as the field trials have been mostly inconsistent; significant management of phytopathogens has been documented mostly in laboratory and greenhouse research. The eradication and prevention of phytopathogens proliferation are two important consequences of microbial inoculants on plants (mostly fungi) (Kaari et al., 2023). For instance, the co-inoculation of potatoes with *Bacillus subtilis* and *Trichoderma harzianum* resulted in the plants developing resistance against *Streptomyces sp.* which causes potato common scab (Wang et al., 2019). According to Kaur, biocontrol may be defined as the employment of microorganisms to limit the occurrence and frequency of plant diseases (Kaur et al., 2019). These bacteria offer reduced toxicity and more selectivity in comparison to conventional chemical insecticides. Numerous bacterial, fungal, nematode, and viral illnesses can be contained or suppressed by PGPR that are endemic to the soil and rhizosphere, making them crucial to the management of phytopathogens. Because they are an environmentally safe substitute for chemical fertilizers and can help plants successfully adapt to difficult climates, which lowers plant performance overall under stress, these bioinoculants are advantageous over chemical fertilizers (Alori et al., 2017). It has been demonstrated that *Pseudomonas* sp, *Azotobacter* sp, *Bacillus* sp, *Klebsiella* sp, *Alcaligenes* sp, and *Enterobacter* sp. enhance plant growth and development (Bashir et al., 2017; Bashir et al., 2018; Bashir et al., 2018a; Bashir et al., 2018b; Bashir et al., 2019; Bashir et al., 2019b; Bashir et al., 2021a; Bashir, 2021b; Hamid & Bashir, 2019a; Minuţ et al., 2022). Many PGPR inoculants, including phytostimulators, bioprotectants, and biofertilizers are presently available on the market that can promote growth via one or more of these methods. Using PGPR in addition to chemical fertilizers and insecticides seems like an intriguing idea. In general, PGPR applications are critical to the creation of novel, effective agricultural inoculants.

9.2 PLANT RHIZOSPHERE

The restricted area of soil called the rhizosphere is where plants and the microorganisms around their roots engage in continuous interaction (Neumann & Ludewig, 2023). One of the three major divisions of the rhizosphere is the endorhizosphere, which is composed of the root tissue containing the endodermis and cortical layers. Soil particles and bacteria adhere to one another on the rhizoplane, or root surface. The vicinity of the root is called the ectorhizosphere (Das et al., 2023). Root exudates initiate the biological and physiological connection between microorganisms and plant roots (Mondal et al., 2023). These exudates provide an extremely selective environment for a wide variety of bacteria, although PGPR are the most prevalent category and are most likely to have a major effect on plant physiology, particularly in root colonization. Plant pathogens can compete with PGPR for root exudates in the rhizosphere, and it is these root exudates that will determine whether their interactions are positive or negative (Santoyo et al., 2021). The efficiency of root colonization by PGPR is determined by microbial competition, expression of several genes, and cell-cell communication via quorum sensing (Liu et al., 2022). Expression of exopolysaccharides forming genes in associative, and endophytic PGPR is responsible for their response to plant exudates (Ashajyothi et al., 2023). Bacteria mostly employ a signaling mechanism for root colonization, and the most prominent signaling molecule in them is acyl homoserine lactone (Knights et al., 2021). The importance of the root zone as an essential ecological unit in how the biosphere operates, discoveries in bacterial taxonomy via metagenomics, and advancements in knowledge of the many mechanisms of action indicated by PGPR (Singh et al., 2020) (Figure 9.1). Rhizobacteria that promote plant growth can do so through a variety of direct processes, such as P and Zn solubilization, Si solubilization, N fixation, and Phytohormone synthesis or by indirect mechanisms, such as lowering pathogenic infection and/or mitigating abiotic challenges (Khan et al., 2021). Figure 9.1 shows

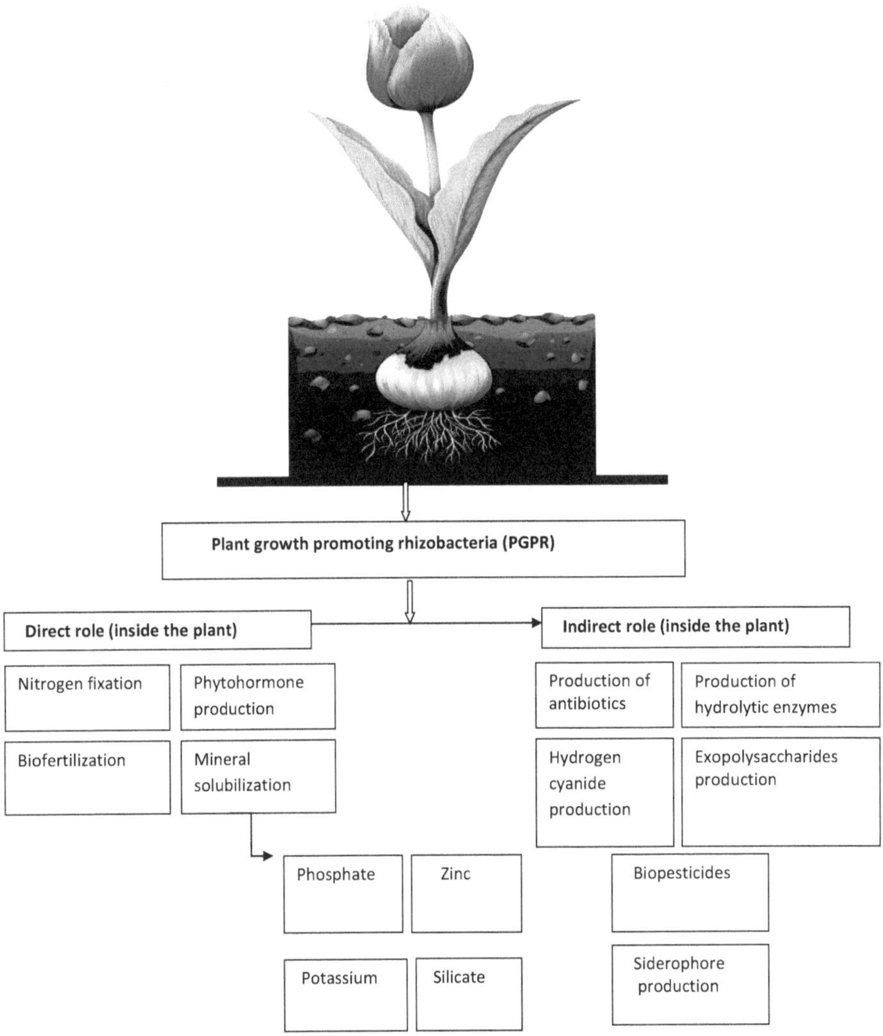

FIGURE 9.1 Direct and indirect mechanism of PGPR for plant growth development.

direct and indirect mechanism of PGPR for plant growth development. At least one mechanism such as phytostimulators, biofertilizers, and biopesticides is used by the many PGPR inoculants now on the market to stimulate growth (Hamid et al., 2021).

9.3 BIOLOGICAL NITROGEN FIXATION BY PGPR

Nitrogen is increasingly turning into one of the main elements limiting plant growth and productivity. This element serves as the fundamental building block of ATP, nucleic acid, and proteins. Although molecular N_2 makes up over 78% of the atmosphere, plants are unable to use it because it is chemically inert. But it's provided to the plant via a process known as biological nitrogen fixation, in which a reductive process by microorganisms converts atmospheric elemental nitrogen into ammonia that plants may consume (Heil et al., 2016). A variety of PGPR species are found in the rhizosphere that can enrich the soil and reduce a significant quantity of nitrogen from the atmosphere. They fall into two categories. The first is made up of symbiotic nitrogen fixers, such *Rhizobium sp.* that infect the roots of leguminous plants and have specificity. The other class of bacteria is non-symbiotic nitrogen fixers, which can be either free-living, associative symbiotic, or endophytes and do not have a plant-specificity. *Azotobacter* sp., *Azospirillum sp.*, *Clostridium sp.*, *Acetobacter diazotrophicus*, and *Cyanobacteria sp.* are a few significant non-symbiotic N_2 fixers (Rana et al., 2021). About two thirds of the nitrogen fixed globally comes from biological nitrogen fixation of which 80% is produced by symbiotic associations, and the remaining portion is derived from free-living associations. Because it can be used as a biofertilizer instead of dangerous agrochemicals and because it increases nodular structure, which promotes more biological nitrogen fixation, the symbiotic relationship between rhizobia and legumes represents the most significant agent in the agricultural system (Mahmud et al., 2020). Inoculating the soil with a mixture of rhizobacterial species can promote nodule formation. Plant signaling chemicals, in particular flavonoids and specialized bacterial-plant recognition, regulate the process of nodule formation. The secondary metabolites called flavonoids are generated by legumes and serve as an attractant for bacteria to colonize the root surface by causing them to release nod factors (Zhang et al., 2019). These nod factors bind to the specific receptors in the root hairs which causes curling in the root hair and subsequent integration of bacteria into the cells of root hair. The nitrogenase enzyme is the major component in the process of nitrogen fixation. It's made of 20 nif genes including both structural genes responsible for the activations of its two metalloproteins such as dinitrogenasereductase and dinitrogenase (Singh & Verma, 2023) while regulatory genes are involved in the synthesis and functioning of nitrogenase enzyme. The members of the associative symbiotic nitrogen fixers fix about 20-40 kg/year nitrogen in the microaerophilic conditions. Different species of *Azospirillum* including *Azospirillum brasilense*, *Azospirillum amazonense*, and *Azospirillum halopraeferens* are the most commonly used associative nitrogen fixers (Guo et al., 2023). Among the several PGPR strains *Azotobacter* sp., including *Azotobacter beijerinckii*, *Azotobacter nigricans*, *Azobacter paspali*, and *Azobacter vinelandii* are the most frequently used free-living aerobic nitrogen fixers (Ranipa et al., 2023). Non-leguminous plants may fix about 15–20 kg of nitrogen annually. Several studies suggest that after the nif genes are found and identified genetic engineering may be used to boost the capacity for nitrogen fixation. On the other hand, experts disagree about the best approaches to maximize and utilize a plant's innate ability to fix nitrogen. Both systems aim to sustain elevated amounts of soil nitrogen and promote plant growth, despite their opposing viewpoints. The importance of PGPR inoculation in agriculture is quickly expanding since it significantly increases the amount of nitrogen in the soil, encourages growth, and reduces the danger of disease (Patel & Patel, 2023).

9.4 PGPR AS PHOSPHATE SOLUBILIZERS

One of the primary necessary macronutrients for the best possible growth and development of plants is phosphorus (P). It plays a part in several plant metabolic activities, including the formation of root stalks, biological nitrogen fixation in legumes, and photosynthesis. It is essential for the manufacture of several cellular components, including phospholipids, nucleic acid, and adenosine triphosphate (Khan et al., 2023). There is a store of both inorganic and organic phosphorus in soil. The inorganic form of phosphorus is attributed to its high reactivity with other soil components, such as calcium in alkaline soil and aluminum in acidic soil. The organic form of phosphorus makes up only 20% of the total amount of phosphorous in the soil and is added by plant, animal, or microbial residue, such as inositol phosphate and monoester (Boitt et al., 2018). Due to its insoluble nature, it cannot be utilized by plants, which can only utilize monobasic and dibasic phosphate, two soluble forms of phosphorus (Jha & Saraf, 2015). The content of phosphorus in soil has altered due to advances in agricultural practices and the dependence of world food supply on phosphorus fertilizers. While the total bioavailability of the phosphorous component is only 10–20%, which is still below the plant's need, the frequent application of agrochemicals in soil is harmful (Yu, 2023). P fertilizers typically cause the maximum phosphorous concentration (0.1–5 mg L-1) to be reached close to the soil surface. The poor phosphorus availability to plants is caused by three dynamic processes: complexation, precipitation, and immobilization. Calcium phosphates are present in calcareous soils, while inorganic phosphates are linked to iron and aluminum in acidic soils (Kumar & Shastri, 2017). It has been shown that certain microorganisms that solubilize phosphate (PSMs) can solubilize phosphorus in its precipitated state, making them an alternative biotechnological approach for meeting plant phosphorus requirements in sustainable agriculture. The main mechanism of mineral phosphate solubilization is dependent on sugar metabolism, in which the bacteria use the exudates from roots for the synthesis of lower molecular weight organic acid. Phosphate solubilizing bacteria (PSB) employ highly efficient mechanisms to mediate the solubilization of insoluble phosphate (Chauhan et al., 2017). Direct oxidation of glucose by the enzyme glucose dehydrogenase is the other way for the synthesis of organic acid particularly gluconic acid. Glucose dehydrogenase (GED) gene and cofactor pyrroloquinoline Quinone (PQQ) is responsible for the activity of dehydrogenase enzyme and mineral phosphate solubilization in the microorganism (Wan et al., 2020). When organic acids like succinic acid, acetic acid, 2-ketogluconic acid, carboxylic acid, malic acid, oxalic acid, succinic acid, and acetic acid are released by bacteria, they lower the pH, challenge P for an adsorption site, and chelate divalent cations bound to phosphate, like calcium phosphate in calcareous soil, which dissolves insoluble phosphate (Patel et al., 2015). Table 9.1 shows different organic acid produced by Phosphate solubilizing bacteria, which play an essential role in making P available to plants. The specificity of phosphate solubilization is that some bacterial species such as *Klebsiella, Enterobacter*, and *pantoea* can solubilize calcium phosphate while other insoluble forms of phosphate are hydrolyzed by the other microorganism (Mendoza-Arroyo et al., 2020). The P-solubilization varies in different soils; calcareous soil with high amount of calcium phosphate shows superior P-solubilization

TABLE 9.1
Organic Acid Produced by Phosphate Solubilizing Bacteria

Phosphate Solubilizing Rhizobacteria	Organic Acids	References
Aspergillus Tubingensis	Succinic acid, gluconic acid, oxalic acid, and citric acid	Ahmad et al. (2023)
Burkholderia sp.	Lactic, citric, oxalic, and tartaric acid	Fitriatin et al. (2022)
Bacillus Aryabhattai	Oxalic acid, malonic acid, citric acid, and succinic acid	Song et al. (2022)
Klebsiella, Burkholderia	Oxalic acid, lactic acid, and gluconic acid	Nacoon et al. (2020)
Paenibacillus sp. *Bacillus cereus*	Gluconic acid, oxalic acid, and malic acid	Chawngthu et al. (2020)
Bacillus Megaterium, Bacillus subtilis	Acetic acid, citric acid, and lactic acid	Do carmo et al. (2019)
Penicillium Oxalicum, Bacillus subtilis	Tartaric acid, citric acid, and lactic acid	Li et al. (2016)
Bacillus Circulans	Gluconic acid and citric acid	Mehta et al. (2015)
Streptomyces Cellulose Aspergillus niger	Malic acid	Jog et al. (2014)

(Ahmad & Kibret, 2014). Some PGPR are also able to mineralize most organic phosphatase (NSAPs) and phosphonatases/carbon-phosphorous (C-P) lyases aiding the release of P from mineral reservoir by H^+ substation for cations bound to phosphate (Arif et al., 2017). Both acidic and alkaline phosphatases are found in soil; acidic phosphatases are found in acidic soil, while alkaline phosphatases are found in alkaline soil. These phosphatases are in charge of dephosphorylating of phoshoester bonds of organic compounds (Shi et al., 2019). The activity of several phosphoesters was discovered to be effective in maize, barley, and wheat. A considerable amount of acid phosphatases were expressed by the inner rhizosphere genus *pseudomonas*, *Enterobacter*, *Rhizobium*, and *Bacillus*. Lyases release the C-P link in the organophosphorous molecule, allowing plants to absorb phosphorous (Hazzoumi et al., 2022). It's possible that two mechanisms (solubilization and mineralization) can be present in same PSB (Wang et al., 2022). Naturally existing microbes in the rhizosphere are called phosphate-solubilizing bacteria (PSB). Some of these strains, which can supply plants with soluble phosphate, have been employed as efficient biofertilizers in the commercial sector. Examples of these strains are *pseudomonas* sp., *Bacillus* spp., *Micrococcus* spp., *Serratia* spp., *Azotobacter* spp., *Microbacterium* spp., *Burkholderia* spp., *Rhizobium* spp., and *Azotobacter* spp. (Bashir et al., 2017; Bashir et al., 2018a; Bashir et al., 2018b; Bashir et al., 2019; Gautam et al., 2021).

9.5 PGPR AS ZINC SOLUBILIZERS

Zinc (Zn) is the most crucial micronutrient for plants. It performs several essential tasks for plants, including resistance to biotic and abiotic stresses, enhancing the activity of carbonic anhydrase for photosynthesis, maintaining membrane integrity, oxidizing and metabolizing carbohydration, and managing the production of hormones and chlorophyll (Noulas et al., 2018). Over 1 billion people in India are fed by the soil used to grow cereals (rice, wheat), oilseeds, and pulses; however, 50% of this soil is zinc deficient, which lowers crop productivity and lowers the nutritional value of grains (Kumar et al., 2020). Different forms of zinc can be found in soil, and the degree of solubility and plant availability varies across different forms. Complexation with silicate minerals, high pH, and interactions with metals (Fe, Al, and Mn) to generate oxides and hydroxides are the main causes of zinc's reduced bioavailability (Natasha et al., 2022). The concentration of zinc in the edible portions of crops and crop yield are increased by agronomic techniques that enhance zinc phytoavailability in the soil through different mineral and organic fertilizers. The most common zinc fertilizers are zinc sulfate ($ZnSO_4$) and zinc oxides (ZnO). One significant disadvantage of using zinc fertilizers in different crops is that they are less soluble in and fixable to the soil matrix, which prevents zinc from being used for plant nutrition (Hefferon, 2019). By a variety of processes, such as the synthesis of organic acid and enzymes, zinc-solubilizing bacteria have the potential to improve the amount of native zinc that is available to plants (Kumar et al., 2019). Acidolysis is the key mechanism employed by ZnSB which decreases the pH of soil and solubilizes insoluble zinc to soluble form. The solubilization of zinc by PGPR strains is linked to a decrease or increase in the pH of the medium. Kumar et al. (2022) stated that Zn bioavailability declines a hundred times with a one-unit increase in pH and vice versa.

In addition to mobilizing soil zinc, rhizobacterial strains belonging to the genera *Acinetobacter*, *Bacillus*, *Cyanobacteria*, *Gluconacetobacter*, *pseudomonas*, and *Serratia* are also effective in soluble zinc oxide and zinc carbonate (ZnCO3), which makes them a cost-effective source in the soil system (Hussain et al., 2015). The most researched and promising candidates are rhizobacteria from the genus *Bacillus*, which not only increases the bioavailability of zinc to plants but also uses other mechanisms to improve plant growth and yield, including the production of growth regulators, biological nitrogen fixation, secretion of antifungal metabolites, and stress tolerance (Jalal, 2022). In addition to the bacteria that dissolve zinc, adding organic matter to the soil can improve its nutritional availability, enzymatic activity, and biological activity. Hence, the applications of zinc solubilizing bacteria may be an innovative method for raising crop quality, production, and growth in soils lacking in zinc (Batool et al., 2021).

9.6 PGPR AS SILICATE SOLUBILIZERS

As silicon is the second most prevalent element in the crust of Earth (Bist et al., 2020), it makes up the majority of soil together with silicates of magnesium, aluminum, iron, calcium, sodium, and potassium. Plants benefit greatly from silicon in terms of growth, yield, and disease resistance (Schaller et al., 2021). The beneficial effects of silicon have primarily been examined in the context of the rice plant (*Oryza sativa*), where it accelerates the pace at which carbon is deposited in panicles and increases the uptake of carbon during photosynthesis (Huang et al., 2021). Similar to silicic acid that is absorbed by plant roots, it mixes with organic substances found in actively dividing tips to strengthen plant cell walls. The need for this element is met by the application of external chemical fertilizers since, although it is abundant in the soil, it is not available for plant nourishment. Water eutrophication and phytotoxicity are two examples of the nutrient cycling disruptions brought on by the widespread use of chemical fertilizers (Zellner et al., 2021).

Silicon solubilizing bacteria are considered as eco-friendly biofertilizers for dissolution and mobilization of naturally available insoluble silicates (Chaganti et al., 2023). Ways by which PGPR solubilize insoluble silicate include production of organic and inorganic acids, exopolysaccharides, and solubilization by ligands (divalent cations) (Kang et al., 2017). Among several ways production of organic acids is one efficient mechanism employed by PGPR to mediate solubilization of silicate minerals (Etesami et al., 2020). The mechanism of solubilization of silica differs depending on the mineral source. Some minerals, like quartz and phytoliths, can be dissolved with merely protons and the exchange of mineral-bound cations, whereas other metal-bound silicates necessitate a combined action of pH shift and ligand assault. Using eight different minerals, the solubilization capacity of a specific bacterial species was investigated. It was shown that *Bacillus* and *pseudomonas* species were more soluble in magnesium trisilicate than in muscovite, quartz, and illite (Li et al., 2022).

The ability of plants to absorb several minerals from soil, including phosphorus, potassium, and silicate, is enhanced by inoculation with bacteria that dissolve silicate. Due to their great reduction in the requirement for potash fertilizers, SSB are now gaining recognition on a global scale (Samantray et al., 2022).

9.7 PGPR AS POTASSIUM SOLUBILIZERS

Potassium (K) enhances crop yield through improving the translocation of sugars and starch, increasing starch-rich grains, forming cellulose, lowering lodging, preserving turgor, photosynthesis, and food formation, activating multiple enzyme systems, decreasing respiration, preventing energy losses, and helping to delay crop diseases (Ahmad et al., 2023). Therefore, the concentration of soluble potassium in soil is often quite low because over 90% of potassium is found in the form of insoluble rock and silicate minerals (Parmar & Sindhu, 2013). A plant with insufficient potassium availability grows slowly, produces few seeds, has poorly developed roots, and produces yields that are below average. It has been observed that several of the microbes that solubilize potassium have significant roles in the natural potassium cycle, including Burkholderia, *Bacillus mucilaginous*, *Acidothiobacillus* sp., *pseudomonas* spp., *Bacillis* spp., Azotobacter spp., *Bacillus edaphicus*, *Ferrooxidans* sp., and *Paenibacillus* sp. (Bashir et al., 2018; Bashir et al., 2019b; Bashir et al., 2021a; Bashir, 2021b; Hamid & Bashir, 2019a). The solubilization of K-bearing minerals (micas and illite) by microorganisms can be attributed to several processes, including lowering pH; synthesis of organic acids like citric acid, oxalic acid, and tartaric acid; chelation of the K cations; and production of capsular polysaccharides that helped dissolve minerals and release potassium (Olaniyan et al., 2022). Consequently, using potassium-solubilizing microorganisms is a viable strategy for crop development that can increase environmentally friendly agricultural production methods and decrease the usage of agrochemicals (Kour et al., 2020).

9.8 PRODUCTION OF PHYTOHORMONES BY RHIZOSPHERIC BACTERIA

The synthesis of several phytohormones, which affect numerous plant developmental processes like cell division, leaf senescence, flower creation, and fruit ripening, is greatly influenced by rhizosphere soil microorganisms, especially the rhizospheric bacteria. One of the most crucial mechanisms used by rhizospheric bacteria for plant growth and development is the production of phytohormones like auxin, gibberellins, and cytokinins (Kudoyarova et al., 2019).

9.8.1 IAA Production by PGPR

Many PGPR strains produce indole-3-acetic acid, a signal molecule that is essential for regulating plant growth and development, including cell elongation, organogenesis, and adaptation to tropical environments. Additionally, it encourages the development of adventitious and lateral roots, apical dominance, and phytopathogen resistance. L-tryptophan, the main building block for the synthesis of IAA, is used by rhizobacteria from exudates from plant roots. There are several ways auxins can encourage plant growth, but the IAA generated by microorganisms has a major impact on root growth because it increases the number of root tips and surface area of the roots, which increases the quantity of nutrients and water that are taken up from the soil (Rehman et al., 2020). Plants that are vulnerable to various diseases and parasites can acquire systemic resistance by inhibiting IAA signaling (Pieterse et al., 2014). More than 80 percent of rhizobacteria produce indole-3-acetic acid, which when injected into crops has a profound effect on soil fertility, crop productivity, and plant growth. Auxin produced by rhizobacteria is one indicator of high-quality PGPR strains. Rhizobium sp. and Microbacterium sp. are two extremely productive producers of IAA (Yadav et al., 2018).

9.9 PGPR AS BIOCONTROL

Using PGPR as biological control is an alternative strategy for shielding plants from various plant diseases caused by phytopathogens. As per Kaur, the utilization of microorganisms to decrease the occurrence and severity of plant illnesses is referred to as biocontrol (Kaur et al., 2019). The PGPR uses a biocontrol mechanism that includes the generation of antibiotics, enzymes that break down cell walls, hydrogen cyanide, siderophores, volatile chemicals, induced systemic resistance, and root colonization efficiency to battle damage caused by phytopathogens. It has been observed that several PGPR species contain inhibitory compounds, and that these compounds have the beneficial effects of eliminating and inhibiting phytopathogens (Saeed et al., 2021). *Pseudomonas* species produce antibiotics such as tensins, cyclic lipopeptides, phenazine, tropolone, pyoluteorin, oomycin A (Meena et al., 2020) while as *Bacillus*, *Streptomyces* and *stenotrophomonas* species produce amphicin, kanosamine, 2,4-diacetylphloroglucinol (DAPG) and oligomycin A. Many studies have been conducted on the activity and specificity of these different antibiotics, and several of these biocontrol strains are now available for purchase (Kour et al., 2019).

9.9.1 Production of HCN

Hydrogen cyanide synthesis by free-living rhizobacteria is one efficient and environmentally safe method for the biological control of phytopathogens. Hydrogen cyanide is a secondary metabolite that keeps phytopathogens from growing by decreasing the activity of enzymes involved in metabolic activities like respiration and the breakdown of carbohydrates (Rehman et al., 2020). To avoid competition or predation, hundreds of organisms, including bacteria, algae, and fungi, produce, release, and metabolize cyanide despite the fact that it is a general metabolic inhibitor. In the rhizospheric soil and plant root nodules, it is found that HCN synthesis is a shared trait of pseudomonas (88.89) and Bacillus (50%) bacteria (Sehrawat et al., 2022). In plants the fungal diseases can be reduced by HCN secreted by *pseudomonas* (Ali et al., 2020). Glycine acts as a precursor for HCN production, and its oxidative decarboxylation by the membrane bound flavoenzyme HCN synthase results in its synthesis (Nandi et al., 2017). While the majority of previous research has defined HCN's function in phytopathogen biocontrol, several more recent studies have shown HCN's function in raising phosphate availability (Rijavec & Lapanje, 2016).

9.9.2 Hydrolytic Enzyme Production

Biopesticides could be used to describe the way that PGPR produces protective enzymes. To produce chemicals with antibiosis and antifungal properties that are used as defensive mechanisms, phytopathogenic agents are inhibited by PGPR, which promotes plant growth. For example, chitinase and glucanase are two hydrolyzing enzymes that would be produced during the process. Because these substances make up the majority of a fungal cell wall, bacteria that generate chitinases and betaglucanases would suppress the growth of fungi. In order to stop *Fusarium udum* from generating wilt, *Sinorhizobium fredii* KCC5 and *Pseudomonas fluorescens* LPK2 produce chitinase and beta-glucanase, respectively (Kumar et al., 2010). By acting on the pollutant and mineralizing it, hydrolytic enzymes enable the plant to readily access nutrients that would otherwise be unavailable. When these macro- and micronutrients are available and in sufficient quantities, amino acids, vitamins, and phytohormones are synthesized, which boosts the plant's nutritional condition and strengthens its defenses against disease invasion (Kaur, 2021). The main role of the enzymes generated by PGPR is thought to be in the biocontrol function, but in reality, they play a more significant role in the degradation of biopolymers, which supplies plants and bacteria with nutrients and causes the mineral cycle, which is crucial for the sustainability of ecosystems (Panicker & Sayyed, 2022).

9.9.3 Siderophore Production

Bacteria in iron-restricted environments produce small compounds called siderophores, which improve the absorption of iron. For all organisms reliant on photosynthesis, iron is essential as an enzyme cofactor for numerous metabolic activities, including photosynthesis, amino acid synthesis, oxygen transfer, nitrogen fixation, and respiration. Iron is a common and abundant chemical component on Earth, usually found in two oxidative states: Fe^{2+} and Fe^{3+}. Because it creates insoluble iron hydroxides and oxides, the latter state is far less accessible to plants (Khoshru et al., 2020).

Certain PGPB have been shown to produce low molecular weight (400–1,500 Da) chemicals that can pull iron out of the earth. Iron-binding substances called siderophores increase the bioavailability of ferric ions (iron) for plant growth (Goswami et al., 2016).

The ability of numerous microorganisms to produce siderophores has been isolated and investigated from both marine and terrestrial environments (Řezanka et al., 2018). The four main chemical structures of siderophores are carboxylates, phenolates, hydroxamates, and pyoverdines. Just 2% of gram-positive bacteria, such as *Bacillus* and *Rhodococcus*, are able to create siderophores, whereas *Enterobacter* and *Pseudomonas* are the most common gram-negative bacteria makers of siderophores (Czarnes et al., 2020). As an essential biological control mechanism that keeps detrimental plant diseases from getting iron supplies by outcompeting them for those resources, beneficial soil and plant-associated microbes manufacture siderophores.

9.9.4 Exopolysaccharides (EPSs) Production

Exopolysaccharides (EPSs) are biodegradable polymers produced by a variety of bacteria, algae, and plants. They're made of glucose residues and their derivatives (Sanalibaba & Cakmak, 2016). EPSs aid in the host's survival under stressful situations (such as salinity, drought, or excessive moisture) by storing water reserves, aggregating soil particles, and promoting essential rhizobacteria-bacteria that interact with plant roots. *Agrobacterium sp., Bacillus drentensis, Rhizobium leguminosarum, Azotobacter, Xanthomonas* sp., and *Rhizobium* sp. are EPS-producing PGPR that can improve soil fertility and support sustainable agriculture (Mahmood et al., 2016). Additionally, EPS play a part in cell recognition, surface adherence, and biofilm formation, all of which aid in the colonization of the soil environment. Aggregation of the soil affects water infiltration, mineral delivery, soil aeration, and organic matter storage. Aggregation of soil is important for soil organic matter storage and fertility recapitalization (Flemming, 2016).

9.9.5 Antibiotic Production

Microbial antagonists have been proposed as a potential substitute for conventional pesticides in the management of plant diseases in agricultural crops. Along with *Bacillus* species and *Pseudomonas* species, PGPR plays a critical role in stopping the spread of disease-causing bacteria by generating antibiotics. In the past 20 years, one of the most promising fields in plant sciences has been the development of antibiotics by PGPR to combat various plant-based diseases and well-studied biocontrol mechanisms (Ulloa-Ogaz et al., 2015). Most *Pseudomonas* species are known to produce antibiotics, including variants of ecomycin, viscosin, cepaciamide A, and oomycin A. Other drugs made by *Pseudomonas* include pyrolynitrin, 2, 4-diacetylphloroglucinol, rhamnolipids, and pyoluteorin. Many lipopeptide antibiotics, including surfactin bacillomycin, are produced by *Bacillus* species in addition to a range of antibiotics and antifungal drugs. The antibiotics are divided into substances that are volatile and non-volatile for further classification. Antibiotics such as polyketides, cyclic lipopeptides, aminopolyols, heterocyclic nitrogenous compounds, etc. are not volatile. Alcohols, aldehydes, ketones, hydrogen cyanide, and other compounds are among the chemicals found in volatile antibiotics (Fouzia et al., 2015).

9.10 COMMERCIALIZATION OF PGPR AS BIOFERTILIZERS

A number of PGPR bacterial strains are accessible as biofertilizers and are gaining traction as an alternative to synthetic agrochemicals and sustainable management in agricultural practices. Table 9.2 shows the different microbial strains and their effect on growth and development of plants. Through a number of processes, including as biological nitrogen fixation, the synthesis of growth regulators, and the solubilization of insoluble elements like phosphorous, zinc, and silicate, biofertilizers improve nutrient availability to the root zone, which in turn stimulates plant development (Ramzan et al., 2019). Use of phosphate solubilizing microorganism such as *Bacillus megaterium* has gained global interest as it can reduce the requirement of phosphate fertilizers of plants up to 75% (Elhaissoufi et al., 2022; Ferguson et al., 2019). The activity and survivability of biofertilizers within the host plant is drastically affected by both biotic and abiotic factors including degree of interaction of PGPR with soil microbiota, extreme temperature, oxygen level in soil, presence of heavy metals, acidity and alkalinity of soil, and deficiency of essential nutrients (Basu et al., 2021). Therefore, a specific climate is needed for inoculants to successfully adapt in agricultural fields (Naamala & Smith, 2020).

9.11 CONCLUSION

Among all the industries in the country, the agricultural sector is crucial to its survival and to satisfying the needs of the expanding populace. The agricultural sector has had multiple technological breakthroughs that have improved crop yields but also caused harm to the land and other environmental issues. In order to maintain food production in an environmentally responsible manner and manage long-term soil fertility, we are searching for simple and affordable operating techniques. Given their many qualities geared toward plant growth and development, natural fertilizers like PGPR are among the most sustainable agricultural practices when it comes to inoculation. Comparing PGPR to conventional chemical-based methods as a plant growth enhancer reveals a number of benefits. Because PGPR lessen the need for synthetic pesticides and fertilizers, they have a less detrimental effect on soil and water resources and are therefore highly environmentally friendly. Furthermore, the application of PGPR can support the advancement of sustainable agricultural methods by enhancing soil biodiversity and health. According to research, PGPR can efficiently boost plant health by a number of processes, including the synthesis of IAA, and the solubilization of various macronutrients and micronutrients like phosphorus, potassium, zinc, and silicates.

TABLE 9.2
Examples of Rhizobacteria That Support Plant Growth

PGPR Strains	Plants	Role in Plant Growth and Development	Reference
Paenibacillus polymyxa	Black pepper (*Piper nigrum*)	Untreated plants acquired systemic resistance to the bacterial spot pathogen *Xanthomonas axonopodis pv*, *Vesicatoria*, and significantly higher biomass was produced by the plants.	Kralova and Jampilek (2023)
Rhizobium strain MRP1	Common pea (*Pisum sativum* L)	The amount of N and P nutrients in plant organs, seed production, protein content, symbiotic properties (nodulation and hemoglobin content), and growth of pea plants were all markedly improved.	Dasgupta et al. (2023)
Bradyrhizobium MRN6	Mung bean (*Vigna radiata*)	Growth metrics were improved at all herbicide doses (quizalafop-p-ethyl and clodinafop) by the herbicide-resistant Rhizobium strain MRP1.	Naz et al. (2022)
Pseudomonas sp. A3R3	Indian mustard (*Brassica juncea* L.)	Increased the biomass of plants grown in Ni-stressed soil (*B. juncea*) and the Ni content (*A. serpyllifolium*) significantly	Pandita (2022)
Psychrobacter sp. SRS8, *Bacillus* sp. SN9 and *Pseudomonas* sp. SR12	Indian mustard (*Brassica oxyrrhina*) and Chinese mustard (*Brassica juncea*)	The test plants showed a rise in biomass, which improved the accumulation of nickel within the plant tissues.	Alves et al. (2021)
Klebsiella pneumonia	Wheat (*Triticum aestivum*)	Increased the lengths of the shoots and roots considerably	Noman et al. (2021)
P. capsici Hydpak1 MF322868 and Hydpk2 MF322869	Chili pepper (*Capsicum annuum*)	Decreased seedling mortality and *P. capsici* disease severity in the chili pepper	Hyder et al. (2020)
Azospirillum amazonense	Rice (*oryza sativa*)	Increased amounts of nitrogen (3.5–18.5%), dry matter accumulation (7–11.6%), and panicle numbers (3–18.6%) during grain maturity	Namdeo et al. (2020)
Bacillus PSB10 SP.	Chickpea (*Cicer arietinum*)	Considerably higher nodulation, leghemoglobin, grain protein, chlorophyll content, and a lower absorption of chromium in the roots, shoots, and grains	Verma et al. (2019)
Burkholderia ambifaria (RZ2MS16) and *Bacillus* sp. (RZ2MS9)	Maize (Zea mays L.) and soybean (Glycine max L.)	For both strains and crops, shoots demonstrated rises of greater than 47%, which was substantially higher than that of uninoculated plants.	Batista et al. (2018)
Azospirillum, *Pseudomonas putida*, and *Azotobacter*	Artichoke (*Cynara cardunculus var. scolymus*)	The combination of bacteria that fix nitrogen and those that solubilize phosphorus enhanced the length and weight of shoots, as well as the coefficient of velocity of germination, seedling vigor index, and mean time of germination in plants.	Allahdadi and Raei (2017)

REFERENCES

Abhilash, P. C., Dubey, R. K., Tripathi, V., Gupta, V. K., & Singh, H. B. (2016). Plant growth-promoting microorganisms for environmental sustainability. *Trends in Biotechnology*, *34*(11), 847–850.

Ahmad, A., Moin, S. F., Liaqat, I., Saleem, S., Muhammad, F., Mujahid, T., & Zafar, U. (2023). Isolation, solubilization of inorganic phosphate, and production of organic acids by individual and co-inoculated microorganisms. *Geomicrobiology Journal*, *40*(1), 111–121.

Ahmad, M., & Kibret, M. (2014). Mechanisms and applications of plant growth promoting rhizobacteria: Current perspective. *Journal of King saud University-Science*, *26*(1), 1–20.

Ali, S., Hameed, S., Shahid, M., Iqbal, M., Lazarovits, G., & Imran, A. (2020). Functional characterization of potential PGPR exhibiting broad-spectrum antifungal activity. *Microbiological Research*, *232*, 126389.

Allahdadi, M., & Raei, Y. (2017). Growth and chlorogenic acid content of artichoke (*Cynara scolymus* L.) affected by bio and chemical fertilizer. *Journal of Biodiversity and Environmental Sciences*, *11*(5), 63–73.

Alori, E. T., Dare, M. O., & Babalola, O. O. (2017). Microbial inoculants for soil quality and plant health. *Sustainable Agriculture Reviews*, 281–307.

Alves, A. R., Oliveira, R. S., Silva, E. F., & Novo, L. A. (2021). Role of bacteria and mycorrhizal fungi in phytomining: Status and future perspectives. In *Microbe mediated remediation of environmental contaminants* (pp. 15–26). Elsevier. https://doi.org/10.1016/B978-0-12-821199-1.00002-X

Arif, M. S., Shahzad, S. M., Yasmeen, T., Riaz, M., Ashraf, M., Ashraf, M. A., & Kausar, R. (2017). Improving plant phosphorus (P) acquisition by phosphate-solubilizing bacteria. In *Essential plant nutrients: Uptake, use efficiency, and management* (pp. 513–556). Springer.

Ashajyothi, M., Balamurugan, A., Patel, A., Krishnappa, C., Kumar, R., & Kumar, A. (2023). Cell wall polysaccharides of endophytic Pseudomonas putida elicit defense against rice blast disease. *Journal of Applied Microbiology*, *134*(2), lxac042.

Attri, I., Awasthi, L. K., Sharma, T. P., & Rathee, P. (2023). A review of deep learning techniques used in agriculture. *Ecological Informatics*, 102217.

Bashir, Z. (2021b). Characterization of Potassium Solubilizing Bacteria from Rhizospheric Soils of Cherry (*Prunusavium*). *Research Journal of Agricultural Sciences*, *12*(2), 494–496.

Bashir, Z., Hamid, B., Zargar, M. Y., & Kundoo, A. A. (2021a). Morphological and Biochemical Characterization of Potassium Solubilizing Bacteria from Rhizospheric Soils of Pinus (*Pinus wallichiana*). *Journal of Research and Development*, *21*, 58–63.

Bashir, Z., Zargar, M. Y., Baba, Z. A., & Mohiddin, F. A. (2018). Solubilization of potassium containing mineral by bacteria from Kashmir iris (*Iris kashmiriana*) rhizosphere. *The Pharma Innovation Journal*, *17*(4), 747–750.

Bashir, Z., Zargar, M. Y., Baba, Z. A., Mohiddin, F. A., Peer, F. A., Mir, S. A., & Najar, G. R. (2018a). Isolation and characterization of phosphorus solubilizing bacteria (PSB) from rhizospheric soils of apple (*Malus Domestica*). *Journal of Pharmacognosy and Phytochemistry*, *7*(2), 2997–3000.

Bashir, Z., Zargar, M. Y., Bhat, T. A., Kousar, S., Baba, Z. A., & Mohiddin, F. A. (2018b). Solubilization of phosphorus containing mineral by bacteria from rhizospheric region of walnut (*Juglans regia*). *International Journal of Current Microbiology and Applied Sciences*, *7*(8), 2391–2398.

Bashir, Z., Zargar, M. Y., Mohiddin, F. A., Baba, Z. A., & Hamid, B. (2019). Isolation and Biochemical Characterization of phosphorus solubilizing bacteria from Rhizospheric region of Apricot and Peach. *Journal of Research and Development University of Kashmir*, *19*, 0972–5407.

Bashir, Z., Zargar, M. Y., Mohiddin, F. A., Kousar, S., Husain, M., & Rasool, F. (2017). Phosphorus solubilizing microorganisms: Mechanism and diversity. *International Journal of Communication Systems*, *5*(5), 666–673.

Bashir, Z., Zargar, M. Y., & Vishwakarma, D. K. (2019b). Potassium solubilizing microorganisms for sustainable agriculture. *Applied Agricultural Practices for Mitigating Climate Change*, *2*, 17–28.

Basu, A., Prasad, P., Das, S. N., Kalam, S., Sayyed, R. Z., Reddy, M. S., & El Enshasy, H. (2021). Plant growth promoting rhizobacteria (PGPR) as green bioinoculants: Recent developments, constraints, and prospects. *Sustainability*, *13*(3), 1140.

Batista, B. D., Lacava, P. T., Ferrari, A., Teixeira-Silva, N. S., Bonatelli, M. L., Tsui, S., & Quecine, M. C. (2018). Screening of tropically derived, multi-trait plant growth-promoting rhizobacteria and evaluation of corn and soybean colonization ability. *Microbiological Research*, *206*, 33–42.

Batool, S., Asghar, H. N., Shehzad, M. A., Yasin, S., Sohaib, M., Nawaz, F., & Uzair, M. (2021). Zinc-solubilizing bacteria-mediated enzymatic and physiological regulations confer zinc biofortification in chickpea (Cicer arietinum L.). *Journal of Soil Science and Plant Nutrition*, *21*, 2456–2471.

Berger, B., Patz, S., Ruppel, S., Dietel, K., Faetke, S., Junge, H., & Becker, M. (2018). Successful formulation and application of plant growth-promoting *Kosakonia radicincitans* in maize cultivation. *BioMed Research International*, *28*, 6439481. https://doi.org/10.1155/2018/6439481

Bhat, R. A., Dervash, M. A., Mehmood, M. A., Skinder, B. M., Rashid, A., Bhat, J. I. A., & Lone, R. (2017). Mycorrhizae: A sustainable industry for plant and soil environment. *Mycorrhiza-Nutrient Uptake, Biocontrol, Ecorestoration*, 473–502.

Bist, V., Niranjan, A., Ranjan, M., Lehri, A., Seem, K., & Srivastava, S. (2020). Silicon-solubilizing media and its implication for characterization of bacteria to mitigate biotic stress. *Frontiers in Plant Science*, *11*, 28.

Boitt, G., Simpson, Z. P., Tian, J., Black, A., Wakelin, S. A., & Condron, L. M. (2018). Plant biomass management impacts on short-term soil phosphorus dynamics in a temperate grassland. *Biology and Fertility of Soils*, *54*, 397–409.

Chaganti, C., Phule, A. S., Chandran, L. P., Sonth, B., Kavuru, V. P. B., Govindannagari, R., & Sundaram, R. M. (2023). Silicate solubilizing and plant growth promoting bacteria interact with biogenic silica to impart heat stress tolerance in rice by modulating physiology and gene expression. *Frontiers in Microbiology*, *14*.

Chauhan, A., Guleria, S., Balgir, P. P., Walia, A., Mahajan, R., Mehta, P., & Shirkot, C. K. (2017). Tricalcium phosphate solubilization and nitrogen fixation by newly isolated Aneurinibacillus aneurinilyticus CKMV1 from rhizosphere of Valeriana jatamansi and its growth promotional effect. *Brazilian Journal of Microbiology*, *48*, 294–304.

Chauhan, A., Saini, R., & Sharma, J. C. (2021). Plant growth promoting rhizobacteria and their biological properties for soil enrichment and growth promotion. *Journal of Plant Nutrition*, *45*(2), 273–299.

Chawngthu, L., Hnamte, R., & Lalfakzuala, R. (2020). Isolation and characterization of rhizospheric phosphate solubilizing bacteria from wetland paddy field of Mizoram, India. *Geomicrobiology Journal*, *37*(4), 366–375.

Czarnes, S., Mercier, P. E., Lemoine, D. G., Hamzaoui, J., & Legendre, L. (2020). Impact of soil water content on maize responses to the plant growth-promoting rhizobacterium Azospirillum lipoferum CRT1. *Journal of Agronomy and Crop Science*, *206*(5), 505–516.

Das, T., Bhattacharyya, A., & Bhar, A. (2023). Linking phyllosphere and rhizosphere microbiome to the plant–insect interplay: The new dimension of tripartite interaction. *Physiologia*, *3*(1), 129–144.

Dasgupta, D., Paul, A., Acharya, K., Minkina, T., Mandzhieva, S., Gorovtsov, A. V., & Keswani, C. (2023). *Bioinoculant mediated regulation of signalling cascades in various stress responses in plants*. Heliyon.

Devi, R., Kaur, T., Kour, D., Yadav, A., Yadav, A. N., Suman, A., & Saxena, A. K. (2022). Minerals solubilizing and mobilizing microbiomes: A sustainable approach for managing minerals' deficiency in agricultural soil. *Journal of Applied Microbiology*, *133*(3), 1245–1272.

Do Carmo, T. S., Cabral, B. V., Dantas, R. C., de Resende, M. M., Cardoso, V. L., & Ribeiro, E. J. (2019). Phosphorus recovery from phosphate rocks using phosphate-solubilizing bacteria. *Geomicrobiology Journal*, *36*(3).

Elhaissoufi, W., Ghoulam, C., Barakat, A., Zeroual, Y., & Bargaz, A. (2022). Phosphate bacterial solubilization: A key rhizosphere driving force enabling higher P use efficiency and crop productivity. *Journal of Advanced Research*, *38*, 13–28.

Etesami, H., & Adl, S. M. (2020). Plant growth-promoting rhizobacteria (PGPR) and their action mechanisms in availability of nutrients to plants. *Phyto-Microbiome in Stress Regulation*, 147–203.

Fahde, S., Boughribil, S., Sijilmassi, B., & Amri, A. (2023). Rhizobia: A promising source of plant growth-promoting molecules and their non-legume interactions: Examining applications and mechanisms. *Agriculture*, *13*(7), 1279.

Fariha, T. I., Buchori, I., & Sejati, A. W. (2021). Modelling industrial growth in the coastal area of semarang metropolitan region using GeOBIA. In *IOP Conference Series: Earth and Environmental Science* (Vol. 887, No. 1, p. 012008). IOP Publishing.

Ferguson, B. J., Mens, C., Hastwell, A. H., Zhang, M., Su, H., Jones, C. H., & Gresshoff, P. M. (2019). Legume nodulation: The host controls the party. *Plant, Cell & Environment*, *42*(1), 41–51.

Fitriatin, B. N., Mulyani, O., Herdiyantoro, D., Alahmadi, T. A., & Pellegrini, M. (2022). Metabolic characterization of phosphate solubilizing microorganisms and their role in improving soil phosphate solubility, yield of upland rice (*Oryza sativa* L.), and phosphorus fertilizers efficiency. *Frontiers in Sustainable Food Systems*, *6*, 1032708.

Flemming, H. C. (2016). EPS-then and now. *Microorganisms*, *4*(4), 41.

Fouzia, A., Allaoua, S., Hafsa, C. S., & Mostefa, G. (2015). Plant growth promoting and antagonistic traits of indigenous fluorescent Pseudomonas spp. isolated from wheat rhizosphere and A. halimus endosphere. *European Scientific Journal*, *11*(24).

Gautam, K., Sirohi, C., Singh, N. R., Thakur, Y., Jatav, S. S., Rana, K., & Parihar, M. (2021). Microbial biofertilizer: Types, applications, and current challenges for sustainable agricultural production. In *Biofertilizers* (pp. 3–19). Woodhead Publishing.

Gomiero, T., Pimentel, D., & Paoletti, M. G. (2011). Environmental impact of different agricultural management practices: Conventional vs. organic agriculture. *Critical Reviews in Plant Sciences*, *30*(1–2), 95–124.

Goswami, D., Thakker, J. N., & Dhandhukia, P. C. (2016). Portraying mechanics of plant growth promoting rhizobacteria (PGPR): A review. *Cogent Food & Agriculture*, *2*(1), 1127500.

Gouda, S., Kerry, R. G., Paramithiotis, S., Shin, H. S., & Patra, L. K. (2018). Revitalization of plant growth promoting rhizobacteria for sustainable development in agriculture. *Microbiological Research*, 206, 131–140.

Guardiola-Márquez, C. E., Santos-Ramírez, M. T., Figueroa-Montes, M. L., Valencia-de Los Cobos, E. O., Stamatis-Félix, I. J., Navarro-López, D. E., & Jacobo-Velázquez, D. A. (2023). Identification and characterization of beneficial soil microbial strains for the formulation of biofertilizers based on native plant growth-promoting microorganisms isolated from northern Mexico. *Plants*, *12*(18), 3262.

Guo, K., Yang, J., Yu, N., Luo, L., & Wang, E. (2023). Biological nitrogen fixation in cereal crops: Progress, strategies, and perspectives. *Plant Communications*, *4*(2).

Hamid, B., & Bashir, Z. (2019a). Potassium solubilization microorganisms: An Alternative technology to chemical fertilizers. *Journal of Research and Development University of Kashmir*, *19*, 0972–5407.

Hamid, B., Zaman, M., Farooq, S., Fatima, S., Sayyed, R. Z., Baba, Z. A., & Suriani, N. L. (2021). Bacterial plant biostimulants: A sustainable way towards improving growth, productivity, and health of crops. *Sustainability*, *13*(5), 2856.

Hazzoumi, Z., Azaroual, S. E., El Mernissi, N., Zaroual, Y., Duponnois, R., Bouizgarne, B., & Meftah Kadmiri, I. (2022). Effect of arbuscular mycorrhizal fungi isolated from rock phosphate mine and agricultural soil on the improvement of wheat plant growth. *Frontiers in Microbiology*, *13*, 881442.

Hefferon, K. (2019). Biotechnological approaches for generating zinc-enriched crops to combat malnutrition. *Nutrients*, *11*(2), 253.

Heil, J., Vereecken, H., & Brüggemann, N. (2016). A review of chemical reactions of nitrification intermediates and their role in nitrogen cycling and nitrogen trace gas formation in soil. *European Journal of Soil Science*, *67*(1), 23–39.

Huang, H., Li, M., Rizwan, M., Dai, Z., Yuan, Y., Hossain, M. M., & Tu, S. (2021). Synergistic effect of silicon and selenium on the alleviation of cadmium toxicity in rice plants. *Journal of Hazardous Materials*, *401*, 123393.

Hussain, A., Arshad, M., Zahir, Z. A., & Asghar, M. (2015). Prospects of zinc solubilizing bacteria for enhancing growth of maize. *Pakistan Journal of Agricultural Sciences*, *52*(4).

Hyder, S., Gondal, A. S., Rizvi, Z. F., Ahmad, R., Alam, M. M., Hannan, A., & Inam-ul-Haq, M. (2020). Characterization of native plant growth promoting rhizobacteria and their anti-oomycete potential against Phytophthora capsici affecting chilli pepper (Capsicum annum L.). *Scientific Reports*, *10*(1), 13859.

Ilangumaran, G., & Smith, D. L. (2017). Plant growth promoting rhizobacteria in amelioration of salinity stress: A systems biology perspective. *Frontiers in Plant Science*, *8*, 1768.

Jalal, A. (2022). *Agronomic biofortification of grain crops with zinc in association with plant growth-promoting bacteria* (pp. 1–10). UNESP.

Jha, C. K., & Saraf, M. (2015). Plant growth promotiong rhizobacteria (PGPR). *Journal of Agricultural Research and Development*, *5*(2), 0108–0119.

Jog, R., Pandya, M., Nareshkumar, G., & Rajkumar, S. (2014). Mechanism of phosphate solubilization and antifungal activity of *Streptomyces spp*, isolated from wheat roots and rhizosphere and their application in improving plant growth. *Microbiology*, *160*(4), 778–788.

Kaari, M., Manikkam, R., Annamalai, K. K., & Joseph, J. (2023). Actinobacteria as a source of biofertilizer/biocontrol agents for bio-organic agriculture. *Journal of Applied Microbiology*, *134*(2), lxac047.

Kang, S. M., Waqas, M., Shahzad, R., You, Y. H., Asaf, S., Khan, M. A., & Lee, I. J. (2017). Isolation and characterization of a novel silicate-solubilizing bacterial strain Burkholderia eburnea CS4-2 that promotes growth of japonica rice (Oryza sativa L. cv. Dongjin). *Soil Science and Plant Nutrition*, *63*(3), 233–241.

Kaur, J. (2021). PGPR in management of soil toxicity. *Rhizobiont in Bioremediation of Hazardous Waste*, 317–344.

Kaur, T., Rani, R., & Manhas, R. K. (2019). Biocontrol and plant growth promoting potential of phylogenetically new Streptomyces sp. MR14 of rhizospheric origin. *AMB Express*, *9*, 1–14.

Khan, F., Siddique, A. B., Shabala, S., Zhou, M., & Zhao, C. (2023). Phosphorus plays key roles in regulating plants' physiological responses to abiotic stresses. *Plants*, *12*(15), 2861.

Khan, N., Ali, S., Shahid, M. A., Mustafa, A., Sayyed, R. Z., & Curá, J. A. (2021). Insights into the interactions among roots, rhizosphere, and rhizobacteria for improving plant growth and tolerance to abiotic stresses: A review. *Cells*, *10*(6), 1551.

Khoshru, B., Mitra, D., Khoshmanzar, E., Myo, E. M., Uniyal, N., Mahakur, B., & Rani, A. (2020). Current scenario and future prospects of plant growth-promoting rhizobacteria: An economic valuable resource for the agriculture revival under stressful conditions. *Journal of Plant Nutrition*, *43*(20), 3062–3092.

Knights, H. E., Jorrin, B., Haskett, T. L., & Poole, P. S. (2021). Deciphering bacterial mechanisms of root colonization. *Environmental Microbiology Reports*, *13*(4), 428–444.

Kour, D., Rana, K. L., Yadav, A. N., Yadav, N., Kumar, M., Kumar, V., & Saxena, A. K. (2020). Microbial biofertilizers: Bioresources and eco-friendly technologies for agricultural and environmental sustainability. *Biocatalysis and Agricultural Biotechnology*, *23*, 101487.

Kour, D., Rana, K. L., Yadav, N., Yadav, A. N., Kumar, A., Meena, V. S., & Saxena, A. K. (2019). Rhizospheric microbiomes: Biodiversity, mechanisms of plant growth promotion, and biotechnological applications for sustainable agriculture. *Plant Growth Promoting Rhizobacteria for Agricultural Sustainability: From Theory to Practices*, 19–65.

Kralova, K., & Jampilek, J. (2023). Applications of nanomaterials in plant disease management and protection. In *Nanotechnology in Agriculture and Agroecosystems* (pp. 239–296). Elsevier.

Kudoyarova, G., Arkhipova, T., Korshunova, T., Bakaeva, M., Loginov, O., & Dodd, I. C. (2019). Phytohormone mediation of interactions between plants and non-symbiotic growth promoting bacteria under edaphic stresses. *Frontiers in Plant Science*, *10*, 1368.

Kumar, A., Dewangan, S., Lawate, P., Bahadur, I., & Prajapati, S. (2019). Zinc-solubilizing bacteria: A boon for sustainable agriculture. *Plant Growth Promoting Rhizobacteria for Sustainable Stress Management: Rhizobacteria in Abiotic Stress Management*, *1*, 139–155.

Kumar, D., Patel, K. P., Ramani, V. P., Shukla, A. K., & Meena, R. S. (2020). Management of micronutrients in soil for the nutritional security. *Nutrient Dynamics for Sustainable Crop Production*, 103–134.

Kumar, H., Bajpai, V. K., Dubey, R. C., Maheshwari, D. K., & Kang, S. C. (2010). Wilt disease management and enhancement of growth and yield of *Cajanus cajan* (L) var. Manak by bacterial combinations amended with chemical fertilizer. *Crop Protection*, *29*(6), 591–598.

Kumar, R., & Shastri, B. (2017). Role of phosphate-solubilizing microorganisms in sustainable agricultural development. In *Agro-Environmental Sustainability: Managing Crop Health* (Volume 1, pp. 271–303). Springer International Publishing.

Kumar, S., Sindhu, S. S., & Kumar, R. (2022). Biofertilizers: An ecofriendly technology for nutrient recycling and environmental sustainability. *Current Research in Microbialogy Sciences*, *3*, 100094.

Kumari, M., Swarupa, P., Kesari, K. K., & Kumar, A. (2022). Microbial inoculants as plant biostimulants: A review on risk status. *Life*, *13*(1), 12.

Li, Z., Bai, T., Dai, L., Wang, F., Tao, J., Meng, S., & Hu, S. (2016). A study of organic acid production in contrasts between two phosphate solubilizing fungi: *Penicillium oxalicum* and *Aspergillus niger*. *Scientific Reports*, *6*(1), 1–8.

Li, Z., Meunier, J. D., & Delvaux, B. (2022). Aggregation reduces the release of bioavailable silicon from allophane and phytolith. *Geochimica ET Cosmochimica Acta*, *325*, 87–105.

Liu, Y., Gao, J., Wang, N., Li, X., Fang, N., & Zhuang, X. (2022). Diffusible signal factor enhances the saline-alkaline resistance and rhizosphere colonization of Stenotrophomonas rhizophila by coordinating optimal metabolism. *Science of the Total Environment*, *834*, 155403.

Mahmood, S., Daur, I., Al-Solaimani, S. G., Ahmad, S., Madkour, M. H., Yasir, M., & Ali, Z. (2016). Plant growth promoting rhizobacteria and silicon synergistically enhance salinity tolerance of mung bean. *Frontiers in Plant Science*, *7*, 876.

Mahmud, K., Makaju, S., Ibrahim, R., & Missaoui, A. (2020). Current progress in nitrogen fixing plants and microbiome research. *Plants*, *9*(1), 97.

Maurya, S., Abraham, J. S., Somasundaram, S., Toteja, R., Gupta, R., & Makhija, S. (2020). Indicators for assessment of soil quality: A mini-review. *Environmental Monitoring and Assessment*, *192*, 1–22.

Meena, M., Swapnil, P., Divyanshu, K., Kumar, S., Harish, Tripathi, Y. N., & Upadhyay, R. S. (2020). PGPR-mediated induction of systemic resistance and physiochemical alterations in plants against the pathogens: Current perspectives. *Journal of Basic Microbiology*, *60*(10), 828–861.

Mehta, P., Walia, A., Kulshrestha, S., Chauhan, A., & Shirkot, C. K, (2015). Efficiency of plant growth-promoting P-solubilizing Bacillus circulans CB7 for enhancement of tomato growth under net house conditions, *Journal of Basic Microbiology*, *55*(1), 33–44.

Mendoza-Arroyo, G. E., Chan-Bacab, M. J., Aguila-Ramírez, R. N., Ortega-Morales, B. O., Canché Solís, R. E., Chab-Ruiz, A. O., Cob-Rivera, K. I., Dzib-Castillo, B., Tun-Che, R. E., & Camacho-Chab, J. C. (2020). Inorganic phosphate solubilization by a novel isolated bacterial strain Enterobacter sp. ITCB-09 and its application potential as biofertilizer. *Agriculture*, *10*(9), 383.

Minuţ, M., Diaconu, M., Roşca, M., Cozma, P., Bulgariu, L., & Gavrilescu, M. (2022). Screening of Azotobacter, Bacillus and Pseudomonas species as plant growth-promoting bacteria. *Processes*, *11*(1), 80.

Mondal, S., Pramanik, K., Pal, P., Mitra, S., Ghosh, S. K., Mondal, T., & Maiti, T. K. (2023). Multifaceted roles of root exudates in light of plant-microbe interaction. In *Unravelling Plant-Microbe Synergy* (pp. 49–76). Academic Press.

Naamala, J., & Smith, D. L. (2020). Relevance of plant growth promoting microorganisms and their derived compounds, in the face of climate change. *Agronomy*, *10*(8), 1179.

Nacoon, S., Jogloy, S., Riddech, N., Mongkolthanaruk, W., Kuyper, T. W., & Boonlue, S. (2020). Interaction between phosphate solubilizing bacteria and arbuscular mycorrhizal fungi on growth promotion and tuber inulin content of Helianthus tuberosus L. *Scientific Reports*, *10*(1), 4916.

Namdeo, V., Mitra, N. G., Jakhar, S. R., & Sahu, R. K. (2020). Increase in growth, productivity, nutritional status of Rice (Oryza sativa L.) and enrichment in soil fertility applied with Azospirillum and nitrogen level. *Journal Homepage URL*, *5*(2), 242–248.

Nandi, M., Selin, C., Brawerman, G., Fernando, W. D., & de Kievit, T. (2017). Hydrogen cyanide, which contributes to Pseudomonas chlororaphis strain PA23 biocontrol, is upregulated in the presence of glycine. *Biological Control*, *108*, 47–54.

Natasha, N., Shahid, M., Bibi, I., Iqbal, J., Khalid, S., Murtaza, B., & Arshad, M. (2022). Zinc in soil-plant-human system: A data-analysis review. *Science of the Total Environment*, *808*, 152024.

Naz, R., Asif, T., Mubeen, S., & Khushhal, S. (2022). Seed application with microbial inoculants for enhanced plant growth. In *Sustainable Horticulture* (pp. 333–368). Academic Press.

Neumann, G., & Ludewig, U. (2023). Rhizosphere chemistry influencing plant nutrition. In *Marschner's Mineral Nutrition of Plants* (pp. 545–585). Academic Press.

Noman, M., Ahmed, T., Shahid, M., Niazi, M. B. K., Qasim, M., Kouadri, F., & Ali, S. (2021). Biogenic copper nanoparticles produced by using the Klebsiella pneumoniae strain NST2 curtailed salt stress effects in maize by modulating the cellular oxidative repair mechanisms. *Ecotoxicology and Environmental Safety*, *217*, 112264.

Noulas, C., Tziouvalekas, M., & Karyotis, T. (2018). Zinc in soils, water and food crops. *Journal of Trace Elements in Medicine and Biology*, *49*, 252–260.

Olaniyan, F. T., Alori, E. T., Adekiya, A. O., Ayorinde, B. B., Daramola, F. Y., Osemwegie, O. O., & Babalola, O. O. (2022). The use of soil microbial potassium solubilizers in potassium nutrient availability in soil and its dynamics. *Annals of Microbiology*, *72*(1), 45.

Pahalvi, H. N., Rafiya, L., Rashid, S., Nisar, B., & Kamili, A. N. (2021). Chemical fertilizers and their impact on soil health. In *Microbiota and biofertilizers: Ecofriendly tools for reclamation of degraded soil environs* (Vol. 2, pp. 1–20). Springer. https://doi.org/10.1071/CP21457

Pandita, D. (2022). Plant growth promoting bacteria (PGPB): Applications and challenges in bioremediation of metal and metalloid contaminated soils. *Metals Metalloids Soil Plant Water Systems*, 485–500.

Panicker, S., & Sayyed, R. Z. (2022). Hydrolytic enzymes from PGPR against plant fungal pathogens. In *Antifungal Metabolites of Rhizobacteria for Sustainable Agriculture* (pp. 211–238). Springer International Publishing.

Parmar, P., & Sindhu, S. S. (2013). Potassium solubilization by rhizosphere bacteria: Influence of nutritional and environmental conditions. *Journal of Microbiology*, *3*(1), 25–31.

Patel, K., Goswami, D., Dhandhukia, P., & Thakker, J. (2015). Techniques to study microbial phytohormones. *Bacterial Metabolites in Sustainable Agroecosystem*, 1–27.

Patel, R., & Patel, A. (2023). Nitrogen fixation in nonlegume plants and genomics of nitrogen-fixing nonrhizobium symbioses. In *The Chemical Dialogue between Plants and Beneficial Microorganisms* (pp. 145–152). Academic Press.

Pieterse, C. M., Zamioudis, C., Berendsen, R. L., Weller, D. M., Van Wees, S. C., & Bakker, P. A. (2014). Induced systemic resistance by beneficial microbes. *Annual Review of Phytopathology*, *52*, 347–375.

Ramzan, S., Pervez, A., Wani, M. A., Jeelani, J., Ashraf, I., Rasool, R., & Maqbool, M. (2019). Soil health: Looking for the effect of tillage on soil physical health. *International Journal of Chemical Studies*, *7*, 1731–1736.

Rana, K. L., Kour, D., Kaur, T., Devi, R., Yadav, A., & Yadav, A. N. (2021). Bioprospecting of endophytic bacteria from the Indian Himalayas and their role in plant growth promotion of maize (Zea mays L.). *Journal of Applied Biology and Biotechnology*, *9*(3), 41–50.

Ranipa, N., Harkhani, K., & Sharma, A. K. (2023). Biofertilizer–A key player in enhancing soil fertility and agricultural sustainability. *International Journal of Agriculture, Environment and Biotechnology*, *16*(02), 121–134. https://doi.org/10.30954/0974-1712.02.2023.10

Rehman, F. U., Kalsoom, M., Adnan, M., Toor, M., & Zulfiqar, A. (2020). Plant growth promoting rhizobacteria and their mechanisms involved in agricultural crop production: A review. *SunText Review of BioTechnology*, *1*(2), 1–6.

Riaz, U., Murtaza, G., Anum, W., Samreen, T., Sarfraz, M., & Nazir, M. Z. (2021). Plant Growth-Promoting Rhizobacteria (PGPR) as biofertilizers and biopesticides. In *Microbiota and biofertilizers: A sustainable continuum for plant and soil health* (pp. 181–196). Springer Nature.

Rijavec, T., & Lapanje, A. (2016). Hydrogen cyanide in the rhizosphere: Not suppressing plant pathogens, but rather regulating availability of phosphate. *Frontiers in Microbiology*, *7*, 1785.

Saeed, Q., Xiukang, W., Haider, F. U., Kučerik, J., Mumtaz, M. Z., Holatko, J., & Mustafa, A. (2021). Rhizosphere bacteria in plant growth promotion, biocontrol, and bioremediation of contaminated sites: A comprehensive review of effects and mechanisms. *International Journal of Molecular Sciences*, *22*(19), 10529.

Samantray, J., Anand, A., Dash, B., Ghosh, M. K., & Behera, A. K. (2022). Silicate minerals-Potential source of potash-A review. *Minerals Engineering*, *179*, 107463.

Sanalibaba, P., & Çakmak, G. A. (2016). Exopolysaccharides production by lactic acid bacteria. *Applied Microbiology Open Access*, *2*(1000115), 10–4172.

Santoyo, G., Guzmán-Guzmán, P., Parra-Cota, F. I., Santos-Villalobos, S. D. L., Orozco-Mosqueda, M. D. C., & Glick, B. R. (2021). Plant growth stimulation by microbial consortia. *Agronomy*, *11*(2), 219.

Schaller, J., Puppe, D., Kaczorek, D., Ellerbrock, R., & Sommer, M. (2021). Silicon cycling in soils revisited. *Plants*, *10*(2), 295.

Sehrawat, A., Sindhu, S. S., & Glick, B. R. (2022). Hydrogen cyanide production by soil bacteria: Biological control of pests and promotion of plant growth in sustainable agriculture. *Pedosphere*, *32*(1), 15–38.

Shi, S., Tian, L., Nasir, F., Bahadur, A., Batool, A., Luo, S., & Tian, C. (2019). Response of microbial communities and enzyme activities to amendments in saline-alkaline soils. *Applied Soil Ecology*, *135*, 16–24.

Shrivastav, P., Prasad, M., Singh, T. B., Yadav, A., Goyal, D., Ali, A., & Dantu, P. K. (2020). Role of nutrients in plant growth and development. In *Contaminants in agriculture: Sources, impacts and management* (pp. 43–59). Springer.

Singh, J., & Verma, P. K. (2023). Role of Nod factor receptors and its allies involved in nitrogen fixation. *Planta, 257*(3), 54.

Singh, M. P., Singh, P., Singh, R. K., Solanki, M. K., & Bazzer, S. K. (2020). Plant microbiomes: Understanding the aboveground benefits. In *Phytobiomes: Current insights and future vistas* (pp. 51–80). Springer.

Sinha, D., & Tandon, P. K. (2020). An overview of nitrogen, phosphorus and potassium: Key players of nutrition process in plants. In *Sustainable solutions for elemental deficiency and excess in crop plants* (pp. 85–117). Springer Nature.

Song, C., Wang, W., Gan, Y., Wang, L., Chang, X., Wang, Y., & Yang, W. (2022). Growth promotion ability of phosphate-solubilizing bacteria from the soybean rhizosphere under maize–soybean intercropping systems. *Journal of the Science of Food and Agriculture, 102*(4), 1430–1442.

Ulloa-Ogaz, A. L., Muñoz-Castellanos, L. N., & Nevárez-Moorillón, G. V. (2015). Biocontrol of phytopathogens: Antibiotic production as mechanism of control. *The Battle Against Microbial Pathogens: Basic Science, Technological Advances and Educational Programs, 1*, 305–309.

Verma, D., Pandey, A., Mohapatra, B., Srivastava, S., Kumar, V., Talukdar, D., Yulianto, R., Jobanputra, A., & Asthir, B. (2019). Plant growth-promoting rhizobacteria: An eco-friendly approach for sustainable agriculture and improved crop production. In *Microbiology for Sustainable Agriculture, Soil Health, and Environmental Protection* (pp. 3–80). Apple Academic Press.

Vidal, A., Klöffel, T., Guigue, J., Angst, G., Steffens, M., Hoeschen, C., & Mueller, C. W. (2021). Visualizing the transfer of organic matter from decaying plant residues to soil mineral surfaces controlled by microorganisms. *Soil Biology and Biochemistry, 160*, 108347.

Wan, W., Qin, Y., Wu, H., Zuo, W., He, H., Tan, J., & He, D. (2020). Isolation and characterization of phosphorus solubilizing bacteria with multiple phosphorus sources utilizing capability and their potential for lead immobilization in soil. *Frontiers in Microbiology, 11*, 752.

Wang, Z., Li, Y., Zhuang, L., Yu, Y., Liu, J., Zhang, L., & Wang, Q. (2019). A rhizosphere-derived consortium of Bacillus subtilis and Trichoderma harzianum suppresses common scab of potato and increases yield. *Computational and Structural Biotechnology Journal, 17*, 645–653.

Wang, Z., Zhang, H., Liu, L., Li, S., Xie, J., Xue, X., & Jiang, Y. (2022). Screening of phosphate-solubilizing bacteria and their abilities of phosphorus solubilization and wheat growth promotion. *BMC Microbiology, 22*(1), 296.

Yadav, A. N., Verma, P., Sachan, S. G., Kaushik, R., & Saxena, A. K. (2018). Psychrotrophic microbiomes: Molecular diversity and beneficial role in plant growth promotion and soil health. *Microorganisms for green revolution: Microbes for sustainable agro-ecosystem* (Vol. 2, pp. 197–240).

Yu, X. (2023). *Role of soil pH on nitrogen and phosphorus fertilizer use efficiency in cereal crops* (Doctoral diss.).

Zellner, W., Tubaña, B., Rodrigues, F. A., & Datnoff, L. E. (2021). Silicon's role in plant stress reduction and why this element is not used routinely for managing plant health. *Plant Disease, 105*(8), 2033–2049.

Zhang, X., Han, L., Wang, Q., Zhang, C., Yu, Y., Tian, J., & Kong, Z. (2019). The host actin cytoskeleton channels rhizobia release and facilitates symbiosome accommodation during nodulation in Medicago truncatula. *New Phytologist, 221*(2), 1049–1059.

10 *Trichoderma* spp. and *Pseudomonas* spp.

An Alternative Technology to Chemical Fertilizers

Mehrun Nisa, Zaffar Bashir, Burhan Hamid, Rehana Rasool, Bilal Ahmad Dar, and Farahanaz Rasool

10.1 INTRODUCTION

Pandemic plant diseases phenomenon is scattering and aggravated by the emergence of plant pathogenic microorganisms in a crop plantation. The living organisms are exposed to the high toxicity of chemical compounds because of excessive use of chemicals. The application of biological control by using *Trichoderma* spp. in agricultural products is the best solution to overcome these problems, and its usage in the crop field is reported to have a significant effect on yield and maintaining soil micro floral population (Susiana et al., 2018; Yatoo et al., 2022). It also suppresses the activity of plant pathogens by enzymatic activities (Vinale et al., 2008).

In some insects and nematodes fungal biological control agents are also found to be effective and helpful. In destruction of certain important crops caused by some pests such as *Tetranychus urticae* and different insects, several *Trichoderma* species have been proven to be effective (Poveda et al., 2020; Poveda, 2021). Some fungi such as Arbuscular mycorrhizal fungi have been extensively studied and have been found to be effective against phytopathogens, such as *Meloidogyne incognita* and other nematodes. Another effective biocontrol agent of phytopathogens (*Arthrobotrys oligospora*) forms tenacious structures to capture nematodes (Liang et al., 2019; Schouteden et al., 2015). In case of abiotic stresses, such as high temperature increase, fungal biopesticides can also provide protection (Kashyap et al., 2020). To increase yield, they are also used as plant defence enhancers due to their ability to induce systemic resistance, protecting them against several pathogens. So, fungi have played important roles in enhancement of plant health and yield (Phour et al., 2020). Fungal species belonging to the genera *Aspergillus*, *Trichoderma*, *Rhizoctonia*, *Colletotrichum*, *Gliocladium*, and others with *Trichoderma* being the most used in the field can incite plant growth (Hossain et al., 2017). Bacteria belongs to the genera *Pseudomonas fluorescence* and *Bacillus* spp. (Cartieaux et al., 2003; Igiehon et al., 2017; Kashyap et al., 2017; Santoyo et al., 2016).

DOI: 10.1201/9781003491460-10

10.2 *TRICHODERMA* SPP.: INTERACTION WITH MICROORGANISMS

Trichoderma strains are used for diverse purposes in agricultural produce as its mechanisms reduce the occurrence of plant disease. *Trichoderma* spp. exhibit antagonistic behaviour against several plant pathogens, but mainly with pathogenic fungi (Harman, 2006). These direct interactions include nutritional competition, hyper parasitism and space and antibiosis or indirectly by improving plant growth and strength (Benítez et al., 2004; Błaszczyk et al., 2014; Druzhinina et al., 2006). The phytopathogens like *Botrytis cinerea* and *Rhizoctonia solani* cause serious damage to economically significant crops and there are various literatures about *Trichoderma* spp. to combat with such plant pathogens (Brunner et al., 2005). It was also found that *Trichoderma* spp. enhanced the activity of *Azotobacter* in maize (Mahato & Neupane, 2018).

10.2.1 Mechanism of Action

Agricultural scientific communities have proposed a few possible mechanisms by which these biocontrol agents suppress the pathogens (Yadav et al., 2017b, 2017c). Figure 10.1 shows *Trichoderma* as biocontrol agent against *Fusarium* spp. Some of the mechanisms are given as follows:

10.2.1.1 Mycoparasitism

Mycoparasitism is an ancestral trait and direct biocontrol method of reducing the pathogen inocula, and the mechanism involves colonizing and parasitizing phytopathogens (Druzhinina et al., 2011; Harman, 2006). Some other fungi have been widely used as mycoparasites for the biological control of agricultural pests. Some species from *Trichoderma* including *Trichoderma harzianum*, *T. asperellum*, *T. atroviride*, and *T. virens* have been often used as biological control agents for plant pathogens. *Trichoderma* spp. as a biocontrol agent holds capacity to inhibit or kill other plant pathogenic fungi (Mukherjee et al., 2012). At least 75 *Trichoderma* species have been identified, with many of them possessing the capacity to inhibit the mycelial growth of *Fusarium* spp., *Botrytis cinerea*, *Rhizoctonia solani*, and *Alternaria alternata* (Druzhinina et al., 2011). The mycoparasitism process involves four sequential steps. The first step is the detection process, and in the second stage there is chemotropic development of antagonistic fungal mycelium towards the fungal plant pathogens which is followed by third and fourth stages. These stages include direct attachment and destruction of cell walls, and finally host fungal cell penetration (Harman et al., 2004). Lytic enzymes chitinases and -1, 3-glucanases are produced by biocontrol *Trichoderma* spp. which function as mycoparasitic agents to break down phytopathogenic fungal cell walls. As per the study by Brunner et al. (2005) pathogenic mycelium can be found by *Trichoderma* in the rhizosphere and grow in the area of pathogens direction.

10.2.1.2 Antibiosis Mechanism

The antagonistic interaction incorporating low-molecular weight diffusible antibiotics that stops the growth of other plant pathogens is known as antibiosis. Various *Trichoderma* spp. are important an source of secondary metabolites for biocontrol of plant pathogens and crucial in interactions with plants. Antimicrobial properties could be the result of several secondary metabolites generated by fungi such as terpenes, polyketides, gliotoxin, and gliovirin. Other antibiotics include gliosoprnins, heptelidic acid, 6-pentyl-pyrone, viridian, Tricholin harzianic acid, and massoilactone (Mukherjee et al., 2012; Vinale et al., 2013; Zafari et al., 2008). *Trichoderma* spp. have the capability to produce peptaibols, which are polypeptide antibiotics that are made of non-proteinogenic amino acids, particularly aminoisobutyric. Peptiabols are from a family of peptides having antibiotic functions and odd characteristic including that their N-terminal is acetylated while their C-terminal contains amino alcohols (Ramachander Turaga, 2020).

10.2.1.3 Nutritional and Space Competition

Trichoderma spp. are excellent competitors for nutritional resources and space (Singh et al., 2013). The pathogen displacement can be attained by the competition between plant pathogens and biocontrols. These avirulent symbionts compete with other fungi for nutrients and space. They inhibit the growth of phytopathogens and modify the rhizosphere through acidification of soil. Several *Trichoderma* spp. are applied to soil directly or by application to seeds (seed dressing) and grow continuously along with growing root system and thus protecting the plants against phytopathogens (Benítez et al., 2004; Howell et al., 2000). The antagonism expressed by various *Trichoderma* spp. against *Chondrostereum purpureum*, *Fusarium oxysporum*, and several other plant pathogens involves competition for carbon and other essential nutrients (Corke & Hunter, 1979; Sivan, 1989). *Trichoderma harzianum* controlled *Fusarium* species on numerous different crops through colonization of the rhizosphere e.g., cotton (Sivan, 1989; Viterbo et al., 2007).

10.2.1.4 Production of Enzymes

Trichoderma spp. produce different enzymes that have a huge role in disease management and plant growth promotion. It was testified that endo and exochitinases produced by *T. Koningii* help in the penetration and destruction of the plant pathogen cells without damaging the tissue of the plant. Biocontrol *Trichoderma* spp. produces numerous hydrolytic enzymes such as glucanase, chitinase, and proteases as well as volatile and non-volatile antibiotics (Elad, 1983). It was also reported that the biocontrol of *Bacillus cinerea* (causal agent of rice blast disease, the most destructive disease of rice worldwide) by the *T. harzianum* is due to the proteases that are produced by the *T. harzianum*. The production of hydrolytic enzymes by *Bacillus cinerea* is inactivated by these proteases (Elad & Kapat, 1999; Metcalf & Wilson, 2001; Mohiddin et al., 2010).

10.2.1.5 Enhancement of Host Defence Response

When rhizosphere is inoculated with *Trichoderma* spp., different mechanisms such as hypersensitive response induced systematic resistance and systematic acquired

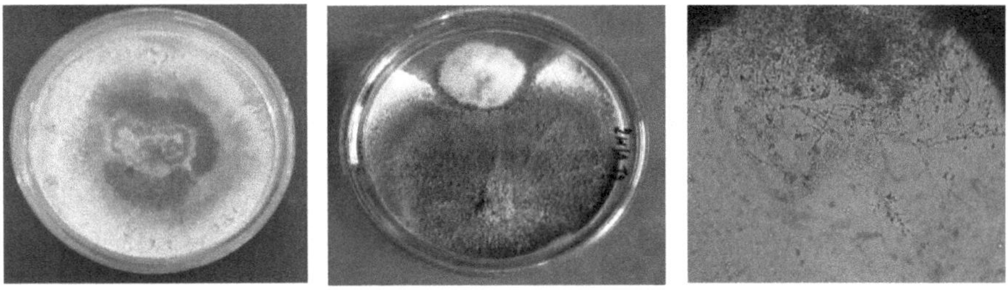

FIGURE 10.1 *Trichoderma* as biocontrol agent against *Fusarium* spp.

resistance protect the plants from microbes (Haggag, 2008). An unidentified *Trichoderma* sp. on cucumber and *Trichoderma virens* on cotton with induced resistance was observed (Koike et al., 2001; De Meyer et al., 1998). It is reported that the peroxidase activity was induced by *T. virens* in the seedlings of cotton more than in the control experiment. In the cells of grapevine, the production of phytoalexin is elicited by ethylene-inducing xylanase produced by the *T. viride* (Calderón et al., 1993; Harman, 2006; Howell et al., 2000). It was noted in cotton that there was stimulation of higher terpenoid in the culture filtrates of few effective strains of *T. virens*. Further the elicitors are mostly glycoproteins or proteins. *Trichoderma harzianum* when applied to seven days old root cucumber seedlings in an aseptic hydroponic system can induce the defensive responses in roots and leaves of this plant (Hanson & Howell, 2004; Yedidia et al., 1999) (Figure 10.1).

10.3 BACTERIA: *PSEUDOMONAS* SPP. AS BIOLOGICAL CONTROL AGENTS

Species of *pseudomonas* are usually found in almost all the environments of nature, such as marine, freshwater, and soil; they are capable of developing profound associations with a variety of plant species (Höfte, 2021). The bacteria pertaining to the *Pseudomonas* spp. are ecologically significant and functionally diverse due to their significant use as plant growth promoters. *Pseudomonas* bacterial strains represent attractive biopesticides due to their phenomenal root-colonization abilities, catabolic adaptability, and ability to secrete various anti-plant pathogenic substances. These bacterial strains perform biocontrol activity on plant pathogens by direct antagonism and induce resistance to disease in the host plant (Cartieaux et al., 2003). *Pseudomonas* species reveal a variety of attributes that make them appropriate as plant growth-promoters and biological control agents. The variety of attributes includes: secretion of plant growth-promoting substances like IAA, siderophores or iron scavengers, hydrolysis of various substances like starch, pectin, lignin, cellulose, production of antibiotics, lytic enzymes and HCN for pathogen suppression, and other necessary substances (Nowak-Thompson et al., 1997; Weller et al., 2002). Figure 10.1 shows the biological control of *Fusarium* spp. by *Pseudomonas*. Beneficial microbes are capable of using various mechanisms to avoid root infections. However, biocontrol agents rely primarily on the production of antibiotics or the impairment of virulent traits of the pathogen. Another significant mechanism is that bacteria can indirectly inhibit the pathogen from proliferating by boosting defence responses in the host plant (Zhang et al., 2020). The capability of bacteria *Pseudomonas* spp. to protect plants from the detrimental effects of phytopathogens at various stages of plant development signifies that *Pseudomonas* plays a crucial role in the development and growth of plants. The aforementioned benefits that *Pseudomonas* delivers depend primarily on their ability to absorb root exudates in an efficient way and to avoid predation by soil predators such as protozoa and nematodes (Abuzar & Haseeb, 2010). These microbial strains have developed an extensive range of antipredatory mechanisms, including one that involves toxin production. Extracellular substances produced by *Pseudomonas* spp. are responsible for the complex interactions with predators, which in turn have a profound impact on the behaviour and physiology of the predators (Gadoury et al., 1989).

Pseudomonas bacterial strains have the capability of deriving energy from a diverse range of organic molecules and produce a wide number of secondary metabolites, among the most prominent of which are hydrogen cyanide, phenazines, 2,4-diacetylphloroglucinol, pyrrolnitrine, lipopeptides, pyochelin, phenazines, antibiotics, and pyrrolnitrine (Keel et al., 1992; Yasmin et al., 2017). *Pseudomonas* spp. that produce the antifungal compound (i.e., DAPG) perform a key role in preventing the spread of various root diseases when delivered through soil or seed into the rhizosphere. Despite its potential to revolutionize plant disease management, biological control of plant diseases by antagonistic microbes is not as popular among farmers compared to other disease control approaches (Reddy et al., 2009).

Other than *Pseudomonas* spp., *Bacillus* spp. have been introduced in agriculture, and they have plant growth-promoting and antipathogenic properties (Bashir et al., 2017); *Bacillus*-based biocontrol agents are usually more competent compared to those based upon other kinds of plant growth-promoting bacteria because *Bacillus* species produce metabolites and form spores more effectively, thereby enhancing the cell viability of commercially manufactured products. Based on various studies, *Bacillus* has been proven to be highly beneficial as a plant growth-promoting bacterium for an array of horticultural and field crops (Hass & Défago, 2005; Etesami et al., 2023; Fan et al., 2018; Santoyo et al., 2016).

Mineralization and solubilization of nutrients including phosphorus, potassium, zinc, silicates and phosphorus; the production of 1-aminocyclo-propne-1-carboxlyic acid (ACC), nitrogen fixation, HCN, antimicrobial compounds, siderophores, hydrolytic enzymes and phytohormones; and abiotic stress tolerance (e.g., by competition for nutrients and space, antibiosis, parasitism, and/or by triggering the defence system in host plant) are various different mechanisms that *Bacillus* sp. utilize for growth of plants and limited proliferation of pathogen (Bashir et al., 2017; Bashir etal., 2018b; Bashir et al., 2018c; Bashir et al., 2019a; Bashir et al., 2018b; Bashir et al., 2019b; Bashir et al., 2021; Bashir, 2021; Shafi et al., 2017; Hamid & Bashir, 2019a; Kohl et al., 2019; Sansinenea et al., 2019).

10.3.1 *Pseudomonas* spp.: Biocontrol Mechanism

The rhizospheric microbes have been identified, characterized, and in several cases, evaluated for action as biocontrol agents against soil-borne plant diseases. In addition to developing systemic resistance in their host plants, these microbes can also produce compounds that may reduce the harm done by phytopathogens, i.e., by producing antibiotics, siderophores, HCN, and various enzymes. These microbes can also compete with pathogens for nutrients, other essential substances, and colonization sites.

The most important mechanisms through which *Pseudomonas* perform biocontrol function are: competition for space acquisition and nutrients, production of antibiotics, and induced systemic resistance.

10.3.1.1 Nutrients and Space Acquisition Competition

Competition in the rhizosphere environment provides a challenging "biological buffering" that usually inhibits exogenous microbes from establishing in the rhizosphere part of the plant. Therefore, it is imperative to determine the potential of introduced pseudomonads to colonize roots and provide protection against soil-borne plant pathogens (Loon et al., 1998). The concept of "rhizosphere competence" refers to the relative ability of a rhizobacterium to colonize the root of a plant. Inoculants consisting of bacteria acquire more strength when these microbes proliferate on the root and subsequently colonize it. Therefore, the development of inoculants is a substantial factor in the disease inhibition resulted by bioinoculants. *Pseudomonas* species have thoroughly demonstrated the process of root colonization, the biotic and abiotic factors that impact colonization, and the genes of bacteria, as well as traits that influence rhizosphere competence (Loper & Buyer, 1991). The exudates of roots containing carbohydrates, amino acids, nucleic acids, sugars, flavonoids, auxins, and organic acids mainly affect the rhizospheric microflora. Biocontrol agents utilized in the soil have to compete with pathogens and injurious microbes for scarce nutrients in root exudates and suitable niches for colonization and eventually overwhelm them. It is believed that the capacity of pseudomonads to establish niches and promptly compete for nutrient acquisition is a general mechanism for antagonistic activity disseminated by biocontrol strains of pseudomonads, which function as probiotics of plant. Fungal contaminants can be eradicated from the soil by increasing competition for nutrients such as carbon, nitrogen, and iron, thereby reducing their ability to proliferate in the soil (Loper & Buyer, 1991).

Pseudomonas spp. established on the roots can function as a sink for nutrients that are accessible and restrict the availability of nutrients for the phytopathogens and its root colonization. Furthermore, in order to successfully remove the pathogens, the pseudomonads compete with indigenous microbes for nutrients in the rhizosphere. Pseudomonads produce siderophores which are also called iron scavengers, sequester the Fe^{3+} available in the rhizosphere, and deprive the pathogens of the iron that they require for survival, thus playing an important role in eradicating pathogens from the same ecological niche (O'Sullivan & O'Gara, 1992). The ability of biocontrol agents to sequester iron under iron-deficient conditions gives them a selective advantage over plant pathogens that do not have efficient iron uptake systems (Bakker et al., 1986).

10.3.1.2 Antibiotic Production

Among the different species of *Pseudomonas* that occupy the rhizosphere, several fluorescent pseudomonad strains have garnered specific attention due to their ability to suppress soil and seed-borne pathogenic fungi. Plant-beneficial microbes aid in the elimination of pathogens from the rhizosphere by secreting antimicrobial metabolites, which enhance the health of plants (Haas & Keel, 2003). The most normally detected classes of antifungal metabolites produced by known *Pseudomonas* spp. are phenazines, pyoluteorin, volatile hydrogen cyanide, and pyrrolnitrin (Nielsen et al., 2001). Besides, new antifungal metabolites, tensins and viscosinamide, have been discovered, and both of these metabolites play a significant role in protecting plants against various phytopathogens (Zhang et al., 2003; Zachow et al., 2015).

The production of antibiotics can be affected by several factors, such as pH, temperature, and the concentrations of different metal ions, especially Zn $^{2+}$ (Duffy & Defago, 1997).

10.3.1.3 Induced Systemic Resistance

Induced systemic response triggered by beneficial microorganisms is highly effective against an extensive range of pathogens, such as viruses, fungi, and bacteria, however the response is considered to be random. Plant hormones such as ethylene and jasmonate are usually considered to be the main regulators of the induced systemic response (Wees et al., 2000). Besides live microbes like Trichoderma, *Pseudomonas*, and *Bacillus*, dead cells of microbes and a number of products from bacterial metabolites, such as lipopolysaccharides, pyocyanin, siderophores, and salicylic acid as well as organelles like flagella, are the primary stimulants of the induced systemic response in plants (Audenaert et al., 2002; Höfte, 2021). A small number of *pseudomonas* strains are able to stimulate the induced systemic response against a wide range of phytopathogens. These immunized plants are able to develop defensive responses quicker and more efficiently as a result of pathogen attacks, which in turn leads to a higher degree of protection (Van Loon & Bakker, 2006).

Induced systemic resistance (ISR) can be described as an array of plant immune responses that are triggered by plant-beneficial microorganisms that live together with plant roots. The *Pseudomonas* as biocontrol agent against *Fusarium* spp. is shown in Figure 10.2.

10.4 ENDOPHYTIC MICROBES AS BIOCONTROL AGENTS

The extensive use of agrochemicals contributes various risks to environment such as change in soil structure, destruction of soil biological communities, and disturbances

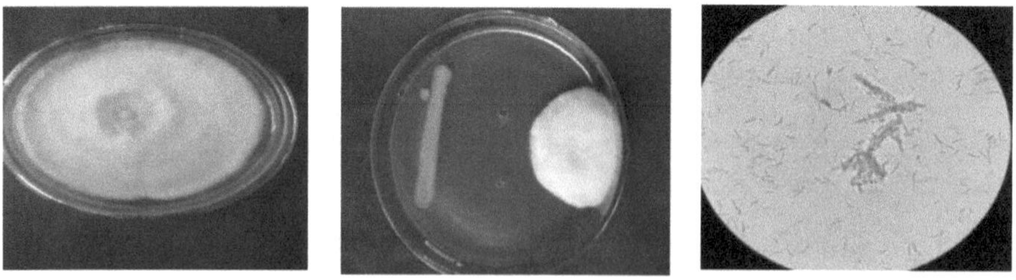

FIGURE 10.2 *Pseudomonas* as biocontrol agent against *Fusarium* spp.

in nutrient cycling that otherwise are key to crop production (Zeng et al., 2022). Similarly, the frequent use of chemical pesticides for management of plant diseases can develop many species that are resistant for these chemicals and has negative consequences on public health since these toxic chemicals persist in the environment and enter the food chain causing serious health concerns. Thus in this era there is a fundamental need of sustainability in agriculture which could limit the spread of plant pathogens preferably by using biological control agents (Santoyo et al., 2017; Vahidinia et al., 2019). Biocontrol may be defined as the use of microorganisms to limit the occurrence and frequency of plant diseases. When compared to traditional chemicals, these agents provide better selectivity and lower toxicity (Kaur et al., 2019).

Endophytes as biological control agents play an important part in plant disease management because they may contain or suppress a number of viral, bacterial, and fungal diseases. The best characteristics of endophytes which make them useful as biocontrol agents include: they can simply colonize inside the roots of plants and do not require the regular doses because of their ability to colonize in significant number and tolerate the drastic climatic variations (Wani et al., 2015). Endophytic bacteria which are native to the area, including *Serratia*, *Pseudomonas*, *Bacillus*, *Streptomyces*, and others, have usage for biocontrol and reduce the requirement to choose bacterial strains with high rhizosphere competence, which are frequently thought to be required for effective seed or root bacterization prior to or during planting (Aravind et al., 2009). Endophytic microbes have plentiful biological benefits and have capacity to enhance the growth of plants by several ways such as production of growth stimulators, management of diseases, production of secondary metabolites, and resistance to different stresses in host plants (Etminani et al., 2018; Hussein & El-Anssary, 2019; Kushwaha et al., 2020).

Endophytes are a different group of naturally occurring organisms (fungi and bacteria) that multiply inside the plants and do not cause any visible disease symptoms. Endophytes penetrate plant tissue predominantly through the roots, but they can also enter through aerial parts of plants such as flowers, stems, and cotyledons, according to studies (Nair & Padmavathy, 2014; Wilson et al., 1995; Garbeva et al., 2001; Gutiérrez-Zamora & Martınez-Romero, 2001; Stoltzfus et al., 1997). Depending on the functionality of endophytes, they are categorized into three main groups, (a). Biological control agents (b). Growth stimulators (c). Stress regulators (Tewari et al., 2019).

Two types of interactions—commensalism and mutualism—are found between plants and endophytes. Through commensalism, the endophyte gets place and nutrients in the host plant, while both endophyte and plant benefit if the interaction is mutualism. Apart from plant-endophyte interaction, plant pathogens within the plant reside in the same environmental condition and thus endophyte compete out these pathogens which reduces the harmful effects of these pathogens (Duijff et al., 1997; Kogel et al., 2006; Mahaffee & Kloepper, 1997). The eradication and prevention of phytopathogens proliferation are two important consequences of inoculating endophytes on plants (mostly fungi). For example, tomato plant when inoculated with *Bacillus amyloliquefaciens* gained resistance against *Alternaria alternata, Fusarium* spp., which are responsible for damping off disease in tomato.

10.5 ENDOPHYTIC BACTERIAL STRAINS AS BIOPESTICIDES

Bacterial endophytic strains as biological control agents have been broadly studied, and the application of these beneficial strains has been used to develop agricultural systems resulting in improved crop development and a reduction in plant disease-causing organisms (Xia et al., 2016). Bacterial endophytes are known to impact plant development in a variety of direct and indirect methods. Directly they can facilitate the development of plants by increasing the uptake of nutrients which take place by some ways such as biological N_2 fixation, nutrient solubilization, and synthesis of growth stimulators like cytokinins, IAA, gibberellins (GA), and ethylene (Long et al., 2008). However indirectly endophytes act as biocontrol agents to neutralize the negative effects of phytopathogens on plants through the production of secondary metabolites, repressive substances such as HCN, lytic enzymes, and antibiotics (Figure 10.3).

Diverse antipathogenic substances produced by bacterial endophytes include peptide, alkaloids, steroids, terpenoids, and phenolic compounds, which can cure the plant diseases caused by bacteria and fungi (Lodewyckx et al., 2002; Yu et al., 2010). Numerous genera of bacteria such as *Pseudomonas*, *Rhizobium*, *Streptomyces*, *Burkholderia*, *Enterobacter*, and *Pseudomonas*, etc. were employed as biocontrol agents with positive results in a variety of sectors. Plant pathogens, such as *F. oxysporum* which causes blight disease in rice, could be effectively removed by the antimicrobial activity of *B. subtilis* MJMP2, which has been isolated from fermented *Brassica campestris* (Cheng et al., 2016; Ibrahim et al., 2020). As per studies different genes have been found in bacterial endophyte such as prnC and phlD which are used by the endophytic bacteria to synthesize antifungal compounds. Endophytic strains from genus *Enterobacter* can also diminish wilt in cotton which is caused by *Verticillium* spp. Antagonistic activity of endophytic bacteria against viruses is not focussed on much but as per some studies, they can effectively suppress the infection

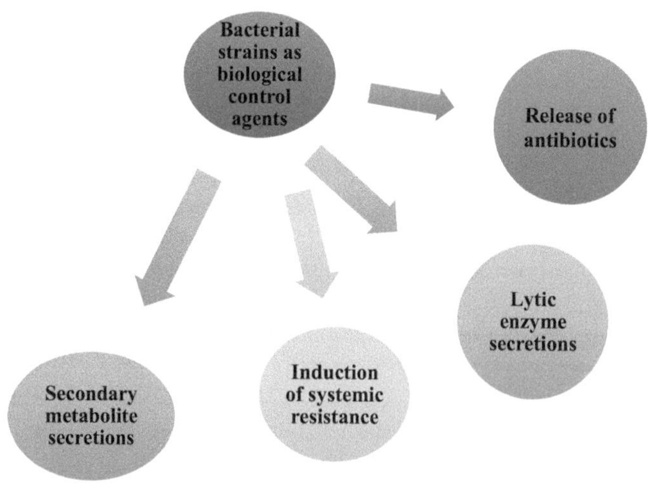

FIGURE 10.3 Ways by which bacterial endophytes act as biocontrol agents.

TABLE 10.1
Bacterial Strain Endophytes as Biological Control Agents against Pathogens

Bacterial Endophytes	Pathogen	Mode of Action	References
B. vallismortis	*Fusarium graminearum*	Antifungal	Zhao et al. (2010)
B. mojavensis	*Fusarium* spp.	Antifungal	Abdallah et al. (2016)
P. aeruginosa	*Colletotrichum truncatum*	Antifungal	Sandani et al. (2019)
B. siamensis	*Acidovorax oryzae*	Antibacterial	Ibrahim et al. (2019)
Bacillus spp. (G4L1 strain)	*Ralstonia solanacearum*	Antibacterial	Fu et al. (2020)
C. vaccinii	*Coleophoma* SP.	Antifungal	Ebadzadsahrai et al. (2020)
B. thuringiensis SY33.3	*Verticillium dahlia*	Antifungal	Azizoglu et al. (2021)
P. fluorescens	*Fusarium* spp.	Antifungal	Yassin et al. (2021)
B. subtilis XZ18-3	*R. cerealis*	Antifungal	Yi et al. (2022)

caused by important viruses (Andreolli et al., 2019; VinodKumar et al., 2018). Endophytes, mostly *Bacillus* sp., are gaining attraction due to their ability to compete out the most dangerous plant pathogen such as tobacco mosaic virus. Growth of this virus is inhibited due to extracellular compound produced by *Bacillus* sp. Some important endophytic bacteria with ability to suppress the growth of phytopathogens are given in Table 10.1.

10.6 FUNGAL–BIOCONTROL AGENTS

Global crops are affected by countless diseases. The common approaches for their prevention are often ineffective, thus fungal endophytes as biocontrol agents are constantly being sought. Fungal endophytes can be found in the internal parts of the plant like stem, leaves, and bark. Interactions of fungal endophytes with plants are economically exploited because they have a positive influence on crop production by suppressing the growth of a wide range of phyto pathogens through the production of inhibitory substances like steroids, terpenoids, alkaloids aliphatic compounds, phenolic compounds, and chromones. As per studies (An et al., 2020; Latz et al., 2018), *Chloranthus japonicus* is colonized by various fungal species, like *Aspergillus flavipes*, *Trichoderma harzianum*, *Aspergillus oryzae*, and *Leptospora rubella,* which were found to be the primary source of resistance against *Bacillus subtilis*, *Bacillus cereus*, *E. coli*, and *Pseudomonas aeruginosa*. Trichodermin, a bioactive compound produced by endophytic fungus *Trichoderma harzianum*, is antagonistic against phytopathogen which are responsible for damping-off disease and early blight of tomato in cucumber plants. *T. harzianum* is also known for the production of siderophore that can chelate iron in the soil, thus making it less available for the growth of pathogens (Chen et al., 2008). In rice *R. solani* causes sheath blight disease, which is the second most deleterious disease of rice and it destroys around 30% of the potential worldwide agricultural output even before harvest, and sometimes can reach up to 50% (Lu et al., 2021). The *T. viride* can suppress the growth of *R. solani* by reducing

TABLE 10.2
Fungal Endophytic Strains as Biological Control Agents

Fungal Endophytes	Pathogen	Mechanism of Action	References
T. viride	*R. solani*	Antifungal	
A. flavus	*P. chrysogenum* and *F. oxysporum*	Antifungal	Santos et al. (2015)
Ngrospora sphaerica	*P. aeruginosa*	Antibacterial	Santos et al. (2015)
Nigrospora sp.	*A. olerace* and *Exserohilum turcicum*	Antifungal	Luo et al. (2017)
C. alienum and *Fusarium* spp.	*Staphylococus aureus*	Antibacterial	Hastuti et al. (2018)
Trichothecium sp.	*Camellia oleifera*	Antifungal	Yu et al. (2018)
T. spirale T76-1	*Curvularia aeria*	Antifungal	Baiyee et al. (2019)
A. pullulans	*Botrytis cinerea*	Antifungal	Di et al. (2020)
Cladosporium spp.	*Fusarium oxysporum.*	Antifungal	Du et al. (2020)
Aspergillus flavipe and *Leptospora rubella*	*C. albican* and *B. cereus*	Antibacterial and antifungal	An et al. (2020)
X. feejeensis	*Fusarium oxysporum*	Antifungal	Brooks et al. (2022)
F. odoratissimum and *M. phaseolina* (USMJMTBR1)	*F. odoratissimum*	Antifungal	Taping et al. (2023)

its ability to synthesize proteins and to absorb amino acids. *Pestalotiopsis adusta*, a fungal endophyte, produces phenolic compound such as pestalachloride, which can suppress the growth of various plant pathogens such as *Gibberella zeae* and *Verticillium albo-atrum* (Li et al., 2008).

Usually in plants *Fusarium oxysporum* causes fungal diseases like *Fusarium* wilt and *Fusarium* root rot, but its biocontrol potential has been also extensively studied by Zimmermann et al. (2016) in maize against parasitic weed *Striga hermonthica.* Antiviral activity also has been seen in some endophytic fungi. For example, *Ephedra aphylla* produces (HSV-2 and VSV) extract which can show inhibition of 40.7% and 15.2% (Selim et al., 2018). Some important fungal endophytes as suppressors of phytopathogens are given in Table 10.2.

10.6.1 Biocontrol Agent Mechanism

10.6.1.1 Release of Secondary Metabolites

Endophytic microbes release various bioactive compounds which stop the growth of numerous disease causing pathogens; these compounds are called secondary metabolites. It includes diverse compounds such as saponins, phenolic, lipids, terpenes, and alkaloids. These defensive molecules are also produced by plants but are insufficient and less effective while as such metabolites from microbial origin have good quality and powerful biocontrol potential. Bioactive secondary metabolites can be antifungal, antibacterial, insecticidal, and antiviral. Fungal endophytes from *Xylaria* sp. produce pili formic acid and cytochalasin D which reduces anthracnose disease in tomato and citrus caused by *Colletotrichum gloeosporioides* (Hussein & El-Anssary, 2019; Lugtenberg et al., 2016; Wei et al., 2017). When a maize plant is colonized by an endophyte *Acremonium zeae*, it can produce an alkaloid derivative pyrrocidines which is antagonistic against *F. verticillioides* and *A. flavus*. So overall, endophytes as biocontrol agents provide an alternate technique for treating a variety of diseases caused by phytopathogens which include bacteria, fungi, and viruses (Wicklow et al., 2005).

10.6.1.2 Hydrolytic Enzyme Secretions

Endophytes can release the lytic enzymes that can degrade these constituents in the cell wall of plant pathogens, thereby suppressing the wide range of plant diseases. Lytic enzymes such as amylase, peptidase, xylanase, cellulase, and chitinases are active polypeptides produced by endophytes that can break the cell wall of pathogenic microbes, thus reducing the chances of infection. One example is the secretion of lytic enzymes such as N-acetylglucosaminidase by *Trichoderma* spp. that can break the mycelium of pathogenic fungi (Loc et al., 2019). Another important example is the cacao witches broom disease which can be reduced by the lytic enzymes from *Streptomyces* (Fadiji & Babalola, 2020; Yang et al., 2021).

10.6.1.3 Resistance Development in Plants

Through use of endophytic microbes plants can acquire induced systemic resistance and systemic acquired resistance to kill the disease-causing pathogens. Resistance through SAR includes the development of hypersensitive reaction in the infected parts of a plant via signalling molecules such as salicylic acid and pathogen related proteins, thus a plant acquires protection against those pathogens. One important example of SAR is the field application of bacterial endophyte such as *Bacillus subtilis* which can reduce the chances of powdery mildew in wheat caused by *Blumeria graminis*. Induced defence mechanisms are due to some signalling molecules such as jasmonic acid, ethylene, and salicylic acid which have inhibitory action against pathogens (Fu et al., 2020; Xie et al., 2021). For example, in grapevine, *Burkholderia* species via ISR inhibits the growth of mycelium and development of spore of *Botrytis cinerea* thus preventing this plant from grey mold disease (Esmaeel et al., 2020; Khare et al., 2016).

10.7 FUTURE PERSPECTIVE AND CONCLUSION

Biocontrol is of greatest importance and is an eco-friendly approach, but is not exploited that much because the research in this area is still conducted in the laboratory and responsiveness for the commercialization of such beneficial microbes is very low. Furthermore, the lack of information among farmers about the ways that are important for the inoculation of biocontrol agents in the field is the second reason for its inefficient use in the field. So to propagate the concept of biocontrol, extension at university level in this direction needs to be improved. To improve biocontrol ability, genetic engineering and other molecular techniques provide a new possibility for the selection and characterization of biocontrol agents. Various approaches that can contribute to increase the efficacy of biocontrol agents include mutation or protoplasm fusion utilizing poly ethylene glycol etc. There is also an exigent need to mass produce the biocontrol agents, understand their mechanism of action, and evaluate the environmental factors that favour the rapid growth of biocontrol agents.

Biological control looks to be the best alternative to plant disease suppression. Biopesticides bring the disease suppression with no environmental issues. It has been proved by various scientists that these biological control agents also trigger the growth of plants by various mechanisms like phosphate solubilization, silicate solubilization, zinc solubilization, and various other important plant growth-promoting secretions like GBA and IAA. Moreover, the novel concept of biocontrol needs a space outside the laboratory to see its fruits in the present production system.

REFERENCES

Abuzar, S., & Haseeb, A. (2010). Plant growth and plant parasitic nematodes in response to soil amendments with plant growth promoting rhizobacteria and inorganic fertilizer in pigeon pea, *Cajanus cajon* L. *World Applied Sciences Journal*, *8*(4), 411–413.

An, C., Ma, S., Shi, X., Xue, W., Liu, C., & Ding, H. (2020). Diversity and antimicrobial activity of endophytic fungi isolated from Chloranthus Japonicus Sieb in Qinling Mountains, China. *International Journal of Molecular Sciences*, *21*(17), 5958.

Andreolli, M., Zapparoli, G., Angelini, E., Lucchetta, G., Lampis, S., & Vallini, G. (2019). Pseudomonas protegens MP12: A plant growth-promoting endophytic bacterium with broad-spectrum antifungal activity against grapevine phytopathogens. *Microbiological Research*, *219*, 123–131.

Aravind, R., Antony, D., Eapen, S. J., Kumar, A., & Ramana, K. V. (2009). Isolation and evaluation of endophytic bacteria against plant parasitic nematodes infesting black pepper (*Piper nigrum* L.). *Indian Journal of Nematology*, *39*(2), 211–217.

Audenaert, K., Pattery, T., Comelis, P., & Hofte, M. (2002). Induction of systemic resistance to Botrytis cinerea in tomato by Pseudomonas aeruginosa 7NSK2: Role of salicylic acid, pyochelin, and pyocyanin. *Molecular Plant-Microbe Interactions*, *15*, 1147–1156.

Azizoglu, Z. B., Yilmaz, S., Azizoglu, U., Karabörklü, S., Temizgul, R., & Ayvaz, A. (2021). Molecular characterization of the chitinase genes of native Bacillus thuringiensis isolates and their antagonistic activity against three important phytopathogenic fungi. *Biologia*, *76*(9), 2745–2755.

Baiyee, B., Pornsuriya, C., Ito, S. I., & Sunpapao, A. (2019). *Trichoderma spirale* T76-1 displays biocontrol activity against leaf spot on lettuce (*Lactuca sativa* L.) caused by *Corynespora cassiicola* or *Curvularia aeria*. *Biological Control*, *129*, 195–200.

Bakker, P. A. H. M., Lamers, J. G., Bakker, A. W., Marugg, J. D., & Weisbeek, P. J. (1986). The role of siderophores in potato tuber yield increase by Pseudomonas putida in a short rotation of potato. *Netherlands Journal of Plant Pathology*, *92*, 249–256.

Bashir, Z. (2021). Characterization of Potassium Solubilizing Bacteria from Rhizospheric Soils of Cherry (*Prunusavium*). *Research Journal of Agricultural Sciences*, *12*(2), 494–496.

Bashir, Z., Hamid, B., Zargar, M. Y., & Kundoo, A. A. (2021). Morphological and biochemical characterization of potassium solubilizing bacteria from rhizospheric soils of pinus (*Pinus wallichiana*). *Journal of Research and Development*, *21*, 58–63.

Bashir, Z., Zargar, M. Y., Baba, Z. A., Mohiddin, F. A., Peer, F. A., Mir, S. A., & Najar, G. R. (2018b). Isolation and characterization of phosphorus solubilizing bacteria (PSB) from rhizospheric soils of apple (*Malus Domestica*). *Journal of Pharmacognosy and Phytochemistry*, *7*(2), 2997–3000.

Bashir, Z., Zargar, M. Y., Bhat, T. A., Kousar, S., Baba, Z. A., & Mohiddin, F. A. (2018c). Solubilization of Phosphorus Containing Mineral by Bacteria from Rhizospheric Region of Walnut (*Juglans regia*). *International Journal of Current Microbiology and Applied Sciences*, *7*(8), 2391–2398.

Bashir, Z., Zargar, M. Y., Mohiddin, F. A., Baba, Z. A., & Hamid, B. (2019). Isolation and Biochemical Characterization of phosphorus solubilizing bacteria from Rhizospheric region of Apricot and Peach. *Journal of Research and Development University of Kashmir*, *19*, 0972–5407.

Bashir, Z., Zargar, M. Y., Mohiddin, F. A., Kousar, S., Husain, M., & Rasool, F. (2017). Phosphorus solubilizing microorganisms: Mechanism and diversity. *International Journal of Communication Systems*, *5*(5), 666–673.

Bashir, Z., Zargar, M. Y., & Vishwakarma, D. K. (2019b). Potassium solubilizing microorganisms for sustainable agriculture. *Applied Agricultural Practices for Mitigating Climate Change*, *2*, 17–28.

Ben Abdallah, R. A., Jabnoun-Khiareddine, H., Nefzi, A., Mokni-Tlili, S., & Daami-Remadi, M. (2016). Endophytic bacteria from Datura stramonium for Fusarium wilt suppression and tomato growth promotion. *Journal of Microbial and Biochemical Technology*, *8*(1), 30–41.

Benítez, T., Rincón, A. M., Limón, M. C., & Codón, A. C. (2004). Biocontrol mechanisms of Trichoderma strains. *International Microbiology*, 7, 249–260.

Błaszczyk, L., Siwulski, M., Sobieralski, K., Lisiecka, J., & Jędryczka, M. (2014). *Trichoderma* spp-Application and prospects for use in organic farming and industry. *Journal of Plant Protection Research*, *54*(4), 309–317.

Brooks, S., Klomchit, A., Chimthai, S., Jaidee, W., & Bastian, A. C. (2022). *Xylaria feejeensis*, SRNE2BP a fungal endophyte with biocontrol properties to control early blight and Fusarium wilt disease in tomato and plant growth promotion activity. *Current Microbiology*, *79*(4), 108.

Brunner, K., Zeilinger, S., Ciliento, R., Woo, S. L., Lorito, M., & Kubicek, C. P. (2005). Improvement of the fungal biocontrol agent *Trichoderma atroviride* to enhance both antagonism and induction of plant systemic disease resistance. *Improvement of the Fungal Biocontrol Agent*, *71*(7), 3959–3965.

Calderón, A. A., Zapata, J. M., Muñoz, R., Pedreño, M. A., & Barceló, A. R. (1993). Resveratrol production as a part of the hypersensitive-like response of grapevine cells to an elicitor from *Trichoderma viride*. *New Phytologist*, *124*, 455–463.

Cartieaux, F., Thibaud, M. C., Zimmerli, L., Lessard, P., Sarrobert, C., David, P., Gerbaud, A., Robaglia, C., Somerville, S., & Nussaume, L. (2003). Transcriptome analysis of Arabidopsis colonized by a plant-growth promoting rhizobacterium reveals a general effect on disease resistance. *The Plant Journal*, *36*, 177–188.

Chen, S. Y., Zhang, C. L., Chen, Y. Z., & Lin, F. C. (2008). Trichodermin (4β-acetoxy-12, 13-epoxytrichothec-9-ene). *Acta Crystallographica Section E: Structure Reports Online*, *64*(4), 702.

Cheng, J., Jaiswal, K. S., Yang, S. H., & Suh, J. W. (2016). Endophytic Bacillus subtilis MJMP2 from Kimchi inhibits Xanthomonas oryzae PV. Oryzae, the pathogen of Rice bacterial blight disease. *Journal of Applied Biological Chemistry*, *59*(2), 149–154.

Corke, A. T. K., & Hunter, T. (1979). Biocontrol of *Nectria galligena* infection of pruning wounds on apple shoots. *Journal of Horticultural Sciences*, *54*, 47–55.

De Meyer, G., Bigirimana, J., Elad, Y., & Höfte, M. (1998). Induced systemic resistance in Trichoderma harzianum T39 biocontrol of Botrytis cinerea. *European Journal of Plant Pathology*, *104*, 279–286.

Di Francesco, A., Zajc, J., Gunde-Cimerman, N., Aprea, E., Gasperi, F., Placì, N., & Baraldi, E. (2020). Bioactivity of volatile organic compounds by Aureobasidium species against gray mold of tomato and table grape. *World Journal of Microbiology and Biotechnology*, *36*, 1–11.

Druzhinina, I. S., Kopchinskiy, A. G., & Kubicek, C. P. (2006). The first 100 Trichoderma species characterized by molecular data. *Mycoscience*, *47*(2), 55–64.

Druzhinina, I. S., Seidl-Seiboth, V., Herrera-Estrella, A., Horwitz, B. A., Kenerley, C. M., Monte, E., Mukherjee, P. K., Zeilinger, S., Grigoriev, I. V., & Kubicek, C. P. (2011). *Trichoderma*: The genomics of opportunistic success. *Nature Reviews Microbiology*, *9*(10), 749–759.

Du, W., Yao, Z., Li, J., Sun, C., Xia, J., & Wang, B. (2020). Diversity and antimicrobial activity of endophytic fungi isolated from *Securinega suffruticosa* in the Yellow River Delta. *PLOS One*, *15*(3), e0229589.

Duffy, B. K., & Defago, G. (1997). Zinc improves biocontrol of Fusarium crown and root rot of tomato by Pseudomonas fluorescens and represses the production of pathogen metabolites inhibitory to bacterial antibiotic biosynthesis. *Phytopathology*, *87*, 1250–1257.

Duijff, B. J., Gianinazzi-Pearson, V., & Lemanceau, P. (1997). Involvement of the outer membrane lipopolysaccharides in the endophytic colonization of tomato roots by biocontrol Pseudomonas fluorescens strain WCS417. *The New Phytologist*, *135*(2), 325–334.

Ebadzadsahrai, G., Higgins Keppler, E. A., Soby, S. D., & Bean, H. D. (2020). Inhibition of fungal growth and induction of a novel volatilome in response to Chromobacterium vaccinii volatile organic compounds. *Frontiers in Microbiology*, *11*, 1035.

Elad, Y. (1983). Parasitism of Trichoderma spp. on *Rhizoctonia solani* and *Sclerotium rolfsii*-scanning electron microscopy and fluorescence microscopy. *Phytopathology*, *73*, 85.

Elad, Y., & Kapat, A. (1999). The role of Trichoderma harzianum protease in the biocontrol of Botrytis cinerea. *European Journal of Plant Pathology*, *105*, 177–189.

Esmaeel, Q., Jacquard, C., Sanchez, L., Clément, C., & Ait Barka, E. (2020). The mode of action of plant associated Burkholderia against grey mould disease in grapevine revealed through traits and genomic analyses. *Scientific Reports*, *10*(1), 19393.

Etesami, B., Jeong, R., & Glick, B. R. (2023). Potential use of Bacillus spp. as an effective biostimulant against abiotic stresses in crops-A review, *Current Research in Biotechnology*, 100128.

Etminani, F., & Harighi, B. (2018). Isolation and identification of endophytic bacteria with plant growth promoting activity and biocontrol potential from wild pistachio trees. *The Plant Pathology Journal*, *34*(3), 208.

Fadiji, A. E., & Babalola, O. O. (2020). Elucidating mechanisms of endophytes used in plant protection and other bioactivities with multifunctional prospects. *Frontiers in Bioengineering and Biotechnology*, *8*, 467.

Fan, B., Wang, C., Song, X., Ding, X., Wu, L., Wu, H., Gao, X., & Borriss, M. (2018). *Bacillus velezensis* FZB42 in: The gram-positive model strain for plant growth promotion and biocontrol. *Frontiers in Microbiology*, *9*, 2491.

Fu, H. Z., Marian, M., Enomoto, T., Hieno, A., Ina, H., Suga, H., & Shimizu, M. (2020). Biocontrol of tomato bacterial wilt by foliar spray application of a novel strain of endophytic Bacillus sp. *Microbes and Environments*, *35*(4), ME20078.

Gadoury, D. M, McHardy, W. E., & Rosenberger, D. A. (1989). Integration of pesticide application schedules for disease and insect control in apple orchards of the northern United States. *Plant Diseases*, *73*, 98–100.

Garbeva, P., Van Overbeek, L. S., Van Vuurde, J. W. L., & Van Elsas, J. D. (2001). Analysis of endophytic bacterial communities of potato by plating and denaturing gradient gel electrophoresis (DGGE) of 16S rDNA based PCR fragments. *Microbial Ecology*, *41*(4), 369–383.

Gutiérrez-Zamora, M. T., & Martınez-Romero, E. (2001). Natural endophytic association between *Rhizobium* etli and maize (*Zea mays* L.). *Journal of Biotechnology*, *91*(2–3), 117–126.

Haas, D., & Défago, G. (2005). Biological control of soil-borne pathogens by fluorescent pseudomonads. *Nature Reviews Microbiology*, *3*, 307–319.

Haas, D., & Keel, C. (2003). Regulation of antibiotic production in root-colonizing Pseudomonas spp. and relevance for biological control of plant disease. *Annual Review of Phytopathology*, *41*, 117–153.

Haggag, W. (2008). Biotechnological aspects of plant resistant for fungal diseases management. *American-Eurasian Journal of Sustainable Agriculture*, *2*, 1–18.

Hamid, B., & Bashir, Z. (2019a). Potassium solubilization microorganisms: An Alternative technology to chemical fertilizers. *Journal of Research and Development University of Kashmir*, *19*, 0972–5407.

Hanson, L. E., & Howell, C. R. (2004). Elicitors of plant defense responses from biocontrol strains of *Trichoderma virens*. *Phytopathology*, *94*, 171–176.

Harman, G. E. (2006). Overview of mechanisms and uses of *Trichoderma* spp. *Phytopathology*, *96*(2), 190–194.

Harman, G. E., Howell, C. R., Viterbo, A., Chet, I., & Lorito, M. (2004). T*richoderma* species—Opportunistic, avirulent plant symbiont. *Nature Reviews Microbiology*, *2*(1), 43–56.

Hastuti, U. S., Rahmawati, D., Sari, R. Y., Fitri, R. D., & Al Asna, P. M. (2018). Antimicrobial activity of endophytic fungi isolated from a medicinal plant, *Hedychium acuminatum. Roscoe*. https://doi.org/10.1063/1.5061895

Höfte, M. (2021). The use of *Pseudomonas* spp. as bacterial biocontrol agents to control plant disease. *Burleigh Dodds Series in Agricultural Science*. https://doi.org/10.19103/AS.2021.0093.11

Hossain, M. M., Sultana, F., & Islam, S. (2017). Plant growth-promoting fungi (PGPF): Phytostimulation and induced systemic resistance. In *Plant-Microbe Interactions in Agro-Ecological Perspectives* (Vol. 2). Springer.

Howell, C. R., Hanson, L. E., Stipanovic, R. D., Puckhaber, L. S., & Wheeler, M. H. (2000). Erratum: Induction of terpenoid synthesis in cotton roots and control of Rhizoctonia solani by seed treatment with *Trichoderma virens*. *Phytopathology*, *90*, 248–252.

Hussein, R. A., & El-Anssary, A. A. (2019). Plants secondary metabolites: The key drivers of the pharmacological actions of medicinal plants. *Herbal Medicine*, *1*(3).

Ibrahim, E., Fouad, H., Zhang, M., Zhang, Y., Qiu, W., Yan, C., & Chen, J. (2019). Biosynthesis of silver nanoparticles using endophytic bacteria and their role in inhibition of rice pathogenic bacteria and plant growth promotion. *RSC Advances*, *9*(50), 29293–29299.

Ibrahim, E., Zhang, M., Zhang, Y., Hossain, A., Qiu, W., Chen, Y., & Li, B. (2020). Green-synthesization of silver nanoparticles using endophytic bacteria isolated from garlic and its antifungal activity against wheat Fusarium head blight pathogen Fusarium graminearum. *Nanomaterials*, *10*(2), 219.

Igiehon, N. O., & Babalola, O. O. (2017). Biofertilizers and sustainable agriculture: Exploring arbuscular mycorrhizal fungi. *Applied Microbiology and Biotechnology*, *101*, 4871–4881.

Kashyap, P. L., Rai, P., Srivastava, A. K., & Kumar, S. (2017). *Trichoderma* for Climate Resilient Agriculture. *World Journal of Microbiology and Biotechnology*, *33*, 1–18.

Kashyap, P. L., Solanki, M. K., Kushwaha, P., Kumar, S., & Srivastava, A. K. (2020). Biocontrol potential of salt-tolerant trichoderma and hypocrea isolates for the management of tomato root rot under saline environment. *Journal of Soil Science and Plant Nutrition*, *20*, 160–176.

Kaur, T., Rani, R., & Manhas, R. K. (2019). Biocontrol and plant growth promoting potential of phylogenetically new Streptomyces sp. MR14 of rhizospheric origin. *AMB Express*, *9*, 1–14.

Keel, C., Schnider, U., Maurhofer, M., Voisard, C., Laville, J., Burger, U., Wirthner, P., Haas, D., & Defago, G. (1992). Suppression of root diseases by *Pseudomonas fluorescens* CHA0: Importance of the bacterial secondary metabolite 2, 4-diacetylphloroglucinol. *Molecular Plant-Microbe Interactions*, *5*, 4–13.

Khare, E., Kim, K., & Lee, K. J. (2016). Rice OsPBL1 (Oryza sativa arabidopsis PBS1-LIKE 1) enhanced defense of Arabidopsis against Pseudomonas syringae DC3000. *European Journal of Plant Pathology*, *146*, 901–910.

Kogel, K. H., Franken, P., & Hückelhoven, R. (2006). Endophyte or parasite–What decides? *Current Opinion in Plant Biology*, *9*(4), 358–363.

Kohl, R., & Kolnaar, W. J. (2019). Ravensberg, mode of action of microbial biological control agents against plant diseases: Relevance beyond efficacy. *Frontiers in Plant Science*, *10*, 845.

Koike, N., Hyakumachi, M., Kageyama, K., Tsuyumu, S., & Doke, N. (2001). Induction of systemic resistance in cucumber against several diseases by plant growth-promoting fungi: Lignification and superoxide generation. *European Journal of Plant Pathology*, *107*, 523–533.

Kushwaha, P., Kashyap, P. L., Kuppusamy, P., Srivastava, A. K., & Tiwari, R. K. (2020). Functional characterization of endophytic bacilli from pearl millet (*Pennisetum glaucum*) and their possible role in multiple stress tolerance. *Plant Biosystems-An International Journal Dealing with all Aspects of Plant Biology*, *154*(4), 503–514.

Latz, M. A., Jensen, B., Collinge, D. B., & Jørgensen, H. J. (2018). Endophytic fungi as biocontrol agents: Elucidating mechanisms in disease suppression. *Plant Ecology & Diversity*, *11*(5–6), 555–567.

Li, E., Jiang, L., Guo, L., Zhang, H., & Che, Y. (2008). Pestalachlorides A–C, antifungal metabolites from the plant endophytic fungus *Pestalotiopsis adusta*. *Bioorganic & Medicinal Chemistry*, *16*, 7894–7899.

Liang, L. M., Zou, C. G., Xu, J., & Zhang, K. Q. (2019). Signal pathways involved in microbe-nematode interactions provide new insights into the biocontrol of plant-parasitic nematodes. *Philosophical Transactions of the Royal Society B: Biological Sciences*, *374*, 20180317.

Loc, N. H., Huy, N. D., Quang, H. T., Lan, T. T., & Thu Ha, T. T. (2019). Characterisation and antifungal activity of extracellular chitinase from a biocontrol fungus, *Trichoderma asperellum* PQ34. *Mycology*, *11*(1), 38–48.

Lodewyckx, C., Vangronsveld, J., Porteous, F., Moore, E. R., Taghavi, S., Mezgeay, M., & der Lelie, D. V. (2002). Endophytic bacteria and their potential applications. *Critical Reviews in Plant Sciences*, *21*(6), 583–606.

Long, H. H., Schmidt, D. D., & Baldwin, I. T. (2008). Native bacterial endophytes promote host growth in a species-specific manner; Phytohormone manipulations do not result in common growth responses. *PloS One*, *3*(7), e2702.

Loon, L. C., Bakker, P. A. H. M., & Pieterse, C. M. J. (1998). Systemic resistance induced by rhizosphere bacteria. *Annual Review of Phytopathology, 36*, 453–483.

Loper, J. E., & Buyer, J. S. (1991). Siderophores in microbial interactions on plant surfaces. *Molecular Plant-Microbe Interactions*, *4*, 5–13.

Lu, Y., Li, Z., Zhao, X., Lv, S., Wang, X., Wang, K., & Ni, H. (2021). Recognition of rice sheath blight based on a backpropagation neural network. *Electronics*, *10*(23), 2907.

Lugtenberg, B. J., Caradus, J. R., & Johnson, L. J. (2016). Fungal endophytes for sustainable crop production. *FEMS Microbiology Ecology*, *92*(12), fiw194.

Luo, H., Zhou, Q., Deng, Y., Deng, Z., Qing, Z., & Sun, W. (2017). Antifungal activity of the extract and the active substances of endophytic Nigrospora sp. from the traditional Chinese medicinal plant Stephania kwangsiensis. *Natural Product Communications*, *12*(12), 1934578X1701201219.

Mahaffee, W. F., & Kloepper, J. W. (1997). Temporal changes in the bacterial communities of soil, rhizosphere, and endorhiza associated with field-grown cucumber (*Cucumis sativus* L.). *Microbial Ecology*, *34*, 210–222.

Mahato, S., & Neupane, S. (2018). Comparative study of impact of *Azotobacter* and *Trichoderma* with other fertilizers on maize growth. *Journal of Maize Research and Development*, *3*(1), 1–16.

Metcalf, D. A., & Wilson, C. R. (2001). The process of antagonism of *Sclerotium cepivorum* in white rot affected onion roots by *Trichoderma koningii*. *Plant Pathology, 50*, 249–257.

Mohiddin, F. A., Khan, M. R., Khan, S. M., & Bhat, B. H. (2010). Why *Trichoderma* is considered super hero (super fungus) against the evil parasites? *Plant Pathology Journal, 9*, 92–102.

Mukherjee, P. K., Horwitz, B. A., & Kenerley, C. M. (2012). Secondary metabolism in *Trichoderma*-A genomic perspective. *Microbiology, 158*(1), 35–45.

Nair, D. N., & Padmavathy, S. (2014). Impact of endophytic microorganisms on plants, environment and humans. *The Scientific World Journal, 22*, 250693. https://doi.org/10.1155/2014/250693

Nielsen, T. H., Thrane, C., Christophersen, C., Anthoni, U., & Srensen, J. (2001). Structure, production characteristics and fungal antagonism of tensin–A new antifungal cyclic lipopeptide from Pseudomonas fluorescens strain. *Journal of Applied Microbiology, 89*, 992–1001.

Nowak-Thompson, B., Gould, S. J., & Loper, J. E. (1997). Identification and sequence analysis of the genes encoding a polyketide synthase required for Pyoluteorin biosynthesis in Pseudomonas fluorescens Pf-5. *Gene, 204*, 17–24.

O'Sullivan, D. J., & O'Gara, F. (1992). Traits of fluorescent Pseudomonas spp. involved in suppression of plant root pathogens. *Microbiology and Molecular Biology Reviews, 56*, 662–676.

Phour, M., Sehrawat, A., Sindhu, S. S., & Glick, B. R. (2020). Interkingdom signaling in plant-rhizomicrobiome interactions for sustainable agriculture. *Microbiological Research, 241*, 126589.

Poveda, J. (2021). *Trichoderma* as Biocontrol Agent against Pests: New Uses for a Mycoparasite. *Biological Control, 159*, 104634.

Poveda, J., Abril-Urias, P., & Escobar, C. (2020). Biological control of plant-parasitic nematodes by filamentous fungi inducers of resistance: Trichoderma, mycorrhizal and endophytic fungi. *Front Microbiology, 11*, 992.

Ramachander Turaga, V. N. (2020). *Peptaibols: Antimicrobial peptides from fung.in bioactive natural products in drug discovery* (pp. 713–730). Springer.

Reddy, B. P., Reddy, M. S., & Krishna, K. V. K. (2009). Characterization of antifungal metabolites of Pf and their effect on mycelia growth of *Magnaporthe grisea* and *Rhizoctonia solani*. *International Journal of Pharmacy Research & Technology*, 1, 1490–1493.

Sandani, H. B. P., Ranathunge, N. P., Lakshman, P. L. N., & Weerakoon, W. M. W. (2019). Biocontrol potential of five *Burkholderia* and *Pseudomonas* Strains against *Colletotrichum truncatum* Infecting Chilli Pepper. *Biocontrol Science and Technology, 29*, 727–745.

Sansinenea, E. (2019). Bacillus spp.: As plant growth-promoting bacteria. *Secondary metabolites of plant growth promoting rhizomicroorganisms: Discovery and applications*, 225–237.

Santos, I. P., Silva, L. C. N., Silva, M. V., Araújo, J. M., Cavalcanti, M. S., & Lima, V. L. M. (2015). Antibacterial activity of endophytic fungi from leaves of *Indigofera suffruticosa* Miller (*Fabaceae*). *Frontiers in Microbiology, 6*, 350.

Santoyo, G., Moreno-Hagelsieb, G., del Carmen Orozco-Mosqueda, M., & Glick, B. R. (2016). Plant growth-promoting bacterial endophytes. *Microbiological Research, 183*, 92–99.

Santoyo, G., Pacheco, C. H., Salmeron, J. H., & Leon, R. H. (2017). The role of abiotic factors modulating the plant-microbe-soil interactions: Toward sustainable agriculture-A review. *Spanish Journal of Agricultural Research, 15*(1), 13.

Schouteden, N., De Waele, D., Panis, B., & Vos, C. M. (2015). Arbuscular mycorrhizal fungi for the biocontrol of plant-parasitic nematodes: A review of the mechanisms involved. *Frontiers in Microbiology*, 6, 1280.

Selim, K., Elkhateeb, W., Tawila, A., El-Beih, A., Abdel-Rahman, T., El-Diwany, A., & Ahmed, E. (2018). Antiviral and antioxidant potential of fungal endophytes of Egyptian medicinal plants. *Fermentation, 4*(3), 49.

Shafi, J., Tian, H., & Ji, M. (2017). *Bacillus* species as versatile weapons for plant pathogens: A review. *Biotechnology & Biotechnological Equipment, 31*, 446–459.

Singh, R., Ong-Abdullah, M., Low, E. T. et al. (2013). Oil palm genome sequence reveals divergence of interfertile species in Old and New worlds. *Nature, 500*, 335–339.

Sivan, A. (1989). The possible role of competition between Trichoderma harzianum and Fusarium oxysporum on rhizosphere colonization. *Phytopathology, 79*, 198.

Stoltzfus, J. R., Malarvithi, P. P., Ladha, J. K., & Bruijn, F. J. D. (1997). Isolation of endophytic bacteria from rice and assessment of their potential for supplying rice with biologically fixed nitrogen. *Plant Soil, 194*, 25–36.

Susiana, P., Achmadi, P., Retno, P. S., Rina, S. K., & Kadarwati, B. (2018). The resistance of potatoes by application of *Trichoderma viride* Antagonists Fungus. *E3S Web of Conferences*, 73.

Taping, J. M. F., Borja, B. T., Bretaña, B. L. P., Tanabe, M. E. N., & Cabasan, M. T. N. (2023). Fungal endophytes as potential biocontrol agent of Panama disease of banana. *Egyptian Journal of Biological Pest Control, 33*, 84.

Tewari, S., Shrivas, V. L., Hariprasad, P., & Sharma, S. (2019). Harnessing endophytes as biocontrol agents. In *Plant Health Under Biotic Stress* (pp. 189–218). Springer.

Vahidinia, A., Samiee, F., Faradmal, J., Rahmani, A., Taravati Javad, M., & Leili, M. (2019). Mercury, lead, cadmium, and barium levels in human breast milk and factors affecting their concentrations in Hamadan, Iran. *Biological Trace Element Research, 187*(1), 32–40.

Van Loon, L. C., & Bakker, P. A. H. M. (2006). Root-associated bacteria inducing systemic resistance. In S. S. Gnanamanickam (Ed.). *Plant-Associated Bacteria* (pp. 269–316). Springer.

Vinale, F., Nigro, M., Sivasithamparam, K., Flematti, G., Ghisalberti, E. L., & Ruocco, M. (2013). Harzianic acid: A novel siderophore from *Trichoderma harzianum. FEMS Microbiology Letters, 347*, 123–129.

Vinale, F., Sivasithamparam, K., Ghisalberti, E. L., Marra, R., Woo, S. L., & Lorito, M. (2008). *Trichoderma*-plant-pathogen interactions. *Soil Biology and Biochemistry, 40*(1), 1–10.

Vinodkumar, S., Nakkeeran, S., Renukadevi, P., & Mohankumar, S. (2018). Diversity and antiviral potential of rhizospheric and endophytic *Bacillus* species and phyto-antiviral principles against tobacco streak virus in cotton. *Agriculture, Ecosystems & Environment, 267*, 42–51.

Viterbo, A., Inbar, J., Hadar, Y., & Chet, I. (2007). Plant disease biocontrol and induced resistance via fungal mycoparasites. In *Environmental and Microbial Relationships* (pp. 127–146). Springer.

Wani, Z. A., Ashraf, N., Mohiuddin, T., & Riyaz-Ul-Hassan, S. (2015). Plant-endophyte symbiosis, an ecological perspective. *Applied Microbiology and Biotechnology, 99*, 2955–2965.

Wees, S. C. M., De Swart, E. A. M., Van Pelt, J. A., Van Loon, L. C., & Pieterse, C. M. J. (2000). Enhancement of induced disease resistance by simultaneous activation of salicylate- and jasmonate-dependent defense pathways in Arabidopsis thaliana. *Proceedings of the National Academy of Sciences of the United States of America, 97*, 8711–8716.

Wei, G., Tan, D., Chen, C., Tong, Q., Li, X. N., Huang, J., Liu, J., Xue, Y., Wang, J., & Luo, Z. (2017). Cytochalasan Alkaloids from *Aspergillus flflavipes. Scientific Reports, 7*, 42434.

Weller, D. M., Raaijmakers, J. M., Gardener, B. B., & Thomashow, L. S. (2002). Microbial populations responsible for specific C soil suppressiveness to plant pathogens. *Annual Review of Phytopathology, 40*, 309–348.

Wicklow, D. T., Roth, S., Deyrup, S. T., & Gloer, J. B. (2005). A protective endophyte of maize: *Acremonium zeae* antibiotics inhibitory to *Aspergillus flavus* and *Fusarium verticillioides. Mycological Research, 109*, 610–618.

Wilson, D. (1995). Endophyte: The evolution of a term, and clarification of its use and definition. *Oikos, 73*, 274–276.

Xia, Z.-K., Ma, Q.-H., Li, S.-Y., Zhang, D.-Q., Cong, L., Tian, Y.-L., & Yang, R.-Y. (2016). The antifungal effect of silver nanoparticles on *Trichosporon asahii*. *Journal of Microbiology, Immunology and Infection*, *49*, 182–188.
Xie, D., Cai, X., Yang, C., Xie, L., Qin, G., Zhang, M., Huang, Y., Gong, G., Chang, X., & Chen, H. (2021). Studies on the control Effect of *Bacillus subtilis* on wheat powdery mildew. *Pest Management Science*, *77*, 4375–4382.
Yadav, A. N., Verma, P., Kour, D., Rana, K. L., Kumar, V., Singh, B., Chauahan, V. S., Sugitha, T., Saxena, A. K., & Dhaliwal, H. S. (2017b). Plant microbiomes and its beneficial multifunctional plant growth promoting attributes. *International Journal of Environmental Sciences & Natural Resources*, *3*, 1–8.
Yadav, A. N., Verma, P., Singh, B., Chauhan, V. S., Suman, A., & Saxena, A. K. (2017c). Plant growth promoting bacteria: Biodiversity and multifunctional attributes for sustainable agriculture. *Advances in Biotechnology and Microbiology*, *5*, 1–16.
Yang, L. N., Liu, H., Wang, Y. P., Seematti, J., Grenville-Briggs, L. J., Wang, Z., & Zhan, J. (2021). Pathogen-mediated stomatal opening: A previously overlooked pathogenicity strategy in the oomycete pathogen *Phytophthora infestans*. *Frontiers in Plant Science*, *12*, 668797.
Yasmin, S., Hafeez, F. Y., Mirza, M. S., Rasul, M., Arshad, H. M. I., Zubair, M., & Iqbal, M. (2017). Biocontrol of bacterial leaf blight of rice and profiling of secondary metabolites produced by rhizospheric Pseudomonas aeruginosa BRp3. *Frontiers in Microbiology*, *8*, 1895.
Yassin, M. T., Mostafa, A. A. F., Al-Askar, A. A., Sayed, S. R. M., Rady, A. M. (2021). Antagonistic activity of *Trichoderma harzianum* and *Trichoderma viride* strains against some fusarial pathogens causing stalk rot disease of maize, in Vitro. *Journal of King Saud University - Science*, *33*, 101363.
Yatoo, A. M., Ali, M. N., Zaheen, Z., Baba, Z. A., Ali, S., Rasool, S., Sheikh, T. A., Sillanpää, M., Gupta, P. K., & Hamid, B. (2022). Assessment of pesticide toxicity on earthworms using multiple biomarkers: A review. *Environmental Chemistry Letters*, *20*(4), 2573–2596.
Yedidia, I., Benhamou, N., & Chet, I. (1999). Induction of defense responses in cucumber plants (*Cucumis sativus* L.) by the biocontrol agent *Trichoderma harzianum*. *Applied and Environmental Microbiology*, *65*, 1061–1070.
Yi, Y., Luan, P., Liu, S., Shan, Y., Hou, Z., Zhao, S., Jia, S., & Li, R. (2022). Efficacy of *Bacillus subtilis* XZ18-3 as a Biocontrol Agent against *Rhizoctonia cerealis* on Wheat. *Agriculture*, *12*, 258.
Yu, H., Zhang, L., Li, L., Zheng, C., Guo, L., Li, W., Sun, P., & Qin, L. (2010). Recent developments and future prospects of antimicrobial metabolites produced by endophytes. *Microbiological Research*, *165*, 437–449.
Yu, J., Wu, Y., He, Z., Li, M., Zhu, K., & Gao, B. (2018). Diversity and antifungal activity of endophytic fungi associated with *Camellia Oleifera*. *Mycobiology*, *46*(2), 85–91.
Zachow, C., Jahanshah, G., De Bruijn, I., Song, C., Ianni, F., Pataj, Z., Gerhardt, H., Pianet, I., Lämmerhofer, M., Berg, G., Gross, H., & Raaijmakers, J. M. (2015). *Molecular Plant-Microbe Interactions*, *28*(7), 800–810.
Zafari, D., Koushki, M. M., & Bazgir, E. (2008). Biocontrol evaluation of wheat take-all disease by Trichoderma screened isolates. *African Journal of Biotechnology*, *7*, 3650–3656.
Zeng, Q., Ding, X., Wang, J., Han, X., Iqbal, H. M. N., & Bilal, M. (2022). Insight into soil nitrogen and phosphorus availability and agricultural sustainability by plant growth-promoting rhizobacteria. *Environmental Science and Pollution Research*, *29*, 45089–45106.
Zhang, Q. X., Kong, X. W., Li, S. Y., Chen, X. J., & Chen, X. J. (2020). Antibiotics of Pseudomonas Protegens FD6 are essential for biocontrol activity. *Australasian Plant Pathology*, *49*, 307–317.

Zhang, Y. L., Tessaro, M. J., Lesner, M., & Li, X. (2003). Knockout analysis of Arabidopsis transcription factors TGA2, TGA5, and TGA6 reveals their redundant and essential roles in systemic acquired resistance. *Plant Cell*, *15*, 2647–2653.

Zhao, Z., Wang, Q., Wang, K., Brian, K., Liu, C., & Gu, Y. (2010). Study of the antifungal activity of *Bacillus Vallismortis* ZZ185 in vitro and identification of its antifungal components. *Bioresource Technology*, *101*(1), 292–297.

Zimmermann, J., Musyoki, M. K., Cadisch, G., & Rasche, F. (2016). Proliferation of the biocontrol agent *Fusarium oxysporum f.* sp. *strigae* and its impact on indigenous rhizosphere fungal communities in maize under different agro-ecologies. *Rhizosphere*, *1*, 17–25.

11 Green Manuring, Vermicomposting, and Composting

A Green Technology for Soil Health

Zaffar Bashi, Bisma Rashi, Gousia Qadir, and Mehrun Nisa

11.1 INTRODUCTION

Sustainable agriculture and green manuring work together to preserve long-term productivity and environmental integrity. "Green manuring" is a kind of sustainable agriculture that emphasizes the utilization of organic materials to improve soil fertility and crop yields. This ancient method has been employed by farmers since the dawn of agriculture. By introducing organic matter to the soil, green manuring enhances soil health, increases water retention, and lessens the demand for inorganic fertilizers. In agriculture, "sustainability" encompasses a broad range of practises, such as organic and ecologically conscious farming, the use of renewable resources, and reducing adverse environmental effects. Sustainable agriculture aims to grow foodstuff in a way that is responsible for the environment, society, and economy. Green manuring is a crucial component of a sustainable farming system. By introducing organic matter to the soil, the practice of "green manuring" assists in maintaining water in the soil, improving soil fertility, and reducing the need for chemical fertilizers. Green manuring not only keeps the soil productive but also reduces soil erosion and the effects of global warming. Additionally, by improving the availability of nutrients in the soil, green manuring can raise crop yields and quality. Composting, or "green manuring," is the process of turning or tilling fresh plant material that hasn't decomposed into the soil's components in order to improve the soil's fertility, physical attributes, and biological makeup. There are several techniques to accomplish green manuring, including cultivating expected short-duration crops prior to mowing them into the soil, adding agricultural residues, or establishing a cover crop (Valizadeh et al., 2023). Since cover crops are planted in between crops to stop soil erosion and replace the soil with organic matter, they can be thought of as a sort of green manure. There are three types of cover crops: biennials, which are planted and grown in the two subsequent growing seasons; annual plants, which

DOI: 10.1201/9781003491460-11

remain in the soil and are cared for over several years; and perennials. It is also possible to sow cover crops in the autumn and till them the next spring. To be more precise, crop residues are the leftovers from harvested crops. These residues might be incorporated into the soil or left on the surface of the ground. This kind of green manuring not only helps the soil restore its nutrients and organic matter, but it also prevents soil erosion. Green manuring is a part of organic farming. "Green manuring" is the practice of tilling under or integrating any crop cultivated with green manure into the soil, either before or soon after flowering. As one of the most crucial elements of the soil for actual soil fertility, improving the state of the soils with micronutrients and organic material is one of the main goals when producing vegetation covers (Valizadeh et al., 2023). The usage of organic fertilizers, particularly green manure crops, has increased due to environmental concerns and a desire to reduce input costs (Pappa et al., 2006). Because agriculture makes use of green and supplemental crops and their integration feeds our soil with organic matter, green manures are required to attain assembly sustainability. To support agricultural sustainability in the long run, many farmers have committed to improving the fertility of the soil management in addition to using green manures. The decomposition of this organic waste affects the accessibility of vitamins and minerals in the soil, even though not all farms utilize green manuring these days. Green manuring has only been practiced by local farmers for a few decades. A green manure crop that is 40–50 days old is predicted to yield 80–100 kg N/ha. Even though crops only use 50% of the nitrogen in fertilizer, an agricultural crop that utilizes green manure could substitute 50–60 kg of nitrogen per hectare (Sharma et al., 2013). For green manuring, legumes such as cowpea, mung, beans, peas, berseem, dhanicha, and sunhemp are suggested. Dhanicha, sunhemp, mungbean, and guar grown in the kharif season have been demonstrated to contribute 8–21 tonnes of green material and 42–95 kg of nitrogen/hectare when grown as green manure crops. Similar to this, when grown in the Rabi season, khesari, cowpea, and berseem can yield 12–29 tonnes of lush matter and 67–68 kg of nitrogen per hectare (Mishra & Nayak, 2004). A beneficial technique for the crop's ideal growth is green manuring. By incorporating nourishment like the nitrogen into the soil, it aids in preserving soil fertility and overall soil health. Table 11.1 shows crops grown for green manuring which maintain and

TABLE 11.1
Crops Grown with Green Manure: Their Nutritional Makeup

Green Manure Crop	N%	P%	K%	Ca%
Dhaincha	0.6	3.7	1.69	0.5
Sunn hemp	0.4–0.8	0.1	0.5	0.4
Cowpea	0.5–0.7	0.2	0.6	0.6
Senji/Indian clover	0.5	0.3	0.7	0.3
Pigeon pea	0.6–0.8	0.2	0.6	0.6
Guar/cluster bean	0.62	0.057	4.32	0.13
Berseem	0.43	0.21	4.59	2.2

TABLE 11.2
Certain Green Manuring Crops That Enhance the Soil with Nitrogen and Organic Matter

			Output in 40–60 days	
Local Name	**Botanical Name**	**Growing Season**	**Green Matter (MT/ha)**	**Nitrogen Contribution (kg/ha)**
Sunhemp	*Crotolaria juncea*	Wet	21.2	91
Dhaincha	*Sesabania aculeata*	Wet	20.2	86
Green gram	*Vigna radiata*	Wet	8.0	42
Cowpea	*Vigna sinensis*	Wet	15.0	74
Guar	*Cyamopsis tetragonoloba*	Wet	20.0	68
Khesari	*Lathyrus sativus*	Dry	12.3	66
Berseem	*Trifolium alexandrium*	Dry	15.5	67

increase soil health. Broadening soil diversity and encouraging the proliferation of beneficial microorganisms are two benefits of green manuring. Numerous beneficial changes are made to the soil environment, enhancing the topsoil's microbial, chemical, and physical characteristics as well as raising crop yield (Singh et al., 2022).

Green manuring was once widely used, but due to increasing requirements for food, the emergence of other lucrative and competitive crops, and the accessibility of both costly and inexpensive artificial fertilizers, it has recently lost favor. The declining levels of soil fertility, challenges with soil accessibility, and public concerns about energy conservation and overuse have led to a recent rise in the importance of green manures for low-input farms using conventional agronomic methods as well as organic producers (Kumar et al., 2014). Table 11.2 represents diverse green manuring crops that enhance the soil with nitrogen and organic matter is as follows.

11.1.1 Green Manuring Crops' Features

Generally speaking, crops made from green manure are quickly maturing crops that have been planted, and must be given time to develop before being incorporated into the soil. The majority of the crops grown with green manure are legumes, including lupins, beans, peas, clover, and alfalfa (Figure 11.1). The aforementioned crops are capable of helping enhance soil structure and fertility (Mandal et al., 2003; Tejada et al., 2008).

Furthermore, green manure crops can enhance water infiltration and retention, lessen weed growth, and lessen soil erosion. Figure 11.1 displays qualities that make green manuring crops desirable for soil and plant health.

The following is a list of the qualities of green manuring crops, point-by-point:

- Legume crops, such as peas, lentils, soybeans, vetch, clover, alfalfa, sunhemp, dhaincha, and so forth, are widely used as green manuring crops.

FIGURE 11.1 Qualities that make green manuring crops desirable.

- These plants are farmed because of their capacity to absorb nitrogen from the environment and retain it in the soil, hence increasing agricultural yields.
- Crops grown through green manuring can also effectively suppress weeds, lowering the necessity for tillage and chemical herbicide application.
- Crops derived from green manure can be grown in a hybridization system or as a type of cover crop.
- Green manure crops offer a viable substitute for mulch and serve as a refuge for beneficial insects and other living organisms.
- They are able to establish and expand rapidly.
- They are intolerant to pests and diseases as well as unfavorable weather conditions like drought, water logging, extreme heat, and low temperatures.
- They are quick to break down and simple to incorporate.

11.1.2 Gains from Green Manuring

The main benefit of green manuring is the improvement of soil fertility and quality. Green manure, which is high in nutrients like potassium, phosphate, and nitrogen, is essential for the health of the soil and plants. The organic matter in green manure adds to the soil's increasing organic matter content, improving the soil's structure and fertility. Consequently, this enhances crop yields and improves produce quality (Dubey et al., 2015). Additionally, drainage and soil structure are enhanced by green manuring. Better air circulation and drainage are made possible by the increased permeability in the soil caused by the decaying organisms in green manure. This makes more water and nutrients available to plants and lowers the chance of water logging. Another benefit of green manuring is that it lessens the necessity for chemical fertilizers. Chemical fertilizers can be costly and environmentally hazardous if overused. The following is a point-by-point list of the advantages of green manuring:

- Increases soil fertility: Introducing organic material to the soil is a technique known as "green manuring" that increases soil fertility. This strengthens the soil's physical composition and expands its capacity to retain water. Additionally, it encourages the growth of beneficial bacteria and earthworms in the soil.
- Boosts soil nutrient content: By including necessary components such as potassium, phosphorus, and nitrogen, green manuring aids in raising the soil's nutrient content. For a plant to grow and develop healthily, these nutrients are necessary.
- Strengthens soil structure: By increasing the soil's tilth and porosity, green manuring aids in strengthening the soil's structure. This aids in improving the soil's airflow, drainage, and water-holding ability.
- Enhances moisture in the soil retention: Green manuring contributes to the soil's increased ability to hold onto moisture. This keeps the soil from drying out and helps to lessen the necessity for frequent irrigation.
- Lowers soil erosion: By raising the soil's organic matter content, green manuring aids in lowering soil erosion. This stops topsoil loss from wind and water erosion and aids in the development of a protective layer on the soil's surface.
- Lowers the growth of weeds: Green manuring aids in lowering the weed population.
- Boosts water permeation: An increased rate for permeation of water into the soil is facilitated by green manuring. This stops soil erosion and reduces the quantity of excess water that goes off.
- Improves soil consolidation: The application of green manuring aids in the better aggregation of soil particles. This enhances the soil's ability to retain water and maintain its structural integrity.
- Enhanced soil physiological activity is a result of green manuring, which also increases the soil's biological activity. This promotes the growth of beneficial microorganisms in the soil, such as earthworms.
- Boosts crop yields: Green manuring increases the amount of nutrients and water-holding capability of the soil, which helps to boost crop yields.
- Decreases the necessity over chemical fertilizers: By naturally raising the soil's nutrient content, green manuring contributes to the reduction of chemical fertilizer use.

- Enhances soil health: By adding green manuring, which adds biological material to the soil, it helps to increase the health of the soil. This promotes the development of advantageous microorganisms in the soil and soil fertility.
- Reduces greenhouse gas emissions: By raising the soil's carbon content, green manuring contributes to a decrease in the emissions of greenhouse gases, including nitrous oxide, carbon dioxide, and methane.
- Boosts carbon sequestration: By raising the soil's organic matter content, green manuring contributes to a boost in carbon sequestration. This lowers the amounts of carbon dioxide in the atmosphere.
- Increases crop diversity: By raising the soil's fertility, green manuring contributes to an increase in crop diversity. This facilitates the growth of numerous crops.
- Boosts soil productivity: By raising the soil's nutrients content and water-holding capacity, green manuring aids in boosting soil productivity.
- Reduces the necessity for synthetic fertilizers: By naturally raising the soil's nutrient content, green manuring contributes to a decrease in the shortage for synthetic fertilizers.
- Boosts the overall number of beneficial microorganisms in soil: The proliferation of the microbes in the soil is facilitated by green manuring. This raises crop yields and increases soil fertility.

11.1.3 The Need for Green Manuring

In order to maintain the productivity of the soil, green manuring is an additional practice that is required because:

- Higher and more diversified cropping per unit area or timeframe has physiologically and chemically deteriorated the soil.
- The high-producing manufactured varieties and hybrids' current spike in inorganic fertilizer prices, sporadic availability issues, and adulteration could not meet the demand for fertilizer.
- Farmyard manures are scarce because farming uses fewer animals as a result of automated cultivation. Their shortage must be addressed with green manuring.

The use of inefficient soil management techniques, extensive agricultural residue burning, hasty adoption of agronomic techniques, and asymmetric nutrient management are all indicators of low crop yields (Florentin et al., 2011). Decreases in the amount of organic nourishment in the soil impair the soil's ability to retain water, interact with water, and absorb nutrients, including the ability to exchange cations. As organic material degrades, all of these eventually lead to decreased nutritional absorption (Bonini & Alves, 2010). Moreover, strict and quick agronomic practices control nutrient release and promote substantial organic depletion, which leads to soil imbalance following crop residue or manure bushing. Applying organic manures is the only way to solve the problem, and it's important to consider the existence of

TABLE 11.3
Amount of Organic Matter and Nitrogen That a Few Crops Utilize while Green Manure Have Buried

Green Manure Crop	N Turned Under (kg ha^{-1})	Organic Matter (kg ha^{-1})
Sunn hemp	75–80	18,500
Guar	60–118	18,500
Dhaincha	75–88	15,950
Senji	113	14,250
Berseem	60	14,000
Arhar	45	–
Cowpea	58	–

green manure when adding organic matter to deficient soils. (Maitra et al., 2018). Table 11.3 depicts that the green manure adds nitrogen and organic matter to soil. Green manuring may have all the right goals, and by giving the soil N and organic matter, it can benefit the soil in many ways.

These difficult times are drawing increased attention to organic farming in contrast to conventional farming. This kind of farming involves managing vitamins and minerals from organic sources, of which green manuring requires special attention, and ensures a recuperative atmosphere for the environment.

11.1.4 Green Manure Application in Sustainable Agriculture

Using green manure in place of synthetic fertilizers is essential for sustainable agriculture.

Organic farming includes sustainable production techniques that help restore the natural balance that has been upset by inappropriate farming practices. Employing green manure on agricultural land has numerous advantages for the environment and the economy. A new agricultural strategy that uses green manure must be implemented in order to achieve sustainable growth. It has been discovered that green manure technologies offer farmers a number of advantages, including higher yields, nitrogen fixation, carbon sequestration, increased SOC content, biodiversity preservation, etc. The high ecological systems service values and low landowner economic rewards of an innovative approach to sustainable farming could lead to a widespread adoption of an infrastructure of ecological compensation. One important tactic for lessening the consequences of salt stress is believed to be the application of green manure. Cultivating environmentally friendly agricultural products increases the amount of labile organic matter (N, P, and K) in the soil, which increases crop production. For this reason, it is essential to include them in crop rotations (Irin & Biswas, 2022). After being added to the soil, legume green manure crops release organic materials such as mucilage, acidic organic matter, amino acids, carbohydrates, and vitamins throughout as well as following their development (Shukla et al., 2011). They are recognized as essential to environmentally friendly farming and have a major impact on soil health (Laishram et al., 2020). The application of green manure improves the ecological, chemical, and physical properties of the soil in salinity-affected areas, which encourages better plant development and growth. Crop productivity and sustainable land use depend on the utilization of green manure technology regarding soil remediation (Wong et al., 2019). Cultivating crops with green manure not only shields the soil but enhances its quality by generating biomass and absorbing solar radiation. Inadvertent benefits of this technique include improved aesthetic value, the potential to create animal feed, decreased compaction, and improved field trafficability.

11.1.5 An Eco-Friendly Method of Green Farming

Adding ecologically friendly organic manure, like green manure, is one of the best strategies to improve soil fertility and soil effectiveness when it comes to water utilization in dry land farming systems (Zhang et al., 2013). The structure of the soil directly affects its aeration, porosity, conductivity, and infiltration. The temperature, aeration, and moisture content of the soil are all improved by green manures. It helps with soil water conservation (Sajjad et al., 2018), which is important in areas that get rain (Martens & Frankenberger, 1992). Management techniques can boost soil water storage. Because soil water storage is strongly correlated with soil physical qualities, conservation of moisture is enhanced. In addition, it aids in erosion control, nematode suppression, and suppression of weed (Chikoye et al., 2001; Osei et al., 2010). Additionally, it aided with integrated weed management (IWM) in maize/cassava intercropping and reduced weed density through the development of various green manure crops (Olorunmaiye, 2010). The possibility of ground water NO^{3-} contamination was reduced by using green manure (GM20) and fertilizer nitrogen (FN60) together (Aulakh et al., 2000).

11.1.6 Increasing the Fertility of the Soil and Enhancing Its Attributes

Green manuring enhances the chemical and physical characteristics of soil by preventing nutrient loss through leaching, increasing the soil's ability to hold water, and increasing its stocks of organic matter (Pandey et al., 2008). The deliberate use of green manure throughout crop cycles enhanced the ability to repurpose nutrients. Several researchers indicated that the advancement of soil characteristics and the widespread abeyant of manure that is green for nitrogen exchange among crops. Because biomass of microbes is an important biological component, it is advised to consider it as an eco-feature when assessing changes in soil usage as well as management characteristics. Several researches have revealed a large increase in soil microbial biomass when legumes are used as green manures. It was recognized that by encouraging soil microbial activity, applying green manure to the soil improved the soil's quality and abundance. Considering the soil's microbiological activity, it is important to highlight that green manures enhance the diversity of soil microorganisms and provide nutrients as a source of carbon for the microbe that transforms bare nutrients in leftover bulb material into nutrients that can be used by crops. Leguminous green manures have the ability to fix a large amount of atmospheric N_2 and retain healthy proportions of organic matter in the soil. The organic nutrients present in soil can only be accessed by non-leguminous green manures (Tejada et al., 2008). Examining modified green blending plants in agricultural systems can help us better understand these conclusive findings on soil microbial populations.

11.1.7 Improving the Availability of Nutrients and Preventing Soil Erosion and Nutrient Loss

Hard pan formation, reduced water retention, surface consolidation, water runoff, and soil erosion are all greatly exacerbated by the ongoing use of heavy machinery and crops that are monoculture in the modern era. Conventional tillage systems provide bare soil dominance, where intense and direct rainfall fractures soil aggregates and blocks soil pores. Soil erosion is prevented as a result of the soil surface becoming sealed and reduced infiltration (Yadav et al., 2021). Green manure and covering crops expect nutrients to be evaporated from the soil as the canopies cover the top soil. It is possible to obtain appropriate agronomical abundance by using the concepts of efficient capacity absorption and land management. Green manure crops help to maintain sustainable productivity by protecting the soil from runoff and allowing nutrients to leach into the deeper layers over time. It also acts as a barrier to water circulation, preventing direct showers from descending on the ground and slowing down surface drainage. Furthermore, by inhibiting roots from penetrating the soil, green manures help to retain the soil in place (Schumann et al., 2000). Green manuring raises the plant nutrient's accessibility viz: nitrogen, phosphorus, potassium, and sulphur by rooting deeply into the soil and absorbing these nutrients into the plant tissue. As a result of their breakdown, the nutrients are readily available to the subsequent crop. Making these minerals readily accessible to crops and minimizing their losses is crucial since other nutrient inputs, such as P and K, are usually limited in organic systems (Yadav et al., 2000).

11.1.8 Reclamation of Problematic Soil

Resolving soil issues and preparing the soil require soil modification for both accessible and high-quality agricultural practices. Sodic soils have limited organic carbon, and if the area is left fallow, the amount of extra organic carbon decreases. Green manuring is highly advantageous in sodic soil because it enhances soil organic content, decreases fallowness, and boosts abundance in the proper places (Khan et al., 2000). Sesbania green manuring considerably improves by strengthening the asphalt-like and actinic backdrop of saline-sodic soils (Baig & Zia, 2006). By producing organic acids and increasing the accessibility of $Ca2^+$, which exchanged with Na^+ of the clay complex to generate an optimal conditions for microbial activity, the addition of organic modifications, such as green manures and crop residues, lowers the pH and ESP of the alkali soils (Mahmoodabadi et al., 2013). The green manure fertilization is used in both alkaline and saline soils to increase the amount of carbon derived from decaying organisms and the accessibility of various minerals for crop production. Shirale et al. (2018) provide examples of how Dhaincha and sunhemp green manures reduce the concentration of particular cations and anions including Cl^-, SO_4^{2-}, HCO_3^-, and CO_3^{2-}. Growers of rice and wheat in a specific rooting zone face ongoing issues related to nutrient extraction and special management. One tactic to solve some of the fertility of the soil problems in this growing cycle is to introduce green manuring. By raising the amount of organic material in the soil and its nitrogen status, this approach can significantly improve soil health. This is made feasible by the use of Sesbania species, which are thought to be perfect for the fallow period among both rice and wheat crops (Tanveer et al., 2019).

11.1.9 Green Manure's Efficacy on Crop Productivity

Green manuring's effects on soil biological matter as well as additional soil characteristics related to improving crop nutrition is reflected in grain production. Several studies have demonstrated the continual delivery of nitrogen (N) by green manures. Since N releases gradually from soils after incorporation, this would correspond with plant N requirements and boost crop yields. *Sesbania aculeata*, when planted 45 days following the harvest of wheat, can yield dry matter ranging from 2.4 to 6.6 t/ha when inserted directly before the primary crop's growing season. Lowland rice with great potential is *Sesbania rostrata*, given enough irrigation. Using Sesbania green manure can increase rice yield by 15% to 60%, depending on the state of the soil. In freshly reclaimed soils, there have been additional instances of these increases. Rice output rose by 15% when phosphorus was added to *Sesbania rostrata*, as stated by Tanveer et al. (2019). Sesbania green manuring seems promising and necessary to boost and maintain the rice-wheat system's output. The timing of the application affects how the green manure affects the rice.

11.1.10 Plant Protection and Green Manure

Two major variables that lower productivity are insect attacks and low soil fertility. Crops grown with green manure, because of their long-term viability and environmental friendliness, are contributing to a reduction in soil-borne disease incidence as well as a limitation on the growth of pests and weeds. Thus, the application of green manuring promotes sustainable crop cultivation (Parajuli, 2011). In comparison, 60 days after sowing, it was 39% in the control field. Several on-farm tests revealed that the early shoot borer occurrence was reduced when dhaincha was sown at 15 kg per ha on the third day of planting cane and integrated at the crop's base where there is significant earthing up. The use of green manure may be helpful in controlling weeds, dangerous soil-borne germs, and nematodes. Blackshaw et al. (2001) claim that green manuring suppresses weeds, increases variety, and reduces the possibility that bulbs would adapt to a particular farming technique, all of which are advantageous for bulb protection. Certain allele-chemicals that prevent edger berry growth are buried in the soil by some green manures, such as rye and clovers. In dry regions, weeds can proliferate quickly; however, green manures protect the soil, facilitate scientific study, and lessen crop-weed competition for nutrients and natural resources. When soil-borne rhizome pathogens use green manuring, their suffering is reduced. Fungal infections in the soil can be controlled with green manuring (Larkin &

Griffin, 2007). Pung et al. (2004) report that *Brassica* green admixture crops were effective in suppressing the growth of the soil-based fungus *Sclerotinia* spp. The green manured soil becomes rich in phosphate and potassium solubilizing microbial community and thus make P and K available to plants (Bashir et al., 2018, 2019).

11.1.11 Vermicomposting: Introduction

Soil forms the foundation of life on earth and is found to be supporting a diverse array of microorganisms. However, urbanization and industrialization and also the growth in technology have led to sustainability challenges. As everyone knows, sustainability is the practice of using natural resources without sacrificing the ability of future generations to meet their own requirements. Vermicomposting has grown in significance and necessity in the post-green revolution age for environmentally friendly farming practices and conservation of the environment (Ahmed et al., 2022). The environmental revolution promoted chemical pesticides and fertilizers to increase crop production, contaminated the environment, and degraded the soil. This meant that after being subjected to so numerous pesticides and chemical fertilizers, the soil lacked organic matter additions, which prevented it from producing a higher yield. Vermicomposting has the potential to significantly contribute to both the enhancement and protection of the environment as well as sustainable agriculture. The delicate equilibrium of natural resources and the delicate ecosystem in our region are well-known, and vermicomposting can lessen the detrimental effects of industrialization and advancements in technology. Vermicomposting can boost soil fertility and health, increase soil biodiversity, and lessen dependency on hazardous agrochemicals by offering an environmentally friendly and sustainable substitute for chemical fertilizers. Vermicompost is organic manure that is widely regarded as the world's best fertilizer (Dominguez et al., 2019). In this earthworms are used to decompose organic matter and produce a mixture of worm casting, organic materials, humus, live earthworms, their cocoons, and other microorganisms. Earthworms as we know are capable of consuming practically any type of organic matter, with one kilogram of worm capable of consuming one kilogram of residue each day. Hence vermicomposting is a cheap and effective way to recycle food scraps, agro-industrial waste, household waste, and animal waste from pigs, chickens, and cattle. Vermicompost is crucial to organic crop production, including cereals, oilseeds, pulses, millets, fruits, vegetables, and also nurseries (Mukherjee et al., 2016). Vermicomposting is an excellent way to lower soil pollution, enhance soil fertility, replace artificial fertilizers, and encourage organic farming. It is cost-effective and easy-to-handle natural manure that is rich in nutrients and growth hormones. Compared to other organic fertilizers, it promotes growth 4 to 5 times more and 30 to 40% more than chemical fertilizers. Vermicomposting provides an excellent opportunity for rural youth that is unemployed to be financially independent by recycling organic waste. At the end, vermicompost is crucial to crop yield, like vegetables and fruits, and is an eco-friendly approach. Participation and implementation of vermicompost projects at the community level can help in adoption of large mass and combat climate change issues (Mukherjee et al., 2016). Waste management presents a number of difficulties, one of which is the vast amount of waste that is generated as a result of population growth and increased consumption of commodities. Because there may not be enough infrastructure, waste management systems may find it extremely difficult to keep up. There are insufficient infrastructure and facilities in many regions of the world to handle trash in an efficient manner. This may result in the incorrect disposal of garbage, which could represent a risk to the environment and cause other issues like:

11.2.1.1 Limited options for disposal: It might be challenging to manage garbage in some places because there aren't many options for getting rid of it. Certain regions could lack sufficient room for landfills or accessibility to recycling centers, for instance. **Exorbitant costs:** Because it necessitates significant investments in infrastructure and technology, properly managing trash can be quite costly. Governments and other entities in charge of trash management may find this difficult.

11.2.1.2 Reluctance to change: People occasionally show resistance to waste management changes, such as the introduction of new recycling initiatives or a decrease in the usage of plastic. Thus, it becomes extremely challenging to put into practice efficient waste management techniques.

11.2.1.3 Environmental effects: Poor waste management and disposal practices can have detrimental effects on the environment, including contaminated land, contaminated water, and contaminated air. It may also result in the atmospheric release of greenhouse gases, which will exacerbate climate change. It can be difficult to manage organic matter, which includes food waste and agro waste, for a number of reasons.

The following are a few particular difficulties in managing organic waste:

11.2.1.4 High volume: A sizeable amount of the waste produced by homes and companies is composed of organic waste. However, given that there is frequently a significant volume of organic matter to collect and treat, it can provide logistical issues for waste management systems.

11.2.1.5 Insufficient infrastructure: Many places lack the facilities—like composting sites—that are required to handle organic waste in an appropriate manner. This may result in the disposal of organic waste in landfills, which may constitute a threat to the environment.

11.2.1.6 Odor and pest issues: Organic waste can attract rodents and flies as well as create offensive odors. Both residential areas and facilities handling organic garbage may have issues with this.

11.2.1.7 Exorbitant costs: Gathering, moving, and processing organic matter can be costly, especially if it has to be carried a great distance to facilities designed to handle it.

11.2.1.8 Restricted applications following processing: There aren't many things that may be done with organic waste. It can be utilized as an energy source or as an amendment to the soil in some circumstances, but there aren't many markets for these goods. Processing organic trash may be challenging if it is contaminated with other waste kinds, including plastic or metal. In order to maintain the health and welfare of the microorganisms and earthworms participating in the process, vermicomposting needs to be carefully managed and observed. Notwithstanding these difficulties, vermicomposting provides a sophisticated, long-term, and ecologically responsible waste management option. By adopting vermicomposting and other sustainable practices, it may support organic farming and waste management in a way that is both sustainable and beneficial for the environment (Tiwari et al., 2020).

11.1.12 Advantages of Vermicomposting

- It has been discovered that vermicomposting enhances the soil's general health and fertility. Vermicomposting is proven to be essential for sustaining an extensive number of beneficial bacteria and fungi, which helps to improve soil structure and health, and also encourage plant growth. It achieves this through preserving a diversified ecology within the compost pile.
- Vermicomposting produces a highly nutrient-dense compost that is also home to a variety of helpful microbes. Our soil can be made healthier and more fertile naturally with the help of this compost through establishing a varied ecology in the composting pile.
- It has been discovered that vermicomposting is essential in the post-green revolution period since it offers a natural and sustainable substitute for artificial fertilizers while also enhancing soil fertility and health. Additionally, it promotes soil biodiversity and lessens the need for chemical fertilizers, all of which contribute to the development of a more environmentally friendly and sustainable agricultural system. As such, it is an efficient means of promoting such a system. Sustainable waste management techniques and organic farming are essential for safeguarding the delicate environment and conserving natural resources.
- Vermicomposting is a bio-oxidative process that helps enhance the biological, chemical, and physical characteristics of soil by combining the efforts of decomposing microorganisms and earthworms to manifest organic waste. By strengthening the soil's fertility and health, this approach makes a greater variety of species of plants and animals more able to thrive there. Earthworms are often called "the farmer's friend" or "nature's ploughman." The earthworm population affects the physical, chemical, and microbiology communities of the soil, which results in the creation of vermicompost, a valuable product. Soil can be enhanced biologically by adding vermicompost. According to recent studies, the biological qualities of soil—such as its organic carbon content, biomass from microbial enzyme activity, population of various beneficial microorganisms, hormone levels, and so on—have significantly enhanced when vermicompost is added (Saha et al., 2022).

11.1.13 Microbiological Aspect of Vermicomposting

The amount of bacteria and "actinomycetes" that are present in the food that earthworms swallow grows and increases by up to 1000 times, and they have been shown to affect and multiply soil microorganisms up to five times. The stomach of E.

foetida has been shown to include a variety of anaerobic N_2-fixing bacteria, such as Clostridium butyricum, C. beijerinckii, and C. paraputrificum. Through their digging activities, earthworms also contribute to improved soil aeration, which in turn increases the soil's microbiota, microbial communities, and biomass, all of which stimulate and accelerate microbial activity (Mathur et al., 2006). Actinomycetes grow quicker and more effectively when soil flows through earthworms since they are bacteria- and earthworm-loving organisms. Actinomycetes were present in casts at a concentration over six times greater than in the initial soil. Thus, organic matter encourages and aids in the production of organic foods, including meat, dairy products (such as milk and cheese), fruits, vegetables, cereals, and so on (Shubha et al., 2021). Earthworms are employed not just to manage solid waste but also to treat sewage. Apart from encouraging the growth of "beneficial decomposer bacteria" in wastewater, earthworms also function as aerators, grinders, crushing plants, biochemical degraders, and biological stimulators. According to research, vermicomposting transforms pathogen-filled biomedical waste, such as *S. aureus*, *P. vulgaris*, *P. pyocyaneae*, and *E. coli*, into a harmless waste that contains commensals, such as C. freundii, and aerobic spore-bearing microorganisms that are normally found in earthworms' alimentary canals and soil. Due to their extraordinary microbiological and antagonistic activity as well as their superior nutritional quality, vermicomposts are currently being used as biofertilizers. Numerous scientists from all over the world have investigated and recorded the wide range of earthworms in the Indian Western Himalayas. There are 51 species of earthworms from the western Himalaya that are divided into 23 genera and seven families, according to researchers' extensive distributional records at the district level for the western Himalayan states of Jammu and Kashmir, Himachal Pradesh, and Uttrakhand. Vermicompost has been shown to have higher dehydrogenase activity of enzymes than commercial media. This is significant since commercial media is often used to measure the respiration activity of various microbial communities. It has been demonstrated that earthworms' ability to increase plant nutrient availability is also influenced by the activity of their gut microbes. Earthworms are beneficial because they decompose litter and have been shown to affect the microflora's activity, which in turn affects the dynamics of the chemical reactions occurring in the soil. Earthworms eat rhizospheric bacteria such as *Pseudomonas*, *Rhizobium*, *Bacillus*, *Azosprillium*, and *Azotobacter* in addition to rhizospheric soil, which promotes plant growth. The right microenvironment in the stomach can cause these bacteria to multiply or become active. Therefore, the amount of PGPR (plant growth-promoting rhizobacteria) increases as a result of earthworm activity. A rich phosphate and potassium-solubilizing microbial community develops in the rhizospheric soil treated with vermicompost (Bashir et al., 2017; Bashir et al., 2018; Bashir et al., 2018a; Bashir et al., 2018b; Bashir et al., 2019; Bashir et al., 2019b; Bashir et al., 2021; Bashir, 2021; Hamid & Bashir, 2019a).

11.1.14 Nutritive Value of Vermicomposting

The nutrients in vermicompost ranges from 0.4 to 0.66% nitrogen, 1.16 to 1.93% phosphorus, and 0.26 to 0.42% potassium. In addition vermicompost includes micronutrients, beneficial soil bacteria, plant growth hormones, and enzymes. Table 11.4

TABLE 11.4
Nutrient Content in Vermicompost

Nutrients	Percent (%)
Phosphorus (P)	1.2–1.8
Potassium (K)	1.5–2.4
Calcium (Ca)	0.5–1.0
Magnesium (Mg)	0.2–0.3
Zinc (Zn)	500–1000 ppm
Iron (Fe)	0.8–1.5
Manganese (Mn)	1000–2000 ppm
Sulphur (S)	0.4–0.5
Copper (Cu)	22–36 ppm
C:N Ratio (Carbon:Nitrogen)	11:1–14:1

depicts the nutrient content in vermicompost. The carbon to nitrogen ratio of vermicompost can range anywhere from 25.0 to 44.9. Vermicomposting has been shown to decrease the carbon-to-nitrogen (C: N) ratio of organic manure or residue, which in turn leads to an increase in the availability of nitrogen, phosphorus, potassium, and total microbial counts (Srivastava et al., 2002). Earthworms digest the plant matter that they eat and then excrete it as a digested form known as worm casts. Worm casts are frequently referred to as "black gold" (Das et al., 2019) (Table 11.4).

11.1.15 Suitable Species Selection for Vermicomposting

In India alone about 3000 species of earthworm are being used for vermicomposting (Sharma et al., 2005). Earthworm can divided broadly into two groups namely Epianecic (the surface living worms) and Epigeic (the burrowing worms) (Ferber et al., 2019). *Epianecic* spp. that burrows and mixes the soil from different layer and *Epigeic* spp. that is reddish-brown earthworms can be found on the surface of soil, and they are very efficient at composting organic waste and also they do not till or process the soil like all other types of earthworms, they play a crucial and important role in breaking down organic matter and enriching the soil with nutrients. The earthworm that lives below the soil is not suitable for vermicompost production. The surface dwelling earthworms only are the ones that should be used in vermicomposting. There are different type of earthworm, for example *Lumbricus rubellus* (redworm), *Lampito mauritii*, *Octochaetona serrata*, *Perionyx excavates*, *Perionyx sansibaricus*, *Pheritima elongataeisenia*, and *Eudrillus enginae*. The African earthworm (*Eudrillus engenial*), red worm (*Eisenia foetida*), and composting worm (*Peronyx excavates*) are promising worms used for vermicompost production (Figure 11.2).

All these three species of worms can be used together for vermicompost production. Red earthworm is the most preferred one because of its very high multiplication rate and the ability of converting the organic matter into vermicompost within 45 to 50 days (Nayak et al., 2014). Vermicompost promotes improvement of soil structure, increases

FIGURE 11.2 Different earthworms.

TABLE 11.5
Characteristics of Red Worm (*Eisenia foetida*)

S. No.	Characteristics	*Eisenia foetida*
1.	Body length	3–10 cm
2.	Body weight	0.4–0.6 gram
3.	Conversion rate	2.0gram/1500worms/2 months
4.	Maturity	50–55 days
5.	Cocoon production	Every 3 months
6.	Incubation of cocoon	20 to 23 days

water retention capacity, and also facilitates promoting the growth of beneficial microorganisms which ultimately leads to healthier and more fertile soil. In order to soil conditioning, apply 5–10% vermicompost by volume to the soil and mix it thoroughly, but for poor soil conditions, higher application rates may be required (Table 11.5).

11.1.16 Methods of Vermicomposting

Vermicomposting can be performed in a number of ways, which includes windrows, beds or bins, flow-through reactors, troughs, towers, pits or trenches, and containers. According to (Pilli et al., 2019), these techniques can be roughly divided into two categories: batch systems and continuous flow systems.

- **Batch systems:** The batch systems involve the production of vermicompost in batches, in which a fixed amount of feedstock is added to the system at the beginning of the process. And then the feedstock is allowed to decompose and that can be consumed by the worms over a particular set of time. After this particular time period, the vermicompost is harvested, and the system is cleaned and made prepared for the next batch. Batch systems are ideal and beneficial for small-scale vermicomposting operations, as they require less investment in equipment and are simpler to manage.
- **Continuous flow systems:** In continuous flow systems, on the other hand, there is constant involvement flow of feedstock into the vermicomposting system, with the vermicompost being harvested on a continuous ongoing

FIGURE 11.3 Vermiculture.

basis. These systems require a more complex design, which includes a heavy mechanism for separating the vermicompost from the worms and a way to control the moisture content of the system. Continuous flow systems are mostly desirable and suited for larger-scale vermicomposting operations, which involve higher volumes of feedstock and require less labor.

- **Windrow vermicomposting**: The windrow method involves building of a long, narrow pile of organic waste and it is being added by earthworms so as to break down the waste into compost. This method is also usually more often used in large-scale vermicomposting operations.
- **Vermiculture:** In order to create vermicompost, the vermiculture process entails cultivating earthworms in an atmosphere of strict regulation, including a worm bin. Vermicompost is made from the worms' excrement, which is fed to them in the form of organic waste.
- The kind of organic waste being composted, the project's unique objectives, and the operation's scale all influence the vermicomposting process selection. As a component of organic farming, vermicompost has demonstrated very encouraging and noteworthy outcomes in terms of promoting plant growth. The final result, known as the vermicast, is also a highly rich source of enzymes and hormones that aid in establishing an environment that is favorable for soil bacteria. Vermicomposting offers an alternative to burning residue and lessens the negative effects of inappropriate handling of organic waste while generating proteins for consumption as animal feed. Vermiculture is seen in Figure 11.3.

11.1.16.1 Preparation Method

11.1.16.1.1 Material Required for Preparation of Vermicompost

All biodegradable garbage—cow dung, residue from crops, weed biomass, veggie trash, leaf litter, house and hotel garbage, garbage from agriculture-related industries, and compostable elements found in rural and urban waste—must be used for the composting process. The inexpensive vermicompost unit at SKUAST-Kashmir is seen in Figure 11.4. The following lists the five fundamental materials or commodities needed to make vermicompost:

11.1.16.1.2 Bedding Material/Organic Residues

Bedding is any appropriate substance that provides the worms with a home and is quite stable. Among the bedding materials that are commonly utilized are peat mosses, cornmeal silage, hay, straw, paper-making sewage, saw dust, bush trimmings, shredding plant stalks, corn stalks, and corn cobs. Because worms require

FIGURE 11.4 Low cost vermicompost unit at SKUAST-Kashmir.

a moist environment to live in and breathe through their skins, their bedding ought to have a significant water absorption capacity. When the bedding breaks down, the worms do consume it, but it's important that this process happens gradually. Put saw dust, husk, coir waste, or sugarcane residue in the bottom of the tubs or container to create a vermiculture bed or worm bed (3 cm). Place a 3 cm layer of fine sand and a 3 cm layer of garden soil over the culture bed. Make sure to use water to hydrate each layer equally.

11.1.16.1.3 Housing or shed facility

It is recommended to use a covered worm culture to protect the worms from excessive light and rain. Cement is used to make the tanks. A wall exists that is split in two. For early decomposition, an additional set of tanks is also constructed. Additionally, a cement tub with a maximum height and breadth of two feet can be constructed. The length can be adjusted at any level, depending on the dimensions of the space. Additionally, the bottom of the tub has a slope-like structure that allows excess water to be drained into the vermicompost unit, also referred to as the vermiwash. It is necessary to gather the drained water in a small sump.

11.1.16.1.4 Worm food

Earthworms, under the right conditions, may eat over their body weight per day. Though nearly any organic food—that is, food originating from plants or animals—can be consumed by them, beef and dairy manures are typically thought of as the best ecological meal for *Eisenia foetida*, as they are the most widely utilized worm feedstock. Cattle dung offers good nutrition, whereas the waste from poultry has a high nitrogen concentration that produces a high-value product in addition to strong nutrition.

11.1.16.1.5 Cow dung/biogas slurry

The animal excrement needs to be dried in the sun before being used to make vermicompost. Any additional waste should be predigested with cow excrement for 20 days using one kilogram of worm (1000 Nos) to fill an area measuring one meter in size, one meter in breadth, and half a meter in height before being placed in a vermi bed during composting. Earthworms do not have to be disposed of in the trash. The earthworm will move about on its own.

11.1.16.1.6 Watering the Vermibed

It's possible that a vermibed doesn't need daily irrigation. On the other hand, the humidity and temperature of the surrounding air may change how often you water. Nonetheless, a consistent 60% level of moisture should be maintained for the duration. Instead of pouring water on the bedding when it becomes necessary, sprinkle it on. Watering must stop prior to vermicompost harvesting.

11.1.16.1.7 Procedure for preparation of vermicompost

All of the strategies being explored here require a comparable procedure or mechanism for the creation of vermicompost.

- The cement rings at the bottom of the vermicomposting area should be covered with a polythene sheet.
- Organic waste should be layered on the sheet, about 15 to 20 cm deep.
- Make slurry out of cow dung.
- Spread an even coating of the slurry over the entire bed.
- Allow the material to break down for 20 days.
- After that time, add dirt or cow dung to the topmost part of the bed.
- Finally, let a few earthworms escape through the crevices.
- Cover the bed with nets or ammunition bags to prevent birds from snatching the earthworms.
- Every three days, water ought to be sprayed to keep the earthworms' bodies sufficiently moist and heated.
- Check the compost after about two months.
- Vermicompost will be available in two to two and a half months.
- It is dark and light, and it has no smell.
- When the compost is done, remove it from the circular container and pile it into a cone.
- Leave the heap alone for two or three hours in order for the earthworms to travel down it gradually.
- Identify the highest point in the stack.
- Sift the bottom portion of the pile to get rid of any worms that can be used to create vermicompost once more.
- Put the bags that hold the decomposing matter in a cool place.

11.1.17 Applications of Vermicompost

Vermicompost, which is considered to be a high source of nutrient-rich organic fertilizer, has numerous applications in the agriculture sector and gardening. Some of its most common applications include soil conditioning:

- Composting: Vermicompost can also be used as a starter material or added to compost piles to boost up the decomposition process and improve the quality content of the final product. Add vermicompost in a 1:1 ratio with other organic materials to start the composting process and boost up the decomposition.

- Fertilization: Vermicompost provides essential nutrients such as nitrogen, phosphorus, and potassium to plants, promoting their growth and development.
- Seed germination: Vermicompost is helpful in improving the germination rate of seeds and enhances their growth, making it an ideal medium for seed starting. Mix vermicompost with soil or growing media in a 1:1 ratio to provide adequate nutrients and moisture to seeds.
- Transplantation: Vermicompost can be used as a soil amendment during transplanting to provide plants with the necessary nutrients and help them establish quickly.
- Plant disease control: Vermicompost is known to have disease-suppressive properties, helping to control soil-borne diseases in plants.
- Lawn care: Vermicompost can be used as a top dressing for lawns to provide nutrients and improve soil health, leading to greener and healthier lawns.
- Hydroponics: Vermicompost can be used as a growing medium in hydroponics systems to provide essential nutrients to plants grown in less soil.

Importance of Vermicompost in Sustainable Crop Production:

- Vermicompost is a tremendous source of plant nutrients, as it contains 0.4–0.66% N, 1.16–1.93% P, and 0.26–0.42% K, along with micronutrients and beneficial soil microorganisms. In addition, vermicompost also contains plant growth promoting hormones and enzymes that enhance both crop and soil efficiency and productivity.
- Vermicompost is a stable, fine-grained organic manure that is excellent for seedlings and for crop development and improvement. It nourishes and enhances the soil's physiochemical and biological qualities. The cast's high nutrient and growth boosting content, favorable for soil microflora, and other factors have a tremendous and promising impact on overall plant growth. It neutralizes soil pH and it also prevents nutrient loss. Earthworms also increase the phosphorus availability in soil.
- Vermicompost has proven to lower the soil compaction and increase the rate of infiltration. Nitrogen of organic matter is either turned into earthworm biomass or excreted as ammonium. In addition 73.3 μg N/gram per soil is present as microbial biomass (Makeschin, 1997). The enhancement of ammonium released in earthworm casts and the conversion of this N toward nitrification and/or reorganization varies from one species of earthworms to another species and also is dependent on their several properties. Vermicomposting results in the generation of a very high-quality compost at room temperature and also in a shorter period of time which is free from pathogens, toxic elements, and weed, and it also minimizes the incidence of harmful pests and microorganism like bacteria, fungi, etc. It contains valuable vitamins and hormones like auxin, and gibberellins which are also beneficial for better crop production.

11.1.18 Vermiwash

When water is allowed to pass through an earthworm action column, the liquid that is produced is known as "vermiwash." When it is used as a foliar spray, it is very beneficial as it contains soil organic molecules, micronutrients from earthworms which are their excretory products and mucus secretions, and also other ingredients. Vermiwash that has been properly collected is translucent and clear and has a light yellow hue. It works as an excellent biopesticide and also has very strong growth-promoting evidence. "Vermiwash" is produced when a large number of earthworms are present. In addition to different enzymes, vitamins, and both micro and macronutrients, it also contains plant growth hormones. The different elements help in promotion of crop productivity and growth while also enhancing the resistance of crops to numerous plant diseases. The frequent and widespread use of inorganic fertilizers, herbicides, and pesticides, as well as the human exploitation of available water resources in modern agricultural systems, has proven to be a significant threat to the sustainability of our agroecosystems. As a result, it is a crucial hour of need to search for effective and eco-friendly alternatives.

11.18.1 Production Techniques of Vermiwash

The earthworm's activity that results in the formation of burrows in the soil is found to be the basis for the process of making vermiwash. Drilospheres are the collective name for the communities of bacteria that can be found in these burrows. The nutrients are carried to the roots of the plants by the water which flows through these channels. The plants then take up these nutrients. The formation of vermiwash takes around 40 to 50 days. A transparent liquid of a brownish colour is obtained by collecting it at the very bottom of the barrel.

11.18.2 Characteristics of Vermiwash

Vermiwash is a coelomic fluid extraction that is high in vitamins, minerals, and macronutrients as well as plant growth hormones such as cytokinins and gibberellins (Buckerfield et al., 1999). Vermiwash is also said to contain some enzymes, growth stimulating hormones, nitrogenous excretory chemicals, and mucus (Tripathi & Bhardwaj 2004). Vermicompost has higher pH and electrical conductivity than vermiwash. Vermicompost has 57% more nitrogen and 79.6% more potassium than vermiwash. Vermiwash has 84% more phosphorus than vermicompost. Vermiwash had 89.1% more Ca and 97.6% more Mg than vermicompost. Vermiwash had 97.8% higher sodium than vermicompost (Manyuchi et al., 2013). Vermiwash makes crops disease-resistant (Yadav et al., 2021). Vermiwash has demonstrated significant potential for sustainable growth in agricultural biotechnology because of its provenance, cost-effectiveness, availability, repeatability, reliability, bio-insecticidal qualities, and eco-friendliness as a soil conditioner.

11.2 COMPOSTING: INTRODUCTION

In the current scenario, environmental health and durability is one of the main apprehensions. The global agricultural, industrial, and metropolitan sectors now generate more trash annually due to the exponential development in human population and activity. Generation of wastes with high magnitude is because of rapid urbanization, industrialization, overpopulation, and indiscriminate disposal of wastes. Different disorganized actions of civilization produce an enormous amount of wastes which causes contamination and pollution (Raj et al., 2018a, 2018b). An active and sustainable waste management (WM) technique is the need of the hour. Waste disposal system is diverse amongst urbanized and unindustrialized nations, but in unindustrialized nations there is a dearth of technical experts, lacking appropriate collection, transportation, and discarding procedures. In order to achieve the Sustainable Development Goals (SDGs), the majority of countries are concentrating on reducing the harmful effects of waste and enhancing waste management practices that will support the maintenance of human and environmental health (Bernal et al., 2017). The scientific community has been paying close attention to waste management (WM) recently as this problem is closely related to sustainability and ecological deprivation (Brewer, 2001). It is a mandatory objective for authorities to provide effective and eco-friendly solid waste management (SWM) approaches comprising: assemblage, carriage, handling, processing, and discarding of generated waste (Raju et al., 2018).

So far frequently applied methods for the discarding of biological trashes or wastes are landfilling and ignition. The procedure of landfilling can be well-preserved by composting (Awomeso et al., 2010). Discharge of methane gas occurs in landfills and decline of carbon footprint can be accomplished by composting (Lights, 2019). In emerging republics like India disposal of waste is predominantly achieved by open dumping, landfilling, incineration, and composting; open disposal of wastes are commonly done (Ghosh et al., 2015). Practices like dumping, landfilling, and incineration of wastes have made extremely threatening impact on the healthiness of our environment. Burning of biological wastes leads to the release of different gases like: nitrous oxide (N_2O), methane (CH_4), and carbon dioxide (CO_2) (Ahmad et al., 2007), whereas landfill and dumping of organic wastes might cause soil pollution and contamination of natural resources. These procedures primarily require large areas for dumping and in big cities it is not possible (Mani & Singh, 2016). So it is an obligation of government and municipalities to provide expertise personnel to find an environment friendly and cost-effective waste management technique (Kaushal et al., 2012). Therefore, a sustainable solution is required to decompose the steadily growing amount of solid waste.

Composting is currently measured as a technique of choice for biodegradation of solid wastes. It is an utmost ideal, environment friendly and economically feasible waste management (WM) practice (Storino et al., 2016; Wei et al., 2014). Composting is an aerobic, biochemical, and microorganism based process that involves the breakdown of organic wastes into humus or in which organic waste is converted into humus like material (Wei et al., 2017). It has been reported that in the future, the scope of market for compost is in floriculture, kitchen gardening, horticulture, agriculture, and construction industries. The compost market will grow faster in upcoming years, with an approximation of $9.2 billion and a Compound Annual Growth Rate (CAGR) of 6.8% (Waliczek et al., 2020). Application of compost will improve soil properties and upsurge the water holding capacity; the organic matter in compost have potential impact on shielding plants from pathogens when utilized as a fertilizer (Somerville et al., 2020). This waste management method has diverse environmentally beneficial results including reduction in emission of greenhouse gases, improvement of soil health, crop productivity, and availability of NPK to crops (Kjerstadius et al., 2016). Decomposition of organic wastes by composting is a substitute solution to non-stop discarding to landfill which roots to negative influence on the environs because of release of greenhouse gases and nasty smell (Al Zuahiri et al., 2015).

The composting process is utilized for conversion and stabilization of organic waste into harmless and steadied form that can be used in numerous farming activities (Garcıá-Gómez et al., 2005). There is unmanaged and limitless use of fertilizer that has spoiled soil health and crop quality. Composting as an eco-friendly approach to waste management is proving to be beneficial and economical. The final product of the composting process is a nutritive substance called humus or compost along with carbon dioxide, water, and heat as derivatives (Abbasi et al., 2000). It's important to note the usage of chemical fertilizers as they are responsible for groundwater pollution (Xu et al., 2020) and have a negative effect on human and environmental health. Incorporation of compost and other organic inputs are reported to improve the growth of crops by making the availability of vital nutrients, enhancing beneficial microbes load, and handling pests and soil-borne diseases (Stewart-Wade, 2020). Therefore, it's evident that application of compost as organic manure will develop soil microbial health and crop productivity saving the environment from chemical alterations.

Nowadays, organic farming is attaining much attention and acceptance due to its diverse beneficial effects on environmental health (Dangour et al., 2010) as well as addressing varied agricultural problems (Nedumaran et al., 2020). It is considered a natural tool to safeguard the environment and promote sustainable development (Skoufogianni et al., 2016). As compared to inorganic and conservative farming practices, organic farming is an ultimate substitute (Agarry et al., 2013). The ruthless and nonstop use of inorganic fertilizers and pesticides has deteriorated soil structure and biodiversity that has resulted in loss of crop productivity and spoilage of ecological health (Brühl & Zaller, 2019; Macit et al., 2007), other than this content of heavy metals in inorganic fertilizers is well recognized (Chuck, 2008). Application of pesticides has developed toxicity among living beings along multiple clinical manifestations and they are greatly injurious to settings (Mahmood et al., 2016). On the other side organic farming and organic products not only maintain healthy environment and balance the sustainability but also develop an affirmative conception about safety among the minds of the customers (Mendon et al., 2020).

11.2.1 Role of Microorganisms in Composting

Composting is an eco-friendly, economical microbial process in which succession of potential microbial communities takes place and converts waste into nutrient rich compost having immense application in the farming sector (Gautam et al., 2010; Rastogi et al., 2020). Diverse microbial flora including bacteria, actinomycetes, yeasts, and fungi are responsible for composting and among them bacteria and fungi are more favorable due to their efficiency in degrading the organic matter (Fan et al., 2017). In the current world scenario recycling, decomposition, and composting are effective management procedures for biodegradable wastes, but little attention has been given towards microbiological features of composting as revealed in literature (Benito et al., 2003; Liu et al., 2011; Ryckeboer et al., 2003a; Tiquia, 2005). To generate good quality compost one must maintain its stability. Microbes have potential applications in the area of waste management, but optimal conditions (including nutrients, air, moisture, pH, and temperature) must be made available for them to harvest suitable end product (Pivato et al., 2013). As mentioned earlier, progression of microorganisms is vital for composting and presence of some microbial communities in compost reveals its quality (Ishii et al., 2000; Ryckeboer et al., 2003b). Among microorganisms, bacteria and fungi are responsible for production of nutrient rich compost also called humus; it's generated by the breaking down of biodegradable organic wastes (Jacoby et al., 2017).

Applications of effective microbial isolates during composting have been recognized as an important biological tool, as they enhance the degradation process by the production of different enzymes (Rastogi et al., 2020). Some prominent microbial communities which are having capacity to degrade solid wastes and colonize in a waste piles are tomesophilies especially bacteria, actinomycetes, fungi, and protozoa (Gajalakshmi & Abbasi, 2008). Such microorganisms have a wide range of temperature (10°C–45°C) for growth and can break down biodegradable wastes efficiently (Cooperband, 2000; Hellmann et al., 1997). Different efficient microorganisms possess the ability to degrade or oxidize complex organic compounds into a simple and stable form (Atalia et al., 2015). A few potential fungal isolates produce extracellular enzymes that can degrade complex organic compounds including polysaccharides and proteins into monomeric forms (Madigan et al., 2009). Wide-ranging bacterial species which can inhabit diverse compost environs have been isolated and reported such as *Pseudomonas*, *Cohnella*, *Cellulomonas*, *Paenibacillus*, and *Bacillus* (He et al., 2013; Ryckeboer et al., 2003). Among gram-positive bacterial species, *Bacillus* and Actinobacteria predominate and flourish during the thermophilic stage of composting process (de Bertolli et al., 1980; Strom, 1985). Besides *Bacillus*, the additional species which were observed at high temperatures (above 70°C) while composting are *Cohnella* sp., *Thermotogae* sp., and *Thermus* sp. respectively (He et al., 2013; Tuomela et al., 2000). The process of composting enhances the development of fungal and bacterial sp. and provides them an ideal environment.

As discussed in an earlier section, composting takes place in multiple phases such as: initial mesophilic, thermophilic, and second mesophilic phase correspondingly. Microbial communities and diversity of micro-flora in pile are reliant on the physiochemical conditions of various phases during the composting process. During the second phase of composting i.e., thermophilic phase, maximum organic material are broken-down into simpler forms and this phase guides the process into quality compost production (Gajalakshmi & Abbasi, 2008; Meena et al., 2018). Because of optimal temperature and pH, microorganisms quicken the process of degradation and rapidly decompose complex organic compounds like amorphous cellulose and hemicellulose (Hu et al., 2011). As compared to chemical procedures, applications of microbial enzymes like cellulase and pectinases to treat biodegradable wastes are appropriate but at industrial level they are expensive. Use of potent thermophilic cellulolytic microbes (fungi and bacteria) to accelerate the process of composting is of better choice (Bohacz, 2017).

Fungi play an important role in decomposition or degradation of wood, paper, cloths, and extra wayward organic material like hemicellulose, cellulose, and lignin (Tuomela et al., 2000). Fungi which are mostly responsible for degradation of waste during composting include *A. fumigatus*, *Malbrancheacinnamomea*, *Ganodermacolossum*, and *Heterobasidionannosum*. Beside these fungi some efficient antagonistic bacterial sp. are being used in bio-fortified compost (Dukare et al., 2011). The antagonistic bacteria overpower several plant pathogens especially pathogenic fungi *Fusarium oxysporum*, *Pythium debaryanum*, *P. aphanidermatum*, and *Rhizoctonia solani*. Accessibility of optimal conditions to microbes during composting results in desirable and better quality end product (Table 11.6).

TABLE 11.6
Compost Stage and Microbial Species

Compost Stage	Microbial Species	References
Mesophilic	*Enterobacter sakazakii*	Chandna et al. (2013)
	Serratia marcescens	Chandna et al. (2013)
	Bacillus flexus, *B. polymyxa*, *B. badius*, *B. cereus*, *B. pumilus*	Partanen et al. (2010), Li et al. (2019), Hefnawy et al. (2013), Chandna et al. (2013), Pane et al. (2020)
	Mycobacterium thermoresistibile, *M. xenopi*	Li et al. (2019)
	Brevibacillus brevis	Hefnawy et al. (2013)
	Staphylococcus sciuri, *S. aureus*, *S. xyloseus*	Hefnawy et al. (2013), Chandna et al. (2013).
	Amycolicicoccus subflavus	Li et al. (2019)
	Penicillium citrinum	Hefnawy et al. (2013)
	Streptomyces griseuss, *S. cinnaborinus*, *S. antibioticus*, *S. roseus*	Hefnawy et al. (2013)
	Fusarium moniliforme, *F. oxysporum*	Hefnawy et al. (2013)
	Aspergillus niger, *A. flavus*	Hefnawy et al. (2013)
	Rhizopus nigricans	Hefnawy et al. (2013)
Thermophilic	*Bacillus flexus*, *B. coagulans*, *B. megaterium*, *B. benzoevorans*, *B. nealsonii*, *B. stearothermophilus*, *B. subtilis*	Partanen et al. (2010), Hefnawy et al. (2013), Chandna et al. (2013), Pane et al. (2020), Antunes et al. (2016)
	Brevibacillus brevis	Antunes et al. (2016)
	Amycolicicoccus subflavus	Li et al. (2019)
	Anoxybacillus flavithermus	Antunes et al. (2016)

(*Continued*)

TABLE 11.6 *(Continued)*
Compost Stage and Microbial Species

Compost Stage	Microbial Species	References
	Comamonas kerstersi	Chandna et al. (2013)
	Gemmatimonas aurantiaca	Antunes et al. (2016)
	Clostridium aciduric, *C. thermocellum*	Partanen et al. (2010), Antunes et al. (2016)
	Klebsiella pneumonia	Hefnawy et al. (2013)
	Kyrpidia tusciae	Antunes et al. (2016)
	Mahella australiensis	Antunes et al. (2016)
	Mycobacterium thermoresistibile, *M. xenopi*	Li et al. (2019)
	Pseudomonas putida, *P. mendocina*	Hefnawy et al. (2013), Pane et al. (2020), Antunes et al. (2016)
	Rhodothermus marinus	Antunes et al. (2016)
	Symbiobacterium thermophilum	Antunes et al. (2016)
	Thermobispora bispora	Antunes et al. (2016)
	Thermobacillus composti	Antunes et al. (2016)
	Solibacillus silvestris	Antunes et al. (2016)
	Thermomonospora curvata	Li et al. (2019), Antunes et al. (2016)
	Thermosediminibacter oceani	Antunes et al. (2016)
	Terribacillus halophilus	Chandna et al. (2013)
	Thermactinomyces sp.	Hefnawy et al. (2013)
	Talaromyces sp.	Hefnawy et al. (2013)
	Aspergillus fumigates	Hefnawy et al. (2013)
	Thermomyces sp.	Partenen et al. (2010), Hefnawy et al. (2013)
	Thermo vulgaris, *T. dichotomicus*	Hefnawy et al. (2013)
Maturation	*Bacillus circulans*, *B. southcampusis*, *B. pumilus*, *B. licheniformis*	Hefnawy et al. (2013)
	Amycolicicoccus subflavus	Li et al. (2019)
	Mycobacterium thermoresistibil, *M. xenopi*	Li et al. (2019)

11.2.2 Application of Compost

Amendment of soil with compost is an old exercise and is still used worldwide. Compost is highly used as fertilizer, but it also helps with pollution prevention, weed control, plant disease prevention, and bioremediation (Coelho et al., 2019; Pane et al., 2019; Ventorino et al., 2019). Application of compost enhances water retention capacity of soil, soil organic carbon content, earthworm count, soil microflora, and crop productivity (D'Hose et al., 2012). It helps to safeguard food security to a boundless range. Good quality compost possesses sufficient nutrients and potential PGPR (Pane et al., 2014). Consumption of compost raises agrarian production and biological stuff of the soil (Luo et al., 2017). Compost possesses its application in the wider areas starting from agriculture, floriculture, pollution remediation, etc., and this organic amendment in soils leads to the overall enhancement of soil health, reduction in the risks of chemical hazards, and maintenance of ecological sustainability.

11.2.3 Composting/Compost and Remediation of Heavy Metals and Pesticides

Environmental and human health are greatly endangered by the enormous discharge of various pollutants into the soil, including heavy metals, pesticides, and polycyclic aromatic hydrocarbons. These pollutants are classified based on their extensive persistence in the environment and how deadly they are, even at very low quantities. Consequently, it's critical to track the percentage of these pollutants in the environment and put prospective removal strategies into action. Using chemical and physical methods, particularly for soil remediation, to clean up contaminated areas is either inefficient or costly. However, composting and compost have been the subject of much research in recent years for soil remediation, and they have been found to be an affordable, environmentally friendly, and suitable method for immobilizing pesticides, eliminating new contaminants, and exclusion of dangerous heavy metals in situ (Zhou et al., 2017). Adsorption by organic matter and degradation by microorganisms are the two ways that composting can remediate contaminated or polluted soil. In general, microbial activity determines how organic contaminants in soil/compost mixtures get decomposed. Throughout this progression enormous amounts of humic acids are produced, which owns the competence to fix with diverse non-degradable contaminants and can modify their assembly.

11.2.4 Heavy Metal Remediation

Among the several non-biodegradable pollutants, heavy metals are conquering significant attention from ecologists due to their detrimental impact on both human and environmental well-being. Agricultural operations, smelting, mining, atmospheric deposition, and other practices can release harmful heavy metals into the environment. Runoff from heavy metal polluted agricultural soils can contaminate surface waters, which can then integrate into groundwater systems (He et al., 2013). Heavy metals are extremely difficult to degrade, and their environmental accumulation poses a serious threat to organisms and bionetworks (Guo et al., 2012; Zhou et al., 2017). By changing the heavy metal's speciation, remediation techniques, particularly composting, can reduce the bioavailability of heavy metals and consequently their toxicity. The integration of compost has a progressive influence on the immobilization and bioavailability of heavy metals. This is because humic acids have the potential to bind with a wide range of metal (loid) s, such as Cd, Cr, Cu, and Pb (Alvarenga et al., 2008). In sandy loam soil, applying 10–20 t ha–1 of farm yard manure (FYM) significantly reduced the percentages of Cd and Pb in amaranth shoots and roots. However, in calcareous polluted soils, the use of green waste compost decreased the uptake of Cu, Pb, and Zn in Greek Cress by 21, 54, and 16%, respectively. Another study found that applying compost effectively reduced the amount of Cd harmfulness to wheat by over 50% by reducing the amount of Cd absorbed by wheat tissue and so improving wheat growth (Liu et al., 2011). After adding 10% compost made from vegetable waste, sewage sludge, and olive husks to soil contaminated with Pb, Cd, and Zn, the content of lead in the aerial portion of the plant was reduced by up to 87% when compared to control samples of white lupin plants (Castaño et al., 2011). In addition, the aerial portions of plants grown in the compost-altered soil had a Zn concentration that was 31% lower than that of the control sample. Compost and red mud have been mentioned as two affordable and active resources for managing heavy metal-polluted soil in situ (Zhou et al., 2017). As a result of their combined use, the accessible heavy metals' efficiency is significantly reduced, which increases the amount of biomass and microbial flora in amended soil. According to the results, modification to the ingredients may lessen the environmental risk of heavy metals in soil (Zhou et al., 2017). After examining the relative effects of compost and biochar on Pb, Cd, and Cr immobilization in soil, photosynthesis, and the growth of maize plants, it was found that both compost and biochar lessen the accessibility of heavy metals in the soil and the toxicity of the plants. However, in comparison to compost, biochar was more effective in reducing the percentage of heavy metals in soil and plants.

11.2.5 Pesticide Remediation

Pesticides are ruthlessly applied in farming or agriculture to switch plant diseases, weeds, and pests and to improve the excellence of the food foodstuffs. But, unselective application of insecticides can lead to undesirable impact on human health, eliminate significant soil micro-flora, and reduce soil fertility. Applications of potential microorganisms to remediate toxic pollutants from polluted sites have been found to be the safest and most sustainable method of decontaminating the environment because bio-magnification and bioaccumulation of dangerous pesticides are the primary cause of loss of microflora, animal biodiversity, and crops.

Compost utilized as soil modifier possesses masses of dynamic microbes to degrade organic impurities like pesticides with the help of accessible nutrients in the compost matrix. In addition, adding compost to contaminated soil increases the amount of organic matter, fostering the growth of the soil's environment and triggering the microbial communities' metabolic activity. The compost to soil considerably boosted the biodegradation of hydrocarbons and increased their removal by 88% after 50 days; prior to this, hydrocarbon biodegradation was negligible without amendment. Secure microbial activity and increased catabolic enzyme abundance were the determining factors for all of this. In their study, Zhou et al. (2017) investigated the breakdown of benzo (a) pyrene in soil composting and found that whereas 46% of the compound was eliminated in unamended soil, in comparison close to 61% was removed in modified composted soil. Pentachlorophenol (PCP) is an organic pollutant that is highly chlorinated with a stable aromatic ring structure and has been discovered to be persistent in soil. Seventy-six per cent of the pentachlorophenol was removed from the polluted soil in a laboratory setting after adding municipal solid waste compost as a nutritional supplement. Fungi capable of mineralizing DDT can be found in cattle manure compost, and soil microbes may accelerate the breakdown of DDT by working in synergism. Similarly, by producing ligninolytic enzymes, basidiomycetes isolated from straw compost can efficiently colonize soil and remove pyrene up to 56% after 28 days (Anastasi et al., 2005). According to reports, a composting heap made of straw compost and soil contaminated with chlorophenols lost more than 90% of its chlorophenols content. As a result, composting is a suitable method for improving soil quality and calming the amount of pesticides in soils due to the degradation of compost microflora.

11.2.6 Compost and Agricultural Sustainability

Widespread use of pesticides and inorganic fertilizers in agriculture has killed beneficial microorganisms, reduced soil fertility, and had a deleterious effect on the ecosystem and public health. Compost and other organic amendments, however, are thought to be a way to achieve sustainable agriculture. Compost application enhances the physicochemical and biological qualities of the soil (Ahmad et al., 2007) and improves its water holding capacity, which in turn boosts plant growth and production. The amendment of compost not only enhances organic matter of the soil but also enhances the amount of accessible macro- and micronutrients for crop growth (Somerville et al., 2020). In terms of soil biological qualities, organic manures increase microbial diversity and their activity in soil, hence boosting nutrient release for crop growth and production. Organic supplements like compost, vermicompost, and biochar have the ability to replace inorganic chemicals while also improving soil structure, plant overall development, and disease management in crops.

11.2.7 Compost and Water Holding Capacity

Organic matter in the form of compost improves soil quality by increasing water-holding capacity, and acts as a sponge in sandy soils by holding water and making it available to plant roots by reducing water wastage through runoff or drainage, as well as assisting plants in dealing with water stress. The yield obtained at 60% irrigation owing to the combined application of biochar and compost was higher than the yield obtained in the control. Obviously, organic matter in soil is primarily responsible for water retention; hence, applying compost is an excellent approach to improve soil organic matter, since it has been proved to affect soil water retention. The porous and spongy nature of compost holds considerable quantity of water, thus minimizing water loss via leaching and evapotranspiration. As a result, compost amendment is a perfect and promising natural fertilizer for dealing with climate change and water scarcity issues, particularly for farmers in dry regions of the world, where soils have low organic matter content and are prone to erosion, deterioration, and desertification processes. The water-holding capacity rose considerably and was 1.57 times higher than in the control treatment. Furthermore, compost was found to improve water-holding capacity and plant-available water by 50% and 34%, respectively, but it was less efficient than cow dung. Therefore, these amendments not only enhance growth and yield of crops but also can help to solve the problem of water shortage.

11.2.8 Compost and Organic Matter/Carbon

Compost is an excellent alternative to chemical fertilizers with a significant impact on soil organic matter and its microbes. By enhancing microbial activity, the organic manures boost enzyme activity, microbial respiration, and consequently nutritional availability for agricultural crops (Meena et al., 2018). Sustaining microbial diversity in soil is also considered a critical component for nutrient cycling and other biological processes in healthy soil (Li et al., 2013). Whilst enzymatic activities are a good indicator of the soil's capacity to execute biochemical processes, when organic fertilizers are applied, the activity of enzymes is increased. In organically treated soil, for example, β-glucosidase activity was reported to be increased by more than 200% when compared to un-treated soil. Furthermore, manure application enhances enzymatic activities and soil organic C, and therefore glucosidase activity was increased as total organic carbon increased (Li et al., 2013). Compost treatment has also been shown to stimulate the activity of a variety of rhizospheric bacteria that aid plant development. According to Ghosh et al. (2015), utilizing organic fertilizers increases organic matter (OM) content of soil while improving its quality and fertility. Improved organic matter in the soil has also been linked to increased structural stability (Tejada et al., 2008), total porosity, and aggregate formation. The OM percentage of compost has a favorable effect on plant development. Previous researches have proven the effect of OM on the physical and chemical characteristics of soil, as well as its general health.

11.2.9 Compost and Soil Structure

Different soil processes rely on the structure of the soil. Amending soils with composts restores the structure of the soil, allowing free gas and water transfer, enhancing seed germination and root growth, as well as lowering the risk of erosion, reducing water evaporation, regulating moisture, and improving drainage (Roman et al., 2015). Compost improves the aggregate stability of the soil and its structure. This is due to the presence of humus in the soil, which binds to the soil and acts as a soil "glue" that holds the soil elements together. In a field study, the application of manure and household waste compost enhanced soil structural stability substantially when compared to the control. Compost application in soil reduces soil loss by 5 to 36%, while soil bulk density decreases by 0.7 to 23%. Somerville et al. (2020) investigated the impact of applying sludge compost to three distinct soils. After three and 15 months of treatment, the bulk density of all three soils fell by 15–26% and 14–25%, respectively.

11.2.10 Compost and Disease Management

The intensive use of inorganic fertilizers and pesticides in the agricultural field has globally destroyed soil fertility, killed beneficial microorganisms, and also decreased natural resistance in crops, thereby making them more vulnerable to diseases besides affecting human health and the environment.

Globally considerable amount of yield losses occurred due to soil-borne diseases, which are difficult to treat, and only limited disease control options are available.

Although chemical pesticides are used, their intensive use in the agricultural field has reduced the natural resistance in crops and made them more susceptible to diseases. Organic amendments, which are widely used in organic and integrated farming have long been known to have a protective impact. Soil-borne pathogens are a critical issue in the agriculture. Although there are numerous strategies to handle these diseases, they are nonetheless costly, have a detrimental impact on soil and plants, damage the environment, and promote pesticide resistance.

Many soil-borne diseases, including *Rhizoctonia solani*, and Fusarium species have been suppressed using compost as an organic disease suppressant, Phytophthora infestans. Several variables can impact compost's suppressive action against Phytophthora infestans; therefore, the effect of compost is not always constant (Pane et al., 2019). Among which feeding substrates, compost composition and most importantly, compost maturity are the important ones. Furthermore, the usage of immature compost may be phytotoxic and have detrimental effects on plant development and soil environment owing to the presence of various chemicals such as phenolic acids, ammonium, salts, and heavy metals. Using mature compost, on the other hand, may enhance soil fertility and crop yield while avoiding phytotoxicity. Despite the fact that the mechanism of plant disease suppression by composts is unknown, the suppressive impact is mostly attributable to indigenous microorganisms. According to Pane et al. (2019), increased populations of fungi and bacteria, as well as more microbial activity, were shown to be related with improved biocontrol ability of composts against Rhizoctonia and Sclerotinia. Different strategies are used by the microorganisms in compost to fight their harmful counterparts which include competition for resources, parasitism, antibiosis, secretion of lytic enzymes, and predation.

Liquid extracts of organic amendments, commonly known as compost/vermicompost teas, have also been identified as a viable alternative to commonly used inorganic chemicals. When applied to plants, these liquid extracts coat the leaves of plants and decrease the available sites for pathogen infection or enhance microbial diversity that can kill harmful phytopathogens. Many researchers have shown that compost teas can prevent the growth of phytopathogenic fungi such as *Pythium ultimum*, *R. solani* (Dionne et al., 2012), Fusarium oxysporum, and Verticillium dahlia (Li et al., 2013). It has been observed that compost extract has a great capacity to inhibit the growth of *Fusarium oxysporum*, *Alternaria alternata*, *Aspergillus niger*, and *Botrytis*. Both vermicompost and compost tea may be utilized as an alternative to pesticides and as plant promoters in crop production to achieve environmental sustainability for farming and food safety. Application of compost extract suppresses F. oxysporum f.sp. Albedinis with an inhibition rate ranging from 20% to 98%. By substituting inorganic fertilizers and pesticides with these eco-friendly organic amendments, diseases and pests can be managed successfully without hurting human health or the environment, and a chemical-free food can be delivered to mankind in the future.

11.2.11 Compost and Plant Growth

Organic fertilizers, such as compost and vermicompost, are an efficient approach to increase and retain soil organic matter, manage diseases and pests, and offer valuable nutrients and other growth promoters like auxins and gibberellins, which plants require. Besides nutrients, compost also includes huge amounts of organic materials that take a long time to degrade and this gradual release of nutrients is highly beneficial to crops. Organic manures increase nutrient availability for agricultural crops by boosting microbial activity, which boosts enzyme activity and microbial respiration. Because compost is made up of decomposed organic waste, its addition enriches the soil with essential nutrients like carbon, phosphorous, nitrogen, and sulphur for crop growth and development. Liu et al. (2011) observed that adding compost to the subsurface increased nutrient cycling and sugarcane growth. The upper layer of compost-amended soil had levels of soil organic carbon and nitrogen that were up to 29.5% higher than in control soil. Compost application over three years increased grain output by roughly 7–15%, according to Zhang and Sun (2016), particularly in the following two to three years. Additionally, it was demonstrated that using compost in place of 30% of N fertilizer is an efficient nutrient management strategy for maintaining N absorption and maize yield (Zhang & Sun, 2016).

In Chinese subtropical arable soils, Cai et al. discovered that manure enhances soil fertility more effectively than synthetic fertilizer, leading to higher crop yields. Agricultural application of compost in an appropriate quantity increased all yield metrics of maize plants by approximately 60%. Furthermore, compost addition enhanced soil fertility, and field studies revealed a 10% increase in radish production without any phytotoxic effect on radish development (Majbar et al., 2018). The application of vermicompost and compost, in conjunction with the prescribed chemical fertilization, significantly promoted crop growth, yield, product quality, and storage life for both tomato and cabbage. The growth of maize in lead-contaminated soil under drought stress and found that the addition of compost and BC to stressed maize crops boosted biomass yield by 50–75% when compared to unamended soil. Similarly, the application of co-composted farm manure and BC increased growth, yield, chlorophyll content, and minimized oxidative stress in the leaves. It also resulted in a decrease in Cd concentrations in wheat tissues, primarily in grains. In contrast, Cd and drought alone reduced wheat growth, increased oxidative stress, and increased Cd contents in wheat tissues.

11.3 CONCLUSION AND FUTURE PROSPECTIVE

Growing and integrating crops made from green manure into the soil to enhance its nutritional value and physical structure is a crucial aspect of organic farming known as "green manuring." Enhancing fertility of the soil by adding nutrients is known as green manuring. It can be applied to lessen erosion, raise the soil's organic matter content, and decrease the demand for external fertilizers. The process of green manuring entails planting carefully chosen cover crops and ploughing them back into the ground. Usually legumes, grasses, or other kinds of vegetation with the capacity to fix the nitrogen from the atmosphere and supply soil with it are used as cover crops. After a brief growth season, the crop is ploughed back into the ground to replenish the organic matter of the soil and nutrition for the following crop. Green manure is a conservation practice that can help produce agriculture in a more sustainable manner. One of the worst examples of land degradation caused by humans is the considerable diminution in soil fertility brought about by incorrect agronomic practices. Numerous factors, including crop yield, soil nutrients content, volume of agricultural production, and environmental effects, are impacted by fertilizer application. Green manures have the potential to significantly increase the fertility and the chemical, physical, and biological qualities of the soil. In addition to improving soil quality, green manuring absorbs nitrogen from the atmosphere in the soil. In terms of weed, disease, and pest insect control, green manure is also safe. To sum up, green manuring could be among the many suitable options for achieving the required ecological and agronomical stability and raising agricultural output.

Vermicomposting is considered to be an eco-friendly and sustainable method for waste management which involves the conversion of organic waste into vermicompost. As compared to traditional composting and manure production, vermicomposting produces vermicompost that has very high nutrient content. Hence, it provides a viable alternative to chemical fertilizers, which can have a very hazardous impact on both crops and human life. It is considered a natural fertilizer as it is produced by using natural processes and contains a variety of essential nutrients, including nitrogen (0.4–0.66%), phosphorus (1.16–1.93%), and potassium (0.26–0.42%), as well as micronutrients, beneficial soil microbes, plant growth hormones, and enzymes. The use of vermicompost can also address the severe and alarming problem of declining soil fertility, which is a major challenge in modern agriculture. Vermicompost can also increase food productivity and nutritional impact while considerably decreasing pollution, soil degradation, and the indiscriminate use of chemical fertilizers.

Vermicomposting is a waste management technique that uses microorganisms and earthworms to manage garbage in an environmentally beneficial way. Among its numerous benefits over traditional composting is that it's a great source of biofertilizers, which improve the biological and physiochemical properties of agricultural soil. Beneficial microbial communities grow in number and variety as a result of vermicomposting. However, due to technological obstacles and a lack of knowledge, it is not frequently used. In order to fully realize the potential of vermicomposting, more development is required.

Indecorous disposal of wastes is having an adverse impact on environmental health; that directly limits the objective of sustainable development. Waste management is a significant part of sustainable development. This arising global problem not only spoils ecological systems but has destroyed natural resources. An authentic and eco-friendly microbial technology that has replaced this unscientific manner of waste disposal is composting. In composting progression of potential microbial community under optimal conditions leads to formation of high quality compost. The composting is known as paramount substitute for waste management and production of highly nutrient rich organic end product. Because it is cost-effective and environmentally benign, farmers can directly implement this waste management technology on their fields. Compost made from organic waste is unquestionably a useful substance with significant advantages for the economy, society, and environment. Composting byproducts can be utilized in horticulture and agriculture sectors to promote sustainable development. Change from inorganic applications to organic applications is the need of the hour. Heterogenous use of pesticides, fertilizers, and weedicides has contaminated our environment, destroyed soil microflora, and reduced fertility level of soil and crop production. Application of compost not only enhances crop production and productivity and disease resistance, but has the potential to safeguard human health and the environs. Due to rapid urbanization and industrialization potential and effective waste management technologies like composting possess huge importance. For a scientific way of waste management and for better quality of compost recent scientific tools must be used. Along with production of qualitative compost, sensors and scientifically updated monitoring systems should be employed to observe critical aspects of the composting process. In the future, the creation of a microbial consortia for rapid composting and waste management techniques should be pursued by researchers. Development of promising psychrophilic microbial consortium for cold climatic zones, where temperature remains below 20°C. Seminars, webinars, conferences, and workshops should be organized to raise audience awareness about environment friendly waste management technologies, especially composting and its application. These parameters will help science and society to achieve goals of sustainable development in the near future.

REFERENCES

Abbasi, S. A., Ramasamy, E. V., Gajalakshmi, S., Khan, F. I., & Abbasi, N. (2000). A waste management project involving engineers and scientists of a university, a voluntary (non-governmental) organization, and lay people-A case study, In *Proceedings of International Conference on Trans Disciplinarily* (pp. 1–3). Swiss Federal Institute of Technology.

Agarry, S. E., Aremu, M. O., & Aworanti, O. A. (2013). Kinetic modelling and half-life study on enhanced soil bioremediation of Bonny light crude oil amended with crop and animal-derived organic wastes. *Journal of Petroleum & Environmental Biotechnology, 4*(2), 137.

Ahmad, R., Jilani, G., Arshad, M., Zahir, Z. A., & Khalid, A. (2007). Bio-Conversion of Organic Wastes for Their Recycling in Agriculture: An Overview of Perspectives and Prospects. *Annals of Microbiology, 57*, 471–479. https://doi.org/10.1007/BF03175343

Ahmed, N., & Al-Mutairi, K. A. (2022). Earthworms effect on microbial population and soil fertility as well as their interaction with agriculture practices. *Sustainability, 14*(13), 7803.

Albrecht, R., Perissol, C., Ruaudel, F., Le Petit, J., & Terrom, G. (2010). Functional changes in culturable microbial communities during a co-composting process: Carbon source utilization and co-metabolism. *Waste Management, 30*(5), 764–770.

Alvarenga, P., Gonçalves, A. P., Fernandes, R. M., de Varennes, A., Vallini, G., Duarte, E., & Cunha-Queda, A. C. (2008). Evaluation of composts and liming materials in the phytostabilization of a mine soil using perennial ryegrass. *Science of the Total Environment, 406*(1–2), 43–56.

Al Zuahiri, F., Pirozzi, D., Ausiello, A., Florio, C., Turco, M., Zuccaro, G., Micoli, L., & Toscano, G. (2015). Biogas production from solid state anaerobic digestion for municipal solid waste. *Chemical Engineering Transactions, 43*, 2407–2412.

Anastasi, A., Varese, G. C., & Marchisio, V. F. (2005). Isolation and identification of fungal communities in compost and vermicompost. *Mycologia, 97*, 33–44.

Antunes, L. P., Martins, L. F., Pereira, R. V., Thomas, A. M., Barbosa, D., Lemos, L. N., Silva, G. M. M., Moura, L. M. S., Epamino, G. W. C., Digiampietri, L. A., Lombardi, K. C., Ramos, P. L., Quaggio, R. B., De Oliveira, J. C. F., Pascon, R. C., Da Cruz, J. B., Da

Silva, A. M., & Setubal, J. C. (2016). Microbial community structure and dynamics in thermophilic composting viewed through metagenomics and metatranscriptomics. *Scientific Reports*, *6*(1), 1–13.

Atalia, K. R., Buha, D. M., Bhavsar, K. A., & Shah, N. K. (2015). Review on composting of municipal solid waste. *IOSR Journal of Environmental Science, Toxicology and Food Technology*, *9*(5), 20–29.

Aulakh, M. S., Khera, T. S., & Doran, J. W. (2000). Yield and nitrogen dynamics in a rice-wheat system using green manure and inorganic fertilizer. *Soil Science Society of America Journal*, *64*, 1867–1876.

Awomeso, J. A., Taiwo, A. M., Gbadebo, A. M., & Arimoro, A. O. (2010). Waste disposal and pollution management in urban areas: A workable remedy for the environment in developing countries. *American Journal of Environmental Sciences*, *6*, 26–32.

Baig, M. B., & Zia, M. S. (2006). Rehabilitation of problem soils through environmental friendly technologies-II: Role of Sesbania (Sesbania aculeata) and gypsum. *Agricultura Tropica et Subtropica*, *39*(1), 26–33.

Bashir, Z. (2021). Characterization of potassium solubilizing bacteria from rhizospheric soils of cherry (*Prunus avium*). *Research Journal of Agricultural Sciences, 12*(2), 494–496.

Bashir, Z., Hamid, B., Zargar, M. Y., & Kundoo, A. A. (2021). Morphological and biochemical characterization of potassium solubilizing bacteria from rhizospheric soils of pinus (*Pinus wallichiana*). *Journal of Research and Development*, *21*, 58–63.

Bashir, Z., Zargar, M. Y., Baba, Z. A., & Mohiddin, F. A. (2018). Solubilization of potassium containing mineral by bacteria from *Kashmir iris* (*Iris kashmiriana*) rhizosphere. *The Pharma Innovation Journal*, *17*(4), 747–750.

Bashir, Z., Zargar, M. Y., Baba, Z. A., Mohiddin, F. A., Peer, F. A., Mir, S. A., & Najar, G. R. (2018a). Isolation and characterization of phosphorus solubilizing bacteria (PSB) from Rhizospheric soils of apple (*Malus domestica*). *Journal of Pharmacognosy and Phytochemistry*, *7*(2), 2997–3000.

Bashir, Z., Zargar, M. Y., Bhat, T. A., Kousar, S., Baba, Z. A., & Mohiddin, F. A. (2018b). Solubilization of phosphorus containing mineral by bacteria from Rhizospheric region of walnut (*Juglans regia*). *International Journal of Current Microbiology and Applied Sciences*, *7*(8), 2391–2398.

Bashir, Z., Zargar, M. Y., Mohiddin, F. A., Baba, Z. A., & Hamid, B. (2019). Isolation and biochemical characterization of phosphorus solubilizing bacteria from Rhizospheric region of apricot and peach. *Journal of Research and Development University of Kashmir*, *19*, 0972–5407.

Bashir, Z., Zargar, M. Y., Mohiddin, F. A., Kousar, S., Husain, M., & Rasool, F. (2017). Phosphorus solubilizing microorganisms: Mechanism and diversity. *International Journal of Communication Systems*, *5*(5), 666–673.

Bashir, Z., Zargar, M. Y., & Vishwakarma, D. K. (2019b). Potassium solubilizing microorganisms for sustainable agriculture. *Applied Agricultural Practices for Mitigating Climate Change*, *2*, 17–28.

Benito, M., Masaguer, A., Moliner, A., Arrigo, N., & Palma, R. M. (2003). Chemical and microbiological parameters for the characterisation of the stability and maturity of pruning waste compost. *Biology and Fertility of Soils*, *37*, 184–189.

Bernal, M. P., Sommer, S. G., Chadwick, D., Qing, C., Guoxue, L., & Michel, F. C. (2017). Current approaches and future trends in compost quality criteria for agronomic, environmental, and human health benefits. In D. L. S. Donald (Ed.), *Advances in agronomy* (Vol. 144, pp. 143–233). Elsevier.

Blackshaw, R. E., Moyer, J. R., Doram, R. C., & Boswell, A. L. (2001). Yellow sweet clover, green manure, and its residues effectively suppress weeds during fallow. *Weed Science*, *49*, 406–413.

Bohacz, J. (2017). Lignocellulose-degrading enzymes, free-radical transformations during composting of lignocellulosic waste and biothermal phases in small-scale reactors. *Science of the Total Environment*, *580*, 744–754.
Bonini, C. S. B., & Alves, M. C. (2010). *Relation between soil organic matter and physical properties of a degraded Oxisol in recovery with green manure, lime and pasture* (pp. 1–6). Paper presented at the 19th World Congress of Soil Science, Soil Solutions for a Changing World, Brisbane, Australia.
Brewer, L. J. (2001). *Maturity and stability evaluation of composted yard debris* (Master's thesis). Oregon State University.
Brühl, C. A., & Zaller, J. G. (2019). Biodiversity decline as a consequence of an inappropriate environmental risk assessment of pesticides. *Frontiers in Environmental Science*, *7*, 464007.
Buckerfield, J. C., Flavel, T., Lee, K. E., & Webster, K. A. (1999). Vermicompost soil and liquid form as plant growth promoter. *Pedobiologia*, *42*, 753–759.
Castaño, R., Borrero, C., & Avilés, M. (2011). Organic matter fractions by SP-MAS 13C NMR and microbial communities involved in the suppression of Fusarium wilt in organic growth media. *Biological Control*, *58*(3), 286–293.
Chandna, P., Nain, L., Singh, S., & Kuhad, R. C. (2013). Assessment of bacterial diversity during composting of agricultural by products. *BMC Microbiology*, *13*(1), 1–14.
Chikoye, D., Ekeleme, F., & Udensi, U. E. (2001). Cogongrass suppression by intercropping cover crops in corn/cassava systems. *Weed Science*, *49*, 658–667.
Chuck, S. (2008). Screening evaluation of heavy metals in inorganic fertilizers. *Minnesota Department of Health* (p. 26). St Paul.
Coelho, L., Osório, J., Beltrão, J., & Reis, M. (2019). Organic compost effects on *Stevia rebaudiana* weed control and on soil properties in the Mediterranean region. *Revista de Ciências Agrárias*, *42*, 109–121.
Cooperband, L. R. (2000). Composting: Art and science of organic waste conversion to a valuable soil resource. Lab Medicine, *31*, 283–289.
D'Hose, T., Cougnon, M., De Vliegher, A., Van Bockstaele, E., & Reheul, D. (2012). Influence of farm compost on soil quality and crop yields. *Archives of Agronomy and Soil Science*, *58*, S71–S75.
Dangour, A. D., Allen, E., Lock, K., & Uauy, R. (2010). Nutritional composition & health benefits of organic foods–using systematic reviews to question the available evidence. *Indian Journal of Medical Research*, *131*(4), 478–480.
Das, A., Layek, J., Yadav, G. S., Babu, S., & Das, S. (2019). Vermicomposting for efficient recycling of biomass for organic crop production in North East India. *Innovative Farming*, *4*(1), 001–014.
de Bertolli, M., Citernesi, U., & Griselli, M. (1980). Bulking agents in sludge composting. *Compost Science & Land Utilization*, *21*, 32–35.
Dominguez, J., Aira, M., Kolbe, A. R., G6mez-Brandón, M., & Pérez-Losada, M. (2019). Changes in the composition and function of bacterial communities during vermicomposting may explain beneficial properties of vermicompost. *Scientific Reports*, *9*(1), 1–11.
Dubey, L., Dubey, M., & Jain, P. (2015). Role of green manuring in organic farming. *Plant Archives*, *15*(1), 23–26.
Dukare, A. S., Prasanna, R., Chandra, D. S., Nain, L., Chaudhary, V., Singh, R., & Saxena, A. K. (2011). Evaluating novel microbe amended composts as biocontrol agents in tomato. *Crop Protection*, *30*(4), 436–442.
Fan, Y. V., Lee, C. T., Ho, C. S., Klemeš, J. J., Wahab, R. A., Chua, L. S., & Sarmidi, M. R. (2017). Evaluation of microbial inoculation technology for composting. *Chemical Engineering Transactions*, *56*, 433–438.
Ferber, T., Slaveykova, V. I., Sauzet, O., & Boivin, P. (2019). Upward mercury transfer by anecic earthworms in a contaminated soil. *European Journal of Soil Biology*, *91*, 32–37.

Florentin, M. A., Penalva, M., Calegari, A., & Derpsch, A. (2011). Green manure/cover crops and crop rotation in conservation agriculture on small farms, *FAO Integrated Crop Management*, *12*, 97.

Gajalakshmi, S., & Abbasi, S. A. (2008). Solid waste management by composting: State of the art. *Critical Reviews in Environmental Science and Technology*, *38*, 311–400.

Garcıá-Gómez, A., Bernal, M. P., & Roig, A. (2005). Organic matter fraction involved in degradation and humification processes during composting. *Compost Science & Utilization*, *13*, 127–135.

Gautam, S. P., Bundela, P. S., Pandey, A. K., Awasthi, M. K., & Sarsaiya, S. (2010). Composting of municipal solid waste of Jabalpur city. *Global Journal of Environmental Research*, *4*, 43–46.

Ghosh, S., Ow, L. F., & Wilson, B. (2015). Influence of biochar and compost on soil properties and tree growth in a tropical urban environment. *International Journal of Environmental Science and Technology*, *12*(4), 1303–1310.

Guo, X., Gu, J., Gao, H., Qin, Q., Chen, Z., & Shao, L. (2012). Effects of Cu on metabolisms and enzyme activities of microbial communities in the process of composting. *Bioresource Technology*, *108*, 140–148.

Hamid, B., & Bashir, Z. (2019a). Potassium solubilization microorganisms: An Alternative technology to chemical fertilizers. *Journal of Research and Development University of Kashmir*, *19*, 0972–5407.

He, Y., Xie, K., Xu, P., Huang, X., Gu, W., Zhang, F., & Tang, S. (2013). Evolution of microbial community diversity and enzymatic activity during composting. *Research in Microbiology*, *164*(2), 189–198.

Hefnawy, M., Gharieb, M., & Nagdi, O. M. (2013). Microbial diversity during composting cycles of rice straw. *International Journal of Advanced Biological and Biomedical Research*, *7*(1), 30–43.

Hellmann, B., Zelles, L., Palojarvi, A., & Bai, Q. (1997). Emission of climate-relevant trace gases and succession of microbial communities during open-windrow composting. *Applied and Environmental Microbiology*, *63*, 1011–1018.

Hu, Z., Liu, Y., Chen, G., Gui, X., Chen, T., & Zhan, X. (2011). Characterization of organic matter degradation during composting of manure–Straw mixtures spiked with tetracyclines. *Bioresource Technology*, *102*(15), 7329–7334.

Irin, I. J., & Biswas, P. K. (2022). Residual effect of green manure on soil properties in green manure-transplant Aman-mustard cropping pattern. *Indian Journal of Agricultural Research*, *1*(6).

Ishii, K., Fukui, M., & Takii, S. (2000). Microbial succession during a composting process as evaluated by denaturing gradient gel electrophoresis analysis. *Journal of Applied Microbiology*, *89*(5), 768–777.

Jacoby, R., Peukert, M., Succurro, A., Koprivova, A., & Kopriva, S. (2017). The role of soil microorganisms in plant mineral nutrition-Current knowledge and future directions. *Frontiers in Plant Science*, *8*, 1167.

Kaushal, R. K., Varghese, G. K., & Chabukdhara, M. (2012). Municipal solid waste management in India-Current state and future challenges: A review. *International Journal of Engineering Science and Technology*, *4*(4), 1473–1489.

Khan, A. R., Ghorai, A. K., & Ingh, S. R. (2000). Improvement of crop and soil sustainability through green manuring in rainfed lowland rice ecosystem. *Agrochimica*, *44*(1–2), 21–29.

Kjerstadius, H., Bernstad Saravia, A., & Spångberg, J. (2016). Can source separation increase sustainability of sanitation management? Impact on nutrient recovery, climate change, and eutrophication of two sanitation systems for a hypothetical urban area in southern Sweden using life cycle assessment (Unpublished manuscript).

Kumar, R., Mahajan, G., Srivastava, S., & Sinha, A. (2014). Green manuring: A boon for sustainable agriculture and pest management - A review. *Agricultural Review*, *35*, 196–206.

Laishram, B., Singh, T. B., Kalpana, A., Wangkheirakpam, M., Chongtham, S. K., & Singh, W. J. (2020). Effect of salicylic acid and potassium nitrate on growth and yield of lentil (*Lens culinaris* L.) under rainfed condition. *International Journal of Current Microbiology and Applied Sciences*, *9*(11), 2779–2791.

Larkin, R. P., & Griffin, T. S. (2007). Control of soil borne potato diseases using Brassica green manures. *Crop Protection*, *26*(7), 1067–1077.

Li, Q., Wang, X. C., Zhang, H. H., Shi, H. L., Hu, T., & Ngo, H. H. (2013). Characteristics of nitrogen transformation and microbial community in an aerobic composting reactor under two typical temperatures. *Bioresource Technology*, *137*, 270–277.

Li, Z., Yang, Y., Xia, Y., Wu, T., Zhu, J., Wang, Z., & Yang, J. (2019). The succession pattern of bacterial diversity in compost using pig manure mixed with wood chips analysed by 16S rRNA gene analysis. *BioRxiv*, 674069.

Lights, Z. (2019). *Reasons why composting is the greenest thing you can do*. One Green Planet.

Liu, J., Xu, X., Li, H., Tao, H., & Xu, Y. (2011). Effect of microbiological inocula on chemical and physical properties and microbial community of cow manure compost. *Biomass and Bioenergy*, *35*, 3433–3439.

Luo, X., Liu, G., Xia, Y., Chen, L., Jiang, Z., Zheng, H., & Wang, Z. J. (2017). Use of biochar-compost to improve properties and productivity of the degraded coastal soil in the Yellow River Delta, China. *Journal of Soils and Sediments*, *17*, 780–789.

Macit, İ., Koç, A., Guler, S., & Deligoz, I. (2007). Yield, quality and nutritional status of organically and conventionally-grown strawberry cultivars. *Asian Journal of Plant Sciences*, *6*(7), 1131–1136.

Madigan, M. T., Martinko, J. M., Dunlap, P. V., & Clark, D. P. (2009). Gram-positive and other bacteria. In *Brock biology of microorganisms* (12th ed., pp. 445–486). Benjamin Cummings.

Mahmood, S., Daur, I., Al-Solaimani, S. G., Ahmad, S., Madkour, M. H., Yasir, M., Hirt, H., Ali, S., & Ali, Z. (2016). Plant growth promoting rhizobacteria and silicon synergistically enhance salinity tolerance of mung bean. *Frontiers in Plant Science*, *7*, 876.

Mahmoodabadi, M., Yazdanpanah, N., Rodriguez Sinobas, L., Pazira, E., & Neshat, A. (2013). Reclamation of calcareous saline sodic soil with different amendments (I): Redistribution of soluble cations within the soil profile. *Agricultural Water Management*, *120*, 30–38.

Maitra, S., Zaman, A., Mandal, T. K., & Palai, J. B. (2018). Green manures in agriculture: A review. *Journal of Pharmacognosy and Phytochemistry*, *7*(5), 1319–1327.

Majbar, Z., Lahlou, K., Ben Abbou, M., Ammar, E., Triki, A., Abid, W., Nawdali, M., Bouka, H., Taleb, M., & El Haji, M. (2018). Co-composting of olive mill waste and wine-processing waste: An application of compost as soil amendment. *Journal of Chemistry*, 7918583.

Makeschin, F. (1997). Earthworms (Lumbricidae: Oligochaeta): Important promoters of soil development and soil fertility. In *Fauna in Soil Ecosystems* (pp. 187–238). CRC Press.

Mandal, U. K., Singh, G., Victor, U. S., & Sharma, K. L. (2003). Green manuring: Its effect on soil properties and crop growth under rice–wheat cropping system. *European Journal of Agronomy*, *19*(2), 225–237.

Mani, S., & Singh, S. (2016). Sustainable municipal solid waste management in India: A policy agenda. *Proceedings of Environmental Science*, *35*, 150–157.

Manyuchi, M. M., Phiri, A., Muredzi, P., & Chitambwe, T. (2013). Comparison of vermicompost and vermiwash bio-fertilizers from vermicomposting waste corn pulp. In *Proceedings of World Academy of Science, Engineering and Technology*, *78*, 346.

Martens, D. A., & Frankenberger, W. T. (1992). Modification of infiltration rates in an organic-amended irrigated soil. *Agronomy Journal*, *84*, 707–717.

Mathur, U. B., Verma, L. K., & Srivastava, J. N. (2006). Effects of vermicomposting on microbiological flora of infected biomedical waste. *Journal of the Indian Society of Hospital Waste Management*, *5*(1), 21–26.

Meena, H., Meena, R. S., Lal, R., Singh, G. S., Mitran, T., Layek, J., Patil, S. B., Kumar, S., & Verma, T. (2018). Response of sowing dates and bio regulators on yield of clusterbean under current climate in alley cropping system in eastern U.P. *Indian Legume Research, 41*(4), 563–571.

Mendon, S., Salins, M., & Aithal, P. S. (2020). Emerging trends in sustainability of organic farming and its impact on purchase intention—A review & research agenda. https://papers.ssrn.com/sol3/papers.cfm?abstract_id=3512860

Mishra, B. B., & Nayak, K. C. (2004). Organic Farming for Sustainable Agriculture. *Orissa Review*, 42–45.

Mukherjee, A., Rakshit, S., Nag, A., Ray, M., Kharbikar, H. L., Shubha, K., & Burman, R. R. (2016). Climate change risk perception, adaptation and mitigation strategy: An extension outlook in mountain Himalaya. *Conservation Agriculture: An Approach to Combat Climate Change in Indian Himalaya*, 257–292.

Nayak, T., Jain, R., Patangray, A. J., & Rathore, A. K. (2014). Vermicompost: Beneficial tool for sustainable farming. *Asian Journal of Multidisciplinary Studies*, 2(8), 255.

Nedumaran, D. G., Prabakaran, V., Kumar, M. A., & Alaguraja, M. (2020). Challanges and Possible of Organic Farming. https://papers.ssrn.com/sol3/papers.cfm?abstractid=3555601

Olorunmaiye, P. M. (2010). Weed control potential of five legume cover crops in maize/cassava intercrop in a Southern Guinea savanna ecosystem of Nigeria. *Amity Journal of Computational Sciences*, *4*(5), 324.

Osei, K., Gowen, S. R., Pembroke, B., Brandenburg, R. L., & Jordan, D. L. (2010). Potential of leguminous cover crops in management of a mixed population of root-knot nematodes (Meloidogyne spp.). *Journal of Nematology*, *42*, 173.

Pandey, D. K., Pandey, R., Mishram, R. P., Kumar, S., & Kumar, N. (2008). *Collection of Dhaincha (Sesbania spp.) variability in Uttar Pradesh, Biodiversity and Agriculture (Souvenir)* (pp. 48–51), Uttar Pradesh Biodiversity Board.

Pane, C., Palese, A. M., Celano, G., & Zaccardelli, M. (2014). Eects of compost tea treatments on productivity of lettuce and kohlrabi systems under organic cropping management. *Italian Journal of Agronomy*, *9*, 153–156.

Pane, C., Sorrentino, R., Scotti, R., Molisso, M., Di Matteo, A., Celano, G., & Zaccardelli, M. (2020). Alpha and beta-diversity of microbial communities associated to plant disease suppressive functions of on-farm green composts. *Agriculture*, *10*(4), 113.

Pane, C., Spaccini, R., Piccolo, A., Celano, G., & Zaccardelli, M. (2019). Disease suppressiveness of agricultural green waste composts as related to chemical and bio-based properties shaped by different on-farm composting methods. *Biological Control*, *137*, 104026.

Pappa, A. V., Rees, R. M., & Watson, C. A. (2006). *Nitrogen transfer between clover and wheat in an intercropping experiment.* 18th World Congress of Soil Science, Philadelphia, PA.

Parajuli, K. J. (2011). Economic impact analysis of mixed-species green manure on organic tomato: Evidence from the northeastern United States (M.Sc. Ag. Thesis). Virginia Polytechnic Institute and State University.

Partanen, P., Hultman, J., Paulin, L., Auvinen, P., & Romantschuk, M. (2010). Bacterial diversity at different stages of the composting process. *BMC Microbiology*, *10*(1), 1–11.

Pilli, K., Jaison, M., & Sridhar, D. (2019). Vermicomposting and its uses in Sustainable Agriculture. *Research Trends in Agriculture Sciences* (pp. 75–88). AkiNik Publications.

Pivato, A., Raga, R., Vanin, S., & Rossi, M. (2013). Assessment of compost quality for its environmentally safe use by means of an eco-toxicological test on a soil organism. *Journal of Material Cycles and Waste Management.*

Pung, H., Aird, P. L., & Cross, S. (2004). The use of Brassica green manure crops for soil improvement and soil borne disease management. In *3rd Australian Soil borne diseases Symposium* (pp. 1–2).

Raj, A., Jhariya, M. K., & Bargali, S. S. (2018a). Climate smart agriculture and carbon sequestration. *Climate change and agroforestry: Adaptation mitigation and livelihood security* (pp. 1–19). New India Publishing Agency (NIPA).

Raj, A., Jhariya, M. K., & Harne, S. S. (2018b). Threats to Biodiversity and Conservation strategies. In K. K. Sood & V. Mahajan (Eds.). *Forests, climate change and biodiversity* (pp. 304–320). Kalyani Publisher.

Raju, K. V., Ravindra, A., Manasi, S., Smitha, K. C., & Srinivas, R. (2018). *Urban Environmental Governance in India*. P. O. Springer International Publishing.

Rastogi, M., Nandal, M., & Khosla, B. (2020). Microbes as vital additives for solid waste composting. *Heliyon*, 6(2).

Roman, P., Martinez, M. M., & Pantoja, A. (2015). *Farmer's compost handbook: Experiences in Latin America*. FAO Rome.

Ryckeboer, J., Mergaert, J., Coosemans, J., Deprins, K., & Swings, J. (2003a). Microbiological aspects of biowaste during composting in a monitored compost bin. *Journal of Applied Microbiology*, *94*(1), 127–137.

Ryckeboer, J., Mergaert, J., Vaes, K., Klammer, S., De Clercq, D., Coosemans, J., Insam, H., & Swings, J. (2003b). A survey of bacteria and fungi occurring during composting and self-heating processes. *Annals of microbiology*, *53*(4), 349–410.

Saha, P., Barman, A., & Bera, A. (2022). Vermicomposting: A step towards sustainability. In *Sustainable Crop Production-Recent Advances*. IntechOpen.

Sajjad, M. R., Subhani, A., Rafique, R., Hussain, T., Naseem, W., Bibi, R., & Malik, S. N. (2018). Green manuring for moisture conservation and wheat yield improvement in rainfed agricultutre. *Journal of Natural Sciences Research*, *8*, 33–41.

Schumann, R. A., Meyer, J. H., & Van Antwerpen, R. (2000). A review of green manuring practices in sugarcane production. *Proceedings South Africa Sugar Technology Assess*, *74*, 93–100.

Sharma, G. D., Thakur, R., Raj, S., Kauraw, D. L., & Kulhare, P. S. (2013). Impact of Integrated Nutrient Management on yield, nutrient uptake, protein content of wheat and soil fertility in a typic haplustert. *The Bioscan*, *8*(4), 1159–1164.

Sharma, S., Pradhan, K., Satya, S., & Vasudevan, P. (2005). Potentiality of earthworms for waste management and in other uses–A review. *The Journal of American Science*, *1*(1), 4–16.

Shirale, A. O., Meena, B. P., Biswas, K., Gourov, P. P., & Kamble, A. L. (2018). *A panacea for the reclamation in saline and sodic soil*. ICAR-Indian Institute of Soil Science.

Shubha, K., Singh, N. R., Mukherjee, A., Dubey, A. K., & Ray, R. K. (2021). Organic vegetable production and its impact on soil, environment and society. In *Advances in Organic Farming* (pp. 191–208). Woodhead Publishing.

Shukla, K. P., Sharma, S., Singh, N. K., Singh, V., Tiwari, K., & Singh, S. (2011). Nature and role of root exudates: Efficacy in bioremediation. *African Journal of Biotechnology*, *10*(48), 9717–9724.

Singh, P. K., Singh, A. P., Singh, V. K., & Singh, O. (2022). Green manuring for sustainable agriculture. In *Encyclopedia of Green Materials* (pp. 1–6). Springer.

Skoufogianni, E., Solomou, A., Molla, A., & Martinos, K. (2016). Organic farming as an essential tool of the multifunctional agriculture. *Organic Farming: A Promising Way of Food Production*, *4*(5), 29.

Somerville, P. D., Farrell, C., May, P. B., & Livesley, S. J. (2020). Biochar and compost equally improve urban soil physical and biological properties and tree growth, with no added benefit in combination. *Science of the Total Environment*, 706, 135736.

Srivastava, A. K., Singh, S., & Marathe, R. A. (2002). Organic citrus: Soil fertility and plant nutrition. *Journal of Sustainable Agriculture*, *19*(3), 5–29.

Stewart-Wade, S. M. (2020). Efficacy of organic amendments used in containerized plant production: Part 1–Compost-based amendments. *Scientia Horticulturae*, *266*, 108856.

Storino, F., Arizmendiarrieta, J. S., Irigoyen, I., Muro, J., & Aparicio-Tejo, P. M. (2016). Meat waste as feedstock for home composting: Effects on the process and quality of compost. *Waste Management*, *56*, 53–62.

Strom, P. F. (1985). Identification of thermophilic bacteria in solid-waste composting. *Applied Environmental and Microbiology*, *50*, 906.

Tanveer, A., Ali, H. H., & Ikram, N. A. (2019). Green Manuring for Soil Health and Sustainable Production of Agronomic Crops. In M. Hasanuzzaman (Eds.). *Agronomic Crops*. Springer.

Tejada, M., Gonzalez, J. L., García-Martínez, A. M., & Parrado, J. (2008). Effects of different green manures on soil biological properties and maize yield. *Bioresource Technology*, *99*(6), 1758–1767.

Tiquia, S. M. (2005). Microbiological parameters as indicators of compost maturity. *Journal of Applied Microbiology*, *99*, 816–828.

Tiwari, A., Kesarwani, K., Ghosh, T., & Rai, S. (2020). Declining nutrients from our plates. *Research Biotica*, *2*(4), 135–140.

Tripathi, G., & Bhardwaj, P. (2004). Comparative studies on biomass production, life cycles and composting efficiency of Eisenia fetida (Savigny) and Lampitomauritii (Kinberg). *Bioresource Technology*, *92*(3), 275–283.

Tuomela, M., Vikman, M., Hatakka, A., & Itavaara, M. (2000). Biodegradation of lignin in a compost environment: A review. *Bioresource Technology*, *72*, 169.

Valizadeh, N., Jalilian, S., Hallaj, Z., Bayat, S. E., Hayati, D., Bazrafkan, K., Kianmehr, N., & Akbari, M. (2023). Encouraging adoption of green manure technology to produce clean rice product. *Scientific Reports*, *13*(8690).

Ventorino, V., Pascale, A., Fagnano, M., Adamo, P., Faraco, V., Rocco, C., Fiorentino, P., & Pepe, O. (2019). Soil tillage and compost amendment promote bioremediation and bio fertility of polluted area. *Journal of Cleaner Production*, *239*, 118087.

Waliczek, T. M., Wagner, N. C., & Guney, S. (2020). Willingness to pay for a specialty blend compost product developed from brown seaweed harvested from coastal regions in Texas. *Hort Technology*, *30*(3), 337–345.

Wei, L., Shutao, W., Jin, Z., & Tong, X. (2014). Biochar influences the microbial community structure during tomato stalk composting with chicken manure. *Bioresource technology*, *154*, 148–154.

Wei, Y., Li, J., Shi, D., Liu, G., Zhao, Y., & Shimaoka, T. (2017). Environmental challenges impeding the composting of biodegradable municipal solid waste: A critical review. *Resources, Conservation and Recycling*, *122*, 51–65.

Wong, V. N. L., Dalal, R. C., & Greene, R. S. B. (2019). Carbon dynamics of sodic and saline soil following gypsum and organic material additions: A Laboratory incubation. *Applied Soil Ecology*, *41*, 29–40.

Xu, Y., Ma, Y., Cayuela, M. L., Sánchez-Monedero, M. A., & Wang, Q. (2020). Compost biochemical quality mediates nitrogen leaching loss in a greenhouse soil under vegetable cultivation. *Geoderma*, *358*, 113984.

Yadav, D., Kumawat, A., Kumar, P., Kumar, S., Singh, D., & Kumar, D. (2021). Chemical fertilization to green manuring: Moving towards sustained production. *Farm Information Bureau*, *9*(1), 15–19.

Yadav, R. L., Dwivedi, B. S., & Pandey, P. S. (2000). Rice-wheat cropping system: Assessment of sustainability under green manuring and chemical fertilizer inputs. *Field Crop Resources*, *65*(1), 15–30.

Zhang, D. B., Yao, P. W., Li, J., Zhao, N., Wang, Z., Yu, C. W., Cao, Q. H., Cao, W. D., & Gao, Y. J. (2013). Effects of two years incorporation of leguminous green manure on soil properties of a wheat field in dryland conditions. *Acta Ecologica Sinica*, *33*, 2272–2281.

Zhang, L., & Sun, X. (2016). Influence of bulking agents on physical, chemical, and microbiological properties during the two-stage composting of green waste. *Waste Management*, *48*, 115–126.

Zhou, G. H., Yu, Y. L., Zhou, X. T., Lu, B. Y., Li, Z. M., & Feng, Y. J. (2017). Effects of drying pretreatment and particle size adjustment on the composting process of discarded flue-cured tobacco leaves. *Waste Management & Research*, *35*(5), 534–540.

12 Impact of the Synthetic Fertilizer Pesticides on Soil Health and Soil Microbiology

Mehnoor Mohmad Bhat, Zaffar Bashir, Burhan Hamid, and Mehrun Nisa

12.1 INTRODUCTION

With the growing economy and advent of new agro technologies throughout the world, particularly China, there is transitioning from a predominantly agricultural nation to a robust agricultural powerhouse (Chen et al., 2020). The COVID-19 pandemic has further impacted agricultural production and socio-economic development, particularly due to challenging circumstances faced by the rural workforce (Wang & Sun, 2018). Consequently, the need for agricultural and rural development has become increasingly apparent. Issues such as lack of advanced agricultural capabilities, limited impact of agricultural industrialization, and weak power of traditional agricultural transformation are becoming increasingly complex. Official data from China's National Statistics Office reveal a continuous decline in rural population, accounting for only 36.11% in 2020. In 2021, agricultural GDP represented only 7.26% of the country's total economic output. Moreover, temporary ratio of agricultural labour productivity was significantly lower than that of secondary and tertiary industries (Jiang et al., 2022). The total number of labourers in 2020 decreased by approximately 5 million compared to the previous year, a year-on-year decrease of 2.7%. Therefore, issues related to agriculture, rural areas, and the rural people are critical in China, necessitating a search for new drivers to support high-quality agricultural growth and promote sustained and stable economic growth in rural areas. The automation of farming and expansion of domesticated animal production, in response to increasing demand for food due to the growing population, creates opportunities and challenges for food safety. As a result, food manufacturers and processors are being held more accountable for preventing food contamination before it reaches consumers (Onyeka Kingsley, 2022). The use of pesticides has increased globally due to advancements in farming. The extensive use of pesticides helps to mitigate crop yield loss and enhance productivity and product quality simultaneously. To protect against the losses caused by pests and diseases and to increase productivity, crops worldwide are treated with an estimated two million tonnes of pesticides every year

 DOI: 10.1201/9781003491460-12

(Antonini & Argilés-Bosch, 2017). Modern agriculture owes much of its success to widespread use of pesticides and fertilizers, which are essential for enhancing crop productivity. This approach has greatly transformed agricultural landscape, by managing pest populations and providing essential nutrients through synthetic fertilizers, thus ensuring crop stability. It has been mainly on maximizing grain yield, without due consideration for long-term effects on soil resources that are crucial for sustainable cereal production. With increasing global food demand to meet nutritional needs of a growing population, it is becoming more important to address complex concerns surrounding human and environmental health. Therefore, there is a need to carefully preserve soil biodiversity and ecological balance.

Among key players in these complex ecosystems are soil microorganisms, whose crucial role lies in the careful regulation of essential soil functions within agroecosystems. These functions include maintenance of soil fertility, efficient nutrient cycling, facilitating nutrient availability, and decomposition of carbon-based and inorganic materials. Moreover, influence of soil microflora—bacteria, fungi, protozoa, and algae—extends to enhancing physical characteristics of soil itself, including porosity, aeration, and water infiltration. This is achieved by their ability to create and stabilize soil aggregates, ranging from macro aggregates to smaller micro aggregates (Srivastava et al., 2016). Furthermore, soil microbial communities actively perform eco-friendly processes, such as bioremediation of contaminated soils, often resulting from anthropogenic activities and biocontrol of plant pathogens. The excessive use of chemical fertilizers and pesticides has led to harmful and unintended consequences, including greenhouse gas emissions from nitrogenous fertilizers, algal blooms in water bodies, and pest resistance (Velthof et al., 1997). Chemical fertilizers and pesticides have been found to have a significant impact on the structural and functional properties of soil microbial communities, according to studies (Bohme et al., 2005). While many adverse effects stem from improper fertilizer use, these can be mitigated through balanced soil management practices. Excessive chemical fertilizer use also leads to nutrient imbalances in soils (Bohme et al., 2005). Soil biodiversity, especially soil microbiome, plays a direct role in maintaining soil fertility and other critical soil functions, which is crucial for ensuring global food security. In light of the current era of global warming and climate change, it is essential to comprehend the link between the soil microbiology and soil functions and to conduct a comprehensive assessment of impact of various human activities on them to promote sustainable agriculture. Moreover, it is of an utmost importance to assess effects of long-term use of pesticides and fertilizers on soil microflora of agroecosystems. To address these challenges, the organic farming practices and biofertilizers containing beneficial microorganisms are gaining importance as sustainable alternatives to traditional chemical fertilizers. "Biofertilizers" enriched with strains like *Paenibacillus, polymyxa*, *Bacillus amyloliquefaciens*, and *Bacillus* species offer increased nutrient bioavailability, improved disease resistance, and enhanced nutrient acquisition by plants (Bashir et al., 2017; Bashir et al., 2019). These microorganisms are also capable of phosphorus solubilization, potassium solubilization, zinc solubilization, and IAA production which is vital for plant development (Bashir et al., 2018; Bashir et al., 2018a; Bashir et al., 2018b; Bashir et al., 2019; Bashir et al., 2019b; Bashir et al., 2021a; Bashir, 2021b; Hamid & Bashir, 2019a). Research shows potential for these "biofertilizers" to positively impact soil microbiological parameters, morphological plant characteristics, and ultimately crop yields. Phosphorus, a vital element in crop production, is often introduced into soil through synthetic fertilizers. However, excessive use of such fertilizers can lead to the "soil degradation," the eutrophication of water bodies, global warming, and a reduction in agricultural soil biodiversity. This move towards biofertilizers aims to strike a balance between increasing agricultural productivity and preserving ecological balance. It implies a shift from traditional agricultural methods to more environmentally friendly and sustainable approaches, ultimately reducing the need for mineral fertilizers. Additionally, an innovative method of enriching conventional mineral fertilizers with microorganisms provides the promising path to mitigate adverse effects of these fertilizers on soil biodiversity. This introduction highlights the critical role of biodiversity in terrestrial ecosystems, emphasizing its importance in maintaining the ecological productivity and regulating climate. In agroecosystems, focus shifts to soil-inhabiting microorganisms, such as bacteria, fungi, and archaea, which play a complex role in below-ground biochemical processes, nutrient cycling, and soil improvement. These microorganisms positively influence plant health, development, and interactions with soil-borne phytopathogens (Frąc et al., 2018).

12.2 SYNTHETIC OR MINERAL FERTILIZERS

Fertilizers, as defined by the Soil Science Society of America, include any organic or inorganic materials, whether of natural or synthetic origin, excluding liming materials or other non-nutrient soil amendments added to soil to provide essential plant nutrients necessary for plant growth. Plants require 16 essential elements for their normal growth and development, with 13 obtained from the soil. Potassium (K), phosphorus (P), and nitrogen (N) are characterized as primary nutrients due to their higher demand (Wang et al., 2017). The intensive and continuous practice of agriculture often depletes these nutrient reserves in soil, necessitating regular replenishment. Synthetic or mineral fertilizers, derived from inanimate sources, include various synthetic and petroleum-based fertilizers, along with viable manures. While they exhibit significant similarities to natural fertilizers, they possess various characteristics. These fertilizers readily dissolve in water, making essential nutrients immediately available to plants for uptake. When used according to recommended procedures, they efficiently supply necessary nutrients for plant growth and are considered normal way. To provide plants with suitable conditions for growth and development, organic fertilizers are commonly used, replacing nutrients in soil. Synthetic fertilizers are particularly prevalent in regions where soils are deficient in basic nutritional elements. Among organic fertilizers, potassium (K), nitrogen (N), and phosphorus (P) are the main ones typically used by farmers in agricultural soils. These fertilizers play a dynamic role in modern agriculture, addressing severe shortage of essential nutrients resulting from intensive and concentrated farming practices. The phosphorus, along with potassium and nitrogen, is one of most important elements in crop production, hence traditional agriculture heavily relies on intensive application of phosphorus, nitrogen, and potassium fertilizers (Pandey, 2004).

12.2.1 Synthetic Nitrogen Fertilizers

Nitrogen, an essential plant nutrient, is abundant in the atmosphere but often deficient in soil. Synthetic nitrogen fertilizers, produced from the Haber-Bosch process, have addressed this issue. This process involves converting atmospheric nitrogen into anhydrous ammonia through high-temperature and high-pressure reactions, serving both agricultural and industrial purposes (Smil, 2001). One prominent synthetic nitrogen fertilizer is ammonium nitrate (NH_4NO_3), containing up to 34% nitrogen in solid form. It provides ammonium and nitrate forms of nitrogen, essential for plant growth. These fertilizers play an active role in modern agriculture, replenishing nitrogen reserves in soil and promoting crop yield. The global consumption of synthetic nitrogen fertilizers has significantly increased over the years, reaching 104 teragrams in 2006 (Faostat, 2009). However, their widespread use has raised concerns, including rising energy costs and fertilizer prices. As a result, there's growing pressure on farmers to maximize nitrogen uptake efficiency, measured by ratio of fertilized yield to unfertilized yield, multiplied by grain nitrogen concentration per unit of the fertilizer nitrogen applied. Ammonium nitrate is a cost-effective and popular choice among farmers due to its high nitrogen absorption. Nitrogen is a key factor in plant growth, enhancing metabolic processes and photosynthesis (Pavlíková et al., 2012). Nitrate fertilization increases amino acids and proteins, promoting growth and stimulating carbon metabolism. It also leads to fluctuations in plant hormone levels and shifts in distribution and phenology, affecting ratios of root-to-shoot and root architecture. Synthetic nitrogen fertilizers, exemplified by ammonium nitrate, are crucial for modern agriculture. They provide an available source of nitrogen, supporting plant growth and crop production. However, their sustainable use requires careful management to maximize productivity and minimize potential environmental impacts (Tripathi & Raghubanshi, 2014).

12.2.2 Synthetic Phosphate Fertilizers

Inorganic phosphate (P), a crucial element for plant growth and crop yield, is often deficient in soils and primarily depends on mined rock phosphate. A scarcity of inorganic phosphate in the soil can lead to a decrease in crop yields (Uchida, 2000). Agriculture is responsible for a significant portion of global inorganic phosphate usage, accounting for approximately 80% to 90% of the total demand. China, Europe, and India collectively contribute to about 60% of the global usage of inorganic phosphate fertilizers, with China being the largest consumer at 34%, followed by India at 19% (Faostat, 2009). The primary synthetic inorganic phosphate fertilizer is typically triple superphosphate (0460-N-P-K), produced by treating rock phosphate with phosphoric acid. Another commonly used inorganic phosphate salt is ammonium phosphate [$(NH_4)3PO_4$], which results from the combination of ammonia and phosphoric acid. This salt forms a crystalline powder. When applied, a small portion of the inorganic phosphate fertilizer is directly available for plants, while the rest remains in the soil with varying degrees of availability. To ensure an adequate supply of phosphorus to crops, farmers often apply excess phosphate (Strickland & Rousk, 2010; Tan et al., 2012). However, this practice raises concerns as it increases the risk of significant phosphorus loss through runoff, leaching, or soil erosion. Ultimately, this excess phosphate may end up in lakes, rivers, and oceans, leading to environmental issues. Considering that the application of balanced amounts of phosphorus mineral fertilizers is a determinant in achieving optimal yields, phosphorus is often present in the soil in forms that are not easily accessible to plants. Phosphorus is a vital element of nucleic acid compounds such as adenosine triphosphate (ATP), guanosine triphosphate (GTP), cytidine triphosphate (CTP), and uridine triphosphate (UTP). Phosphorus is also involved in chemical processes responsible for maintaining the proper functioning of plant cells, cell division, the activation or inactivation of intracellular enzymes, and the development of the morphological structures of plants such as roots, stems, and stalks. Therefore, phosphorus is required at all stages of plant development (Moghaddam, 2011).

12.2.3 Synthetic Potassium Fertilizers

Potassium (K) fertilizers play a crucial role in replenishing essential minerals in soil that have been depleted due to intensive agricultural practices. Potassium was initially discovered by farmers centuries ago when they observed positive effects of wood ash from metal pots on plant growth, known as potassium hydroxide. Potassium is available in two primary forms: muriate (meaning chloride) of potash or MOP, and sulfate of potash or SOP. These synthetic potassium fertilizers are essential for addressing potassium deficiencies in agricultural soils and ensuring optimal crop yield and plant growth.

12.3 AGRICULTURAL PESTICIDES—IMPLICATIONS AND CONCERNS

Pesticides, including fungicides, herbicides, and insecticides, are biochemical substances used to control, mitigate, or eliminate pests in agriculture. They play a significant role in modern farming, as evidenced by the global pesticide market's estimated value of $61.8 billion in 2014, projected to rise to $83.7 billion by 2019. India alone consumes around 80,000 tonnes of pesticides annually, primarily used in cotton production (45% of total pesticide use), rice cultivation (23%), and the growth of fruits, vegetables, cereals, oilseeds, pulses, and millets (6–7%). Pesticides are crucial in reducing crop loss due to plant diseases (Pinstrup-Andersen, 2001), which can result in a 10–15% yield drop in major crops worldwide. Herbicides are the most commonly used pesticides in agriculture. With the need to control phytopathogens and meet increasing food demands, chemical pesticides have become indispensable. As a result, pesticide consumption has been steadily rising in agricultural soils, making Asia the world's largest pesticide consumer, with China leading in production and consumption. The increased use of fertilizers and insecticides since World War II has led to a surge in insect pests and diseases in agricultural soils. Overuse of agrochemicals and monoculture practices have exacerbated pest problems. Insecticides, being bioactive and toxic substances, directly or indirectly affect soil productivity and the quality of the agroecosystem. Pesticides can be classified based on various criteria, such as biochemical composition, target pests, persistence in soil, toxicity, and volatilization behavior. Pesticides encompass a wide range of chemicals, including insecticides, fungicides, herbicides, and more (Lehr, 2014). Ideally, insecticides should have minimal impact on non-target soil organisms, low soil persistence, cost-effectiveness, and biodegradability (Altieri & Nicholls, 2003). However, many pesticides exhibit severe and long-lasting toxicity, harming numerous life forms beyond their intended targets (Imfeld & Vuilleumier, 2012). Most pesticides can penetrate the cell walls of non-target soil microorganisms, disrupting their normal metabolism and leading to cell death (Zacharia, 2011). Pesticides pose a significant threat to soil microbiota and overall soil health, disrupting natural soil habitats (Zacharia, 2011). Moreover, they cause soil and water contamination, enter the food chain, affect human and environmental health, and contribute to the development of pesticide-resistant varieties (Sattler et al., 2006). The impact of pesticide toxicity varies, with assessment being particularly challenging due to low-level contamination from continuous use of poorly degradable pesticides or high-level contamination from disposal or accidental releases (Imfeld & Vuilleumier, 2012). Soil microorganisms play a key role in the conversion of accumulated toxic pesticides in the soil. Therefore, it is crucial to assess the long-term and continuous use of biochemical pesticides on soil microbial structure and function. The extensive use of pesticides in agriculture has raised concerns about their impact on soil health, non-target organisms, and environmental sustainability, underscoring the importance of careful pesticide management and sustainable agricultural practices.

12.3.1 The Interplay of Environmental Factors, Microflora, and Health

Starting with two primary issues, the limitation of land resources and the increase in global population, we see a cascade of effects:

1. Overuse of Chemicals: The scarcity of land and the growing population lead to the excessive use of pesticides and chemicals in an attempt to maximize agricultural output. These chemicals can have a detrimental effect on soil microflora, including fungi, protozoa, bacteria, and cyanobacteria, disrupting their normal functioning.
2. Short-Term Consequences: The immediate effects of this disruption can be seen in the form of pollution of soil, air, and water, and an imbalance in the natural ecosystem.
3. Long-Term Consequences: Over time, these effects can lead to more serious problems, damaging the environment and human health.

This flow of information illustrates the interconnectedness of environmental factors, microflora, and human health, highlighting the importance of sustainable practices. Figure 12.1 depicts impact of agrochemicals on microflora.

12.4 SOIL HEALTH—A CRITICAL ASPECT OF ENVIRONMENTAL HEALTH

Soil health is considered a fundamental component of ecosystem health, reflecting the integration of nutrient cycles, energy flows, stability, and resilience in an

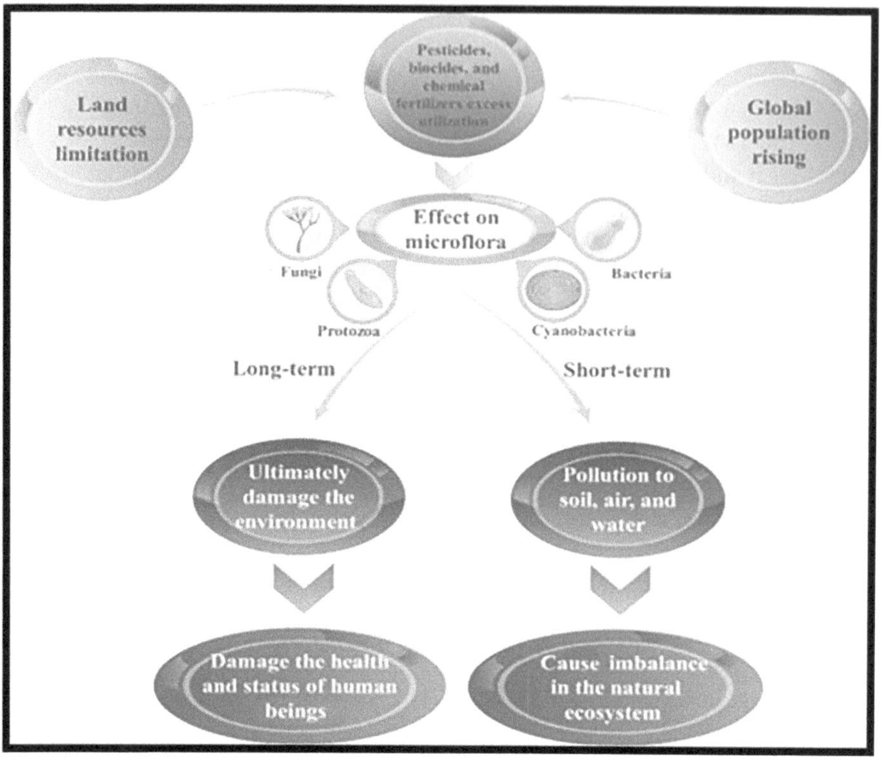

FIGURE 12.1 Impact on microflora.

ecosystem. A healthy ecosystem is characterized by its ability to sustain biological productivity and maintain environmental quality even in the face of disturbances or stress. Soil health is closely related to biodiversity and stability. It is often used interchangeably with term "soil quality," which is defined as the soil's capacity to function within ecosystem boundaries, sustaining biological productivity, promoting environmental quality, and supporting plant and animal health. Fundamentally, soil health reflects the continued capacity of soil to function as a vital living system within the context of ecosystems and land use. This includes sustaining biological productivity, enhancing air and water quality, and supporting the health of plants, animals, and humans (Avidano et al., 2005). The occurrence of disease outbreaks in plants can serve as an indicator of volatility in soil ecosystem strength. Soil health also influences the ability of a biological community to suppress plant pathogens, the population density of plant pathogens in soil, and the severity of disease outbreaks. Therefore, the ability of soil to suppress diseases is an indicator of a stable and healthy soil ecosystem (Van Bruggen & Semenov, 2000). Soil is considered a living and dynamic ecosystem, hosting an immense diversity of microorganisms that play an essential role in regulating its properties. Soil health also influences the ability of the biological community to suppress plant pathogens, the population density of plant pathogens in the soil, and the severity of disease outbreaks. In broad terms, soil functions include life support processes such as anchoring and nutrient supply to plants, water retention, and conductivity, support for food systems, and regulatory functions similar to gas exchange, nutrient cycling, microbial diversity, pollutant detoxification, carbon sequestration, and heavy metal retention (Larson & Pierce, 1991). Therefore, managing soil health is crucial for ensuring sustainable agricultural production and maintaining soil biodiversity, including microbial diversity. Basic indicators for assessing soil health include physical parameters like texture and water-holding capacity, chemical parameters such as pH, salinity, and soil organic matter content, and biological parameters like bacterial activity, nitrogen mineralization, and "soil respiration." These indicators offer a valuable understanding of the health of the soil and overall ecosystem. The deterioration of soil health can be attributed to intensive agricultural practices and excessive use of synthetic fertilizers and pesticides. These practices affect soil functions, physical and biochemical environment, and bacterial activity, which together, impact soil health. Measuring soil microflora, its structure, function, and distribution is essential to gaining insights into soil health and for applying strategies to restore and maintain soil quality.

12.4.1 Soil Health: A Tripartite Analysis

Soil health is a complex concept that encompasses three main categories: physical, chemical, and biological. Each of these categories contributes to the overall health and productivity of the soil.

Physical Factors: These include the texture of the soil, its capacity to hold water, and the degree of compaction. Each of these factors plays a role in determining how well plants can grow in the soil (Figure 12.2).

1. **Chemical Factors**: The chemical health of the soil is determined by its pH level, the availability of nutrients, and the amount of organic matter present in the soil. These factors influence the soil's fertility and its ability to support plant growth.
2. **Biological Factors**: The biological health of the soil is influenced by the ratio of carbon to nitrogen (C:N) and the level of microbial and enzymatic activity. Factors contribute to the soil's ability to cycle nutrients and support plant life.

All these factors are interconnected and influence each other, contributing to the overall health of the soil. Understanding these interactions is crucial for effective soil management.

12.5 SOIL MICROBIOLOGY—THE UNSEEN WORLD OF MICROBIAL DIVERSITY

Soil microorganisms play a crucial role in soil fertility. They include *bacteria*, *archaea*, *algae*, *protozoa*, *fungi*, *rotifers*, *tardigrades*, and *nematodes*, and perform

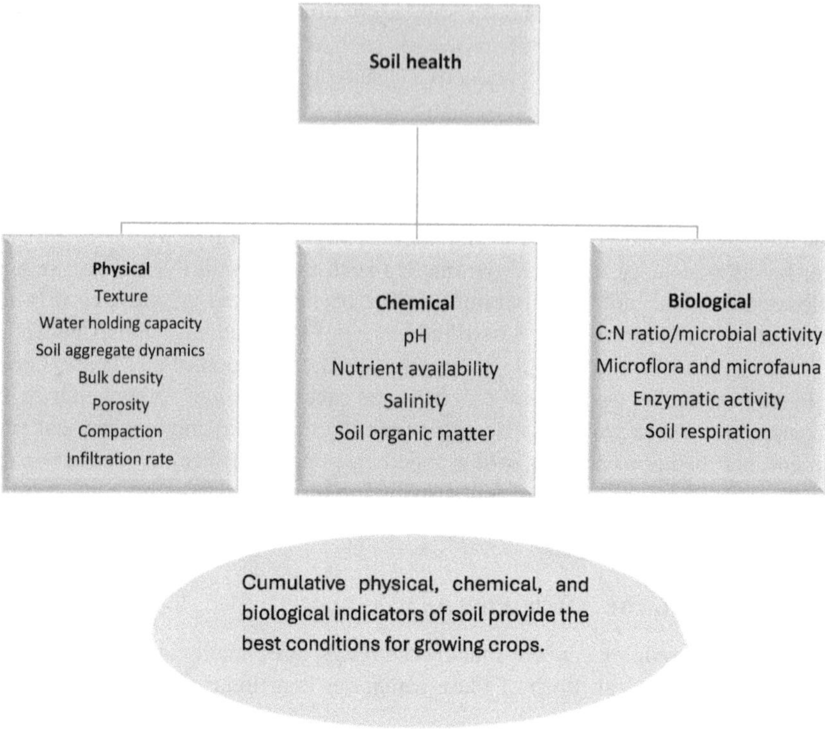

FIGURE 12.2 The categories that contribute to the overall health and productivity of the soil.

essential functions such as nutrient cycling and organic matter decomposition (Li et al., 2007; Tejada et al., 2015). Soil microbial mutualism offers nutritional benefits to plant species by receiving sugars from them. In particular, *rhizobia* and *fungi* have a positive impact on legumes and *ectomycorrhizal* tree species respectively, as compared to non-legumes and non-*ectomycorrhizal* plants.

The functioning of terrestrial ecosystems heavily relies on soil microbes, which serve as catalysts for promoting plant diversity. Soil microorganisms that are not dependent on a host organism can increase the variety of plants in an area by increasing the amount of nutrients and changing which plants are dominant due to both mutually beneficial and harmful interactions. Plant diversity can also support the growth of soil microbes. This is because it can provide them with food resources like exudates and litter, as well as physical microhabitats and favourable environmental conditions. Ecologists are aware of the importance of microbes in soil functions. Nevertheless, it is the progress in technology that has allowed scientists to extract and describe microbial communities to evaluate for roles in the soil. The rhizosphere soil is home to interactions between plants and microbes that play an important role in various processes, such as plant growth, pollutant degradation, and nutrient mobilization. These interactions are responsible for atmospheric nitrogen fixation, nutrient solubilization, and controlled movements. Additionally, they regulate the availability of pollutants to plants through the production of chelating agents, biosurfactants, acidification, redox changes, and phosphate/potassium solubilization (Bashir et al., 2017; Bashir et al., 2018; Bashir et al., 2018a; Bashir et al., 2018b; Bashir et al., 2019; Bashir et al., 2019b; Bashir et al., 2021a; Bashir, 2021b; Hamid & Bashir, 2019a). Different types of pollutants, including those that originate from agriculture, can be transformed and degraded by bacteria and fungi. The microorganisms undertake a variety of processes, collectively known as degradation, to achieve this. These processes can result in complete mineralization of the pollutants or partial transformation into non-toxic metabolites that are excreted in the soil. Once excreted, the non-toxic metabolites can become immobilized by bonding to matrix components such as humic acid. The earth is home to an incredible variety of tiny living things, and they form a complex and diverse microbial community in the soil. These microorganisms are responsible for performing most of the enzymatic processes in the soil and are essential for conserving energy and nutrients in their living biomass. The fertility of the soil and the yield of crops are greatly impacted by the abundance of microorganisms residing in the soil. It has been discovered that a mere gram of soil holds nearly 10,000 distinct microbial species. Sophisticated analytical techniques have indicated that every gram of soil could contain as many as one million prokaryotic genomes. Bacteria are the most commonly found microbes in soil, followed by fungi and protozoa, actinomycetes, and algae. The prevalence of fungal-bacterial dominance can fluctuate in cultivated soils (Sylvia et al., 1998). The ratios of fungal-bacterial biomass in agricultural soils can be affected by agricultural practices and environmental factors. If soil is altered organically, it tends to promote higher fungal/bacterial biomass ratios, while a use of mineral fertilizers may result in the opposite trend. These differences demonstrate the beneficial impact of organic fertilizers on soil ecosystems and microflora, which helps to support sustainable agriculture.

The introduction of advanced molecular tools for microbial detection, identification, enumeration, and characterization has revolutionized our ability to study soil microorganisms without the need for cultivation. Techniques like ARDRA, denaturing gradient gel electrophoresis (Berg, 2000), phospholipid fatty acid analysis, and fatty acid methyl ester analysis have enabled researchers to explore the structural and functional diversity of unculturable soil microbes. This has opened new paths for understanding the complex world of soil microorganisms. Numerous studies have been conducted to investigate the effects of biotic and abiotic factors on microbial activity and diversity in soils. These factors include land-management practices, nutrient modifications (organic manures or inorganic fertilizers), land-use practices, and environmental conditions. The influence of chemical fertilizers and pesticides on soil microorganisms has been widely studied, revealing variable effective patterns. These effects can be both direct due to the action of agrochemicals on microorganisms, and indirect resulting from changes in the soil environment caused by these substances (Seymour, 2005). Soil microbiology is a fascinating domain characterized by its incredible microbial diversity. These microorganisms play a fundamental role in soil health, agricultural productivity, and food security, making them a subject of intense research and interest. Figure 12.3 depicts the functions of various soil microflora.

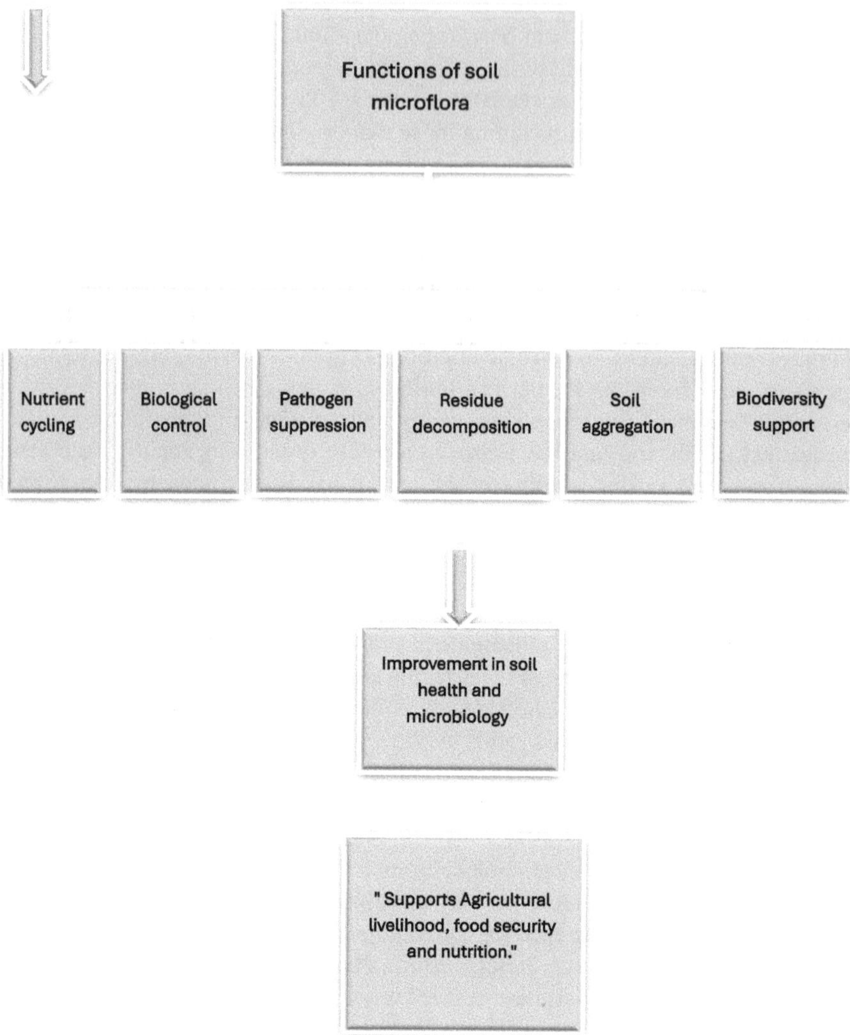

FIGURE 12.3 Functions of soil microflora.

Advanced techniques have expanded our ability to explore this hidden assembly of microorganisms, shedding light on their structure and function and their responses to various agricultural practices and environmental conditions.

12.6 THE INFLUENCE OF SYNTHETIC FERTILIZERS ON SOIL HEALTH

Synthetic or inorganic fertilizers have a significant role in contemporary agriculture, enhancing crop yield. The prolonged use of these fertilizers, particularly in diverse crop types, has led to alterations in soil organic matter (SOM), nitrogen content, pH, and moisture levels (Bohme et al., 2005). Their impact on soil health and ecosystem functions is closely tied to primary productivity. Instead of causing direct toxicity, inorganic fertilizers tend to affect the manufacturing process. SOM, particularly soil organic carbon (SOC), is a crucial parameter for assessing soil health. Inorganic fertilizers often contribute to a decline in the use of organic manure. Some studies have shown that excessive application of mineral nitrogen can either have a neutral or inhibitory effect on the decomposition of low-nitrogen plant resources. In contrast, long-term organic fertilizer uses tend to increase SOM, particularly through adding higher litter and root biomass (Ladha et al., 2011). The application of mineral nitrogen fertilizers can have complex interactions with carbon changes in soil. It can also induce soil acidification, especially if the soil lacks buffering capacity, causing a release of base cations like calcium and magnesium. Continued nitrogen addition can decrease soil base cation content, leading to the release of toxic aluminium from soil reserves, which can negatively impact plant nutrient availability. A severe case of soil acidification in China due to excessive synthetic nitrogen fertilizer use has been reported, leading to decreased crop production. Soil acidification not only affects microbial nitrogen immobilization, breakdown, and mineralization of SOM but also contributes to condensed water quality due to increased nitrate and cation leaching to groundwater, lakes, and rivers. Leached nitrate can further contribute to the oxidation of pyrite in the undersoil, releasing various trace elements (Venterea et al., 2004). Microbial communities in soil are highly delicate to elevated levels of reactive nitrogen. Soil acidification can trigger the production of nitrous oxide, a strong greenhouse gas, through nitrification and denitrification processes (Velthof et al., 2012). The excessive supply of reactive nitrogen and phosphorus can lead to ammonia and nitrogen oxide releases into the air and losses of nitrate and phosphorus into water bodies. Excess phosphorus can cause a buildup of soil phosphorus beyond its sorption capacity, resulting in increased losses to surface waters through various pathways, including overland flow, erosion, and subsurface leaching (Velthof et al., 2012). While synthetic fertilizers contribute to increased crop production, their overuse can have harmful effects on soil wellbeing. These effects include soil acidification, release of toxic elements, reduced microbial activity, water quality degradation, and increased greenhouse gas emissions. It is vital to manage the use of synthetic fertilizers carefully to diminish these bad impacts and uphold soil health.

12.7 THE INFLUENCE OF SYNTHETIC FERTILIZERS ON SOIL MICROBIOLOGY

Fertilizers are known to significantly enhance the nutrient content of soil, ultimately improving crop productivity. This enhancement occurs through mechanisms such as increased root turnover, rhizodeposition, and integration of crop residues into soil, all of which contribute to sharp microbial activity. So boost in microbial activity, in turn, has implications for SOC content and microbial communities (Lupwayi et al., 2012). Soil microbiology plays a vital role in understanding complicated relationships between soil health, agricultural practices, and the application of fertilizers. The functional diversity of soil microbial communities is intricately linked to the availability of key resources, namely phosphorus (P), carbon (C), and nitrogen (N). Various studies have shown that the presence and abundance of these essential nutrients play an essential role in determining the composition and activity of soil microbes. Consequently, agricultural practices, such as excessive use of mineral fertilizers, pesticides, and power tillage, can have opposing effects on living groups in the soil. Table 12.1 depicts a summary of recent research on the impact of pesticides on soil microbial community and enzymatic activity. These practices can disrupt the habitats and functions of these soil organisms, thereby affecting the overall soil ecosystem. The direct effects of synthetic fertilizer application on soil microbiota can be both positive and negative. For instance, the application of high concentrations of ammonia fertilizer typically results in short-term damage to soil microorganisms, primarily within the application zone (Angus et al., 2014). The outcomes depend on numerous factors, including type amount, and the way fertilizer is applied. However, more complete studies have shown that mineral fertilizer application can lead to a distinguished increase in microbial biomass, as observed in various long-term experiments conducted around the world (Geisseler & Scow, 2014). Long-term studies on the impact of fertilizers with urea and anhydrous ammonia have recorded minimal effects on soil microbial populations and soil quality, it is essential to note that response to fertilizers can vary depending on specific soil conditions. For example, soils with a pH below 5 may experience a reduction in microbial biomass due to fertilizer application, while soils with a pH of at least 7 may see an increase in microbial biomass. The application of phosphorus (P) fertilizers is known to decrease the presence of mycorrhizal fungi, although the degree of this decrease depends on the fungal species involved and the level of plants present in the soil. Comparing the effects of organic and inorganic fertilizers, several studies have reported that organically modified soils tend to exhibit significantly higher organic carbon content, increased microbial populations, and enhanced microbial activities. The bacterial community structure in organically edited soils also more closely looks like that of natural soils. Additionally, certain bacterial communities, such as Pseudomonas, which are beneficial for plant growth, can adversely affect long-term synthetic fertilizer application, whereas the opposite inclination is observed in organically amended soils (Wu et al., 2012). Moreover, the sensitivity of soil microorganisms to heavy metals and pesticides, such as cadmium, copper, and cypermethrin, appears to be higher in soils treated with mineral fertilizers compared to organically amended soils. This sharp sensitivity is an important consideration when assessing the general impact of different fertilizer practices on soil health. The relationship between fertilizers and soil microbiota is complicated, and influenced by various factors, and quantity of fertilizer applied, manner of application, soil features, and other agricultural practices. Understanding these intricate relations is crucial for achieving sustainable agriculture and soil health management. Further research is needed to provide conclusive insights into the complex effects of nitrogen (N) and phosphorus (P) on soil microbiota, as contradictory reports exist in this regard (Li et al., 2013). The enduring impact of pesticides on diverse soil types is governed by multiple factors, including biochemical and microbial changes, volatilization, sorption, uptake by plants or organisms, and groundwater leaching. Among these, sorption and biodegradation are the most crucial processes. Agricultural soils encounter a variety of pesticides through

TABLE 12.1
A Summary of Recent Research on the Impact of Pesticides on Soil Microbiology and Enzymatic Activity

S. No	Location	Pesticide	Impact on Microbial Species	Biochemical Reaction Impact	Affected Soil Enzymes	Citation
1.	Spain	Oxyfluorfen herbicide, chlorpyrifos insecticide	Microbial population increased	Inhibition of dehydrogenase, urease, beta-glucosidase, and phosphatase activity increased	Dehydrogenase, urease, beta-glucosidase, and phosphatase	Franco Andreu et al., 2016
2.	Spain	Cypermethrin insecticide	Microbial diversity decreased with increasing insecticide dose	Activities of dehydrogenase, urease, beta-glucosidase, phosphatase, and arylsulfatase decreased	Dehydrogenase, urease, beta-glucosidase, phosphatase, and arylsulfatase	Tejada et al., 2015
3.	Poland	Imidacloprid insecticide	The number of bacteria decreased	The concentration of NO_3^- decreased in treated soils, and the concentration of NH4 increased in 103 FR treated soils	Dehydrogenase, phosphatase	Cynon & Piotrowski, 2015
4.	Tunisia	Glyphosate herbicide	Positive effect on bacteria diversity	Enzymatic activities	Enzymatic activities responded differently to the action of glyphosate	Cherni et al., 2015
5.	India	Chlorpyrifos insecticide with control	Decrease in bacterial and fungal colonies	The number of colony-forming units of bacteria and fungi decreased in soil amended with insecticide	Colony forming units	Supreeth et al., 2016
6.	Tunisia	Glyphosate herbicide	Positive effect on bacteria diversity	Enzymatic activities	Enzymatic activities responded differently to the action of glyphosate	Cherni et al., 2015
7.	Spain	Mesotrione herbicide	Soil biomass increased with higher doses	Soil biomass, dehydrogenase	Soil dehydrogenase activity decreased with increasing herbicide dose	Pose-Juan et al., 2015
8.	Spain	Chlorpyrifos pesticide under the influence of earthworm	Strong inhibition of cholinesterase and carboxylesterase activities of earthworms	Inhibited cholinesterase and carboxylesterase	Cholinesterase, carboxylesterase	Sanchez Hernandez et al., (2018)
9.	China	Dimethomorph	High doses of dimethomorph resulted in varying degrees of soil enzymatic activity	Varying degrees of soil enzymatic activity	Soil enzyme activity	Wang et al. (2017)

direct application. The destiny of pesticides in the soil can significantly differ due to factors such as application frequency, spatial and temporal variations, crop rotation, and climatic conditions. For instance, certain studies have observed the evolution of microflora capable of rapidly degrading specific pesticides in soils where those pesticides have been applied for prolonged periods. Pesticides exert different effects on soils compared to plants or animals. Biochemical residues tend to decompose more slowly in soil, and this persistence can be influenced by various factors. These factors include a biochemical structure of the pesticide, its instability, water solubility, preparation method, and application technique. In addition, soil-related factors like soil type, the organic matter content, clay content, pH levels, the microbial diversity, and the presence of invertebrates play significant roles in pesticide behaviour. The United States Environmental Protection Agency (US EPA) defines "half-life" as the rate at which a pesticide decomposes in soil or water. The pesticides that persist in their original form in soil for extended periods are more likely to leach through soil. Pesticides are categorized based on their "half-life" with groups ranging from low persistence (half-life less than 30 days) to very high persistence (half-life greater than 100 days). Organochlorine pesticides, which contain five or more chlorine atoms per molecule, are known to be highly persistent in the environment. Persistent bioaccumulate toxic (PBT) pollutants are toxic substances that endure in the environment and accumulate in food chains, posing risks to human health and ecosystems. Many of these pesticides have been classified by the US EPA as persistent bioaccumulate toxic (PBT). Pesticides that are highly persistent in soil have long-lasting effectiveness in pest control. Their prolonged presence in the soil can have detrimental and toxic effects on soil microflora and wildlife, potentially leading to environmental contamination. The ideal scenario for pesticide degradation is not to release toxic particles into the soil. The residual concentration of pesticides in soil is influenced by factors such as soil type, application quantity, and plant growth stage. The effects of pesticides on soil health can vary based on their concentrations, for higher amounts of pesticides can inhibit soil microbial activity and crop yield, while lower concentrations may have no adverse effects or may even be less toxic to soil health. This concentration is dependent on the importance of precise application and responsible pesticide use. Several studies have examined the influence of various pesticides on soil microbiology and enzymatic activities. These studies have utilized different techniques to measure microbial parameters, including microbial biomass, genetic diversity, and catabolic activities. The findings suggest that changes in these microbial parameters are more pronounced when pesticides like methamidophos, oxadiazon, glyphosate, chlorpyrifos, quinalphos, and fungicides are applied in high doses. Soil characteristics also play a significant role influencing toxic effects of the pesticides on microflora. The physical and biochemical properties of soil such as pH, carbon-based carbon content, and clay content can impact the response of soil microorganisms to pesticide exposure. Additionally, the stage of pesticide application, frequency of action, use of organic amendments in soil, and crop age can impact pesticide toxicity on the soil health. Pressed treatments, for example, can alter the functional structure and reduce the functional diversity of soil bacteria. Repeated applications of pesticide mixtures, including in crop treatments, may also have similar effects. The impact of pesticides on soil health is complex and dependent on multiple factors, including pesticide concentration, soil type, and agricultural practices. To mitigate potential damage to soil ecosystems, it is essential to apply pesticides carefully, considering factors such as pesticide interaction, application frequency, and soil characteristics, while monitoring their effects on soil microbiology and enzymatic activities.

12.8 IMPACT OF PESTICIDES ON SOIL MICROBIOLOGY

Pesticides, a diverse group of biochemical agents used in agriculture to combat harmful biological agents, include bactericides, herbicides, insecticides, and fungicides. While these are designed to be species-specific, repeated use can inadvertently affect beneficial soil microorganisms. These organisms may undergo genetic changes that make them less beneficial to the soil ecosystem and may develop resistance to the pesticides intended to control them. This implies that damage or modification of these primary microbes can have cascading negative effects on the entire soil ecosystem, affecting organisms up to the largest mammals. Soil microbes exhibit varied responses to different biochemical pesticides applied in agricultural soils, and these responses are influenced by several factors. These factors encompass the nature of the pesticide, soil properties, and composition of groups within the soil. The total number of bacteria, fungi, protozoa, and algae in the soil may either increase or decrease depending on factors like the toxicity and nutrient availability of pesticides. However, the overall structural and functional diversity of soil microbial populations may change following pesticide applications, as studies have shown that populations of sensitive microbial communities may decrease in response to pesticide exposure, while populations of microbes capable of withstanding higher pesticide concentrations may increase (Wang et al., 2017). This can occur because these resilient microbes may use carbon-based compounds released from dead microbial cells or even the pesticide itself as an energy or carbon source. Reduced competition in the soil due to pesticide-induced changes can also be a contributing factor. Although some studies have reported an overall increase in microbial biomass following pesticide application, there may be an immediate decrease in functional microbial diversity (Lupwayi et al., 2009). Long-term pesticide applications can lead to the dominance of a few specific microbial groups in the soil, thereby affecting the overall community structure and related biological processes. The type of pesticide used plays a critical role in determining how soil microbial populations are affected. Some pesticides like organophosphates and herbicides may have little to no negative effects on soil microbial biomass while others such as pesticides, fungicides, and specific herbicides can significantly decrease microbial biomass (Wang et al., 2017). The effects of pesticides can be complex. For example, a specific pesticide may not directly impact the microbial population but can have opposing effects when its degradation products are considered. In addition, the type and diversity of microbial groups affected by pesticides may vary and the specific effects of pesticides can differ between bacterial, fungal, and actinomycetes populations (Moghaddam et al., 2011). The behavior of soil microbial inhabitants in response to pesticide exposure is complex and can be influenced by the type of pesticide, soil properties, and the composition of the microbial community. Understanding these dynamics is vital for sustainable agricultural practices and maintaining soil health. It is crucial to use pesticides judiciously and consider their long-term impacts on soil microbial ecosystems and related biological processes. The impact of pesticide toxicity on agricultural soil and its complex components, widely used to protect crops and increase agricultural yields, has inadvertently led to the accumulation of residues within the soil, thereby posing direct and indirect threats to the diverse biology and vegetation within this vital ecosystem. The research findings showed that certain pesticides, including Cypermethrin, Thiamethoxam, and Glyphosate, cause a decline in microbial activity and enzymatic activity, particularly Phosphatase and Dehydrogenase enzymes. This reduction extends to the production of carbon dioxide (CO_2) in the soil, indicative of reduced microbial respiration and consequently a disadvantaged decay of organic matter. Additionally, some fungicides such as azoxystrobin and Falcon 460 EC have been observed to overpower enzyme activity and exert harmful effect on the soil-dwelling microbial communities. It is worth noting that these impacts are intricately linked to pesticide concentrations and they are strengthened when pesticides are used in combination with other substances or when they are applied at levels exceeding recommended amounts. Consequently, disruption of soil chemical processes and damage to microbial growth and nutrient cycling arise as dangerous challenges, underscoring the urgency of sensible pesticide management to preserve soil health and promote the sustainability of agricultural practices.

12.9 SUMMARY

With the world population projected to reach 9.7 billion by 2050, there is a growing need to produce 70% more food to ensure global food safety (Faostat, 2009). The maintenance of soil health is critical for both current agricultural productivity and the long-term sustainability of agriculture. Soil health harnesses the power of biology to benefit human needs by working in harmony with natural processes. However, the modernization of agriculture has led to excessive use of capital, chemicals, and technology. While this method has been successful in increasing food production and pest control, it has also resulted in a range of economic, environmental, and social challenges. One of the most significant features affected by these practices is soil health. The excessive and often unnecessary application of artificial fertilizers and pesticides has led to severe soil pollution, harming soil biodiversity and microbial communities living within. The soil microbiota is exposed to high levels of toxic and persistent chemical inputs, which negatively affect their structural and functional diversity. These impacts extend to vital soil properties, such as nutrient content, predominant soil species, and microbial population diversity and activity (Tilman, 2011). The effects of chemical inputs on soil health can range from short-term variations to long-lasting, potentially irreversible changes. In contrast, organic alterations and the use of biocontrol agents have shown promise in maintaining overall soil quality and fertility, promoting sustainable agricultural practices. These alternatives offer cost-effective and environmentally friendly options for sustainable agriculture (Oliveira et al., 2014). Understanding the impact of pesticides on soil ecosystems remains a complex and evolving field, with varying research findings. Some pesticides can serve as a carbon or energy source for microorganisms, contributing to their degradation and assimilation, while others can be toxic to microbial populations. Soil properties, nature and concentration of pesticides, their activity, and production of metabolites during metabolism in soil all play a role in determining the impact of pesticides on soil biological properties. Long-term pesticide applications can disrupt the biochemical equilibrium of soil, significantly risking soil quality and fertility, and leading to reduced crop productivity. Future research should focus on understanding the molecular responses of microbes to pesticide applications, aiding in risk assessments related to pesticide contamination and its impact on soil microbial diversity and enzymatic activities. Furthermore, more studies are needed to explore "the impact of specific agricultural practices," including organic alterations, crop rotations, tillage, and pesticide use on soil regenerative capacity. By adopting a broader and ecologically informed approach, agriculture can better harness abundance and diversity to achieve more sustainable and productive farming systems.

REFERENCES

Altieri, M. A., & Nicholls, C. I. (2003). Soil fertility management and insect pests: Harmonizing soil and plant health in agroecosystems. *Soil Tillage Research*, *72*(2), 203–211.

Angus, J. F., Gupta, V. V. S. R., Pitson, G. D., & Good, A. J. (2014). Effects of banded ammonia and urea fertilizer on soil properties and the growth and yield of wheat. *Crop Pasture Science*, *65*(4), 337–352.

Antonini, C., & Argilés-Bosch, J. M. (2017). Productivity and environmental costs from intensification of farming: A panel data analysis across EU regions. *Journal of Cleaner Production*, *140*, 796–803.

Avidano, L., Gamalero, E., Cossa, G. P., & Carraro, E. (2005). Characterization of soil health in an Italian polluted site by using microorganisms as bioindicators. *Applied Soil Ecology*, *30*, 21–33.

Bashir, Z. (2021b). Characterization of Potassium Solubilizing Bacteria from Rhizospheric Soils of Cherry (*Prunusavium*). *Research Journal of Agricultural Sciences*, *12*(2), 494–496.

Bashir, Z., Hamid, B., Zargar, M. Y., & Kundoo, A. A. (2021a). Morphological and biochemical characterization of potassium solubilizing bacteria from rhizospheric soils of pinus (*Pinus wallichiana*). *Journal of Research and Development*, *21*, 58–63.

Bashir, Z., Zargar, M. Y., Baba, Z. A., Mohiddin, F. A., Peer, F. A., Mir, S. A., & Najar, G. R. (2018a). Isolation and characterization of phosphorus solubilizing bacteria (PSB) from rhizospheric soils of apple (*Malus Domestica*). *Journal of Pharmacognosy and Phytochemistry*, *7*(2), 2997–3000.

Bashir, Z., Zargar, M. Y., Bhat, T. A., Kousar, S., Baba, Z. A., & Mohiddin, F. A.(2018b). Solubilization of Phosphorus Containing Mineral by Bacteria from Rhizospheric Region of Walnut (*Juglans regia*). *International Journal of Current Microbiology and Applied Sciences*, *7*(8), 2391–2398.

Bashir, Z., Zargar, M. Y., Baba, Z. A., & Mohiddin, F. A. (2018). Solubilization of potassium containing mineral by bacteria from Kashmir iris (*Iris kashmiriana*) rhizosphere. *The Pharma Innovation Journal*, *17*(4), 747–750.

Bashir, Z., Zargar, M. Y., Mohiddin., F. A., Baba, Z. A., & Hamid, B. (2019). Isolation and Biochemical Characterization of phosphorus solubilizing bacteria from Rhizospheric region of Apricot and Peach. *Journal of Research and Development University of Kashmir*, *19*, 0972–5407.

Bashir, Z., Zargar, M. Y., Mohiddin, F. A., Kousar, S., Husain, M., & Rasool, F. (2017). Phosphorus solubilizing microorganisms: Mechanism and diversity. *International Journal of Communication Systems*, *5*(5), 666–673.

Bashir, Z., Zargar, M. Y., & Vishwakarma, D. K. (2019b). Potassium solubilizing microorganisms for sustainable agriculture. *Applied Agricultural Practices for Mitigating Climate Change*, *2*, 17–28.

Berg, G. (2000). Diversity of antifungal and plant-associated Serratia plymuthica strains. *Journal of Applied Microbiology*, *88*, 952–960.

Bohme, L., Langer, U., & Bohme, F. (2005). Microbial biomass, enzyme activities and microbial community structure in two European long-term field experiments. *Agriculture, Ecosystems & Environment*, *109*, 14–152.

Chen, J., Zhang, F., & Yang, X. (2020). Analysis of regional entrepreneurship, spatial spillover and economic growth. *Science & Technology Progress and Policy*, *37*, 44–50.

Cherni, D. T., & Chebil, S. (2015). Effect of glyphosate on enzymatic activities, rhizobiaceae and total bacterial communities in an agricultural tunisian soil. *Water Air and Soil Pollution*, *226*(5).

Cycoń, M., & Piotrowska-Seget, Z. (2015). Biochemical and microbial soil functioning after application of the insecticide Imidacloprid. *Journal of Environmental Sciences*, *27*, 147–158.

Faostat, F. A. O. (2009). *Statistical databases*. Food and Agriculture Organization of the United Nations.

Frąc, M., Hannula, S. E., Belka, M., & Jędryczka, M. (2018). Fungal biodiversity and their role in soil health. *Frontiers in Microbiology*, *9*, 707.

Franco-Andreu, L., Gómez, I., Parrado, J., García, C., Hernández, T., & Tejada, M. (2016). Behavior of two pesticides in a soil subjected to severe drought. Effects on soil biology. *Applied Soil Ecology*, *105*, 17–24.

Geisseler, D., & Scow, K. M. (2014). Long-term effects of mineral fertilizers on soil microorganisms-A review. *Soil Biology and Biochemistry*, *75*, 54–56.

Hamid, B., & Bashir, Z. (2019a). Potassium solubilization microorganisms: An Alternative technology to chemical fertilizers. *Journal of Research and Development University of Kashmir*, *19*, 0972–5407.

Imfeld, G., & Vuilleumier, S. (2012). Measuring the effects of synthetic pesticides on bacterial communities in soil: A review. *European Journal of Soil Biology*, *49*, 1–4.

Jiang, S., Zhou, J., & Qiu, S. (2022). Digital agriculture and urbanization: Mechanism and empirical research. *Technological Forecasting and Social Change*, *180*, 121724.

Ladha, J. K., Reddy, C. K., Padre, A. T., & Van Kessel, C. (2011). Role of nitrogen fertilization in sustaining organic matter in cultivated soils. *Journal of Environmental Quality*, *40*(6), 1756–1766.

Larson, W. E., & Pierce, F. J. (1991). Conservation and enhancement of soil quality. In *IBSRAM Proceedings 2, Evaluation for Sustainable Land Management in the Developing World*, *12*, 175–203.

Lehr, P. (2014). *Global markets for biopesticides* (Report Code CHM029E). BCC Research.

Li, F., Liu, M., Li, Z., Jiang, C., Han, F., & Che, Y. (2013). Changes in soil microbial biomass and functional diversity with a nitrogen gradient in soil columns. *Applied Soil Ecology*, *64*, 1–6.

Li, Z., Wu, X., & Chen, B. (2007). Changes in transformation of soil organic C and functional diversity of soil microbial community under different land uses. *Agricultural Science China*, *6*, 1235–1245.

Lupwayi, N. Z., Harker, K. N., Dosdall, L. M., Turkington, T. K., Blackshaw, R. E., O'Donovan, J. T., Cárcamo, H. A., Otani, J. K., & Clayton, G. W. (2009). Changes in the functional structure of soil bacterial communities due to fungicide and insecticide applications in canola. *Agricultural Ecosystems & Environment*, *130*, 109–114.

Lupwayi, N. Z., Lafond, G. P., Ziadi, N., & Grant, C. A. (2012). Soil microbial response to nitrogen fertilizer and tillage in barley and corn. *Soil Tillage Research, 118*, 139–146.
Moghaddam, N. S., Zakaria, M. P., Omar, D., Sijam, K., & Khakvar, R. (2011). Effects of imidacloprid on the biodiversity of soil microbes in selected soils of Malaysia. In T. B. Moorman & C. C. Dowler (Eds.). *Proceedings of the Second International Conference on Environmental Science and Development IPCBEE* (Vol. 4). IACSIT Press.
Oliveira, C. M., Auad, A. M., Mendes, S. M., & Frizzas, M. R. (2014). Crop losses and the economic impact of insect pests on Brazilian agriculture. *Crop Protection, 56*, 50–54.
Onyeka Kingsley, N., & Ayibapreye, J. (2022). Chemical pesticides and food safety. In R. Ramón Eduardo Rebolledo (Ed.). *Insecticides* (Ch. 3). IntechOpen.
Pandey, S., & Singh, D. K. (2004). Total bacterial and fungal population after chlorpyrifos and quinalphos treatments in groundnut (Arachis hypogaea L.) soils. *Chemosphere*, 55, 283–290.
Pavlíková, D., Neuberg, M., Žižková, E., Motyka, V., & Pavlík, M. (2012). Interactions between nitrogen nutrition and phytohormone levels in Festulolium plants. *Plant Soil Environment, 58*(8), 367–372.
Pinstrup-Andersen, P. (2001). *The future world food situation and the role of plant diseases.* The Plant Health Instructor. https://doi.org/10.1094.PHI-I-2001-0425-01
Pose-Juan, E., Sánchez-Martín, M. J., Herrero-Hernández, E., & Rodríguez-Cruz, M. S. (2015). Application of mesotrione at different doses in an amended soil: Dissipation and effect on the soil microbial biomass and activity. *Science of the Total Environment, 536*, 31–38.
Sanchez-Hernandez, J. C., Del Pino, J. N., Capowiez, Y., Mazzia, C., & Rault, M. (2018). Soil enzyme dynamics in chlorpyrifos-treated soils under the influence of earthworms. *Science of the Total Environment, 612*(15), 1407–1416.
Sattler, C., Kachele, H., & Verch, G. (2006). Assessing the intensity of pesticide use in agriculture. *Agriculture, Ecosystems & Environment, 119*, 299–304.
Seymour, N. (2005). Impacts of pesticides and fertilizers on soil. In *Proceedings, Soil Biology-Soil Health Conference.* Condamine Catchment Management Association Inc.
Smil, V. (2001). *Enriching the earth: Fritz haber, carl bosch, and the transformation of food production.* MIT Press.
Srivastava, P., Singh, P. K., Singh, R., Bhadouria, R., Singh, D. K., & Singh, S. (2016). Relative availability of inorganic N-pools shifts under land use change: An unexplored variable in soil carbon dynamics. *Ecological Indicators, 64*, 228–236.
Strickland, M. S., & Rousk, J. (2010). Considering fungal: Bacterial dominance in soils-methods, controls, and ecosystem implications. *Soil Biology and Biochemistry, 42*, 1385–1395.
Supreeth, M., Chandrashekar, M. A., Sachin, N., & Raju, N. S. (2016). Effect of chlorpyrifos on soil microbial diversity and its biotransformation by Streptomyces sp. HP-11. *3 Biotech, 6*, 1–6.
Sylvia, D. M., Fuhrmann, J. J., Hartel, P. G., & Zuberer, D. A. (1998). *Principles and applications of soil microbiology.* Prentice Hall.
Tan, H., Barret, M., Rice, O., Dowling, D. N., Burke, J., & Morrissey, J. P. (2012). Long-term agrichemical uses leads to alterations in bacterial community diversity. *Plant Soil Environ, 58*, 452–458.
Tejada, M., García, C., Hernández, T., & Gómez, I. (2015). Response of soil microbial activity and biodiversity in soils polluted with different concentrations of cypermethrin insecticide. *Archives of Environmental Contamination and Toxicology, 69*(1), 8–19.
Tilman, D., Balzer, C., Hill, J., & Befort, B. L. (2011). Global food demand and the sustainable intensification of agriculture. *Proceedings of the National Academy of Sciences of the United States of America, 108*, 20260–20264.

Tripathi, S. N., & Raghubanshi, A. S. (2014). Seedling growth of five tropical dry forest tree species about light and nitrogen gradients. *Journal of Plant Ecology*, *7*(3), 250–263.

Uchida, R. (2000). Essential nutrients for plant growth: Nutrient functions and deficiency symptoms. In J. A. Silva & R. Uchida (Eds.), *Plant nutrient management in Hawaii's soils, approaches for tropical and subtropical agriculture* (pp. 31–55). College of Tropical Agriculture and Human Resources, University of Hawaii at Manoa.

Van Bruggen, A. H., & Semenov, A. M. (2000). In search of biological indicators for soil health and disease suppression. *Applied Soil Ecology*, *15*(1), 13–24.

Velthof, G. L., Lesschen, J. P., Webb, J., Pietrzak, S., Kros, J., & Oenema, O. (2012). The impact of the nitrates directive on gaseous N emissions effects of measures in nitrates action programme on gaseous N emissions. DG Environment, European Commission. http://ec.europa.eu/environment/water/water-nitrates/pdf/Final Report Impact Nitrates Directive def.pdf

Velthof, G. L., Oenema, O., Postma, R., & van Beusichem, M. L. (1997). Effects of type and amount of applied nitrogen fertilizer on nitrous oxide fluxes from intensively managed grassland. *Nutrient Cycling in Agroecosystems*, *46*, 257–267.

Venterea, R. T., Groffman, P. M., Verchot, L. V., Magill, A. H., & Aber, J. D. (2004). Gross nitrogen process rates in temperate forest soils exhibiting symptoms of nitrogen saturation. *Forest Ecology and Management*, *196*(1), 129–142.

Wang, C., Zhang, Q., Wang, F., & Liang, W. (2017). Toxicological effects of dimethomorph on soil enzymatic activity and soil earthworm (Eisenia fetida). *Chemosphere*, *169*, 316–323.

Wang, J., & Sun, J. M. (2018). Talent development in China: Current practices and challenges ahead. *Advances in Developing Human Resources*, *20*, 389–409.

Wu, F., Gai, Y., Jiao, Z., Liu, Y., Ma, X., & An, L. (2012). The community structure of microbial in arable soil under different long-term fertilization regimes in the Loess Plateau of China. *African Journal of Microbiology Research*, *6*, 6152–6164.

Zacharia, J. T. (2011). Identity, physical and chemical properties of pesticides. In M. Stoytcheva (Ed.). *Pesticides in the modern world trends in pesticides analysis* (pp. 1–18). Intech Publisher.

13 Biotechnological Interventions in Organic Farming

Gulab Khan Rohela, Pawan Saini, Mudasir Gani, Plabani Roy, Syam S, Gulzar Ahmad Khan, and Zaffar Bashir

13.1 INTRODUCTION

World human population is expected to grow by 50% before 2050, which would put further strain on natural resources, particularly non-arable agricultural lands and availability of fresh water (Bahar et al., 2020). Studies show that before 2050, global agricultural output must rise by 100% to keep up with the growing human population. To meet the demand of higher productivity, chemical based fertilizers and pesticides found their way into the agricultural system in the 19th century (Pretty & Bharucha, 2014). However, in the earlier times, agriculture was practiced not by using synthetic chemical based fertilizers, herbicides, and pesticides (Gamage et al., 2023). Based on the increased demand for higher productivity, the mid-19th century saw the first developments in manufacturing and usage of chemical fertilizers in agriculture (Hemathilake & Gunathilake, 2022). These early fertilizers were potent, affordable, and biodegradable in nature. Later, based on the resistant nature of pests towards simple chemicals; highly aromatic and toxic natured chemicals were being utilized as synthetic pesticides to get rid of pests and microbial pathogens (Singh et al., 2023).

Although the introduction of more complex based chemical inputs were proven as advantageous in the short run for higher yield, but they also had major long-term negative consequences, including soil compaction, erosion, and reductions in soil fertility overall, reduced soil micro biota, as well as health risks related to harmful chemicals finding their way into the food chain through biomagnification (Parizad & Bera, 2023). Scientists engaged in soil biology have started looking for solutions to these negative consequences in an effort to keep the agricultural production levels higher without affecting the sustainability nature of the soil fertility (Raimi et al., 2022).

Among the various alternatives, organic farming is considered as a potent, recyclable, and sustaining practice, where natural fertilizers made of organic materials, such as green manure and compost manure were used (Akanmu et al., 2023). It is well acknowledged that the most economically viable and sustainable agricultural system is organic farming which enables the farmers to produce the agricultural products in

DOI: 10.1201/9781003491460-13

a better and safer way (Thakur et al., 2022). Organic farming also includes microbes based maintenance of soil fertility which is considered as a good strategy for utilizing the soil based nutrients, adding bone meal as a source of vital mineral nutrients, and also farming with crop rotation and companion planting (Gamage et al., 2023).

Organic farming is also known as biological farming or ecological farming. It was begun in the early 1900s in reaction to rapidly evolving farming methods. Utilizing diverse crops, biological pest control, and cultivating insect predators are recommended (Mani, 2022). It is still being advanced by a number of groups' new methodologies and strategies. The goal of organic farming systems is to strictly prohibit or limit the use of synthetic materials while allowing the use of naturally occurring components present in soil and in the surrounding environments in a sustainable and eco-friendly way (Varga et al., 2022). Supporters of organic farming assert benefits for food safety, openness, food security, self-sufficiency, as well as for sustainability, openness, and autonomy (Rudolph & Muchesa, 2023). Organic farming is a creative and sustainable method that improves many farmers' lives and agricultural output while also being environmentally beneficial (Singh et al., 2023). According to the most recent report, organic farming is on the rise globally and is primarily viewed as an alternative to agrochemical-free farming, which will improve the water reusability from agricultural sources and support nature through the development of sustainable ecosystem, reducing the toxicity of metals and other chemicals in agricultural products (Babajani et al., 2023). It can be stated simply as using natural resources to improve and sustainably produce food products.

While organic farming is seen to be environmentally beneficial, there are certain drawbacks such as improving the soil fertility through vermicomposting, biofertilizers, biosolutions, and other means is a time consuming process and it needs to be monitored regularly (Parizad & Bera, 2023). In certain nutrient deficient land forms, it becomes mandatory to go for crop rotation and intercropping patterns for improving the soil fertility or to interrupt the life cycle of pests. Similarly, lower yields, troubles in maintaining soil fertility levels, and market access are just a few of the numerous difficulties faced by organic growers (Jouzi et al., 2017). Certain organic and plant based pesticides, such as rotenone and pyrethrum, which are plant extracts that are frequently employed in organic fields are effective against pests but have also been shown to be harmful to fish, people, and animals (Sivaranjani & Rakshit, 2019). As a result, overuse of organic and plant based pesticides may cause runoff from the field to harm surrounding water bodies and enter the food chain. To lessen the constraints and difficulties associated with organic farming, biotechnology based applications should be combined with organic farming for wide applicability of organic farming. This chapter will provide information on biotechnological applications for improving the efficiency of organic farming.

13.2 SOIL MANAGEMENT IN ORGANIC FARMING

Soil management is an essential component of organic farming as healthy soil was the basis of effective organic crop production. Sustainable agricultural methods that support soil fertility, biodiversity, and the general health of ecosystems are the foundation of organic farming (Das et al., 2023). Organic soil management aims to

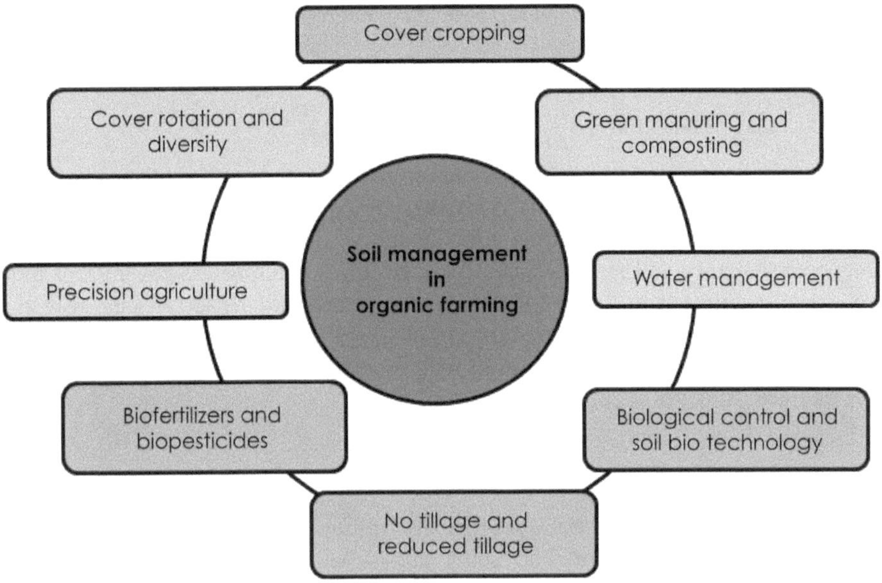

FIGURE 13.1 Different soil management practices in organic farming.

improve the natural processes within the soil ecosystem, promoting long-term productivity and resilience (Kumar et al., 2022). Biotechnology interventions are often approached cautiously in the context of organic farming, and its principles emphasize the use of natural and sustainable practices (Alduvín, 2022). The following are some key practices for managing soil in organic farming (Figure 13.1):

13.2.1 Crop Rotation

In organic farming, where synthetic inputs are limited, crop rotation becomes an integral part of maintaining soil health and sustainable agricultural practices. By incorporating diversity into the farming system, organic farmers can decrease the reliance on external inputs while promoting long-term soil fertility and resilience (Fenz et al., 2023). Crop rotation interrupts the cycles of pests and diseases and keeps soil nutrients from being depleted. Rotation aids in maintaining a balanced nutrient profile in the soil since various crops have varying nutrient requirements (Niether et al., 2023).

13.2.2 Cover Cropping

Cover cropping is a sustainable agricultural practice widely used in organic farming. It involves planting specific crops, known as green manure or cover crops, to protect and cover the soil during periods when the main cash crops are not growing (Crystal-Ornelas et al., 2021). Cover crops are planted in the off-season to enhance soil structure, protect the soil from erosion, and prevent weed growth. Additionally fixing nitrogen from the atmosphere, legume cover crops improve soil fertility (Seitz et al., 2023).

13.2.3 Green Manure

Growing particular crops, usually legumes or cover crops, and then adding them to the soil while they are still green and actively growing is known as "green manure" (Jalilian et al., 2022). Incorporating green manure crops into the soil is an excellent practice to enhance soil fertility and structure. Green manures are cover crops that are grown specifically to be ploughed and incorporated into the soil (Rautaray et al., 2020). Leguminous green manure crops, including clover and vetch, have a unique ability to form a symbiotic relationship with nitrogen-fixing bacteria and fix atmospheric nitrogen (Carr et al., 2020).

13.2.4 Organic Matter and Compost

Adding compost is an excellent practice for improving soil health and fertility. Compost is a rich source of organic matter, and its benefits extend to various aspects of soil quality. When using compost, it's important to ensure that it is well-rotted and fully decomposed to avoid potential issues with nutrient availability and the introduction of pathogens (Nakasaki et al., 2019). Additionally, incorporating compost as part of a holistic soil management strategy, along with other practices like cover cropping and proper irrigation, can yield the best results for sustainable and productive agriculture (Krause et al., 2022).

13.2.5 Mulching

In organic farming, mulching is the process of adding an organic or inorganic layer to the soil. Mulch, this layer, benefits crops and the farming system as a whole in a number of ways. Mulching is especially crucial in organic farming since it supports the ideas of weed control, soil quality, and sustainability (Amipara & Makwana, 2023). Utilizing organic mulches is a beneficial and sustainable practice for maintaining soil health and promoting optimal conditions for plant growth. Organic mulches hold soil moisture, regulate soil temperature, and suppress weeds (straw, hay, or cover crop residues). Mulching also aids in the prevention of soil erosion and compaction (Khan et al., 2022).

13.2.6 No-Till or Reduced Tillage

No-till and reduced tillage are sustainable farming practices that minimize soil disturbance compared to conventional tillage (Krauss et al., 2022). Choosing between no-till and reduced tillage depends such as climate, type of soil, crop rotation, and farmer preferences. Minimize soil disturbance as much as possible to protect the beneficial organisms and soil structure (Zhang et al., 2020). Reduced tillage practices preserve soil biodiversity, lessen erosion, and help to retain moisture (Lowry & Brainard, 2019).

13.2.7 Biological Control

Encouraging natural predators and beneficial organisms is an ecologically friendly and sustainable approach to pest and disease control in agriculture (Elnahal et al., 2022). By fostering a healthy soil ecosystem, farmers can harness the power of nature to maintain a balance between pests and their natural enemies. A diverse range of organisms that contribute to pest and disease control can be found in healthy soil ecosystems (Tariq et al., 2020).

13.2.8 Minimize Synthetic Inputs

This practice aims to reduce the reliance on synthetic or chemical inputs such as fertilizers, pesticides, and herbicides, promoting environmental health, biodiversity, and long-term sustainability (Richard et al., 2022). Organic farming depends on natural inputs and biological processes to sustain soil fertility and address pest issues (Bellone et al., 2023).

13.2.9 pH Management

Regularly check and regulate soil pH to ensure it aligns with the needs of the crops being cultivated. Employ organic amendments such as lime or sulphur to organically adjust soil pH levels as necessary (Dai et al., 2019).

13.2.10 Water Management

Sustainable water use in agriculture requires a combination of methods and strategies that are adapted to local conditions. In order to achieve effective water management and ensure the resilience of agricultural systems in the metamorphosis of environmental conditions, farmers, researchers, policymakers, and local communities all play critical roles (Klaimala et al., 2022). Adopt water-efficient irrigation methods to conserve water and mitigate soil erosion. Additionally, apply mulching and cover crops to retain soil moisture effectively (Parizad & Bera, 2023).

13.2.11 Crop Diversity

The range of plant species and varieties grown in agricultural systems is referred to as crop diversity, or agrobiodiversity. It is a fundamental aspect of environmentally, economically, and socially responsible agriculture. Diversify crops to promote a variety of root structures and nutrient requirements, contributing to overall soil health (Gentsch et al., 2020).

13.2.12 Biopesticides and Biofertilizers

Biopesticides and Biofertilizers are eco-friendly alternatives to pesticides and chemical fertilizers, respectively. Biopesticides and Biofertilizers contribute to sustainable agriculture by reducing the reliance on synthetic inputs and minimizing environmental impacts. When integrated into well-designed agricultural systems, they can enhance soil fertility, improve crop health, and support overall ecosystem resilience (Gupta et al., 2023). The development of biotechnology has made it possible to produce biopesticides and fertilizers (Parewa et al., 2021). These products may include natural compounds or advantageous microorganisms that support soil fertility and organic, sustainable pest control (Kamboj et al., 2019).

13.2.13 Soil Microbial Biotechnology

Soil microbial communities can be improved through the use of biotechnological instruments. This could entail applying microbial amendments or inoculants, which encourage the development of advantageous microorganisms and enhance soil health and nutrient cycling (Hesham et al., 2021).

13.2.14 Precision Agriculture

Precision agriculture can benefit from biotechnology, which includes the use of sensors, drones, and other cutting-edge technologies (Raj et al., 2022). Farmers can increase overall productivity, minimize their pact on the environment, and make their best use of resources with precision agriculture (Roy & George, 2020). The key is to adopt biotechnological interventions that are compatible with organic farming principles, promote sustainability, and minimize potential risks to the nature. Regulatory frameworks and certification standards for organic products often have guidelines in place to ensure the compatibility of any biotechnological interventions with organic farming practices.

Different observations of studies of soil management practices related to organic farming are listed in Table 13.1.

13.3 WEED MANAGEMENT IN ORGANIC FARMING

Improper weed management in agricultural systems has its negative impact with regard to nutrient unavailability, disease incidence, and overall loss in crop yield (Peerzada et al., 2019). Weeds generated the greatest potential loss globally, accounting for approximately 34% of losses, compared to other biotic variables of pests and microbial pathogens (Mahawar et al., 2023). In conventional agriculture, chemical based herbicides and weedicides are being used, which has several disadvantages with regard to deprivation of soil nutrient status. The application of herbicides and weedicides can cause the soil to become either alkaline or acidic (Srivastava et al., 2020). Weedicides have the potential to cause rashes, allergies, and severe skin and/or eye irritation to farm workers (Wakeil et al., 2013). Additional drawbacks are that when they are consumed by people without being sterilized and completely cleaned with water, they could cause cancer. Because the chemicals are soluble in water, an overabundance of water can wash them into neighboring bodies of water, where they can spread water-borne illnesses like diarrhea (Mishra, 2023). They have the ability to eradicate the beneficial soil's microorganisms which results in alteration of soil fertility. Herbicide/weedicide exposure in crop plants can result in decreased

TABLE 13.1
Observations of Studies in Different Soil Management Practices

#	Particular	Observation	Reference
1	Crop rotation and cover crop	Cover cropping is a key practice to increase system net primary productivity and increase the quantity of high-quality plant residues available for integration into soil organic matter (SOM). Crop rotation practices and local environmental conditions, however, influence the magnitude of soil C stock change.	McClelland et al. (2021)
2	Green manuring and composting	It was observed that green manure's primary effects on soil quality improvement related to an increase in soil organic matter (SOM) levels, nutrient availability and cation exchange capacity (CEC), organic acid formation, a decrease in Al^{3+} levels, and the recycling and mobilization of nutrients from deeper soil layers.	Iderawumiand Kamal (2022)
3	No tillage and reduced tillage	The experiment demonstrated that reduced tillage decreased soil erosion in organic agriculture compared to conventional agriculture and was mainly driven by soil surface cover and SOM.	Seitz et al. (2019)
4	Biofertilisers and biopesticides	Improve soil fertility and control various insects, pests, and disease in agricultural crops	Boudhand Singh (2019)
5	Precision agriculture	Adopting this technology-intensive system on a large scale could increase productivity and resource savings and reduce environmental impact. It does, however, require assistance from the Global Positioning System, various sensors for monitoring soil moisture, nutrients, etc., and geo-referenced maps for different soil properties.	Roy and George (2020)
6	Biological control	IPM and organic are compatible approaches to agricultural production that both rely upon biological control as one tool for producers to use.	Baker et al. (2020)

production of amino acids, reduced cell division, and increased internal herbicide concentrations (Kalia & Gosal, 2011).

As compared to conventional agriculture, in organic agriculture, cultural practices typically have a slower manifesting influence on crop-weed interactions (Thakur et al., 2023). Therefore, rather than focusing solely on crop wise based weed control strategy, weed management should be approached over a longer period of time and require thorough integration with other cultural practices in order to improve the entire cropping system. Given its effects on soil, nutrient, pest, and weed control, cover crop management is significant in this regard (Nordblom et al., 2023). It is emphasized that preventive and cultural weed management must be used in conjunction with physical based direct weed control to reduce weed emergence by choosing the right crop sequence, cover crops, tillage etc. (Cordeau, 2022). The adoption of organic farming has been hampered by worries about possible increases in weed populations in the absence of herbicide use. But as the public's desire for organic food and the awareness of organic farming has grown recently, so too has the selection of weed control strategies (Casagrande et al., 2016). Overall, the chemical based weedicides and herbicides have a detrimental effect on soil fertility, living organisms, and the surrounding environment.

In the context of negative impact of chemical based weedicides/herbicides, scientists are forced to come up with novel solutions for safe food production and environmentally friendly weed management trends. New herbicides with safer environmental and toxicological characteristics are becoming more and more necessary. The term bioherbicides refers to a class of herbicides that are on basis of natural substances and have the efficiency to be a safe alternative for the environment (de Souza Barros et al., 2021). Similarly, products made from living things or their secondary metabolites that decrease weed populations without endangering the environment are referred to as bioherbicides (Soltys et al., 2013; Maes et al., 2021). Research in certain parts of the world is on phytopathogen based bioherbicides which are mixed with adjuvants and made as a formulation that can boost herbicidal efficacy, which is another exciting area of study (Souza et al., 2021). Since chemical weed management is still the most common agronomic practice, it is imperative that new herbicides be developed as the effectiveness of the presently available ones is rapidly declining (Duke et al., 2018).

Biological herbicides can minimize the effects of synthetic herbicides on the environment and minimize their use. Bioherbicides need to be consistently effective, reasonably priced, and efficient in order to be successful. The most well-known examples of natural bioherbicides are phytotoxic compounds found in the extracts of mustard (*Brassica* sp.), sunflower (*Helianthus annuus* L.), green radish (*Raphanussativus*), annual wormwood (*Artemisia annua L.*), bottle brush (*Callistemon citrinus*), eucalyptus (*Eucalyptus* sp.), and sorghum (*Sorghum bicolor* (L.) Moench.), which can be used to protect crop plants without reducing the yield (Motmainna et al., 2021; Soltys et al., 2013). The identified plant based compounds with herbicidal activity are artemisinin, limonene, glucosinolates, isothiocyanates, leptospermone, momilactone, sarmentine, sorgoleone, sarmentine, and essential oils of different plant species (Pantovic & Secanski, 2023).

Apart from plant based herbicides/weedicides, there are currently 24 bioherbicides registered worldwide that are based on fungi, bacteria, or viruses (Aneja et al., 2017). A number of additional microbial candidates are being evaluated as potential bioherbicides at different stages. There are many articles discussing the use of essential oils as herbicides; for example, manuka oil, which comes from the bush *Leptospermnumscoparium*, is used as bioherbicide (Duke et al., 2022). Plant, bacterial, fungal, and viral based bioherbicides/bioweedicides which were reported from the parts of the world are represented in Table 13.2.

13.4 BIOFERTILIZERS IN ORGANIC FARMING

Biofertilizers are substances containing living microorganisms which, when applied to seeds, plant surfaces, or soil, colonize the rhizosphere or interior of the plant and

TABLE 13.2
Bacterial, Fungal, and Plant Based Bioherbicides

#	Group of Organism	Source of Bioherbicide	Action	Weed Control	Reference
1	**Bacteria**	*Pseudomonas aeruginosa*	Reduced germination; prevention of development of seedlings	*Sorghum Amaranthus hybridus; Pennisetum purpureum Echinocholoa crus-galli*	Adetunji et al. (2023), Lawrance et al. (2019)
		Pseudomonas fluorescens	Reduced seed germination and root growth	*Aegilops cylindrica*, *Taeniatherum Caput-Medusae Bromus tectorum*	Kennedy (2018), Reinhart et al. (2020), Tekiela (2020)
		Streptomyces sp.	Growth inhibition	Broad leaved weeds	Aneja et al. (2017)
		Curtobacterium spp.	Changes in protein metabolism	*Petunia* spp.	Pugazhendhi et al. (2019)
		Xanthomonas campestris	Inhibition of growth	*Poaannua*	Aneja et al. (2017)
2	**Fungi**	*Acremonium diospyri*	Inhibition of growth	*Diospyrosvirginiana*	Aneja et al. (2017)
		Alternaria eichhorniae	Capacity to lower water hyacinth biomass and growth rate	*Eichhorniacrassipes*	Shabana (2005)
		Alternaria cassaie	Inhibition of growth	*Cassia* spp.	Aneja et al. (2017)
		Alternaria alternata	Blocks electron transport in photosystem II	Monocots and dicots	Chen and Qiang (2017)
		Ascochyta agropyrina	Inhibition of root growth	*Cirsium arvense, Mercurialis annua Chenopodium album, Sonchus oleraceus, Setaria viridis*	Cimmino et al. (2013)
		Colletotrichum gloeosporiodes sp. malvae	Inhibition of growth	*Malvapusilla*	Aneja et al. (2017)
		Diaporthegulyae	Necrosis	*Papaverrhoes*	
		Drechslera avenacea	Inhibition of Growth and Death	*Avenafatua*	Hetherington et al. (2002)
		Fusarium oxysporum	Causing the death of weeds	*Orobanche* spp.	Aybeke (2017)
		Fusarium fujikuroi	Growth inhibition with reduced plant height	*Sorghum bicolour Cucumissativus*	Daniel et al. (2018)
		Gibbago trianthemae	Inhibition of growth	*Trainthemaportulacastrum*	Aneja et al. (2017)

(*Continued*)

TABLE 13.2 *(Continued)*
Bacterial, Fungal, and Plant Based Bioherbicides

#	Group of Organism	Source of Bioherbicide	Action	Weed Control	Reference
		Phoma macrostoma	Photodegradation chlorosis and mortality	*Taraxacum officinale Cirsium arvense*	Bailey et al. (2011), Hubbard et al. (2016)
		Phytphtherapalmivora	Inhibition of growth	*Morreniaodorata*	Aneja et al. (2017)
		Sprisoriumophiuri	Control of weeds	*Rottboelliacochinchinensis*	Babu et al. (2004)
		Trichoderma koningiopsis	Yellowing of leaves	*Euphorbia heterophylla*	Bordin et al. (2018)
3	**Plants**	*Achilleasantolina*	Inhibition of growth and changes in the metabolic process	*Medicagopolymorpha*	Algandaby et al. (2018)
		Ailanthus altissima	Inhibition of germination and growth	*M. sativa*	Tsao et al. (2002)
		Brassica napus	Inhibition of germination of weed plants by reducing root growth	*Convolvulus arvensis, Phalaris minor,*	Aliki et al. (2014)
		Carum carvi	Leaf damage through biochemical changes	*Echinochloa crusgalli*	Synowiec et al. (2019)
		Chrysanthemum coronarium	Inhibition of seed germination and seedling growth	*Sinapis arvensis Phalaris canariensis,*	Hosni et al. (2013)
		Citrus sinensis	Inhibition of growth	*Solanumnigrum*	Verdeguer et al. (2020)
		Cynara cardunculus	Inhibition of seed germination of weed plantys	*Phalaris minor Trifolium incarnatum, Silybum marianum*	Kaab et al. (2020)
		Eucalyptus citriodora	Inhibition of plant growth and germination	*Cenchrus echinatus*	Ootani et al. (2017)
		Cymbopogon nardus	Inhibition of seed germination	*Digitaria horizontalis*	Ootani et al. (2017)
		Juglans nigra	Growth inhibition	*Conyza canadensis*	Shrestha (2009)
		Pinus densiflora	Root and shoot growth inhibition	*Lolium multiflorum, Digitaria sanguinalis*	Kato-Noguchi et al. (2017)
		Mimosa pigra	Root growth retardation	*Lactuca sativa, Ruellia tuberosa*	Koodkaew et al. (2018)
		Rosmarinus officinalis L. and *Thujaoccidentalis*	Inhibition of seed germination	*P. oleracea., Acroptilon repens*	Ramezani et al. (2008)
		Syzygium aromaticum	Inhibition of growth	*E. crus*galli	Verdeguer et al. (2020)

promote growth by increasing the availability of essential nutrients to the host plant (Ahamad et al., 2023). Unlike chemical fertilizers, biofertilizers work symbiotically with plants, enhancing nutrient uptake, improving soil fertility, and promoting sustainable agricultural practices. They play a key role in biological nitrogen fixation, phosphorus solubilization, and other processes that contribute to soil health and plant growth (Rohela & Saini, 2022). The use of biofertilizers is an environmentally friendly and sustainable approach to enhance crop productivity while minimizing the negative impact on ecosystems. Biofertilizers play an important role in organic farming by enhancing soil fertility and promoting sustainable agricultural practices (Bhardwaj et al., 2014). Unlike chemical fertilizers, biofertilizers are derived from living organisms and work in harmony with the natural processes of the soil. Here are some key types of biofertilizers commonly used in organic farming:

13.4.1 Nitrogen-Fixing Biofertilizers

13.4.1.1 *Rhizobium* spp.

These bacteria form symbiotic relationships with leguminous plants, such as peas and beans. They colonize the plant's roots, fix atmospheric nitrogen into a useful form that the plants might use for their growth. The symbiotic relationship between *Rhizobium* and leguminous plants is an excellent example of biological nitrogen fixation, a natural process that benefits both the bacteria and the plants involved (**Raza et al., 2020**). This mutualistic association is essential for sustainable agriculture and has been widely exploited to improve soil fertility and crop production.

13.4.1.2 *Azotobacter* spp.

These free-living bacteria convert atmospheric nitrogen into ammonia, making it available for plants. They are versatile and can associate with a variety of crops. Unlike Rhizobium, which forms a symbiotic relationship with leguminous plants, Azotobacter lives freely in the soil and contributes to nitrogen fixation in a non-symbiotic manner (Reinhart et al., 2020; Aasfar et al., 2021). It is a versatile and beneficial bacterium that enhances soil fertility and supports plant growth. Its ability to fix nitrogen in a non-symbiotic manner makes it valuable for a wide range of crops and agricultural systems.

13.4.1.3 *Azospirillum* spp.

Azospirillum is a genus of nitrogen-fixing bacteria that forms associative symbiotic relationships with the roots of certain plants, which primarily are grasses and cereals (Pedraza et al., 2020). These bacteria colonize the non-leguminous plant roots and facilitate nitrogen uptake from the soil which produce growth-promoting substances. These bacteria are known for their ability to enhance plant growth and development through various mechanisms that improve nutrient availability and uptake. Its use as a biofertilizers contributes to sustainable and environmentally friendly agricultural practices.

13.4.2 Phosphate-Solubilizing Biofertilizers

13.4.2.1 Phosphate-Solubilizing Bacteria (PSB)

Those are groups of microorganisms that play a crucial role in accessing phosphorus more available to plants by solubilizing insoluble forms of phosphate. Phosphorus is an essential nutrient for plant growth and development, but it is often found in forms that are not readily accessible to plants. They help in the conversion of these insoluble phosphates into forms that plants can absorb (Arif et al., 2017). This enhances phosphorus availability, crucial for plant growth and improving soil fertility. Diverse microbial communities from rhizospheric soil samples from Kashmir valley, India were isolated and characterized with high P solubilizing potential (Bashir et al., 2017, 2018b, 2018c, 2019a). Their use as biofertilizers is an eco-friendly approach that contributes to sustainable agriculture.

13.4.2.2 Mycorrhizal Fungi

They are the group of beneficial fungi that form symbiotic associations with the roots of most plants such as mulberry (Ozturk et al., 2024). These symbiotic relationships, known as mycorrhizae, are mutually beneficial, as both the fungi and the plants receive advantages from the association. There are different types of mycorrhizal fungi, with the two main types being ectomycorrhizal (ECM) and arbuscular mycorrhizal (AM) fungi. These fungi form symbiotic relationships with plant roots, extending their reach into the soil and improving nutrient absorption, including phosphorus (Khaliq et al., 2022). Mycorrhizal fungi form essential symbiotic relationships with plants, providing a range of benefits that contribute to nutrient cycling, improved water absorbance, soil health, disease resistance, stress tolerance, carbon sequestration, promotion of biodiversity, and the overall sustainability of ecosystems (Wahab et al., 2023).

13.4.3 Potassium-Mobilizing Biofertilizers

Potassium-mobilizing bacteria (KMB): Potassium-mobilizing bacteria (KMB) are a group of bacteria that play a key role in making potassium more available to plants by solubilizing or mobilizing potassium from mineral sources in the soil (Bashir et al., 2018a, 2019a, 2019b, 2021; Bashir, 2021; Gautam et al., 2022). Potassium is an essential nutrient for plants, contributing to various physiological processes, such as osmoregulation, enzyme activation, and the maintenance of cell turgor (Pandey & Mahiwal, 2020). Potassium-mobilizing bacteria have a beneficial role in improving potassium availability to plants, supporting plant growth, and contributing to sustainable agriculture. Their use as biofertilizers is one of the strategies aimed at optimizing nutrient management in agricultural systems.

13.4.4 Sulphur-Oxidizing Biofertilizers

Sulphur-oxidizing biofertilizers are a type of biofertilizer that involves sulphur-oxidizing bacteria, which play a crucial role in making sulphur more available to plants in a form that they can easily assimilate. Sulphur is an essential nutrient for plants, playing a critical role in various biochemical processes, including amino acid and protein synthesis. Sulphur-oxidizing bacteria contribute to the sulphur cycle in the soil, making this nutrient more accessible to plants. *Thiobacillus* bacteria have the ability to oxidize sulphur, making it available to plants. Sulphur-oxidizing biofertilizers play a crucial role in improving sulphur availability to plants, supporting plant growth, and contributing to sustainable and environmentally friendly agricultural practices.

13.4.5 Biofertilizers for Disease Suppression

Biofertilizers have a role in disease suppression by promoting the growth of beneficial microorganisms that antagonize or compete with plant pathogens (El-Saadony et al., 2022). The use of biofertilizers for disease suppression is part of a broader approach known as biological control or biocontrol. Here are some biofertilizers and mechanisms through which they contribute to disease suppression:

a. ***Trichoderma* spp.** (Abdul-Halim et al., 2023):
   - *Trichoderma* is a genus of fungi that can act as a biofertilizer and has biocontrol properties.
   - Some *Trichoderma* species produce enzymes that can degrade the cell walls of pathogenic fungi, suppressing their growth.
   - *Trichoderma* can also outcompete pathogenic fungi for nutrients and space.

b. ***Bacillus subtilis*** (Luo et al., 2022):
   - *Bacillus subtilis* is a well-known biofertilizer that has biocontrol properties.
   - It produces antibiotics and antifungal compounds, inhibiting the growth of certain plant pathogens.
   - *Bacillus subtilis* can also induce systemic resistance in plants.

c. ***Pseudomonas fluorescens*** (Gade & Koche, 2022):
   - *Pseudomonas fluorescens* is a beneficial bacterium that acts as a biofertilizer and has biocontrol capabilities.
   - It produces siderophores, which sequester iron and limit its availability to pathogenic bacteria.
   - *Pseudomonas fluorescens* can also produce antifungal compounds.

d. ***Streptomyces* spp.** (Du et al., 2022):
   - *Streptomyces* species are soil bacteria that can act as biofertilizers and produce antibiotics.
   - Some *Streptomyces* strains have antifungal properties, suppressing the growth of pathogenic fungi in the rhizosphere.

e. **Combinations of Beneficial Microorganisms:**
   - Formulations containing a mix of beneficial microorganisms, such as bacteria and fungi, can provide a broader spectrum of disease suppression.
   - These combinations may include mycorrhizal fungi, PGPR, and other biocontrol agents.

f. **Biofertilizer Seed Coating:**
   - Applying biofertilizers as seed coatings can introduce beneficial microorganisms directly to the rhizosphere, enhancing disease suppression from the early stages of plant growth.

The effectiveness of biofertilizers in disease suppression can depend on various factors, including the specific pathogens present, environmental conditions, and the compatibility of the biofertilizer with the target crop. Integrated pest management (IPM) strategies often incorporate the use of biofertilizers for sustainable and environmentally friendly disease control in agriculture.

13.4.6 Azolla as a Nitrogen Fixer

Azolla is a small aquatic fern that has gained attention for its remarkable ability to fix atmospheric nitrogen through a symbiotic relationship with the nitrogen-fixing cyanobacterium *Anabaena azollae* (Akhtar et al., 2021). This unique partnership allows Azolla to thrive in nitrogen-deficient water bodies and provides a sustainable source of nitrogen for agriculture. Azolla is adaptable to various climates and can be cultivated in tropical, subtropical, and temperate regions, making it a versatile nitrogen-fixing option for different agricultural settings (Adetunji et al., 2023). Azolla's ability can fix atmospheric nitrogen through its symbiotic relationship with *Anabaena azollae* which makes it an attractive and sustainable option for enhancing nitrogen availability in agricultural ecosystems, particularly in rice cultivation. Its use as a nitrogen-fixing biofertilizer aligns with environmentally friendly and resource-efficient agricultural practices.

13.4.7 Advantages of Biofertilizers

Biofertilizers are advantageous in agriculture for several reasons. Biofertilizers contribute to sustainable agriculture by improving soil structure, increasing nutrient availability, and promoting plant growth without relying on synthetic chemicals. Integrating biofertilizers into organic farming practices supports ecological balance and reduces environmental impact. Here are some key advantages of using biofertilizers:

- **Environmentally Friendly**: Biofertilizers are derived from natural sources, often consisting of beneficial microorganisms like bacteria, fungi, and algae. Unlike chemical fertilizers, they do not introduce harmful synthetic chemicals into the soil and water, reducing environmental pollution and promoting sustainable agriculture.
- **Nutrient Cycling**: Biofertilizers play a role in nutrient cycling. They help in the breakdown of organic matter, making essential nutrients more available to plants. This enhances soil fertility over time and reduces the need for external inputs.
- **Nitrogen Fixation**: Certain types of biofertilizers, such as nitrogen-fixing bacteria like *Rhizobium* and *Azotobacter*, can convert atmospheric nitrogen into a form that plants can use. This reduces the reliance on synthetic nitrogen fertilizers, which are energy-intensive to produce and can contribute to environmental issues.
- **Improved Soil Structure**: Biofertilizers can enhance soil structure by promoting the growth of beneficial microorganisms. This improves soil

aeration, water retention, and overall soil health, leading to better plant growth.

- **Disease Suppression**: Some biofertilizers contain microorganisms that can suppress the growth of pathogenic organisms. This helps in disease control and reduces the dependence on chemical pesticides, contributing to sustainable and eco-friendly agricultural practices.
- **Cost-Effectiveness**: While there may be initial costs associated with the production and application of biofertilizers, in the long run, they can be cost-effective. Reduced dependence on chemical fertilizers and pesticides can lead to lower input costs for farmers.
- **Promotion of Beneficial Microbial Activity**: Biofertilizers stimulate the activity of beneficial microorganisms in the soil. This microbial activity contributes to the breakdown of organic matter, the release of nutrients, and the overall improvement of soil health.
- **Adaptability to Organic Farming**: Biofertilizers are compatible with organic farming practices. As they are derived from natural sources, their use aligns with organic certification standards and helps farmers meet the growing demand for organic produce.
- **Reduced Soil Degradation**: Continuous use of chemical fertilizers can lead to soil degradation and loss of fertility over time. Biofertilizers promote sustainable agriculture by maintaining soil health and preventing long-term damage.
- **Biodiversity Conservation**: The use of biofertilizers supports the conservation of biodiversity in the soil. The diverse microbial communities in biofertilizers contribute to a balanced and healthy ecosystem within the soil.

It's important to note that the effectiveness of biofertilizers can vary depending on factors such as climate, soil type, and crop species. Additionally, proper application methods and management practices are essential to maximize the benefits of biofertilizers.

13.5 BIOPESTICIDES IN ORGANIC FARMING

The extensive utility of synthetic fertilizers and chemical pesticides' subsequent damage to the humans, other non-target organisms, and environment led to the concept of organic farming with the logic to repair, maintain, and improve the ecological balance. The major components of organic farming include crop rotation, crop residues, organic manures, biopesticides, biofertilizers etc. Biopesticides are basically naturally occurring organisms that regulate the population density of insect pests and are an important component of organic farming which not only keeps the humans, animals, and environment safe but also enhances agro-ecosystem health and maintains ecological balance. The three main categories of biopesticides include the Plant Incorporated Protectants (PIPs), biochemical biopesticides, and microbial biopesticides. The use of entomopathogens viz. viruses, bacteria, fungi, and nematodes as microbial biopesticides has increased over the years, and a number of biopesticides based on entomopathogens have been developed commercially.

The microbial biopesticides that are available include *Nucleopolyhedrovirus* (NPV), *Granulovirus* (GV), *Metarhizium anisopliae*, *Steinernema carpocapsae*, *Heterorhabditis bacteriophora*, *Bacillus thuringiensis*, *Beauveria bassiana*, etc (Table 13.3). They are compatible with other pest management strategies and can be isolated, mass multiplied, and applied under field conditions or developed as commercial formulations with inert ingredients to increase their efficacy and persistence (Gani et al., 2022). They cause lethal (infecting and killing) and sub-lethal effects (interfere with the growth, development, reproductive attributes, and behavior) on the host populations and are highly specific in their action. Their targeted approach and natural degradation are important for preserving biodiversity and maintaining

TABLE 13.3
Entomopathogens as Biopesticides with Their Trade Name and Target Pest

Entomopathogen	Trade Name	Target Pests
Entomopathogenic Bacteria		
Bacillus thuringiensis subsp. *aizawai*	Florbac, Agree, Design, Xentari	Lepidoptera
B. thuringiensis subsp. *israelensis*	Tekar, Bactimos, Acrobe	Diptera
B. thuringiensis subsp. *kurstaki*	Biobit, Thuricide, Bactospeine, Javelin, Delfin, Costar, Able, Condor, Costar, CRYMAX, Cutlas, Futura, Lepinox, Steward, Halt, Biolep, Bioasap	Lepidoptera
B. thuringiensis subsp. *galleriae*	Spicturin	Lepidoptera
B. thuringiensis subsp. *tenebrionis*	Novodor, Trident	Coleoptera
Paenibacilluspopilliae	Milky Spore Powder	Japanese beetle
Entomopathogenic Fungi		
Beauveriabassiana	Naturalis, Conidia, Botegha, Botanigard, Ostrinil, Boveril, Koppert, Betal	Hemiptera, Thysanoptera, Coleoptera, Mites
Hirsutellathompsonii	Almite, Agrilife Lancer	Mites
Isariafumosorosea	NoFly and PFR-97	Whiteflies, Thrips, Aphids, Mites
Lecanicilliumlecanii	VERTALEC, MYCOTAL, VERTI-SIN	Hemiptera, Thysanoptera
Lecanicilliumlongisporum	Vertalec	Aphids
Metarhiziumanisopliae	MET52 G, BioMagic, Green Muscle, BioBlast, Betel, MET52 EC	Orthoptera, Coleoptera, Hemiptera
Verticilliumlecanii	Vertimax, Shock, Biofire, S-Vertilec, Varunastra, Verticill, Simridhi	Hemiptera, Thysanoptera, mites
Metarhiziumbrunneum	Met52	Ticks
Paecilomyceslilacinus	MeloCon	Plant-parasitic nematodes
Pochoniachlamydosporia	Pochar	Plant-parasitic nematodes

TABLE 13.2 ***(Continued)***
Entomopathogens as Biopesticides with Their Trade Name and Target Pest

Entomopathogen	Trade Name	Target Pests
Entomopathogenic Nematodes		
Heterorhabditisbacteriophora	Nemasys and Terranem	Soil born insect pests
Steinernemacarpocapsae	Ecomask and NemAttack	
Steinernemafeltiae	Entonem, Fungus Gnat and Rootknot Exterminator and Scanmask	
Steinernemacarpocapsae and *Heterorhabditisheliothidis*	Double-Death	
Entomopathogenic viruses		
*Anticarsiagemmatalis*MNPV	Coopervirus SC, Protege, Multigen Baculo-soja, BaculovirusNitral	*Anticarsiagemmatalis*
*Autographacalifornica*NPV	Loopex	*Autographacalifornica, Trichoplusiani*
*Autographacalifornica*MNPV + *Spodopteraalbula*NPV	VPN-ULTRA	*Heliothisvirescens, Spodopteraexigua, Estigmeneacrea, Plutellaxylostella Autographacalifornica, Trichoplusiani, Pseudoplusiaincludens,*
*Spodopterasunia*NPV	VPN 82	*Spodoptera* spp.
*Cydia Pomonella*GV	Carpovirusine, Madex Twin, Madex Plus, Madex Top	*Cydiapomonella, Grapholitamolesta*
*Adoxophyesorana*GV	Capex	*Adoxophyesorana*
*Erinnyisello*GV	*Baculoviruserinnyis,*	*Erinnyisello*
*Helicoverpazea*SNPV	Gemstar, HzNPVCCAB	*Heliothis*and *Helicoverpa*spp.
*Helicoverpaarmigera*NPV	Diplomata, Helicovex	*Heliothis*and *Helicoverpa*spp.
*Lymantriadispar*MNPV	Gypchek	*Lymantriadispar*
*Phthorimaeaoperculella*GV	BaculovirusCorpoica, PTMbaculovirus, Tutavir	*Phthorimaeaoperculella, Teciasolanivora, Tutaabsoluta*
*Phthorimaeaoperculella*GV + *Bacillusthuringiensis*	Matapol Plus, Bacu-Turin	*Phthorimaeaoperculella, Teciasolanivora, Symmetrischematangolias*
*Spodopteraexigua*NPV	SPOD-X LC, Spexit	*Spodopteraexigua*
SpodopterafrugiperdaMNPV	Spodovir	*Spodopterafrugiperda*
CryptophlebialeucotretaGV	Cryptex	*Thaumatotibialeucotreta*
*Spodopteralittoralis*NPV	Littovir	*Spodopteralittoralis, Spodopterafrugiperda*
*Neodiprionabietis*NPV	Abietiv	*Neodiprionabietis*

ecological balance. The biopesticides are the suitable alternatives to chemical pesticides and the growing awareness about the harmful effects of chemical pesticides on human health, environment, and other non-target organisms has led to a rising demand for biopesticides in recent years. Although biopesticides have gained significant importance in recent years they still account for just 4.2% of the total pesticide market in India (Chakraborty et al., 2023). The biopesticides market is anticipated to reach 13.9 billion USD by 2028 from 6.7 billion USD by 2023 at a CAGR of 15.9 % (Anonymous, 2023). The prominent biopesticides companies include the BASF SE, Bayer AG, Syngenta, Koppert, Nufarm, Isagro S. P.A., Certis USA L.L.C., Novozymes, UPL Limited, FMC Corporation etc.

The awareness about the biopesticides has increased in recent years as basic and applied research has greatly enhanced their effectiveness. Also, techniques for their isolation, mass production, storage, and application have been improved in recent years. Thus, there is a great scope for setting up more biopesticide production units to meet the demand in the future. The biopesticide production units should be set up in areas which have appropriate climatic conditions and away from farming, industrial, and urban areas. Further, for better efficacy of biopesticides, the local organisms should be preferred in any area as they are already acclimatized to the environmental conditions prevalent in that area (Manzoor et al., 2023). The research on the development and evaluation of novel biopesticide formulations with improved insecticidal efficacy through the inclusion of different compounds, more virulent isolates, and also through genetic engineering and nanotechnology approaches is going on to expand their use in insect pest management programs for organic farming.

The biopesticides are being registered and regulated under the Insecticides Act, 1968, and only 12 different types of biopesticides have been registered in India under this act (Rajni et al., 2022). They include the following:

- *Bacillus thuringiensis* var. galleriae
- *Pseudomonas fluorescens*
- Neem based pesticides
- *Cymbopogon* NPV of *Helicoverpaarmigera*
- NPV of *Spodopteralitura*
- *Bacillus thuringiensis* var. israelensis
- *Beauveriabassiana*
- *Trichodermaviride*
- *Trichodermaharzianum*
- *Bacillus sphaericus*
- *Bacillus thuringiensis* var. kurstaki

13.5.1 Advantage of Biopesticides

- Safe to the humans, environment, and other non-target organisms including beneficial insects
- No harmful/toxic residues
- Ability to multiply in the target hosts and remain active for a long period of time

- Host specific
- Compatible with pest management strategies and can be easily incorporated into IPM programs thereby causing lesser usage of pesticides
- Cost effective
- Lower probability of hosts to develop resistance against biopesticides
- Reduce the selection pressure for the evolution of pesticide resistance in host populations
- Effective in small quantities and decompose quickly

13.5.2 Disadvantages of Biopesticides

- Slow speed of kill and low efficacy
- Susceptibility to adverse environmental conditions
- Need for additional management strategies
- Difficulty in mass production

13.6 CROP DIVERSITY/CROP DIVERSIFICATION IN ORGANIC FARMING

Currently, agriculture is mired in a myriad of issues, primarily brought about by farming's biased development that placed a strong focus on intensification. Developing new fundamental ideas for farming is the first step toward more sustainable farming practices. Integrated and organic agricultural systems are now the two main ideologies that best exemplify such integrative methods. In order to create and execute these more environmentally friendly farming practices, new ideas in science, technology, and education are required (Hufnagel et al., 2020). It is necessary to construct a hierarchy of objectives for either ecological (long-term alternative) farming systems or an integrated (short-term alternative) based on the examination of the flaws of existing farming and the prospects in the future (Figure 13.2).

The relationship between the crop rotation and crop protection is shown in Figure 13.2. To achieve the associated goals, the agricultural practices must be

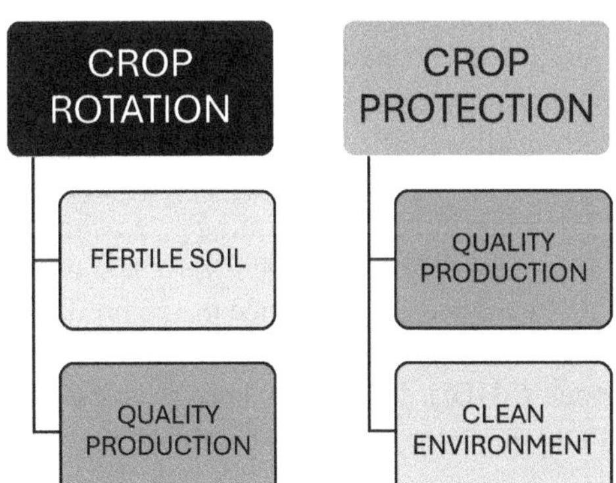

FIGURE 13.2 Relationship between the crop rotation and crop protection.

revised. As a result, each approach requires both the overall strategy and the concretization of a combination of techniques and procedures. This frequently calls for the creation of new technologies or a critical synthesis of the body of current knowledge. Overcoming goals that appear to be at odds with one another is the challenge in this process. In the end, a new method is created as a logical, multi-objective approach that is safe, adaptable, and makes use of a wide range of tools, depending on the growing season and the particulars of the farm.

13.6.1 Centralized Role of Crop Rotation in Organic Farming

Tanveer et al. (2019) described crop rotation as the practice of switching around the crop species grown on the same plot of land. In general, rotation breaks the cycles of disease and weed growth and improves soil quality, which increases crop output. Rotation is a crucial part of sustainable agriculture. For instance, in comparison to monoculture, barley dry weights and seed yields were higher in the barley-forage rotation in all years by 29 and 26%, respectively. In India, carrying out organic farming has grown in importance. A nationwide network focused on innovation and the transition to organic farming was the farms' natural next step. It will be easier for the groups to collaborate closely on REC and practice knowledge, expertise, and innovation in organic farming. Moreover, a sharp and clear picture will emerge of the threats and opportunities for organic farming. Opportunities will be used and threats counteracted and solved (Mukherjee et al., 2022).

Crop rotation has two aspects: the crops cultivated this year and their split over the available area, and the crops grown throughout time in a set order (succession of crops in time). The notion of crop rotation may be reinforced by leveraging the interplay between spatial and temporal dimensions. A variety of crops are produced in a certain order as part of the crop rotation process. If a precise rotational plan is devised, with an ideal squad and lineup in mind, the value contributed by this group of players (crops) grows. Some of the players can only score well if their performance is carefully prepared by others (preceding crop). The target of the crop rotation is offering appropriate, optimal, and homogenous conditions to all the players in the team. This is the basis for sustainability quality production.

As illustrated in the Figure 13.3, the main purposes of crop rotation are to maintain and increase soil fertility and to prevent and manage weeds, diseases, and pests (Wijnands, 1999). With the least amount of external input pesticides, fossil fuels, machines, and fertilizers an optimally high-quality crop may be ensured. Crop rotation determines which crops are healthy and vigorous, which is the foundation of quality output.

Diversification of cropping in organic farming is a central agroecological principle, which may improve resource use efficiency, reduce pests and diseases, diversify income sources, and enhance the resilience of the production (Rodriguez et al., 2021). Crop diversification between organic and conventional farming systems use Shannon Diversity Index (SHDI). It captures both richness (number) and evenness (abundance) of crops and analyzes impact of livelihood assets on it using ordinary least square model.

Crop rotation plays a central and crucial role in the basic design of an organic farm. It is not only the major weapon to prevent and control pests, diseases, and

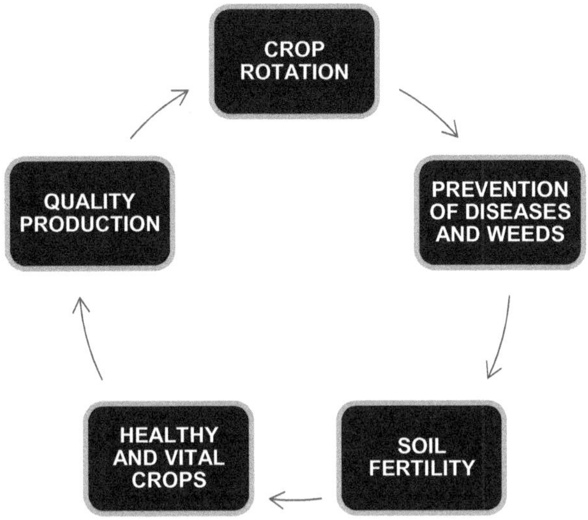

FIGURE 13.3 The main purposes of crop rotation.

weeds, but it is also the basis for maintenance and improvement of soil fertility. Crop rotation appears to most farmers to be merely an idea or an abstract concept rather than a tangible, meticulously thought out and executed reality. Tscharntke et al. (2021), in their practice crops in general are changed a lot, too often crops change from preceding crop and the crops grown together in one course in the rotation are not only often not of comparable demands and characteristics but they also keep on changing. This leads to fragmented parcels in terms of soil fertility and also leads to heterogeneity and thus endangers the sustainability of quality production.

13.7 ROLE OF GENOME EDITING TECHNOLOGY IN MAKING ORGANIC AGRICULTURE MORE SUSTAINABLE

CRISPR-Cas9 (Clustered Regularly Interspaced Short Palindromic Repeats and CRISPR-associated protein 9) is a revolutionary genome editing technology that allows precise modification of DNA within living organisms (Asmamaw & Zawdie, 2021). It has transformed genetic engineering and offers a powerful tool for manipulating the genetic information of various organisms, including plants, animals, and microorganisms. CRISPR-Cas9 has democratized genome editing, making it more accessible and versatile, with implications for various scientific, medical, and agricultural advancements (Aljabali et al., 2024). Researchers continue to refine and expand the technology for enhanced precision and safety. CRISPR-Cas9 based genome editing technology has the potential to play a significant role in making organic agriculture more sustainable. Organic agriculture aims to minimize the use of synthetic inputs, promote ecological balance, and enhance soil fertility. It's important to note that while CRISPR-Cas9 provides powerful tools for crop improvement, ethical and regulatory considerations are essential to ensure the responsible use of this technology. Transparency, public engagement, and adherence to ethical guidelines contribute to building trust and acceptance of genetically edited crops (Gordon et al., 2021). Genome editing can contribute to these goals in several ways:

13.7.1 Crop Improvement through Genome Editing: CRISPR-Cas9 genome editing technology has revolutionized the field of crop improvement by providing a precise and efficient way to modify specific genes within the plant genome (Kaur et al., 2021). This technology allows scientists to edit the DNA of crops with unprecedented accuracy, enabling the development of improved varieties with desirable traits (Iqbal et al., 2020). Here are some ways in which CRISPR-Cas9 contributes to crop improvement:

- **Disease Resistance:** CRISPR-Cas9 can be used to introduce or enhance resistance to various diseases in crops. By targeting specific genes associated with susceptibility to pathogens, scientists can create crops that are more resilient and require fewer chemical inputs for disease control (Bisht et al., 2019).
- **Pest Resistance:** Targeting genes involved in interactions with pests allows for the development of crops with increased resistance to insect damage (Bisht et al., 2019). This reduces the need for synthetic pesticides, aligning with sustainable and environmentally friendly agricultural practices.
- **Abiotic Stress Tolerance:** CRISPR-Cas9 can be employed to enhance a plant's ability to withstand abiotic stresses such as drought, salinity, and extreme temperatures (Li et al., 2022). By editing genes involved in stress response pathways, crops can be made more resilient to challenging environmental conditions.
- **Improved Nutritional Content:** Researchers can use CRISPR-Cas9 to modify genes related to nutritional content, resulting in crops with enhanced levels of vitamins, minerals, and other beneficial compounds (Priti et al., 2023). This process, known as biofortification, addresses nutritional deficiencies and contributes to improved public health.
- **Optimized Crop Yield:** CRISPR-Cas9 technology enables the modification of genes involved in plant growth and development, leading to increased crop yields (Sami et al., 2021). By fine-tuning these genetic factors, scientists can create different crop varieties with improved productivity and resource-use efficiency.
- **Reduced Allergenicity:** Genetic modifications using CRISPR-Cas9 can be applied to reduce or eliminate allergenic components in certain crops (Camerlengo et al., 2020). This can be particularly important for enhancing the safety and acceptance of specific food products.
- **Faster Breeding Process:** Traditional breeding methods can be time-consuming. CRISPR-Cas9 accelerates the breeding process by allowing precise modifications to be made in a shorter timeframe (Ahmar et al., 2023). This speedier development of improved crop varieties is crucial for addressing global food security challenges.

- **Precision in Trait Modification:** CRISPR-Cas9 offers unparalleled precision in gene editing (Aljabali et al., 2024). This precision allows scientists to modify specific traits without introducing undesired changes, resulting in more predictable and controllable outcomes.

13.7.1 Nutritional Enhancement through Genome Editing

Genome editing in the context of organic farming has the potential to contribute to nutritional enhancement by improving crop traits such as nutrient content, resistance to diseases, and overall crop performance (Kumar et al., 2021). While genetic modification has been a topic of debate, genome editing techniques, such as CRISPR-Cas9, offer precise and targeted modifications to the plant genome without introducing foreign genes. Here are some ways in which genome editing can enhance the nutritional profile of organic crops:

- **Increased Nutrient Content:** Genome editing can be used to enhance the levels of essential nutrients in crops, such as vitamins, minerals, and antioxidants (Gaikwad et al., 2020). For example, researchers can modify the genes responsible for nutrient synthesis or accumulation to increase the nutritional content of organic crops.
- **Biofortification**: Biofortification involves enhancing the nutrient levels in crops to address specific nutritional deficiencies in human diets. Genome editing can be employed to increase the levels of key nutrients like iron, zinc, and vitamin A in organic crops, making them more nutritionally valuable (Qiao et al., 2016; Bhambhani et al., 2021).
- **Improved Crop Resistance**: By editing the genes associated with disease resistance, organic crops can be made more resilient to pests and pathogens (Ntui et al., 2023). This can reduce the need for chemical pesticides, aligning with the principles of organic farming while maintaining the crop's overall health and nutritional content.
- **Optimized Plant Metabolism**: Genome editing allows for the modification of metabolic pathways within plants (Dellero, 2020). This can be used to improve the efficiency of nutrient utilization and assimilation, ensuring that the organic crops make better use of available resources and nutrients.
- **Enhanced Stress Tolerance**: Genome editing can help develop crops that are more resilient to environmental stressors such as drought, salinity, or extreme temperatures (Yadav et al., 2023). This ensures a more stable yield and helps maintain the nutritional quality of organic crops even under challenging conditions.
- **Reduced Anti-Nutritional Factors**: Some plants naturally contain anti-nutritional factors that can interfere with the absorption of nutrients in the human body. Genome editing can be employed to reduce or eliminate these factors, improving the bioavailability of nutrients in organic crops (Bhambhani et al., 2021).
- **Faster Crop Development**: Traditional breeding methods for improving crop traits can be time-consuming. Genome editing allows for the

acceleration of crop development by directly targeting specific genes, resulting in crops with improved nutritional profiles in a shorter time frame (Zhang et al., 2018).

13.7.2 Reduced Environmental Impact through Genome Editing

Genome editing in organic farming has the potential to reduce environmental impact in several ways, promoting sustainable agricultural practices. Here are some ways in which genome editing can contribute to a more environmentally friendly organic farming system:

- **Reduced Need for Pesticides:** Genome editing can enhance the resistance of crops to pests and diseases, reducing the reliance on chemical pesticides in organic farming (Gossen & McDonald, 2020). This approach aligns with the principles of organic agriculture, which emphasizes natural and holistic methods of pest and disease management.
- **Improved Resource Use Efficiency:** By editing genes related to nutrient uptake and utilization, crops can be optimized for better resource use efficiency (Hawkesford & Griffiths, 2019). This can lead to reduced nutrient runoff into water bodies, minimizing the risk of water pollution and improving the overall sustainability of organic farming practices.
- **Enhanced Stress Tolerance:** Genome-edited crops with improved tolerance to environmental stressors, such as drought or salinity, can thrive in challenging conditions (Li et al., 2022). This can lead to more reliable crop yields, even in regions prone to adverse environmental conditions, contributing to the resilience of organic farming systems.
- **Decreased Land Requirements:** Through genome editing, crops can be modified to achieve higher yields on existing agricultural land (Hüdig et al., 2022). This can reduce the need for expanding agricultural areas into natural habitats, helping to preserve biodiversity and protect ecosystems.
- **Lower Greenhouse Gas Emissions:** Increased crop resilience and improved resource use efficiency can contribute to lower greenhouse gas emissions associated with organic farming (Niggli et al., 2009). For example, optimizing nitrogen utilization in crops can reduce the need for nitrogen based fertilizers, which are associated with greenhouse gas emissions.
- **Mitigation of Climate Change Impacts:** Genome editing can be employed to develop crops that are better adapted to changing climatic conditions (Yadav et al., 2023). This adaptability can help mitigate the impacts of climate change on organic farming, ensuring the continued success of agricultural practices in the face of environmental challenges.
- **Conservation of Soil Health:** Genome editing can contribute to the development of crops with improved root structures and nutrient acquisition capabilities (Li et al., 2016). This can enhance soil health, reduce erosion, and minimize the need for intensive soil management practices, promoting sustainable organic farming.

- **Minimized Crop Wastage:** Genome editing can lead to the development of crops with extended shelf life and improved post-harvest characteristics (Kumari et al., 2022). This can contribute to reducing food waste, which is an important environmental consideration.

13.7.3 Enhanced Adaptability through Genome Editing

Genome editing offers a powerful tool for enhancing the adaptability of crop plants to changing environmental conditions. Through precise modifications to the plant's genome, researchers can confer traits that improve resilience, stress tolerance, and overall adaptability. Here are several ways in which genome editing can contribute to enhanced adaptability in crop plants:

- **Climate Adaptation:** Genome editing can be employed to introduce or enhance traits that help crops adapt to specific environmental conditions, such as temperature and soil variations (Yadav et al., 2023). This can be crucial for sustainable organic agriculture in diverse geographical regions.
- **Adaptation to Altered Growing Conditions**: As climate patterns shift, crops need to adapt to new environmental conditions. Genome editing facilitates the introduction of specific genetic changes to help crops adjust to altered growing conditions, ensuring stable yields (Bailey-Serres et al., 2019).
- **Improved Water Use Efficiency:** Genome editing can target genes involved in water uptake and utilization, leading to crops that use water more efficiently (Sathee et al., 2022). This is important in regions facing water scarcity or where irrigation is a common practice.
- **Enhanced Adaptability to Sustainable Agricultural Practices:** Genome editing can be employed to develop crops that adapt well to sustainable farming practices, such as no-till agriculture or organic farming (Vanderschuren et al., 2023). This promotes environmentally friendly and resource-efficient cultivation methods. It's important to note that while CRISPR-Cas9 provides powerful tools for crop improvement, ethical and regulatory considerations are essential to ensure the responsible use of this technology. Transparency, public engagement, and adherence to ethical guidelines contribute to building trust and acceptance of genetically edited crops. While genome editing holds promise for reducing the environmental impact of organic farming, it's essential to approach its implementation with caution, considering ethical, social, and regulatory aspects. Open communication and engagement with stakeholders, including farmers, consumers, and environmental organizations, are crucial for the responsible adoption of these technologies in organic agriculture. It's important to note that the use of genome editing in organic farming should align with the principles and regulations of organic agriculture. Transparency, ethical considerations, and public engagement are crucial aspects when implementing such technologies to ensure acceptance within the organic farming community and among consumers.

13.8 CLIMATE RESILIENT AGRICULTURE THROUGH BIOTECHNOLOGICAL INTERVENTIONS

Agriculture resilience to climate change interventions is a serious matter that should be the focus of all economic, political, scientific, and public interventions to assure sustainable life on this planet (Iqbal et al., 2023). Climate resilient agriculture refers to the development and implementation of agricultural practices and technologies that enable crops and farming systems to withstand and adapt to the challenges posed by climate change. It involves adopting practices and implementing technologies that enable the agricultural sector to cope with and adapt to the challenges posed by climate change (Munawar et al., 2020). The goal is to enhance the capacity of farming systems to maintain or improve productivity, sustainability, and food security in the face of changing climatic conditions. Biotechnological interventions play a crucial role in achieving climate resilience in agriculture. Here are some key ways in which biotechnology can contribute:

13.8.1 Drought-tolerant crops

Biotechnology can be used to develop crops that are more resistant to drought conditions. Genetic engineering techniques, such as the introduction of drought-responsive genes, can help create plants that require less water to thrive (Varshney et al., 2011). Biotechnological interventions for drought-resistant crops by

13.8.1.1 Genetic Modification for Drought Tolerance

Scientists can identify and incorporate specific genes responsible for drought tolerance into crop plants. For instance: Drought-Responsive Genes: Genes that regulate water use efficiency, water uptake, and water retention in plants can be identified and introduced into crops. These genes might include those involved in the synthesis of osmoprotectants, which help plants cope with water stress.

13.8.2 Transgenic Approaches

- **Overexpression of Drought-Responsive Genes**: Certain genes naturally produced by plants in response to drought stress can be overexpressed to enhance the plant's ability to withstand water scarcity. Examples include genes encoding for aquaporins (water channel proteins) and late embryogenesis abundant (LEA) proteins (Rani et al., 2021).
- **Introducing Drought-Responsive Transgenes**: Researchers can introduce genes from other drought-tolerant organisms, such as desert plants or bacteria, into crop plants to confer improved drought resistance.

13.8.3 Marker-Assisted Breeding

Identifying and Selecting Drought-Resistant Traits: Biotechnology tools can be used to identify genetic markers associated with drought resistance (Varshney et al., 2011). This information is then used in conventional breeding programs to select plants with improved drought resistance.

13.8.4 Crops with Altered Root Systems

Enhanced Root Architecture: Biotechnology can be employed to modify the root systems of crops, promoting deeper and more extensive root structures. This helps plants access water from deeper soil layers, making them more resilient to drought conditions.

13.8.5 Genome Editing Techniques

CRISPR-Cas9 and Other Editing Tools: Genome editing tools like CRISPR-Cas9 allow for precise modification of specific genes associated with drought tolerance. Researchers can edit the plant genome to enhance its ability to cope with limited water availability.

13.8.6 Synthetic Biology Approaches

Designing Artificial Pathways: Synthetic biology techniques enable the creation of artificial pathways in plants that enhance their ability to tolerate drought stress. This could involve the introduction of novel genes or pathways not naturally found in the plant.

13.8.7 Biochemical Approaches

Metabolic Engineering: Biotechnological methods can be used to modify the metabolic pathways in plants to enhance their ability to produce stress-responsive compounds, such as osmoprotectants or antioxidants, which help combat the effects of drought (Roy et al., 2021). It's important to note that the development and deployment of drought-resistant crops should be approached cautiously, considering environmental, ethical, and regulatory aspects. While biotechnology offers promising solutions, it should be integrated with other sustainable agricultural practices for comprehensive and responsible crop improvement.

13.8.8 Heat-tolerant and Flood Resistant Crops

Rising temperatures due to climate change can negatively impact crop yields. Biotechnology can be employed to develop crops that are more heat-tolerant. Genetic modifications can enhance the ability of plants to withstand high temperatures and maintain optimal growth under stressful conditions. In areas prone to flooding, the introduction of genes that enhance a plant's ability to survive in flooded soils can be used to develop crops with improved tolerance to waterlogged conditions.

13.8.9 Salt-tolerant Crops

Salinity in soil is a major concern for agriculture, particularly in coastal regions. Biotechnological interventions can help develop crops that can thrive in saline soils. This involves the introduction of genes that enable plants to tolerate high levels of salt.

13.8.10 Improved Nutrient Utilization

Biotechnology can be used to enhance the nutrient uptake efficiency of crops. This includes the development of plants that can better utilize nutrients present in the soil, leading to improved yields and reduced environmental impact (Iqbal et al., 2023).

13.8.11 Carbon Sequestration

Certain crops and plants can be engineered to capture and store more carbon dioxide, helping to mitigate the effects of climate change. This can contribute to the reduction of greenhouse gas levels in the atmosphere (Garland & Curry, 2022).

13.8.12 Precision Farming

Biotechnological tools, such as genetically modified organisms (GMOs) and gene editing techniques can be integrated with precision agriculture practices. Conservation and sustainable use of genetic resources and traditional knowledge of germplasm is essential. Advances in biotechnology, especially in the areas of *in vitro* culture techniques, cryopreservation, and molecular biology, have generated new opportunities for improved conservation, utilization, and management of climate resilient agriculture. Biomolecurization is considered the basis for sustaining life on the planet, playing as important of a role as bioeconomy. It has been predicted that the 21st century will be an era of bio-economy driven by advances in bioscience and biotechnology. Bio-economy may become the fourth economy after agricultural, industrial, and information technology, having far reaching impacts on sustainable development in agriculture, forestry, environmental protection, light industry, food supply, health care, and other micro-economy aspects. Thus, a strategic and forward vision for climate resilient agriculture in the 21st century and 22nd century is of far reaching significance for sustainable development.

13.9 ORGANIC FARMING AND REBUILDING OF CROPS BY NEW BREEDING TECHNIQUES

New breeding methods and organic farming are two different approaches to agriculture, each with its own set of values and methods. But they all aim to advance environmentally friendly and sustainable agriculture. Let's examine each idea and how it helps to rebuild crops. The foundation of organic farming is an emphasis on biodiversity, soil health, and ecological balance. Genetically modified organisms (GMOs) and synthetic pesticides and herbicides are not used in this method (Reeve et al., 2016; Röös et al., 2018). Rather, in order to preserve soil fertility and manage pests, organic farmers rely on organic fertilizers, companion planting, crop rotation, and other natural techniques (Lakner & Breustedt, 2017). By promoting healthier and more resilient ecosystems, organic farming aids in the rebuilding of agricultural production. Beneficial soil organisms, including fungi, bacteria, and earthworms, are protected by organic practices, which refrain from using synthetic chemicals. This raises the availability of nutrients and improves soil structure, which benefits crops' overall health (Dhiman, 2020).

13.9.1 Attributes of Organic Farming (Sustainable Practices)

13.9.1.1 No Synthetic Chemicals

Organic farming avoids synthetic pesticides, herbicides, and fertilizers, relying on natural alternatives to control pests and enhance soil fertility (Singh, 2021). Organic farmers rely on natural methods and materials to manage pests and diseases, improve soil fertility, and promote overall crop health. Instead of synthetic fertilizers, organic farmers use natural fertilizers like compost, manure, and other organic matter to enhance soil fertility (McGuire, 2017).

13.9.1.2 Crop Rotation and Polyculture

Organic farmers often use crop rotation and polyculture to maintain soil health and reduce the risk of pests and diseases. Rotating crops helps break pest and disease cycles, reducing the need for synthetic interventions. Planting complementary crops can help deter pests and promote a healthier growing environment (Adamczewska-Sowińska & Sowiński, 2020).

13.9.2 Soil Health

13.9.2.1 Compost and Organic Matter

Organic farming emphasizes the use of compost and organic matter to improve soil structure, water retention, and nutrient content (Aldebron et al., 2020). According to Cooper et al. (2020), organic matter in the form of plant residues, cover crops, and other decaying materials contributes to the formation of humus. Humus is a stable organic component that enhances soil fertility and structure. Compost is a rich source of essential nutrients, including nitrogen, phosphorus, potassium, and various micronutrients. These nutrients are released slowly, providing a sustained and balanced supply to plants (Cooper et al., 2020).

13.9.2.2 No Soil Erosion

Practices like cover cropping and reduced tillage help prevent soil erosion, maintaining long-term soil health. Cover crops protect the soil from erosion by preventing raindrop impact, reducing surface runoff, and enhancing soil structure. Mulching helps control soil erosion by reducing water runoff, minimizing the impact of raindrops on the soil, and improving water infiltration (Werf et al., 2020). Organic farming often promotes reduced tillage or no-till practices to minimize soil disturbance (Martínez-Mena et al., 2020).

13.9.3 Biodiversity

13.9.3.1 Preservation of Ecosystems

Organic farms typically support higher levels of biodiversity, preserving natural ecosystems and promoting a healthier environment. Organic farming avoids the use of synthetic pesticides, herbicides, and fertilizers. Instead, it relies on natural alternatives, such as compost, cover crops, and beneficial insects. This helps prevent the pollution of soil and water, protecting aquatic ecosystems and soil-dwelling organisms (Wittwer et al., 2021). According to Das et al. (2020), organic farmers often practice crop rotation, which involves planting different crops in a specific sequence over time. This helps break pest and disease cycles, improves soil fertility, and promotes biodiversity in the agricultural landscape.

13.9.3.2 No Genetically Modified Organisms

Organic farming prohibits the use of genetically modified organisms (GMOs), supporting natural biodiversity (Husaini & Sohail, 2018). Organic farming standards and certifications typically prohibit the use of genetically modified seeds or organisms in the cultivation process. The aim is to promote natural, traditional breeding methods and maintain the integrity of the genetic makeup of crops (Kothamasi & Vermeylen, 2011).

13.9.4 Reduced Environmental Impact

13.9.4.1 Lower Carbon Footprint

Organic farming tends to have a lower carbon footprint as it avoids the use of synthetic chemicals that contribute to greenhouse gas emissions. Organic farming practices such as cover cropping, crop rotation, and reduced tillage help improve soil structure and increase organic matter content. This leads to enhanced carbon sequestration in the soil, acting as a carbon sink and mitigating the release of greenhouse gases into the atmosphere (Adewale et al., 2018).

13.9.4.2 Water Conservation

Organic practices often involve water conservation methods, reducing the environmental impact of agriculture. Organic farmers often use cover crops, which are grown during periods when the main cash crop is not in the field (Sivaranjani & Rakshit, 2019). Cover crops help reduce soil erosion, improve soil structure, and increase water infiltration. By protecting the soil, cover crops contribute to water conservation. The use of compost and mulch in organic farming helps improve soil water retention (Mandal et al., 2021).

13.9.5 New Breeding Techniques (NBTs)

New breeding techniques (NBTs) encompass a range of advanced methods for developing plant varieties with desired traits. These techniques include CRISPR-Cas9 gene editing, RNA interference, and other precision breeding methods. Unlike traditional genetic modification, NBTs allow for more precise modifications at the molecular level, often without introducing foreign genes (Crespo-Herrera & Ortiz, 2015). With the help of NBTs, crops with better qualities like greater yield, improved nutritional value, and resistance to pests and diseases may develop more quickly. Breeders can overcome obstacles more quickly by precisely targeting particular genes (Bharadwaj, 2016). In order to rebuild agricultural systems in the face of climate change, NBTs can be useful tools for creating crops that are more suited to face climate change (Lammerts van Bueren et al., 2018).

New breeding methods and organic farming are essential for sustainable agriculture, addressing environmental issues, and ensuring food security.

13.9.6 New Breeding Techniques

13.9.6.1 Precision Breeding

13.9.6.1.1 CRISPR and Gene Editing

CRISPR (Clustered Regularly Interspaced Short Palindromic Repeats) technology has revolutionized the field of genetic engineering and gene editing, including its application in precision breeding (Ahmad et al., 2021). New breeding techniques like CRISPR-Cas9 enable precise modification of plant genomes, allowing for targeted improvements in traits such as disease resistance, drought tolerance, and nutritional content (Enfissi et al., 2021).

13.9.6.1.2 Accelerated Breeding

Accelerated breeding refers to the use of advanced technologies and techniques to speed up the traditional breeding process in plants and animals. The goal is to achieve desired traits in a shorter time frame compared to conventional breeding methods. Techniques like marker-assisted selection speed up the breeding process by identifying desirable traits more efficiently (Cordovil et al., 2020).

13.9.7 Improved Crop Varieties

13.9.7.1 Resilience to Climate Change

Resilience involves enhancing adaptive capacity, which refers to the ability of systems to adjust to changing climate conditions. This includes the capacity to modify practices, make informed decisions, and implement strategies that reduce vulnerability to climate-related impacts. New breeding techniques allow the development of crops better adapted to changing climatic conditions, helping farmers mitigate the impacts of climate change (Garland & Curry, 2022).

13.9.7.2 Disease Resistance

Traits like resistance to pests and diseases can be incorporated more efficiently using advanced breeding methods. New Breeding Techniques (NBTs), including advanced genetic engineering methods like CRISPR-Cas9, offer innovative ways to enhance disease resistance in crops and livestock (Deng et al., 2020). These techniques enable precise modifications of the genetic material, allowing for the development of plants and animals with improved resistance to diseases. NBTs make it feasible to stack multiple resistance genes within the same organism. Stacking genes with different modes of action provides a broader and more durable defense against a range of pathogens, reducing the risk of resistance development in the pathogens (Qaim, 2020).

13.9.8 Increased Productivity

13.9.8.1 Enhanced Yield

By selectively breeding crops with higher yields and improved agronomic traits, new breeding techniques contribute to increased agricultural productivity. NBTs can be used to modify genes responsible for growth and development, influencing factors like plant height, branching patterns, and fruit development (Saini et al., 2020). These modifications can result in plants that are more efficient in utilizing resources and energy for enhanced yield.

13.9.8.2 Reduced Resource Requirements

Crops engineered for better resource efficiency can contribute to sustainable agriculture by requiring fewer inputs such as water and fertilizers. NBTs can be employed to enhance the efficiency of photosynthesis in crops (Maitra & Gitari, 2020). Improved photosynthetic efficiency can lead to increased biomass production using the same or lower amounts of sunlight and carbon dioxide. According to Richard et al. (2022) NBTs offer high precision in introducing specific traits, minimizing unintended effects on the plant's physiology or unintended environmental consequences. This precision contributes to more targeted resource utilization.

13.9.9 Addressing Global Challenges

13.9.9.1 Nutritional Enhancement

New breeding techniques offer the potential to enhance the nutritional content of crops, addressing global issues related to malnutrition and food quality. NBTs can be used for biofortification, which involves enhancing the nutritional content of crops (Chandra et al., 2020). According to Sarkar et al. (2021) specific genes associated with the production of vitamins and minerals can be targeted for modification using NBTs. This approach has been applied to enhance the levels of key nutrients such as vitamin A, iron, zinc, and folate in crops. NBTs can be used to modify the amino acid composition of crops, improving the nutritional quality of plant proteins.

13.9.9.2 Reduced Food Waste

Crops with improved shelf life and resistance to post-harvest losses can contribute to reducing food waste. NBTs can be used to develop crops with improved resistance to pests and diseases, reducing the likelihood of post-harvest losses due to spoilage. Crops with enhanced resistance can potentially have a longer shelf life, providing more time for distribution and consumption. NBTs enable the customization of traits related to post-harvest storage requirements. For example, modifications can be made to control ripening processes, extend freshness, or improve resistance to bruising during handling (Hörz, 2022).

13.9.10 Integration of Organic Farming and NBTs

Organic farming and new breeding techniques both play important roles in the sustainable development of agriculture. By adopting practices that prioritize environmental health and utilizing advanced breeding methods, farmers can work towards rebuilding crops that are resilient, productive, and adapted to future challenges (Yadav et al., 2021). Combining organic farming practices with NBTs can be a holistic approach. For instance, using NBTs to enhance disease resistance in organic crops could reduce the need for chemical interventions while maintaining organic farming principles (Sarkar et al., 2021). The integration of these approaches should be approached with caution, considering ethical, environmental, and regulatory aspects. Striking a balance between innovation and sustainability is crucial for the long-term success of agriculture. The combination of organic farming practices and new breeding techniques offers a holistic approach to sustainable agriculture. By emphasizing environmentally friendly practices and leveraging cutting-edge technologies, these approaches aim to ensure long-term food security while minimizing negative

TABLE 13.4
Rebuilding Crops with New Breeding Methods and Using Organic Farming Practices

Particulars	Observations	Reference
1. Organic interventions conferring stress tolerance and crop quality	A high demand exists for nutrient-rich organic food items. Modern organic agriculture has demonstrated benefits in terms of food/biomass production, soil health, biodiversity, climate resilience, nutritional security, and a high quality of life.	Sarkar et al. (2021)
2. Molecular techniques from simple breeding to speed breeding in crop plants	From traditional to modern speed breeding (clustered regularly interspaced short palindromic repeats (CRISPR), and CRISPR-associated (Cas) proteins) combined with genome editing tools, which aim to produce crop generations with desired traits annually	Ahmar et al. (2020)
3. New plant breeding technologies for food security and sustainable agricultural development	New plant breeding technologies (NPBTs), including genetically modified and gene-edited crops, offer great potential for sustainable agricultural development and food security while resolving inadequacies of the Green Revolution.	Qaim (2020)
4. Crop biotechnology in food production	Through agricultural biotechnology, researchers can improve food crop features by utilizing site-directed nucleases and other cutting-edge new plant breeding methods, such as genomic analysis.	Steinwand and Ronald (2020)

impacts on the environment (Paramesh et al., 2022). Some of the experimental observations are mentioned in Table 13.4.

13.10 ADVANTAGES OF ORGANIC FARMING

13.10.1 High Premium on Products

Realizing the ill effects of consuming chemical foods, consumers prefer to buy organic food products, have higher perception of these products, and are willing to pay additional money for them (Rahman et al., 2021). Further, the growing awareness about the advantages of organic products over the conventional agricultural products have not only increased the consumer base but also marketing opportunities as the organic market is now the fastest growing market all over the world including India (Das et al., 2020). The demand of organic food is more than that of its supply and this factor boosts up the price of organic food. Therefore, when the farmers sell organic foods in the market then they earn more income. The profits are more due to organic premiums (Smith et al., 2019) and are normally priced 10–40% higher than conventional food (Winter & Davis, 2006). This premium is very important for a farmer whose income is just sufficient to feed his/her family with one meal. This is an important advantage of organic farming due to which many farmers turn toward this method. This also makes organic farming more competitive in addition to its efficiency in production (Suharyati et al., 2016).

13.10.2 Low Investment in Farming

Organic farming is very cost effective and does not involve capital investment as high as that required in chemical farming. The costs of production are reduced in comparison to traditional farming (Gaur, 2016; Shepherd, 2003) and relying mainly on internal resources and limiting the input of external auxiliary materials. This results in less fossil and mineral resources being consumed (Kumar et al., 2021). Resources like fertilizers and pesticides are produced locally by recycling various farm resources such as cattle shed wastes, cow dung, cow urine, droppings of goats and sheep, and fish wastes as manures, and crop wastes such as sugar cane trashes, dried leaves, green manure, green leaves manure, and weeds as a mulching material as well as manure. Hence farmers realize reduced cost on external inputs as one of the major advantages (Sudheer, 2023).

13.10.3 Livelihood Opportunities

In developing countries, labour is relatively inexpensive in comparison to input costs, favouring the transformation to less input dependent, but more labour intensive production processes (Soni et al., 2022). As organic farming requires more labour, countries like India where the agriculture is highly influenced by the vagaries of various biotic and abiotic factors, organic farming is capable enough to provide economic security to the mediocre farmers as well (Roychowdhury et al., 2014), resulting in more income-generating employment per farm (Halberg, 2008). Additionally, market oriented organic agriculture projects create more options for paid work locally therefore benefitting women also by providing livelihood opportunities within the vicinity of their households (Figure 13.4).

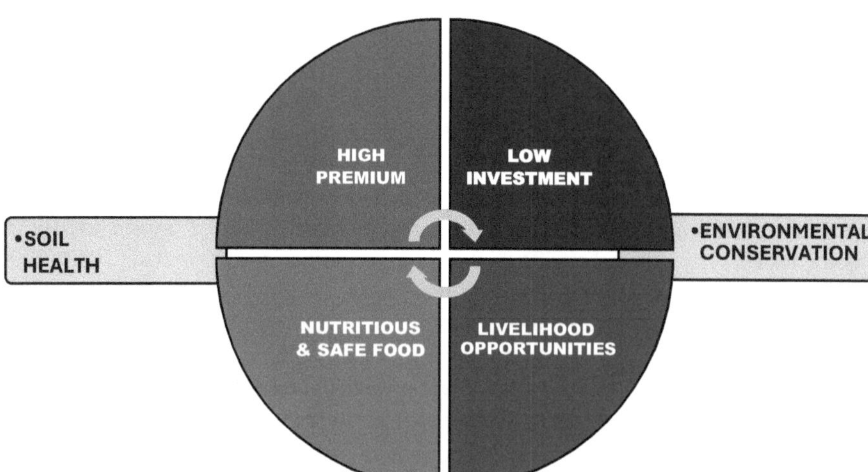

FIGURE 13.4 The main purposes of crop rotation.

13.10.4 Nutritious and Safe Food

As advantages of organic agriculture go beyond the purely financial ones (Gulkazi et al., 2014) organic foods are being preferred over conventionally grown foods (Suharyati et al., 2016). Preference relies on the nutritional superiority and safety standards of food grown under organic conditions (Ullah et al., 2015; Zahedipour et al., 2019) as use of herbicides, fungicides, synthetic fertilizers, and pesticides is completely restricted in case of organic farming (Reganold & Wachter, 2016). This gives an advantage to organic farming over conventional farming as the food produced does not contain any harmful chemicals (Yanakittkul & Aungvaravong, 2020), artificial flavors, or preservatives, and contains more of vitamin C and phosphorus (Forman et al., 2012), and thereby the people consuming organic foods reduce their risks of physical ailments (Gaur, 2016).

13.10.5 Sustainable Soil Health

Maintaining and improving the long-term soil fertility is one of the priorities of sustainable agriculture (Colom-Gorgues, 2009). Organic farming is one of the solutions which retain the fertility of the soil (Aceleanu, 2016) for a longer period of time and thus allow the farmers to use the land for a longer period for cultivation (Mendez et al., 2019). This starts with minimal tillage, which improves and preserves the soil structure, helps to preserve the microorganisms, prevents the decomposition of humus by avoiding deep and intensive loosening, and improves the elasticity and protection against soil erosion and the water balance (Boltianska et al., 2021). Further use of green manures, green leaves manures, dried leaves, vegetable by-products, algae, and composts (Bergstrand, 2022) increase soil organic matter, soil biochemical and ecological properties along with water holding capacity and prevents the entering of sunlight into the soil and makes the soil sufficient with moisture and improved water holding capacity. Because of its higher ability to store carbon in the soil, organic agriculture represents a means to improve CO_2 abatement if adopted on a large scale (Gomiero et al., 2011).

13.10.6 Environmental Conservation

Overuse of chemical pesticides is a major contributor to biodiversity loss in conventional agricultural systems. As organic agricultural practices strictly avoid the spraying of any synthetic chemicals and pesticides therefore it does not pollute soil, air, and water leading to markedly lower eco-toxicity potentials on the one hand and higher richness of biodiversity on other (Kumar et al., 2021; Oquist et al., 2007; Sushma & Dinesh, 2020). Further, by encouraging the development of native plants and utilizing crop rotation, organic fields encourage biodiversity. This promotes the preservation of robust environments that are home to a variety of species, including pollinators and helpful insects and are expressed in the form of floral diversity, faunal diversity, habitat diversity, soil biological activity etc (Khadda, 2021). Aside from curbing the number of harmful chemicals polluting the water supply, organic farming also reduces the number of greenhouse gasses coming from farms (Holka et al., 2022). In addition, using organic pesticides has reduced the amount of nitrous oxide that gets introduced into the atmosphere. Nitrous oxide is considered one of the most damaging greenhouse gasses and one of the main ingredients in chemical pesticides (Jezierska-Tys et al., 2021; Ntinyari & Gweyi-Onyango, 2021)

13.11 CONCLUSION AND FUTURE PROSPECTS

Intensive farming with unsustainable agricultural practices (chemical based) can lead to reduced biodiversity, disruption of natural soil micro flora, chemical based food contaminations leading to biomagnification of toxic metals in food chains and its health complications. Low-input and organic farming could both reduce this detrimental effect in order to preserve biodiversity and use natural methods to control weeds, insects, and other pests. Increased biodiversity on organic farms leads to improved insect pest control, weed control, and maintenance of soil fertility and with improved yield and quality. Furthermore, in areas where conventional agriculture has historically been used, organic farming may aid in reversing the reductions of habitual species. One of the benefits of organic farming that is most commonly mentioned when contrasting it with the traditional production method is its favourable impact on restoration of biodiversity. The synthesis of bio compounds possessing herbicidal properties ought to ensure their impact on specific weeds, while also taking into account the conservation of the biodiversity in the areas where they are utilized, and promoting increased food and environmental security and quality.

Because organic farming preserves biodiversity, produces highly nutritious crops, and improves soil fertility through crop rotation, all of which can have a significant impact on the agricultural sector by lowering poverty, increasing productivity, and guaranteeing food security, it has become a prominent research topic and progressively increasing trend of implemented agricultural practice in the last ten years. The bioformulations (biopesticides, biofertilizers, and bioherbicides) based organic farming also has certain drawbacks which need to be addressed in achieving the sustainable agricultural practices.

To accomplish the goals of sustainable and eco-friendly agricultural practices, the following list of vital actions should be followed:

- Biotechnology based research should be carried out for climate resilient agriculture through creation of crops which can resist and grow well in changing environments under biotic and abiotic stresses.
- Thorough research should be carried out for identifying the potential plant metabolites with biopesticidal or bioherbicidal properties.
- Assess the products of bioweedicides/bioherbicides/biofertilizers and their formulations, timing, application techniques, and other factors to reduce or eliminate the entry of microorganisms or their by-products into the food chains of animals and humans and its impact on environment.
- Create high-throughput assays to assess the toxicological and biochemical mechanisms of action of plant secondary metabolites which were to be or are being utilized as bioherbicides. This will help with pathogenesis/antagonism role determination, strain selection, and biocontrol range prediction.
- Additionally, protocols for tracking the competency, stability, and effects of bioherbicidal microbes discharged into the environment ought to enhance our comprehension and evaluation of threats to the microbial ecosystem as a whole.

- Biotechnology based products which are applied for improving soil fertility, eradicating pests, and prevention and control of diseases should be produced in a safer and economically viable way.
- To lessen the constraints and difficulties of organic farming, a combination of biochar, enhanced sludge, with vermicompost, biofertilizers, organominerals, and digital technology is essential.
- Soil, season and altitude based crop rotation, inter cropping, mixed cropping, and cover cropping practices should be developed for minimizing the disease and pest infestations.
- Further research on artificial intelligence should be carried out and implemented for real time monitoring of soil health status, irrigation, detection of pests and pathogenic microbial load in crop, vegetative, and floral stage based inputs of mineral nutrients, ripening, and harvest related strategies for better yield.

In conclusion, the future of agriculture majorly depends on a combination of organic farming, biotechnology based safe formulations and genome edited crops, and artificial intelligence based real time monitoring of agricultural practices.

REFERENCES

Aasfar, A., Bargaz, A., Yaakoubi, K., Hilali, A., Bennis, I., Zeroual, Y., & Meftah Kadmiri, I. (2021). Nitrogen fixing *Azotobacter* species as potential soil biological enhancers for crop nutrition and yield stability. *Frontiers in Microbiology*, *12*, 628379.

Abdul-Halim, A. M. A. A., Shivanand, P., Krishnamoorthy, S., & Taha, H. (2023). A review on the biological properties of *Trichoderma* spp. as a prospective biocontrol agent and biofertilizer. *Journal of Applied Biology and Biotechnology*, *11*(5), 34–46. https://doi.org/10.7324/JABB.2023.11504

Aceleanu, M. I. (2016). Sustainability and competitiveness of Romanian farms through organic agriculture. *Sustainability*, *8*(3), 1–19.

Adamczewska-Sowińska, K., & Sowiński, J. (2020). Polyculture management: A crucial system for sustainable agriculture development. *Soil Health Restoration and Management*, 279–319.

Adetunji, C. O., Akram, M., Adetuyi, B. O., Khan, F. S., Rashid, A., Anwar, H., & Yerima, M. B. (2023). Biofertilizer derived from cyanobacterial: Recent advances. In *Next generation algae: Vol. I: Applications in agriculture, food and environment* (pp. 295–320). Wiley Publisher. https://doi.org/10.1002/9781119857839.ch13

Adewale, C., Reganold, J. P., Higgins, S., Evans, R. D., & Carpenter-Boggs, L. (2018). Improving carbon foot printing of agricultural systems: Boundaries, tiers, and organic farming. *Environmental Impact Assessment Review*, *71*, 41–48.

Ahamad, L., Shahid, M., & Danish, M. (2023). Microbial biofertilizers: An environmentally-friendly approach to sustainable agriculture. In *Microbiomes for the Management of Agricultural Sustainability* (pp. 167–182). Springer Nature Switzerland.

Ahmad, S., Tang, L., Shahzad, R., Mawia, A. M., Rao, G. S., Jamil, S., & Tang, S. (2021). CRISPR-based crop improvements: A way forward to achieve zero hunger. *Journal of Agricultural and Food Chemistry*, *69*(30), 8307–8323.

Ahmar, S., Gill, R. A., Jung, K. H., Faheem, A., Qasim, M. U., Mubeen, M., & Zhou, W. (2020). Conventional and molecular techniques from simple breeding to speed breeding in crop plants: Recent advances and future outlook. *International Journal of Molecular Sciences*, *21*(7), 2590.

Ahmar, S., Hensel, G., & Gruszka, D. (2023). CRISPR/Cas9-mediated genome editing techniques and new breeding strategies in cereals-current status, improvements, and perspectives. *Biotechnology Advances*, 108248.

Akanmu, A. O., Olowe, O. M., Phiri, A. T., Nirere, D., Odebode, A. J., Karemera Umuhoza, N. J., & Babalola, O. O. (2023). Bioresources in organic farming: Implications for sustainable agricultural systems. *Horticulturae*, *9*(6), 659.

Akhtar, M., Sarwar, N., Ashraf, A., Ejaz, A., Ali, S., & Rizwan, M. (2021). Beneficial role of Azolla sp. in paddy soils and their use as bioremediators in polluted aqueous environments: Implications and future perspectives. *Archives of Agronomy and Soil Science*, *67*(9), 1242–1255.

Aldebron, C., Jones, M. S., Snyder, W. E., & Blubaugh, C. K. (2020). Soil organic matter links organic farming to enhanced predator evenness. *Biological Control*, *146*, 104278.

Alduvín, C. (2022). Cisgenesis and organic farming. In *Cisgenic crops: Potential and prospects* (pp. 121–143). Springer International Publishing.

Algandaby, M. M., & Salama, M. (2018). Management of the noxious weed; *Medicago polymorpha* L. via allelopathy of some medicinal plants from Taif region, Saudi Arabia. *Saudi Journal of Biological Sciences*, *25*(7), 1339–1347.

Aliki, H. M., Reade, J. P., & Back, M. A. (2014). Effects of concentrations of *Brassica napus* (L.) water extracts on the germination and growth of weed species. *Allelopathy Journal*, *34*(2).

Aljabali, A. A., El-Tanani, M., & Tambuwala, M. M. (2024). Principles of CRISPR-Cas9 technology: Advancements in genome editing and emerging trends in drug delivery. *Journal of Drug Delivery Science and Technology*, 105338.

Amipara, R. P., & Makwana, B. D. (2023). Efficacy of organic inputs on nutrient management and cost-benefit ratio on pulse crops in organic farming. *Journal of Agriculture and Ecology*, *16*, 5–10.

Aneja, K. R., Khan, S. A., & Aneja, A. (2017). Bioherbicides: Strategies, challenges and prospects. *Developments in Fungal Biology and Applied Mycology*, 449–470.

Anonymous. (2023). https://www.marketsandmarkets.com/Market-Reports/biopesticides-267.html

Arif, M. S., Shahzad, S. M., Yasmeen, T., Riaz, M., Ashraf, M., Ashraf, M. A., & Kausar, R. (2017). Improving plant phosphorus (P) acquisition by phosphate-solubilizing bacteria. *Essential Plant Nutrients: Uptake, Use Efficiency, and Management*, 513–556.

Asmamaw, M., & Zawdie, B. (2021). Mechanism and applications of CRISPR/Cas-9-mediated genome editing. *Biologics: Targets and Therapy*, 353–361.

Aybeke, M. (2017). Fusarium infection causes genotoxic disorders and antioxidant-based damages in *Orobanche* spp. *Microbiological Research*, *201*, 46–51.

Babajani, A., Muehlberger, S., Feuerbacher, A., & Wieck, C. (2023). Drivers and challenges of large-scale conversion policies to organic and agro-chemical free agriculture in South Asia. *International Journal of Agricultural Sustainability*, *21*(1), 2262372.

Babu, R. M., Sajeena, A., Seetharaman, K., Vidhyasekaran, P., Rangasamy, P., Prakash, M. S., & Biji, K. R. (2004). Advances in bioherbicides development-An overview. *Crop Protection*, *23*(7), 655–655.

Bahar, N. H., Lo, M., Sanjaya, M., Van Vianen, J., Alexander, P., Ickowitz, A., & Sunderland, T. (2020). Meeting the food security challenge for nine billion people in 2050: What impact on forests. *Global Environmental Change*, *62*, 102056.

Bailey, K. L., Pitt, W. M., Leggett, F., Sheedy, C., & Derby, J. (2011). Determining the infection process of *Phoma macrostoma* that leads to bioherbicidal activity on broad leaved weeds. *Biological Control*, *59*(2), 268–276.

Bailey-Serres, J., Parker, J. E., Ainsworth, E. A., Oldroyd, G. E., & Schroeder, J. I. (2019). Genetic strategies for improving crop yields. *Nature*, *575*(7781), 109–118.

Baker, B. P., Green, T. A., & Loker, A. J. (2020). Biological control and integrated pest management in organic and conventional systems. *Biological Control*, *140*, 104095.

Bashir, Z. (2021). Characterization of potassium solubilizing bacteria from rhizospheric soils of cherry (*Prunusavium*). *Research Journal of Agricultural Sciences*, *12*(2), 494–496.

Bashir, Z., Hamid, B., Zargar, M. Y., & Kundoo, A. A. (2021). Morphological and biochemical characterization of potassium solubilizing bacteria from rhizospheric soils of pinus (*Pinus wallichiana*). *Journal of Research and Development*, *21*, 58–63.

Bashir, Z., Zargar, M. Y., Baba, Z. A., & Mohiddin, F. A. (2018a). Solubilization of potassium containing mineral by bacteria from Kashmir iris (*Iris kashmiriana*) rhizosphere. *The Pharma Innovation Journal*, *17*(4), 747–750.

Bashir, Z., Zargar, M. Y., Baba, Z. A., Mohiddin, F. A., Peer, F. A., Mir, S. A., & Najar, G. R. (2018b). Isolation and characterization of phosphorus solubilizing bacteria (PSB) from rhizospheric soils of apple (*Malus Domestica*). *Journal of Pharmacognosy and Phytochemistry*, *7*(2), 2997–3000.

Bashir, Z., Zargar, M. Y., Bhat, T. A., Kousar, S., Baba, Z. A., & Mohiddin, F. A. (2018c). Solubilization of phosphorus containing mineral by bacteria from rhizospheric region of walnut (*Juglans regia*). *International Journal of Current Microbiology and Applied Sciences*, *7*(8), 2391–2398.

Bashir, Z., Zargar, M. Y., Mohiddin, F. A., Baba, Z. A., & Hamid, B. (2019a). Isolation and biochemical characterization of phosphorus solubilizing bacteria from rhizospheric region of apricot and peach. *Journal of Research and Development University of Kashmir*, *19*, 0972–5407.

Bashir, Z., Zargar, M. Y., Mohiddin, F. A., Kousar, S., Husain, M., & Rasool, F. (2017). Phosphorus solubilizing microorganisms: Mechanism and diversity. *International Journal of Communication Systems*, *5*(5), 666–673.

Bashir, Z., Zargar, M. Y., Vishwakarma, D. K. (2019b). Potassium solubilizing microorganisms for sustainable agriculture. *Applied Agricultural Practices for Mitigating Climate Change*, *2*, 17–28.

Bellone, D., Jeuffroy, M. H., Bertrand, M., Mistou, M. N., Barbu, C., Ballini, E., Morison-Valantin, M., Gauffreteau, A., & Pashalidou, F. G. (2023). Are innovative cropping systems less dependent on synthetic pesticides to treat Septoria leaf blotch (*Zymoseptoria tritici*) than conventional systems? *Crop Protection*, *170*, 106266.

Bergstrand, K. (2022). Organic fertilizers in greenhouse production systems—A review. *Scientia Horticulturae*, *295*.

Bhambhani, S., Kondhare, K. R., & Giri, A. P. (2021). Advanced genome editing strategies for manipulation of plant specialized metabolites pertaining to biofortification. *Phytochemistry Reviews*, 1–19.

Bharadwaj, D. N. (2016). Sustainable agriculture and plant breeding. *Advances in Plant Breeding Strategies: Agronomic, Abiotic and Biotic Stress Sraits*, 3–34.

Bhardwaj, D., Ansari, M. W., Sahoo, R. K., & Tuteja, N. (2014). Biofertilizers function as key player in sustainable agriculture by improving soil fertility, plant tolerance and crop productivity. *Microbial Cell Factories*, *13*, 1–10.

Bisht, D. S., Bhatia, V., & Bhattacharya, R. (2019). Improving plant-resistance to insect-pests and pathogens: The new opportunities through targeted genome editing. *In Seminars in Cell & Developmental Biology Academic Press*, *96*, 65–76.

Boltianska, N., Boltianskyi, O., & Manita, I. (2021). *Environmental benefits of organic agricultural production* (pp. 280–284). ХНТУСГ. https://repo.btu.kharkov.ua//handle/123456789/39749

Bordin, E. R., Frumi Camargo, A., Rossetto, V., Scapini, T., Modkovski, T. A., Weirich, S., Carezia, C., Barretta Franceschetti, M., Balem, A., Golunski, S. M., & Treichel, H. (2018). Non-toxic bioherbicides obtained from *Trichoderma koningiopsis* can be applied to the control of weeds in agriculture crops. *Industrial Biotechnology*, *14*(3), 157–163.

Boudh, S., & Singh, J. S. (2019). Pesticide contamination: Environmental problems and remediation strategies. *Emerging and Eco-Friendly Approaches for Waste Management*, 245–269.

Camerlengo, F., Frittelli, A., Sparks, C., Doherty, A., Martignago, D., Larré, C., & Masci, S. (2020). CRISPR-Cas9 multiplex editing of the α-amylase/trypsin inhibitor genes to reduce allergen proteins in durum wheat. *Frontiers in Sustainable Food Systems*, *4*, 104.

Carr, P. M., Cavigelli, M. A., Darby, H., Delate, K., Eberly, J. O., Fryer, H. K., Gramig, G. G., Heckman, J. R., Mallory, E. B., Reeve, J. R., Silva, E. M., & Woodley, A. L. (2020). Green and animal manure use in organic field crop systems. *Agronomy Journal*, *112*(2), 648–674.

Casagrande, M., Peigné, J., Payet, V., Mäder, P., Sans, F. X., Blanco-Moreno, J. M., Antichi, D., Bàrberi, P., Beeckman, A., Bigongiali, F., Cooper, J., & David, C. (2016). Organic farmers' motivations and challenges for adopting conservation agriculture in Europe. *Organic Agriculture*, *6*, 281–295.

Chakraborty, N., Mitra, R., Pal, S., Ganguly, R., Acharya, K., Minkina, T., Sarkar, A., & Keswani, C. (2023). Biopesticide consumption in India: Insights into the current trends. *Agriculture*, *13*(3), 557.

Chandra, A. K., Kumar, A., Bharati, A., Joshi, R., Agrawal, A., & Kumar, S. (2020). Microbial-assisted and genomic-assisted breeding: A two way approach for the improvement of nutritional quality traits in agricultural crops. *3 Biotech*, *10*, 1–15.

Chen, S., & Qiang, S. (2017). Recent advances in tenuazonic acid as a potential herbicide. *Pesticide Biochemistry and Physiology*, *143*, 252–257.

Cimmino, A., Zonno, M. C., Andolfi, A., Troise, C., Motta, A., Vurro, M., & Evidente, A. (2013). Agropyrenol, a phytotoxic fungal metabolite, and its derivatives: A structure–activity relationship study. *Journal of Agricultural and Food Chemistry*, *61*(8), 1779–1783.

Colom-Gorgues, A. (2009). The challenges of organic production and marketing in Europe and Spain: Innovative marketing for the future with quality and safe food products. *Journal of International Food and Agribusiness Marketing*, *21*(2–3), 166–190.

Cooper, J., Greenberg, I., Ludwig, B., Hippich, L., Fischer, D., Glaser, B., & Kaiser, M. (2020). Effect of biochar and compost on soil properties and organic matter in aggregate size fractions under field conditions. *Agriculture, Ecosystems & Environment*, *295*, 106882.

Cordeau, S. (2022). Conservation agriculture and agroecological weed management. *Agronomy*, *12*(4), 867.

Cordovil, C. M., Bittman, S., Brito, L. M., Goss, M. J., Hunt, D., Serra, J., & Hutchings, N. (2020). Climate-resilient and smart agricultural management tools to cope with climate change-induced soil quality decline. In *Climate Change and Soil Interactions*, 613–662.

Crespo-Herrera, L. A., & Ortiz, R. (2015). Plant breeding for organic agriculture: Something new? *Agriculture & Food Security*, *4*(1), 1–7.

Crystal-Ornelas, R., Thapa, R., & Tully, K. L. (2021). Soil organic carbon is affected by organic amendments, conservation tillage, and cover cropping in organic farming systems: A meta-analysis. *Agriculture, Ecosystems & Environment*, *312*, 107356.

Dai, X., Zhou, W., Liu, G., Liang, G., He, P., & Liu, Z. (2019). Soil C/N and pH together as a comprehensive indicator for evaluating the effects of organic substitution management in subtropical paddy fields after application of high-quality amendments. *Geoderma*, *337*, 1116–1125.

Daniel, Jr, J. J., Zabot, G. L., Tres, M. V., Harakava, R., Kuhn, R. C., & Mazutti, M. A. (2018). Fusarium Fujikuroi: A novel source of metabolites with herbicidal activity. *Biocatalysis and Agricultural Biotechnology*, *14*, 314–320.

Das, S., Chatterjee, A., & Pal, T. K. (2020). Organic farming in India: A vision towards a healthy nation. *Food Quality and Safety*, *4*(2), 69–76.

Das, S., Ray, M. K., Panday, D., & Mishra, P. K. (2023). Role of biotechnology in creating sustainable agriculture. *PLOS Sustainability and Transformation*, *2*(7), e0000069.

Dellero, Y. (2020). Manipulating amino acid metabolism to improve crop nitrogen use efficiency for a sustainable agriculture. *Frontiers in Plant Science*, *11*, 602548.

Deng, Y., Ning, Y., Yang, D. L., Zhai, K., Wang, G. L., & He, Z. (2020). Molecular basis of disease resistance and perspectives on breeding strategies for resistance improvement in crops. *Molecular Plant, 13*(10), 1402–1419.

De Souza, V. M., Pedrosa, J. L. F., Gonçalves, D. R., Medeiros, F. C. L. D., Carvalho, G. R., Gonçalves, A. H., & Teixeira, P. V. V. Q. (2021). Herbicides of biological origin: A review. *The Journal of Horticultural Science and Biotechnology, 96*(3), 288–296.

Dhiman, V. (2020). Organic farming for sustainable environment: Review of existed policies and suggestions for improvement. *International Journal of Research and Review, 7*(2), 22–31.

Du, Y., Wang, T., Jiang, J., Wang, Y., Lv, C., Sun, K., & Huang, L. (2022). Biological control and plant growth promotion properties of Streptomyces albidoflavus St-220 isolated from Salvia miltiorrhiza rhizosphere. *Frontiers in Plant Science, 13*, 976813.

Duke, S. O., Owens, D. K., & Dayan, F. E. (2018). *Natural product-based chemical herbicides* (pp. 153–165). Weed Control, CRC Press.

Duke, S. O., Pan, Z., Bajsa-Hirschel, J., & Boyette, C. D. (2022). The potential future roles of natural compounds and microbial bioherbicides in weed management in crops. *Advances in Weed Science, 40.*

Elnahal, A. S., El-Saadony, M. T., Saad, A. M., Desoky, E. S. M., El-Tahan, A. M., Rady, M. M., AbuQamar, S. F., & El-Tarabily, K. A. (2022). The use of microbial inoculants for biological control, plant growth promotion, and sustainable agriculture: A review. *European Journal of Plant Pathology, 162*(4), 759–792.

El-Saadony, M. T., Saad, A. M., Soliman, S. M., Salem, H. M., Ahmed, A. I., Mahmood, M., & AbuQamar, S. F. (2022). Plant growth-promoting microorganisms as biocontrol agents of plant diseases: Mechanisms, challenges and future perspectives. *Frontiers in Plant Science, 13*, 923880.

Enfissi, E. M., Drapal, M., Perez-Fons, L., Nogueira, M., Berry, H. M., Almeida, J., & Fraser, P. D. (2021). New plant breeding techniques and their regulatory implications: An opportunity to advance metabolomics approaches. *Journal of Plant Physiology, 258*, 153378.

Fenz, S., Neubauer, T., Heurix, J., Friedel, J. K., & Wohlmuth, M. L. (2023). AI-and data-driven pre-crop values and crop rotation matrices. *European Journal of Agronomy, 150*, 126949.

Forman, J., Silverstein, J., Committee on Nutrition, Council on Environmental Health, & American Academy of Pediatrics. (2012). Organic foods: Health and environmental advantages and disadvantages. *Pediatrics, 130*(5), e1406–e1415.

Gade, R. M., & Koche, M. D. (2022). Enhancing the growth and disease suppression ability of Pseudomonas fluorescens. In *New and future developments in microbial biotechnology and bioengineering* (pp. 351–368). Elsevier.

Gaikwad, K. B., Rani, S., Kumar, M., Gupta, V., Babu, P. H., Bainsla, N. K., & Yadav, R. (2020). Enhancing the nutritional quality of major food crops through conventional and genomics-assisted breeding. *Frontiers in Nutrition, 7*, 533453.

Gamage, A., Gangahagedara, R., Gamage, J., Jayasinghe, N., Kodikara, N., Suraweera, P., & Merah, O. (2023). Role of organic farming for achieving sustainability in agriculture. *Farming System, 1*(1), 100005.

Gani, M., Shafi, I., Hassan, T., Mantoo, M. A., Guroo, M. A., Gupta, R. K., & Bali, K. (2022). Baculoviruses for insect pest management: Applications, challenges and opportunities. In T. H. Askary (Ed.). *Pest management: Methods, applications and challenges*. Nova Science Publishers Inc.

Garland, S., & Curry, H. A. (2022). Turning promise into practice: Crop biotechnology for increasing genetic diversity and climate resilience. *PLoS Biology, 20*(7), e3001716.

Gaur, M. (2016). Organic farming in India: Status, issues and prospects. *SOPAAN-II, 1*(1), 26–36.

Gautam, S. S., Gondwal, M., Soni, R., & Gautam, B. P. S. (2022). Potash biofertilizers: Current development, formulation, and applications. In *Trends of Applied Microbiology for Sustainable Economy Academic Press*, 481–500.

Gentsch, N., Boy, J., Batalla, J. D. K., Heuermann, D., von Wirén, N., Schweneker, D., Feuerstein, U., Groß, J., Bauer, B., Reinhold-Hurek, B., Hurek, T., & Guggenberger, G. (2020). Catch crop diversity increases rhizosphere carbon input and soil microbial biomass. *Biology and Fertility of Soils*, *56*, 943–957.

Gomiero, T., Pimentel, D., & Paoletti, M. G. (2011). Environmental impact of different agricultural management practices: Conventional vs. organic agriculture. *Critical Reviews in Plant Sciences*, *30*(1–2), 95–124.

Gordon, D. R., Jaffe, G., Doane, M., Glaser, A., Gremillion, T. M., & Ho, M. D. (2021). Responsible governance of gene editing in agriculture and the environment. *Nature Biotechnology*, *39*(9), 1055–1057.

Gossen, B. D., & McDonald, M. R. (2020). New technologies could enhance natural biological control and disease management and reduce reliance on synthetic pesticides. *Canadian Journal of Plant Pathology*, *42*(1), 30–40.

Gul Kazi, A., Siddique, S., Hamid, M., & Tariq, A. (2014). Organic farming: The return to nature. In *Improvement of Crops in the Era of Climatic Changes* (pp. 249–281). Springer.

Gupta, S., Sharma, S., Aich, A., Verma, A. K., Bhuyar, P., Nadda, A. K., Mulla, S. I., & Kalia, S. (2023). Chicken feather waste hydrolysate as a potential biofertilizer for environmental sustainability in organic agriculture management. *Waste and Biomass Valorization*, 1–17.

Halberg, N. (2008). Energy use and greenhouse gas emission in organic agriculture. In *Proceedings of the International Conference Organic Agriculture and Climate Change* (pp. 1–6). ENITA of Clermont.

Hawkesford, M. J., & Griffiths, S. (2019). Exploiting genetic variation in nitrogen use efficiency for cereal crop improvement. *Current Opinion in Plant Biology*, *49*, 35–42.

Hemathilake, D. M. K. S., & Gunathilake, D. M. C. C. (2022). *Agricultural productivity and food supply to meet increased demands* (pp. 539–553). Future Foods Academic Press.

Hesham, A. E. L., Kaur, T., Devi, R., Kour, D., Prasad, S., Yadav, N., Singh, C., Singh, J., & Yadav, A. N. (2021). Current trends in microbial biotechnology for agricultural sustainability: Conclusion and future challenges. *Current Trends in Microbiology Biotechnology for Sustainable Agriculture*, 555–572.

Hetherington, S. D., Smith, H. E., Scanes, M. G., & Auld, B. A. (2002). Effects of some environmental conditions on the effectiveness of *Drechslera avenacea* (Curtis ex Cooke) Shoem: A potential bioherbicidal organism for Avena fatua L. *Biological Control*, *24*(2), 103–109.

Holka, M., Kowalska, J., & Jakubowska, M. (2022). Reducing carbon footprint of agriculture-can organic farming help to mitigate climate change? *Agriculture*, *12*(9), 1383.

Hörz, P. F. (2022). Organic agriculture as crisis management: A cultural anthropologist's approach to a phenomenon of "reflexive modernity". *Das culturas da alimentação ao culto dos alimentos*, *I*(15), 269.

Hosni, K., Hassen, I., Sebei, H., & Casabianca, H. (2013). Secondary metabolites from *Chrysanthemum Coronarium* (garland) flowerheads: Chemical composition and biological activities. *Industrial Crops and Products*, *44*, 263–271.

Hubbard, M., Taylor, W. G., Bailey, K. L., & Hynes, R. K. (2016). The dominant modes of action of macrocidins, bioherbicidal metabolites of *Phoma macrostoma*, differ between susceptible plant species. *Environmental and Experimental Botany*, *132*, 80–91.

Hüdig, M., Laibach, N., & Hein, A. C. (2022). Genome editing in crop plant research—Alignment of expectations and current developments. *Plants*, *11*(2), 212.

Hufnagel, J., Reckling, M., & Ewert, F. (2020). Diverse approaches to crop diversification in agricultural research. A review. *Agronomy for Sustainable Development*, *40*, 1–17.

Husaini, A. M., & Sohail, M. (2018). Time to redefine organic agriculture: Can't GM crops be certified as organics? *Frontiers in Plant Science*, *9*, 423.

Iderawumi, A. M., & Kamal, T. O. (2022). Green manure for agricultural sustainability and improvement of soil fertility. *Farming and Management*, *7*(1), 1–8.

Iqbal, N., Hayat, M. T., Tahir, A. A., Akhtar, S., Bangash, N., Nazeer, A., Saleem, A. R., & Riyazuddin, R. (2023). Global prospects of climate-resilient agriculture. In N. Iqbal, M. T. Hayat & A. A. Tahir (Eds.). *Climate-Resilient Agriculture, Vol 1: Crop Responses and Agroecological Perspectives* (pp. 1–25). Springer International Publishing.

Iqbal, Z., Iqbal, M. S., Ahmad, A., Memon, A. G., & Ansari, M. I. (2020). New prospects on the horizon: Genome editing to engineer plants for desirable traits. *Current Plant Biology*, *24*, 100171.

Jalilian, S., Mondani, F., Fatemi, A., & Bagheri, A. (2022). Evaluation of animal-and green manures co-application on soil micronutrients and yield of sesame (*Sesamum indicum* L.) in organic farm condition. *Journal of Plant Nutrition*, *45*(15), 2348–2359.

Jezierska-Tys, S., Joniec, J., Bednarz, J., & Kwiatkowska, E. (2021). Microbiological nitrogen transformations in soil treated with pesticides and their impact on soil greenhouse gas emissions. *Agriculture*, *11*(8), 787.

Jouzi, Z., Azadi, H., Taheri, F., Zarafshani, K., Gebrehiwot, K., Van Passel, S., & Lebailly, P. (2017). Organic farming and small-scale farmers: Main opportunities and challenges. *Ecological Economics*, *132*, 144–154.

Kaab, S. B., Rebey, I. B., Hanafi, M., Hammi, K. M., Smaoui, A., Fauconnier, M. L., De Clerck, C., Jijakli, M. H., & Ksouri, R. (2020). Screening of Tunisian plant extracts for herbicidal activity and formulation of a bioherbicide based on *Cynara cardunculus*. *South African Journal of Botany*, *128*, 67–76.

Kalia, A., & Gosal, S. K. (2011). Effect of pesticide application on soil microorganisms. *Archives of Agronomy and Soil Science*, *57*(6), 569–596.

Kamboj, P., Garg, R., & Khokher, K. (2019). Chapter-1: Impact of bio pesticides and bio fertilizers to conserve nutrient and disease management. *Advances*, *1*.

Kato-Noguchi, H., Kimura, F., Ohno, O., & Suenaga, K. (2017). Involvement of allelopathy in inhibition of understory growth in red pine forests. *Journal of Plant Physiology*, *218*, 66–73.

Kaur, H., Pandey, D. K., Goutam, U., & Kumar, V. (2021). CRISPR/Cas9-mediated genome editing is revolutionizing the improvement of horticultural crops: Recent advances and future prospects. *Scientia Horticulturae*, *289*, 110476.

Kennedy, A. C. (2018). Selective soil bacteria to manage downy brome, jointed goat grass, and medusa head and do no harm to other biota. *Biological Control*, *123*, 18–27.

Khadda, B. S. (2021). Prospects of organic farming in India. *A Voice for Agriculture, Mirror*, *2*(7), 27–34.

Khaliq, A., Perveen, S., Alamer, K. H., Zia Ul Haq, M., Rafique, Z., Alsudays, I. M., & Attia, H. (2022). Arbuscular mycorrhizal fungi symbiosis to enhance plant–soil interaction. *Sustainability*, *14*(13), 7840.

Khan, B. A., Nijabat, A., Khan, M. I., Khan, I., Hashim, S., Nadeem, M. A., & Ikram, M. (2022). Implications of mulching on weed management in crops and vegetable. In *Mulching in agroecosystems: Plants, soil & environment* (pp. 199–213). Springer Nature Singapore.

Klaimala, P., Khunlert, P., Chuntib, P., Pundee, R., Kallayanatham, N., Nankongnab, N., Kongtip, P., & Woskie, S. (2022). Pesticide residues on children's hands, home indoor surfaces, and drinking water among conventional and organic farmers in Thailand. *Environmental Monitoring and Assessment*, *194*(6), 427.

Koodkaew, I., Senaphan, C., Sengseang, N., & Suwanwong, S. (2018). Characterization of phytochemical profile and phytotoxic activity of *Mimosa pigra* L. *Agriculture and Natural Resources*, *52*(2), 162–168.

Kothamasi, D., & Vermeylen, S. (2011). Genetically modified organisms in agriculture: Can regulations work? *Environment, Development and Sustainability*, *13*, 535–546.

Krause, H. M., Stehle, B., Mayer, J., Mayer, M., Steffens, M., Mäder, P., & Fliessbach, A. (2022). Biological soil quality and soil organic carbon change in biodynamic, organic, and conventional farming systems after 42 years. *Agronomy for Sustainable Development*, *42*(6), 117.

Krauss, M., Wiesmeier, M., Don, A., Cuperus, F., Gattinger, A., Gruber, S., Haagsma, W. K., Peigné, J., Palazzoli, M. C., Schulz, F., van der Heijden, M. G., & Steffens, M. (2022). Reduced tillage in organic farming affects soil organic carbon stocks in temperate Europe. *Soil and Tillage Research*, *216*, 105262.

Kumar, A., Anju, T., Kumar, S., Chhapekar, S. S., Sreedharan, S., Singh, S., & Lim, Y. P. (2021). Integrating omics and gene editing tools for rapid improvement of traditional food plants for diversified and sustainable food security. *International Journal of Molecular Sciences*, *22*(15), 8093.

Kumar, A., Singh, K., Verma, P., Singh, O., Panwar, A., Singh, T., Kumar, Y., & Raliya, R. (2022). Effect of nitrogen and zinc nanofertilizer with the organic farming practices on cereal and oil seed crops. *Scientific Reports*, *12*(1), 6938.

Kumari, C., Sharma, M., Kumar, V., Sharma, R., Kumar, V., Sharma, P., & Irfan, M. (2022). Genome editing technology for genetic amelioration of fruits and vegetables for alleviating post-harvest loss. *Bioengineering*, *9*(4), 176.

Lakner, S., & Breustedt, G. (2017). Efficiency analysis of organic farming systems a review of concepts, topics, results and conclusions. *German Journal of Agricultural Economics*, *66*(2), 85–108.

Lammerts van Bueren, E. T., Struik, P. C., van Eekeren, N., & Nuijten, E. (2018). Towards resilience through systems-based plant breeding. A review. *Agronomy for Sustainable Development*, *38*, 1–21.

Lawrance, S., Varghese, S., Varghese, E. M., & Asok, A. K. (2019). Quinoline derivatives producing *Pseudomonas aeruginosa* H6 as an efficient bioherbicide for weed management. *Biocatalysis and Agricultural Biotechnology*, *18*, 101096.

Li, X., Xu, S., Fuhrmann-Aoyagi, M. B., Yuan, S., Iwama, T., Kobayashi, M., & Miura, K. (2022). CRISPR/Cas9 technique for temperature, drought, and salinity stress responses. *Current Issues in Molecular Biology*, *44*(6), 2664–2682.

Li, X., Zeng, R., & Liao, H. (2016). Improving crop nutrient efficiency through root architecture modifications. *Journal of Integrative Plant Biology*, *58*(3), 193–202.

Lowry, C. J., & Brainard, D. C. (2019). Organic farmer perceptions of reduced tillage: A Michigan farmer survey. *Renewable Agriculture and Food Systems*, *34*(2), 103–115.

Luo, L., Zhao, C., Wang, E., Raza, A., & Yin, C. (2022). Bacillus amyloliquefaciens as an excellent agent for biofertilizer and biocontrol in agriculture: An overview for its mechanisms. *Microbiological Research*, *259*, 127016.

Maes, C., Meersmans, J., Lins, L., Bouquillon, S., & Fauconnier, M. L. (2021). Essential oil-based bioherbicides: Human health risks analysis. *International Journal of Molecular Sciences*, *22*(17), 9396.

Mahawar, H., Bajpai, A., Sreekanth, D., Pawar, D., & Barman, K. K. (2023). Emerging weeds under climate change and their microbial management. In *Bioinoculants: Biological Option for Mitigating Global Climate Change* (pp. 57–86). Springer Nature Singapore.

Maitra, S., & Gitari, H. (2020). Scope for adoption of intercropping system in organic agriculture. *Indian Journal of Natural Sciences*, *11*(63), 28624–28631.

Mandal, A., Dhaliwal, S. S., Mani, P. K., & Toor, A. S. (2021). Conservation agricultural practices under organic farming. In *Advances in organic farming* (pp. 17–37). Woodhead Publishing.

Mani, M. (2022). Organic pest management in horticultural crops. *Trends in Horticultural Entomology*, *16*, 211–241.

Manzoor, S., Gani, M., Hassan, T., Shafi, I., Wani, F. J., Mumtaz, S., Eroglu, G. B., Yaqoob, M., & Mantoo, M. A. (2023). Comparative evaluation of temperate, subtropical, and tropical isolates of nucleopolyhedrovirus against tomato fruit borer, *Helicoverpa armigera* (Hubner) (Lepidoptera: Noctuidae). *Egyptian Journal of Biological Pest Control, 33*(40), 1–7.

Martínez-Mena, M., Carrillo-López, E., Boix-Fayos, C., Almagro, M., Franco, N. G., Díaz-Pereira, E., & De Vente, J. (2020). Long-term effectiveness of sustainable land management practices to control runoff, soil erosion, and nutrient loss and the role of rainfall intensity in Mediterranean rainfed agroecosystems. *Catena, 187*, 104352.

McClelland, S. C., Paustian, K., & Schipanski, M. E. (2021). Management of cover crops in temperate climates influences soil organic carbon stocks: A meta-analysis. *Ecological Applications, 31*(3), e02278.

McGuire, A. M. (2017). Agricultural science and organic farming: Time to change our trajectory. *Agricultural & Environmental Letters, 2*(1), 170024.

Mendez, E. K., Orrego, C. E., & Giraldo, G. I. (2019). Organic versus conventional: A comparative study on the shelf life of passion fruit (*Passiflora edulis Sims*) crops. *International Food Research Journal, 26*(2), 393–399.

Mishra, R. K. (2023). Fresh water availability and its global challenge. *British Journal of Multidisciplinary and Advanced Studies, 4*(3), 1–78.

Motmainna, M., Juraimi, A. S., Uddin, M. K., Asib, N. B., Islam, A. K. M. M., & Hasan, M. (2021). Assessment of allelopathic compounds to develop new natural herbicides: A review. *Allelopathy Journal, 52*, 21–40.

Mukherjee, K., Konar, A., & Ghosh, P. (2022). Organic farming in India: A brief review. *International Journal of Research in Agronomy, 5*(2), 113–118.

Munawar, S., ul Qamar, M. T., Mustafa, G., Khan, M. S., & Joyia, F. A. (2020). Role of biotechnology in climate resilient agriculture. *Environment, Climate, Plant and Vegetation Growth*, 339–365.

Nakasaki, K., Hirai, H., Mimoto, H., Quyen, T. N. M., Koyama, M., & Takeda, K. (2019). Succession of microbial community during vigorous organic matter degradation in the primary fermentation stage of food waste composting. *Science of the Total Environment, 671*, 1237–1244.

Niether, W., Macholdt, J., Schulz, F., & Gattinger, A. (2023). Yield dynamics of crop rotations respond to farming type and tillage intensity in an organic agricultural long-term experiment over 24 years. *Field Crops Research, 303*, 109131.

Niggli, U., Fließbach, A., Hepperly, P., & Scialabba, N. (2009). Low greenhouse gas agriculture: Mitigation and adaptation potential of sustainable farming systems. *Ökologie & Landbau, 141*, 32–33.

Nordblom, T., Gurusinghe, S., Erbacher, A., & Weston, L. A. (2023). Opportunities and challenges for cover cropping in sustainable agriculture systems in Southern Australia. *Agriculture, 13*(3), 688.

Ntinyari, W., & Gweyi-Onyango, J. P. (2021). Greenhouse gases emissions in agricultural systems and climate change effects in sub- Saharan Africa. In N. Oguge, D. Ayal, L. Adeleke & I. da Silva (Eds.). *African Handbook of Climate Change Adaptation.* Springer.

Ntui, V. O., Tripathi, J. N., Kariuki, S. M., & Tripathi, L. (2023). Cassava molecular genetics and genomics for enhanced resistance to diseases and pests. *Molecular Plant Pathology*, 1–14. https://doi.org/10.1111/mpp.13402

Ootani, M. A., dos Reis, M. R., Cangussu, A. S. R., Capone, A., Fidelis, R. R., Oliveira, W., Barros, H. B., Portella, A. C. F., de Souza Aguiar, R., & dos Santos, W. F. (2017). Phytotoxic effects of essential oils in controlling weed species *Digitaria horizontalis* and *Cenchrus echinatus*. *Biocatalysis and Agricultural Biotechnology, 12*, 59–65.

Oquist, K. A., Strock, J. S., & Mulla, D. J. (2007). Influence of alternative and conventional farming practices on subsurface drainage and water quality. *Journal of Environmental Quality*, *36*(4), 1194–1204.

Ozturk, M., Kamili, A. N., Altay, V., & Rohela, G. K. (2024). Mulberry: Mycorrhizal interactions and development related to other organisms. In *Mulberry*. Springer. https://doi.org/10.1007/978-3-031-49117-74

Pandey, G. K., & Mahiwal, S. (2020). *Role of potassium in plants* (pp. 45–49). Springer.

Pantovic, J. G., & Secanski, M. (2023). Weed control in organic farming. *Contemporary Agriculture*, *72*(1–2), 43–56.

Paramesh, V., Ravisankar, N., Behera, U., Arunachalam, V., Kumar, P., Solomon Rajkumar, R., & Rajkumar, S. (2022). Integrated farming system approaches to achieve food and nutritional security for enhancing profitability, employment, and climate resilience in India. *Food and Energy Security*, *11*(2), e321.

Parewa, H. P., Joshi, N., Meena, V. S., Joshi, S., Choudhary, A., Ram, M., Meena, S. C., & Jain, L. K. (2021). Role of biofertilizers and biopesticides in organic farming. *Advances in Organic Farming*, 133–159.

Parizad, S., & Bera, S. (2023). The effect of organic farming on water reusability, sustainable ecosystem, and food toxicity. *Environmental Science and Pollution Research*, *30*(28), 71665–71676.

Pedraza, R. O., Filippone, M. P., Fontana, C., Salazar, S. M., Ramírez-Mata, A., Sierra-Cacho, D., & Baca, B. E. (2020). Azospirillum. In *Beneficial microbes in agro-ecology* (pp. 73–105). Academic Press.

Peerzada, A. M., Bukhari, S. A. H., Dawood, M., Nawaz, A., Ahmad, S., & Adkins, S. (2019). Weed management for healthy crop production. *Agronomic Crops. Management Practices*, *2*, 225–256.

Pretty, J., & Bharucha, Z. P. (2014). Sustainable intensification in agricultural systems. *Annals of Botany*, *114*(8), 1571–1596.

Priti, Kamboj, D., Chaudhary, V., Baliyan, N., Rani, R., & Jangra, S. (2023). Nutritional enhancement in horticultural crops by CRISPR/Cas9: Status and future prospects. *Sustainable Agriculture in the Era of the OMICs Revolution*, 399–430.

Pugazhendhi, A., Radhakrishnan, R., & Duc, P. A. (2019). Curtobacterium sp. MA01 generates oxidative stress to inhibit the plant growth. *Biocatalysis and Agricultural Biotechnology*, *20*, 101274.

Qaim, M. (2020). Role of new plant breeding technologies for food security and sustainable agricultural development. *Applied Economic Perspectives and Policy*, *42*(2), 129–150.

Qiao, Y., Halberg, N., Vaheesan, S., & Scott, S. (2016). Assessing the social and economic benefits of organic and fair trade tea production for small-scale farmers in Asia: A comparative case study of China and Sri Lanka. *Renewable Agriculture and Food Systems*, *31*(3), 246–257.

Rahman, S. M. E., Mele, M. A., Lee, Y. T., & Islam, M. Z. (2021). Consumer preference, quality, and safety of organic and conventional fresh fruits, vegetables, and cereals. In *Foods*, *10*(1).

Raimi, M. O., Iyingiala, A. A., Sawyerr, O. H., Saliu, A. O., Ebuete, A. W., Emberru, R. E., Sanchez, N. D., & Osungbemiro, W. B. (2022). Leaving no one behind: Impact of soil pollution on biodiversity in the global south: A global call for action. In *Biodiversity in Africa: Potentials, Threats and Conservation* (pp. 205–237). Springer Nature.

Raj, E. F. I., Appadurai, M., & Athiappan, K. (2022). Precision farming in modern agriculture. In *Smart Agriculture Automation Using Advanced Technologies: Data Analytics and Machine Learning, Cloud Architecture, Automation and IoT* (pp. 61–87). Springer Singapore.

Rajni, Y., Singh, S., & Singh, A. N. (2022). Biopesticides: Current status and future prospects. *Proceedings of the International Academy of Ecology and Environmental Sciences*, *12*, 211–233.

Ramezani, S., Saharkhiz, M. J., Ramezani, F., & Fotokian, M. H. (2008). Use of essential oils as bioherbicides. *Journal of Essential Oil Bearing Plants*, *11*(3), 319–327.

Rani, S., Kumar, P., & Suneja, P. (2021). Biotechnological interventions for inducing abiotic stress tolerance in crops. *Plant Gene*, *27*, 100315.

Rautaray, S. K., Pradhan, S., Mohanty, S., Dubey, R., Raychaudhuri, S., Mohanty, R. K., Mishra, A., & Ambast, S. K. (2020). Energy efficiency, productivity and profitability of rice farming using *Sesbania* as green manure-cum-cover crop. *Nutrient Cycling in Agroecosystems*, *116*, 83–101.

Raza, A., Zahra, N., Hafeez, M. B., Ahmad, M., Iqbal, S., Shaukat, K., & Ahmad, G. (2020). Nitrogen fixation of legumes: Biology and Physiology. *The Plant Family Fabaceae: Biology and Physiological Responses to Environmental Stresses*, 43–74.

Reeve, J. R., Hoagland, L. A., Villalba, J. J., Carr, P. M., Atucha, A., Cambardella, C., & Delate, K. (2016). Organic farming, soil health, and food quality: Considering possible links. *Advances in Agronomy*, *137*, 319–367.

Reganold, J. P., & Wachter, J. M. (2016). Organic agriculture in the twenty-first century. *Nature Plants*, *2*, 15221.

Reinhart, K. O., Carlson, C. H., Feris, K. P., Germino, M. J., Jandreau, C. J., Lazarus, B. E., Mangold, J., Pellatz, D. W., Ramsey, P., Rinella, M. J., & Valliant, M. (2020). Weed-suppressive bacteria fail to control *Bromus tectorum* under field conditions. *Rangeland Ecology & Management*, *73*(6), 760–765.

Richard, B., Qi, A., & Fitt, B. D. (2022). Control of crop diseases through integrated crop management to deliver climate-smart farming systems for low-and high-input crop production. *Plant Pathology*, *71*(1), 187–206.

Rodriguez, C., Dimitrova Mårtensson, L. M., Zachrison, M., & Carlsson, G. (2021). Sustainability of diversified organic cropping systems—challenges identified by farmer interviews and multi-criteria assessments. *Frontiers in Agronomy*, *3*, 698968.

Rohela, G. K., & Saini, P. (2022). Nitrogen-Fixing biofertilizers and biostimulants. In *Microbial Biostimulants for Sustainable Agriculture and Environmental Bioremediation* (pp. 83–100). CRC Press.

Röös, E., Mie, A., Wivstad, M., Salomon, E., Johansson, B., Gunnarsson, S., Wallenbeck, A., Hoffmann, R., Nilsson, U., Sundberg, C., & Watson, C. A. (2018). Risks and opportunities of increasing yields in organic farming-A review. *Agronomy for Sustainable Development*, *38*(2).

Roy, A., Purkaystha, S., & Bhattacharyya, S. (2021). Advancement in molecular and fast breeding programs for climate-resilient agriculture practices. Harsh environment and plant resilience. *Molecular and Functional Aspects*, 73–98.

Roy, T., & George, K. J. (2020). Precision farming: A step towards sustainable, climate-smart agriculture. *Global Climate Change: Resilient and Smart Agriculture*, 199–220.

Roychowdhury, R., Abdelgawwad, M. R., & Banerjee, U. (2014). Status, trends and prospects of organic farming in India: A review. *Journal of Plant Biology Research*, *2*(2), 38–48.

Rudolph, M., & Muchesa, E. (2023). A Review of the agroecological farming system as a viable alternative food production approach in South Africa. *South African Journal of Agricultural Extension (SAJAE)*, *51*(2), 43–76.

Saini, P., Saini, P., Kaur, J. J., Francies, R. M., Gani, M., Rajendra, A. A., & Chauhan, S. S. (2020). Molecular approaches for harvesting natural diversity for crop improvement. *Rediscovery of Genetic and Genomic Resources for Future Food Security*, 67–169.

Sami, A., Xue, Z., Tazein, S., Arshad, A., He Zhu, Z., Ping Chen, Y., & Jin Zhou, K. (2021). CRISPR–Cas9-based genetic engineering for crop improvement under drought stress. *Bioengineered*, *12*(1), 5814–5829.

Sarkar, D., Dubey, P. K., Chaurasiya, R., Sankar, A., Shikha, Chatterjee, N., & Rakshit, A. (2021). Organic interventions conferring stress tolerance and crop quality in agroecosystems during the United Nations Decade on Ecosystem Restoration. *Land Degradation & Development*, *32*(17), 4797–4816.

Sathee, L., Jagadhesan, B., Pandesha, P. H., Barman, D., Adavi, B. S., Nagar, S., & Chinnusamy, V. (2022). Genome editing targets for improving nutrient use efficiency and nutrient stress adaptation. *Frontiers in Genetics*, *13*, 900897.

Seitz, D., Fischer, L. M., Dechow, R., Wiesmeier, M., & Don, A. (2023). The potential of cover crops to increase soil organic carbon storage in German croplands. *Plant and Soil*, *488*(1–2), 157–173.

Seitz, S., Goebes, P., Puerta, V. L., Pereira, E. I. P., Wittwer, R., Six, J., van Der Heijden, M. G., & Scholten, T. (2019). Conservation tillage and organic farming reduce soil erosion. *Agronomy for Sustainable Development*, *39*, 1–10.

Shabana, Y. M. (2005). The use of oil emulsions for improving the efficacy of Alternaria eichhorniae as a mycoherbicide for waterhyacinth (*Eichhornia crassipes*). *Biological Control*, *32*(1), 78–89.

Shepherd, M., Pearce, B., Cormack, B., Philipps, L., Cuttle, S., Bhogal, A., Costigan, P., & Unwin, R. (2003). An assessment of the environmental impacts of organic farming, A review for Defra project OF0405. London: Department for Environment, Food and Rural Affairs (with ADAS, EFRC and IGER).

Shrestha, A. (2009). Potential of a black walnut (*Juglans nigra*) extract product (NatureCur®) as a pre-and post-emergence bioherbicide. *Journal of Sustainable Agriculture*, *33*(8), 810–822.

Singh, M. (2021). Organic farming for sustainable agriculture. *Indian Journal of Organic Farming*, *1*(1), 1–8.

Singh, R., Umapathi, A., Patel, G., Patra, C., Malik, U., Bhargava, S. K., & Daima, H. K. (2023). Nanozyme-based pollutant sensing and environmental treatment: Trends, challenges, and perspectives. *Science of the Total Environment*, *854*, 158771.

Sivaranjani, S., & Rakshit, A. (2019). Organic farming in protecting water quality. *Organic Farming: New Advances Towards Sustainable Agricultural Systems*, 1–9.

Smith, O. M., Cohen, A. L., Rieser, C. J., Davis, A. G., Taylor, J. M., Adesanya, A. W., Jones, M. S., Meier, A. R., Reganold, J. P., Orpet, R. J., Northfield, T. D., & Crowder, D. W. (2019). Organic farming provides reliable environmental benefits but increases variability in crop yields: A global meta-analysis. *Frontiers in Sustainable Food Systems*, *3*, 82.

Soltys, D., Krasuska, U., Bogatek, R., & Gniazdowska, A. (2013). Allelochemicals as bioherbicides-present and perspectives. In *Herbicides-Current Research and Case Studies in Use*. IntechOpen.

Soni, R., Gupta, R., Agarwal, P., & Mishra, R. (2022). Organic farming: A sustainable agricultural practice. *Vantage*, *3*(1), 21–44.

Srivastava, P., Balhara, M., & Giri, B. (2020). Soil health in India: Past history and future perspective. *Soil Health*, 1–19.

Steinwand, M. A., & Ronald, P. C. (2020). Crop biotechnology and the future of food. *Nature Food*, *1*(5), 273–283.

Sudheer, P. S. K. (2023). Economics of organic versus chemical farming for three crops in Andhra Pradesh, India. *Journal of Organic Systems*, *8*(2), 37–49.

Suharyati, A., Hartono, S., & Waluyati, L. R. (2016). Competitive and comparative advantages analysis of organic rice farming in Karanganyar regency, Central Java Province. *Ilmu Pertanian (Agricultural Science)*, *1*(1), 25.

Sushma, R., & Dinesh, G. M (2020). Organic farming and conventional farming: A comparative study. *Journal of Emerging Technologies and Innovative Research*, *7*(11), 1044–1052.

Synowiec, A., Możdżeń, K., Krajewska, A., Landi, M., & Araniti, F. (2019). Carum carvi L. essential oil: A promising candidate for botanical herbicide against *Echinochloa crus-galli* (L.) P. Beauv. In maize cultivation. *Industrial Crops and Products*, *140*, 111652.

Tanveer, A., Ikram, R. M., & Ali, H. H. (2019). Crop rotation: Principles and practices. *Agronomic Crops: Management Practices* (Vol. 2, pp. 1–12).

Tariq, M., Khan, A., Asif, M., Khan, F., Ansari, T., Shariq, M., & Siddiqui, M. A. (2020). Biological control: A sustainable and practical approach for plant disease management. *Acta Agriculturae Scandinavica, Section B—Soil & Plant Science*, *70*(6), 507–524.

Tekiela, D. R. (2020). Effect of the bioherbicide *Pseudomonas fluorescens* D7 on downy brome (Bromus tectorum). *Rangeland Ecology & Management*, *73*(6), 753–755.

Thakur, N., Kaur, S., Kaur, T., Tomar, P., Devi, R., Thakur, S., Mehta, D. K., & Yadav, A. N. (2022). Organic agriculture for agro-environmental sustainability. In *Trends of Applied Microbiology for Sustainable Economy* (pp. 699–735). Academic Press.

Thakur, N., Nigam, M., Mann, N. A., Gupta, S., Hussain, C. M., Shukla, S. K., Shah, A. A., Casini, R., Elansary, H. O., & Khan, S. A. (2023). Host-mediated gene engineering and microbiome-based technology optimization for sustainable agriculture and environment. *Functional & Integrative Genomics*, *23*(1), 57.

Tsao, R., Romanchuk, F. E., Peterson, C. J., & Coats, J. R. (2002). Plant growth regulatory effect and insecticidal activity of the extracts of the tree of heaven (*Ailanthus altissima* L.). *BMC Ecology*, *2*, 1–6.

Tscharntke, T., Grass, I., Wanger, T. C., Westphal, C., & Batáry, P. (2021). Beyond organic farming–harnessing biodiversity-friendly landscapes. *Trends in Ecology & Evolution*, *36*(10), 919–930.

Ullah, A., Shah, S. N. M., Ali, A., Naz, R., Mahar, A., & Kalhoro, S. A. (2015). Factors affecting the adoption of organic farming in Peshawar-Pakistan. *Agricultural Sciences*, *6*(6), 587–593. https://doi.org/10.4236/as.2015.66057

Vanderschuren, H., Chatukuta, P., Weigel, D., & Mehta, D. (2023). A new chance for genome editing in Europe. *Nature Biotechnology*, *41*(10), 1378–1380.

Varga, K., Fehér, J., Trugly, B., Drexler, D., Leiber, F., Verrastro, V., Magid, J., Chylinski, C., Athanasiadou, S., Thuerig, B., László, A., & Tamm, L. (2022). The state of play of copper, mineral oil, external nutrient input, anthelmintics, antibiotics and vitamin usage and available reduction strategies in organic farming across Europe. *Sustainability*, *14*(6), 3182.

Varshney, R. K., Bansal, K. C., Aggarwal, P. K., Datta, S. K., & Craufurd, P. Q. (2011). Agricultural biotechnology for crop improvement in a variable climate: Hope or hype? *Trends in Plant Science*, *16*(7), 363–371.

Verdeguer, M., Sánchez-Moreiras, A. M., & Araniti, F. (2020). Phytotoxic effects and mechanism of action of essential oils and terpenoids. *Plants*, *9*(11), 1571.

Wahab, A., Muhammad, M., Munir, A., Abdi, G., Zaman, W., Ayaz, A., & Reddy, S. P. P. (2023). Role of arbuscular mycorrhizal fungi in regulating growth, enhancing productivity, and potentially influencing ecosystems under abiotic and biotic stresses. *Plants*, *12*(17), 3102.

Wakeil, N., Shalaby, S., Abdou, G., & Sallam, A. (2013). Pesticide-residue relationship and its adverse effects on occupational workers. *Insecticides-Development of Safer and More Effective Technologies*, 57–81.

Werf, H. M., Knudsen, M. T., & Cederberg, C. (2020). Towards better representation of organic agriculture in life cycle assessment. *Nature Sustainability*, *3*(6), 419–425.

Wijnands, F. W. T. (1999). Crop rotation in organic farming: Theory and practice. In *Designing and Testing Crop Rotations for Organic Farming. Proceedings from an International Workshop* (pp. 21–35). Danish Research Centre for Organic Farming.

Winter, C. K., & Davis, S. F. (2006). Organic foods. *Journal of Food Science*, *71*(9), R117–R124.

Wittwer, R. A., Bender, S. F., Hartman, K., Hydbom, S., Lima, R. A., Loaiza, V., & Van Der Heijden, M. G. (2021). Organic and conservation agriculture promote ecosystem multifunctionality. *Science Advances*, *7*(34), eabg6995.

Yadav, A., Bhuj, B. D., Ram, S., Singh, C. P., Dhar, S., Yadav, R. K., & Ashok, K. (2021). Organic farming for sustainable agriculture: A review. *Annals of the Romanian Society for Cell Biology*, *25*(6), 8088–8123.

Yadav, R. K., Tripathi, M. K., Tiwari, S., Tripathi, N., Asati, R., Chauhan, S., & Payasi, D. K. (2023). Genome editing and improvement of abiotic stress tolerance in crop plants. *Life*, *13*(7), 1456.

Yanakittkul, P., & Aungvaravong, C. (2020). A model of farmer's intentions towards organic farming: A case study on rice farming in Thailand. *Heliyon*, *6*(1), e03039.

Zahedipour, Z., Asghari, M., Abdollahi, B., Alizadeh, M., & Danesh, Y. R. R. (2019). A comparative study on quality attributes and physiological responses of organic and conventionally grown table grapes during cold storage. *Scientia Horticulturae*, *247*, 86–95.

Zhang, Y., Massel, K., Godwin, I. D., & Gao, C. (2018). Applications and potential of genome editing in crop improvement. *Genome Biology*, *19*, 1–11.

Zhang, Y., Xie, D., Ni, J., & Zeng, X. (2020). Conservation tillage practices reduce nitrogen losses in the sloping upland of the Three Gorges Reservoir area: No-till is better than mulch-till. *Agriculture, Ecosystems & Environment*, *300*, 107003.

14 Water Movement in Organic Soils

Mahrukh Qureshi, Rohitashw Kumar, and Munjid Maryam

14.1 INTRODUCTION

The formation of organic soil takes place when the amount of biomass produced by plants surpasses the rate of decomposition, primarily due to strong constraints on the activity of microbes. These conditions include poor oxygen availability due to permanent saturation with water, low temperatures, excessive precipitation, and a variety of physicochemical characteristics including low pH or low nutritional content (Silva Neto et al., 2023). Despite making up just 3% of the earth's surface, organic soils such as high-latitude bogs, mires, and tropical peatlands store a disproportionately large amount of soil organic carbon worldwide—roughly 28% (Silva Neto et al., 2023). These soils cover around 4.23 million km^2 worldwide (Xu et al., 2018). These soils consist of 20% to 35% organic matter at minimum (Mukherjee & Lal, 2015; Grenon et al., 2021). Organic soils possess an acidic pH of less than 4, a low dry bulk density, elevated C:N ratios, a significant tendency for shrinkage, and an extraordinarily high ability to hold soil water (Hewitt et al., 2021).

According to the IPCC (2014), organic soils are generally classified as histosols under the FAO-UNESCO categorization. The histosols are soils that form primarily from "moss peat in boreal, arctic, and subarctic regions, via moss peat, reeds or sedge peat, and forest peat in temperate regions to mangrove peat and swamp forest peat in the humid tropics." "Peat soils," "muck soils," "bog soils," and "organic soils" are among common terms for histosols. Organic soils provide several ecological benefits (Figure 14.1). Apart from their role as carbon sinks, these soils are crucial for preserving the hydrological cycle since the areas containing these soils control the water flow in watersheds. Furthermore, organic soils are extremely relevant to biodiversity because they are linked to a variety of phyto-physiognomies, in which the ecological processes of various fauna and flora species are directly linked to the movement of the water table in organic soils. This, in turn, affects the availability of resources for biodiversity and the preservation of interspecies interactions (Silva Neto et al., 2023).

14.2 ORGANIC SOILS IN AGRICULTURE

Organic soils are often highly valuable for agriculture once they have been drained. They have a high nutritional status and are typically simple to operate. Worldwide,

DOI: 10.1201/9781003491460-14

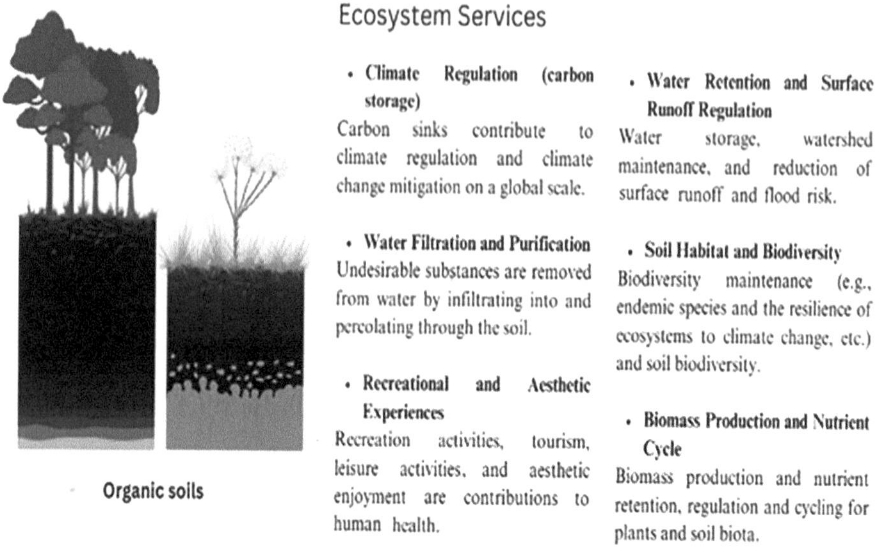

FIGURE 14.1 Illustration of a few primary ecosystem services provided by organic soils (Source Silva Neto et al., 2023).

organic soils are extremely productive cropping systems, especially when it comes to producing vegetables. Vegetable crops may be grown on organic soils with the benefit of abundant water availability and minimal tillage energy requirements. While organic soils are still used for grazing in many parts of Europe, they are also widely utilized for horticultural produce, especially when they are well-drained and include a mineral component in the topsoil (Armstrong & Castle, 1999). The biggest organic soil agricultural region in the United States is the Florida Everglades, covering approximately 2300 km^2 (Kolka et al., 2016), whereas 170,000 km^2 of organic soils are utilized for agriculture in Canada (Oleszczuk et al., 2008). Approximately 125,000 km^2 of peatland are utilized for agriculture in Europe (Grenon et al., 2021). Subtropical and tropical regions present in Asia are used for agriculture; with 2610 km^2 of peatland utilized for agriculture in China (Laine et al., 2009), and 42,000 km^2 of peatland are used for agriculture in Indonesia (Dommain et al., 2018; Jauhiainen et al., 2014; Konecny et al., 2016; Oleszczuk et al., 2008).

Climate, vegetation, peat stratigraphy, mire geomorphology, stage of moorsh formation, air water regime, the need to preserve the environment, and soil physical and chemical characteristics are some of the variables that affect how productive organic soils are for farming. To restore pH and nutrient availability to proper levels, growers can amend organic soils with significant volumes of fertilizer and lime (Castillo & Wright, 2008; Ewing et al., 2012). In Western Europe, sand cover aids arable farming on organic soils by increasing soil carrying capacity (Ilnicki, 2002). For sustainable usage, shallow organic soils should be farmed or afforested, depending on biophysical and economic factors (Kreshtapova et al., 2002).

14.3 IMPACT OF SOIL ORGANIC MATTER ON SOIL WATER

Organic matter boosts soils' capacity to hold onto water, reduces runoff losses, enhances aeration, and, by encouraging granulation, improves soil structure. The amount of water in the soil is directly impacted by soil organic matter (SOM). It influences infiltration and aggregation as well, which indirectly impacts the water content of the soil. It is believed that the quantity and composition of soil organic matter directly affect the properties of soil water. Elevated concentrations of organic matter result in higher soil water contents (Murphy, 2015; Védère et al., 2022). Compared to fine-textured soils, sandy soils' soil water content responds more to an increase in organic matter (Bonfante et al., 2020; Libohova et al., 2018). Water circulation and air aeration in the soil are marginally improved by the granularity produced by combining organic materials, such as humus (a substance produced by the breakdown of organic matter), with a layer of clay (Khusnida et al., 2024). Organic matter in the soil affects the amount of water in the soil indirectly through infiltration, its impact on vegetation, and microbial activity, which results in soil structure and aggregation (Figure 14.2). Through these mechanisms, soil organic matter affects the properties of the soil, reducing bulk density and surface sealing to increase soil porosity (Rawls et al., 2004) and, eventually, improving soil water penetration and retention. Soil organic matter provides aggregate stability against wetting and raindrop impact which in turn can help the soil maintain its infiltration capacity and avoid surface crusting. This minimizes runoff, particularly during periods of heavier rainfall. Reducing runoff ensures that rainwater reaches the soil to be retained for agricultural and pasture growth while also minimizing water erosion

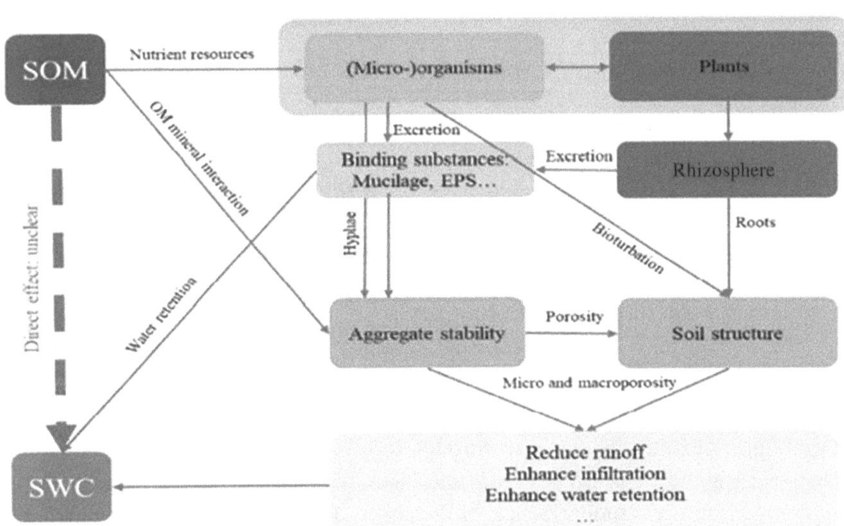

FIGURE 14.2 Diagrammatic depiction of the feedback processes resulting from the interactions between soil, plants, and microbes that indirectly affect SOM on SWC (Source: Védère et al., 2022).

and runoff (Murphy, 2015). As a source of nutrients, soil organic matter affects soil life, influencing the growth and features of the detritusphere and rhizosphere. These zones' processes, in addition to organic matter itself, contribute to the development of aggregates, soil structure, and stability—all of which are essential for regulating soil water content (Védère et al., 2022).

Water-holding capacity is an essential indicator of soil physical fertility since it indicates the soil's ability to store and deliver water for plant development. There is a considerable interplay between bulk density, water-holding capacity, aggregate stability, and soil structure because of the connections to soil porosity. Numerous pedotransfer functions have been developed to forecast the link between soil organic matter and accessible soil water (Murphy, 2015). A collection of linear regression equations encompassing a broad range of soil textures and SOM levels has been developed by Rawls et al. (1992) using data from over 2500 soil strata. The formulas that Rawls employed are:

At 10 kPa

At 1500 kPa

$$\theta = 0.4118 - 0.0030 \times \text{sand} + 0.0023 \times \text{clay} + 0.0317 \times \text{SOM} \quad (14.1)$$

$$\theta = 0.0260 + 0.0050 \times \text{clay} + 0.0158 \times \text{SOM} \quad (14.2)$$

Similarly, Hudson (1994) developed equations for available water capacity (AWC) utilizing surface-horizon data from agricultural soils in the United States using sands, silty loams, and silty clay loams:

$$\text{Sand: AWC}(\text{mm}/\text{mm}) = 0.038 + 0.022(\text{SOM}\%) \quad (14.3)$$

$$\text{Silty loam: AWC}(\text{mm}/\text{mm}) = 0.09 + 0.097(\text{SOM}\%) \quad (14.4)$$

$$\text{Silty clay loam: AWC}(\text{mm}/\text{mm}) = 0.063 + 0.028(\text{SOM}\%) \quad (14.5)$$

The pore geometry that governs soil water dynamics is defined by the structure of the soil. In addition, pore volume, pore size distribution, pore connectivity, and soil aggregation all affect water flow and solute transport in soils. Comprehending the hydraulic characteristics of organic soils is crucial for enhancing irrigation efficiency and managing water quality in farmed areas.

14.4 HYDRAULIC CHARACTERISTICS OF ORGANIC SOIL

The movement of water and nutrients through soils is greatly influenced by the hydraulic characteristics of the soil. These characteristics, which are necessary for a variety of water management operations, include bulk density, available water, permanent wilting point, infiltration rate, field capacity, water holding capacity, and water table depth. They also include soil saturated and unsaturated hydraulic conductivity. The hydraulic properties of soil are highly dynamic and vary due to various factors such as disturbances in land use and management, seasonal and diurnal variations (Zhou et al., 2008), the influence of environmental conditions like temperature and precipitation variations (Hu et al., 2009; Novak et al., 2009), organic activity, and root dynamics (Rasse et al., 2000). Soil texture, soil structure, bulk density, pore volume, pore size distribution, and organic matter are of particular importance and have been shown to influence water transport in soils (Azuka et al., 2022; Strudley et al., 2008; Zhou et al., 2008).

There are three phases in soil: solid, liquid, and gaseous. Plant fibers, humus, and mineral matter, including varying-sized grains (from sand to clay) and amorphous materials like carbonates, phosphates, and hydroxides, comprise the solid phase of organic soils. The primary factor influencing bulk density, water retention, hydraulic conductivity, and drainable porosity in organic soils is their fiber content due to its indirect influence on volume weight, water absorption, and even tillage in the case of surface horizons. Subcellular and amorphous (not fiber), organic material with a particle size of less than 0.1 mm provides significant cation exchange and water retention (Verry et al., 2011), but it also severely restricts hydraulic conductivity due to an increase in the proportion of amorphous material.

14.4.1 Bulk Density and Porosity

An organic soil's bulk density, also known as volume weight, is the weight of a certain volume of soil, often stated on a dry weight basis. In general, there is a strong correlation between this characteristic of organic soils and the moisture-absorbing values (Farnham & Finney, 1965). The bulk density of organic soils is substantially lower than that of mineral soils. Bulk density frequently increases as the degree of the decomposition increases. In general, bulk density decreases with increasing organic matter content. According to Mobilian and Craft (2022), the bulk density of organic soils varies between 0.1 g cm^{-3} for fibrists and 0.2–0.3 g cm^{-3} for hemists and saprists. In very fibrous, undecomposed materials, values vary from 0.05 g/cm^3 to less than 0.5 g/cm^3 in well-decomposed materials (Andriesse, 1988). There is a high correlation between the bulk densities of organic/peat soils and their organic carbon content (Hossain et al., 2015). Because organic soils have a lower bulk density than mineral soils, they often have a larger porosity and a faster rate of water transport. According to Boelter (1974), sapric materials often have less than 85% of pores, but fibric peats typically have a total porosity of 90% by volume. Larger macro and mesopores (diameters ranging from 1×10^{-5} to 1×10^{-3} m) are crucial for the fast movement of water in and out of organic soils.

14.4.2 Water Retention Capacity

Numerous studies have documented the water retention capacity of organic soils (Hallema et al., 2015; Kennedy et al., 2018; Wallor et al., 2018). The creation of soil aggregates and the expansion of soil porosity, which improves water infiltration and retention for plant growth, are encouraged by soil organic matter. The physical, chemical, and hydraulic characteristics of soil are altered as a result of the accelerated decomposition and humification processes that occur when organic soils are cultivated (Kechavarzi et al., 2010; Kroetsch et al., 2011). For instance, due to increased overall porosity and a higher proportion of macropores, undecomposed peat can

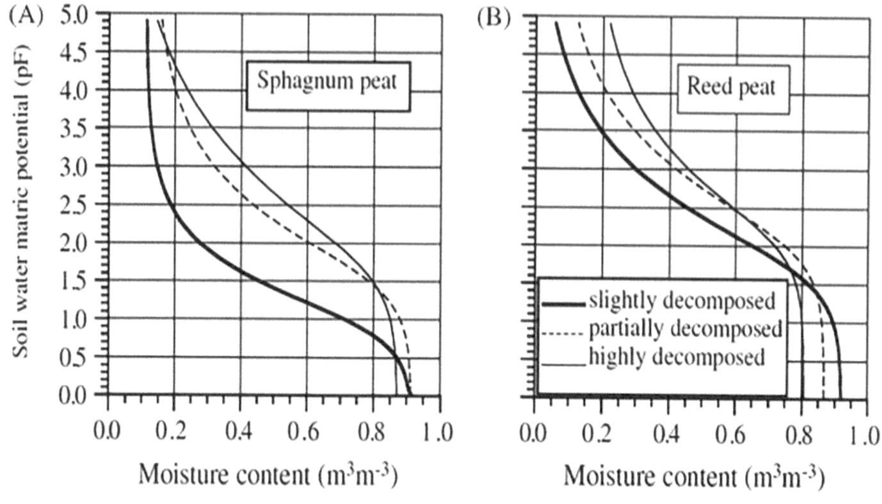

FIGURE 14.3 Water retention curves for Sphagnum peat and reed peat (Source: Brandyk et al., 2002).

retain water better than decomposed peat (Verry et al., 2011). The degree of decomposition and botanical makeup of peat residues determine the connection between matric potential and soil moisture content in organic soils. The water retention curve is the connection between soil water content and matric potential. The Sphagnum peat and reed peat water retention curves (Figure 14.3) illustrate the soil water properties of partially or highly degraded organic matter, with a loss of 11–19% and a moisture content of 4–6%, respectively. In both situations, the moisture content near the wilting point increased as the degree of decomposition increased (Brandyk et al., 2002).

14.4.3 Hydraulic Conductivity

Just like fractured geological media, organic soils contain a dual porosity matrix made up of inactive pores made up of closed, dead-end voids created by the remnants of plant cells and active holes that carry water through the soil matrix (Carey et al., 2007). These soils can have extremely high conductivities when they have a high proportion of coarse roots and tubular stems since they have developed relatively recently and are mostly undecomposed. On the other side, hydraulic conductivity may be extremely low in organic soils that lack structure and have severely degraded organic matter. In contrast to the Darcian behavior on mineral soils, it has been discovered that the saturated hydraulic conductivity (Ksat) in peat soils increases with pore water pressure (Lafond et al., 2014). Pore water pressure also affects the unsaturated hydraulic conductivity, K(h), in peat soils (Holden & Burt, 2003). Furthermore, substantial anisotropy in organic soil profiles has been verified by earlier research. At the surface horizon and at a depth of around 30 cm, it was discovered that the horizontal Ksat in a cutover peat soil was more than four times larger than the vertical conductivities (Surridge et al., 2005). In managed peat soils, Kechavarzi et al. (2010) also noted comparable changes in hydraulic conductivities. Hydraulic conductivity levels in organic soils have been found to decline with time. This happens because peat colloids swell and because fluid movement causes peat particles to migrate, which blocks pores and lowers local porosity (Brandyk et al., 2002). Between the ground surface and a depth of 50 cm or more, the saturated hydraulic conductivity of peat usually decreases by several orders of magnitude (Hoag & Price, 1995). Moreover, some studies indicate that layering of peat increases the hydraulic conductivity heterogeneity throughout the depth profile (Beckwith et al., 2003). This drop much surpasses the predicted decrease resulting from a decrease in the geometric mean pore diameter within the same depth zone (Rezanezhad et al., 2016). In peat soils, water flow is significantly impacted by the separate and combined impacts of hydraulic conductivity anisotropy and heterogeneity (Beckwith et al., 2003).

Water repellency, also known as hydrophobicity, is a frequent property of peat soils and has a substantial impact on the hydrologic processes occurring inside the soil profile. It is especially related to the initial wetness and the degree of material breakdown (Caron et al., 2015). The high hysteretic qualities of these organic layers are partially due to their water repellency (Naasz et al., 2008). Water content and hydraulic conductivity are often higher after drainage than on rewetting for the same matric potential (Caron et al., 2015). Because unstable wetting fronts form in non-structured soils, preferential flow is frequently seen. Fingers or preferred flow routes have been seen in peat soils as a result of many variables, including high water repellency in the soil profile and increased hydraulic conductivity with depth (Dekker & Ritsema, 2000).

14.5 SOIL WATER MOVEMENT IN ORGANIC SOILS

Soil water adheres to the same fundamental principle of seeking equilibrium. It is always flowing toward a lower potential energy. The moving force responsible for flow actually corresponds to the rate at which potential energy decreases with distance. Water has a natural inclination to flow through soil due to variations in its potential energy between different points. Thus, the total potential of soil water may be conceptualized as the product of the individual contributions of many components (Hillel, 2012), such as:

$$\psi_t = \psi_g + \psi_p + \psi_o + \psi_n + \psi_m \tag{14.6}$$

Where

ψ = total potential
ψ = gravitational potential
ψ = pressure (or matric) potential
ψ = osmotic potential
ψ = pneumatic potential
ψ = matric potential

Darcy's Law provides the foundation for water flow in saturated porous material. When the Darcy equation is extended to include the idea of water potentials, it may be expressed as:

$$q = -K_{sat} \frac{d\psi_t}{dz} \tag{14.7}$$

Where

q = water flow density (m/s)

ψ_t = total potential (m)

z = vertical position coordinate (m)

$d\psi/dz$ = total potential gradient

The water flow is explained under unsaturated soil conditions using the following equation by Darcy-Buckingham and Richards:

$$q = -K(\theta) \frac{d\psi_t}{dz} \tag{14.8}$$

$$\frac{d\theta}{dt} = \frac{d}{dz}\left[K(\theta)\frac{d\psi_t}{dz}\right] \tag{14.9}$$

Where

θ = water content

The type and composition of the soil are extensively linked to the composition and flow of water in the soil. Significant variations may be seen in the physical characteristics of various soil types. The biogeochemical processes in drained lands are mostly driven by the flow of water, and assessing the impact of enhanced drainage on drainage water quality requires a detailed grasp of how it affects field and landscape hydrology (Skaggs et al., 1994). Additionally, organic soils exhibit non-Darcian behavior, meaning that during a pumping test, the apparent hydraulic conductivity increases, suggesting that hydraulic conductivity rises with hydraulic gradient (Rycroft et al., 1975). Because soil organic matter has a significant water-holding capacity, it can also alter the availability of water. Thus, a rise in organic matter will cause the amount of water that is accessible to increase. It has been noted that the soil surface layer has greater FC and PWP because of the layer's higher SOM content (Costa et al., 2013). Numerous factors, including land use, crop management, and natural soil qualities, influence how SOM affects water availability (Lal, 2020).

The dual-porosity medium found in organic soils consists of a "mobile region" where nutrients and water flow through relatively easily and an "immobile region" where fluid flow velocity is minimal (Rezanezhad et al., 2016). Drying and shrinking cause changes in the paths that water takes while moving through the unsaturated zone of organic soils. Water flow in soils can be directed by the structure of the peat and the direction of the decomposing plant material (Wang et al., 2020). Compared to decomposed peat, undigested peat has a higher drainage affinity because of its lower air-entry pressure (Kechavarzi et al., 2010). According to Kechavarzi et al. (2010), humification and decomposition cause the soil to shrink and deteriorate, which can reduce the amount of water that can be retained, transmitted, and stored. Predicting the drainage characteristics of cultivated organic soils becomes more challenging because varying levels of decomposition can be discovered within the same soil profile. This results in stratification, where a finer, more humic layer sits on top of a coarser, more fibric layer (Lafond et al., 2014). As hydraulic conductivity rises with depth in these soil profiles (fine-textured profiles over coarse-textured profiles), preferential soil water movement develops (Dekker & Ritsema, 2000). Moreover, organic soils' hydrophobicity can cause erratic flow patterns and irreversible drying processes in OM, which make it harder to rewet the soil (Dekker & Ritsema, 2000).

Peatlands are extremely important as water reserves due to two key reasons. Peatlands, like wetlands, are found in landscape depressions that receive fluxes of surface and subsurface water from the upslope direction. Peatlands are made of organic components that absorb a large amount of water from the environment and may store up to 17 times their weight in water (Campos et al., 2012). Since it enables peatlands to retain water during the rainy season and release it during dry times, this "sponge-like" behavior (Silva et al., 2013) is especially significant in regions with substantial seasonality in rainfall distribution. On peatlands with somewhat deeper water tables, such as those below 15 cm, water retention features have a significant impact on the specific yield (change in water table depth resulting from precipitation) (Dettmann & Bechtold, 2016). Given the shallow water table, hydraulic equilibrium may be presumed, and if the soil's water retention properties are known, the water table depth can be utilized to estimate the amount of water present (Dimitrov & Lafleur, 2021).

In general, hydraulic conductivity (K) drops from 0.5 to 0.1 over a 50 cm distance; occasionally, it drops even more (Hoag & Price, 1995). Specific yield (Sy) also reduces over this same range from 0.5 to 0.1. In contrast, there is often less change in hydraulic structure with depth in the bottom layer of more degraded peat, known as the catotelm (Ingram, 1978). Since the more degraded peat in the catotelm has fewer pores than the peat in the acrotelm, the flowpath is more convoluted and there are more frictional effects on the flow, resulting in a lower K. This bilayer structure is essential to the control of water storage and release in intact bogs. Bogs easily release "excess" water because flow occurs through and over the more porous top layers of the acrotelm when the water table is high (Bay, 1969). With just a little dip in the water table, this efflux is made possible by the high Sy in the near-surface strata. But prolonged drainage combined with evapotranspiration losses lowers the water table, allowing flow through lower permeability layers (Boelter, 1974). A very minimal capillary rise is required to guarantee a sufficient moisture supply to the developing section of the plant because of the high storativity supplied by these loosely packed partially or undecomposed mosses, which maintain a high and stable water table.

14.6 DRAINAGE OF ORGANIC SOILS

The moist circumstances that lead to the creation of organic soils sometimes restrict their agricultural value. Therefore, in order to use them for agriculture, drainage is frequently needed. However, the deterioration of organic soils is also a result of their agricultural usage. Because these soils were originally wet, they nearly always need artificial drainage to reach their full potential. Drainage encompasses both the management of floods (via arterial drainage) and the management of moisture inside the field (through traditional under-drainage) (Armstrong & Castle, 1999). When water is withdrawn from organic soils, the underlying organic matter undergoes accelerated oxidation and breakdown rates, which results in large-scale atmospheric emissions of carbon dioxide (CO_2) and nitrous dioxide (N_2O) (Conchedda & Tubiello, 2020). Due to the introduction of perennial crops like oil palm in Southeast Asia after the 1990s, agriculture has been identified as a key contributor to the draining of organic soils worldwide. According to calculations using AR5 GWP for the N_2O emissions, the worldwide greenhouse gas emissions from drained cropland and grassland organic soils in 2019 were calculated as 833 MtCO_2eq (IPCC, 2014) (Figure 14.4). Nearly 25 Mha or about 7.5% of the 328 Mha of worldwide histosols had been drained for agriculture with a limited increase since 1990 (Conchedda & Tubiello, 2020).

The volume of undrained organic soil is reduced as water is removed, either by drainage or evapotranspiration from the surface (Fuchsman, 1986). Undrained organic soil has a high water content, usually between 80 and 90%. The oxidation of the organic matter and the loss of water cause the surface of an organic soil to shrink and the ground level to drop with drainage. Organic soils that are drained for farming or other uses always result in land subsidence. Given that organic soils have a high porosity, or saturated water content (up to 80–90%), and an exceptionally low density, the impacts of compaction and desiccation during first drainage can be substantial. When peat bogs are drained, the equilibrium between the buildup and breakdown of organic material is drastically altered. While organic soil naturally

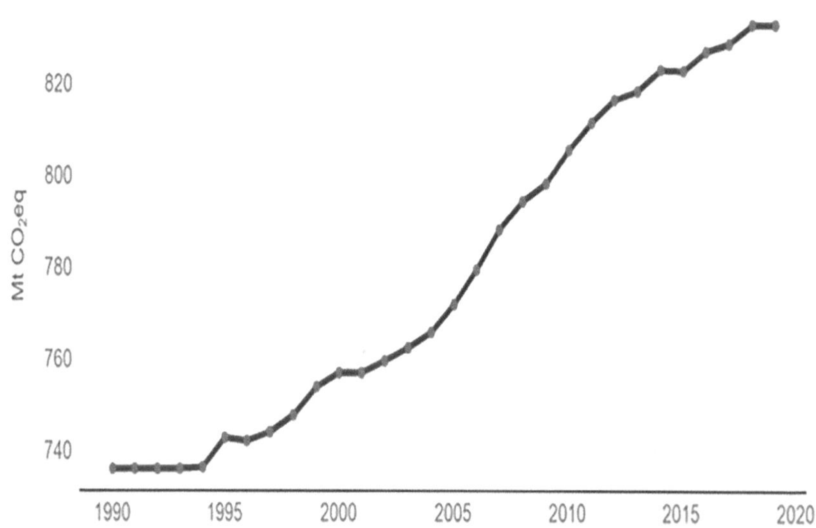

FIGURE 14.4 Global emissions from drained organic soils (Source: Conchedda & Tubiello, 2020).

accumulates a few inches every 100 years, the pace at which drained organic soil is lost can be 100 times higher—in severe circumstances, up to a few inches annually. Consequently, deposits that have accumulated over a very long period that is highly important to human activities may eventually evaporate (Galloway, 2016).

14.7 CONCLUSION

Physical, chemical, and biological elements interact in a complicated way to determine how water moves in organic soils. Soil organic matter (SOM) plays a crucial role in soil water management, enhancing its capacity to hold water, reduce runoff losses, enhance aeration, and improve soil structure. It directly impacts soil water content, with higher concentrations resulting in higher soil water contents. Mineral soils and organic soils are not the same in many ways. In comparison to mineral soils, organic soils have varied soil porosities due to "mobile" and "immobile" areas in the soil matrix. This results in higher preferential flow, hydrophobicity, and variable expansion and contraction characteristics. Organic soils have a lower bulk density than mineral soils, so water moves through them more quickly and with greater porosity. Water retention capacity in organic soils is influenced by soil organic matter, which encourages soil aggregates and expands porosity. Peatlands are crucial water reserves due to their ability to absorb large amounts of water from the environment. However, organic soils often require artificial drainage to reach their full potential due to their moist conditions. Agriculture has been identified as a key contributor to the draining of organic soils worldwide, with high global greenhouse gas emissions from drained cropland and grassland organic soils. Further, drainage of organic soil reduces the volume resulting in land subsidence. We can more effectively manage and make use of these special soils for a range of uses, from conservation to agriculture, if we comprehend their processes. To adjust to changes and guarantee the sustainable use of organic soils in the face of changing environmental challenges, ongoing study and monitoring are crucial.

REFERENCES

Andriesse, J. P. (1988). *Nature and management of tropical peat soils* (No. 59). Food & Agriculture Organization.

Armstrong, A. C., & Castle, D. A. (1999). Drainage of organic soils. *Agricultural Drainage*, *38*, 1083–1105.

Azuka, C. V., Umeugokwe, C. P., & Ajoagu, G. M. (2022). Hydraulic properties of different soil types and its implication on koupendri catchment hydrology. *Bulgarian Journal of Agricultural Science*, *28*(4).

Bay, R. R. (1969). Runoff from small peatland watersheds. *Journal of Hydrology*, *9*(1), 90–102.

Beckwith, C. W., Baird, A. J., & Heathwaite, A. L. (2003). Anisotropy and depth-related heterogeneity of hydraulic conductivity in a bog peat. I: Laboratory measurements. *Hydrological Processes*, *17*(1), 89–101.

Boelter, D. H. (1974). The hydrologic characteristics of undrained organic soils in the Lake States. *Histosols: Their Characteristics, Classification, and Use*. *6*, 33–46.

Bonfante, A., Basile, A., & Bouma, J. (2020). Exploring the effect of varying soil organic matter contents on current and future moisture supply capacities of six Italian soils. *Geoderma*, *361*, 114079.

Brandyk, T., Szatylowicz, J., Oleszczuk, R., & Gnatowski, T. (2002). Water-related physical attributes of organic soils. *Organic Soils and Peat Materials for Sustainable Agriculture*, 33–66.

Campos, J. R. D. R., Silva, A. C., & Vidal-Torrado, P. (2012). Mapping, organic matter mass and water volume of a peatland in Serra do Espinhaço Meridional. *Revista Brasileira de Ciência do Solo*, *36*, 723–732.

Carey, S. K., Quinton, W. L., & Goeller, N. T. (2007). Field and laboratory estimates of pore size properties and hydraulic characteristics for subarctic organic soils. *Hydrological Processes: An International Journal*, *21*(19), 2560–2571.

Caron, J., Price, J. S., & Rochefort, L. (2015). Physical properties of organic soil: Adapting mineral soil concepts to horticultural growing media and histosol characterization. *Vadose Zone Journal*, *14*(6), 1–14.

Castillo, M. S., & Wright, A. L. (2008). Soil phosphorus pools for Histosols under sugarcane and pasture in the Everglades, USA. *Geoderma*, *145*(1–2), 130–135.

Conchedda, G., & Tubiello, F. N. (2020). Drainage of organic soils and GHG emissions: Validation with country data. *Earth System Science Data Discussions*, *2020*, 1–47.

Costa, A. D., Albuquerque, J. A., Costa, A. D., Pértile, P., & Silva, F. R. D. (2013). Water retention and availability in soils of the State of Santa Catarina-Brazil: Effect of textural classes, soil classes and lithology. *Revista Brasileira de Ciência do Solo*, *37*, 1535–1548.

Dekker, L. W., & Ritsema, C. J. (2000). Wetting patterns and moisture variability in water repellent Dutch soils. *Journal of Hydrology*, *231*, 148–164.

Dettmann, U., & Bechtold, M. (2016). Deriving effective soil water retention characteristics from shallow water table fluctuations in peatlands. *Vadose Zone Journal*, *15*(10), vzj2016-04.

Dimitrov, D. D., & Lafleur, P. M. (2021). Revisiting water retention curves for simple hydrological modelling of peat. *Hydrological Sciences Journal*, *66*(2), 252–267.

Dommain, R., Frolking, S., Jeltsch-Thömmes, A., Joos, F., Couwenberg, J., & Glaser, P. H. (2018). A radiative forcing analysis of tropical peatlands before and after their conversion to agricultural plantations. *Global Change Biology*, *24*(11), 5518–5533.

Ewing, J. M., Vepraskas, M. J., Broome, S. W., & White, J. G. (2012). Changes in wetland soil morphological and chemical properties after 15, 20, and 30 years of agricultural production. *Geoderma*, *179*, 73–80.

Farnham, R. S., & Finney, H. R. (1965). Classification and properties of organic soils. *Advances in Agronomy*, *17*, 115–162.

Fuchsman, C. H. (Ed.). (1986). *Peat and water: Aspects of water retention and dewatering in peat*. Springer Science & Business Media.

Galloway, K. E. (2016). *Natural treatment systems for stormwater cleanup in Southern California: A biofilter coastal case study*. University of California.

Grenon, G., Singh, B., De Sena, A., Madramootoo, C. A., von Sperber, C., Goyal, M. K., & Zhang, T. (2021). Phosphorus fate, transport and management on subsurface drained agricultural organic soils: A review. *Environmental Research Letters*, *16*(1), 013004.

Hallema, D. W., Périard, Y., Lafond, J. A., Gumiere, S. J., & Caron, J. (2015). Characterization of water retention curves for a series of cultivated Histosols. *Vadose Zone Journal*, *14*(6), 1–8.

Hewitt, A. E., Balks, M. R., Lowe, D. J., Hewitt, A. E., Balks, M. R., & Lowe, D. J. (2021). Organic soils. *The Soils of Aotearoa New Zealand*, 113–132.

Hillel, D. (2012). *Applications of soil physics*. Elsevier.

Hoag, R. S., & Price, J. S. (1995). A field-scale, natural gradient solute transport experiment in peat at a Newfoundland blanket bog. *Journal of Hydrology*, *172*(1–4), 171–184.

Holden, J., & Burt, T. P. (2003). Hydraulic conductivity in upland blanket peat: Measurement and variability. *Hydrological Processes*, *17*(6), 1227–1237.

Hossain, M. F., Chen, W., & Zhang, Y. (2015). Bulk density of mineral and organic soils in the Canada's arctic and sub-arctic. *Information Processing in Agriculture*, *2*(3–4), 183–190.

Hu, W., Shao, M., Wang, Q., Fan, J., & Horton, R. (2009). Temporal changes of soil hydraulic properties under different land uses. *Geoderma*, *149*(3–4), 355–366.

Hudson, B. D. (1994). Soil organic matter and available water capacity. *Journal of Soil and Water Conservation*, *49*(2), 189–194.

Ilnicki, P. (2002). Agricultural production systems for organic soil conservation. In *Organic soils and peat materials for sustainable agriculture* (pp. 187–200). CRC Press.

Ingram, H. A. P. (1978). Soil layers in mires: Function and terminology. *Journal of Soil Science*, *29*(2), 224–227.

IPCC. (2014). *Mitigation of climate change: Contribution of working group III to the fifth assessment report of the intergovernmental panel on climate change* (O. Edenhofer, R. Pichs-Madruga, Y. Sokona, E. Farahani, S. Kadner, K. Seyboth, A. Adler, I. Baum, S. Brunner, P. Eickemeier, B. Kriemann, J. Savolainen, S. Schlömer, C. von Stechow, T. Zwickel & J. C. Minx, (Eds.). Cambridge University Press.

Jauhiainen, J., Kerojoki, O., Silvennoinen, H., Limin, S., & Vasander, H. (2014). Heterotrophic respiration in drained tropical peat is greatly affected by temperature—A passive ecosystem cooling experiment. *Environmental Research Letters*, *9*(10), 105013.

Kechavarzi, C., Dawson, Q., & Leeds-Harrison, P. B. (2010). Physical properties of low-lying agricultural peat soils in England. *Geoderma*, *154*(3–4), 196–202.

Kennedy, C. D., Alverson, N., Jeranyama, P., & DeMoranville, C. (2018). Seasonal dynamics of water and nutrient fluxes in an agricultural peatland. *Hydrological Processes*, *32*(6), 698–712.

Khusnida, A., Shohrukhbek, B., Pakhlovonjon, K., & Javohirbek, M. (2024). Composition and movement of water in soil. In *Proceedings of International Conference on Educational Discoveries and Humanities* (Vol. 3, No. 1, pp. 13–17).

Kolka, R., Bridgham, S. D., & Ping, C. L. (2016). Soils of peatlands: Histosols and gelisols. *Wetland soils: Genesis, hydrology, Landscapes, and Classification*, *2*.

Konecny, K., Ballhorn, U., Navratil, P., Jubanski, J., Page, S. E., Tansey, K., & Siegert, F. (2016). Variable carbon losses from recurrent fires in drained tropical peatlands. *Global Change Biology*, *22*(4), 1469–1480.

Kreshtapova, V. N., Krupnov, R. A., & Uspenskaya, O. N. (2002). Quality of organic soils for agricultural use of cutover peatlands in Russia. In *Organic Soils and Peat Materials for Sustainable Agriculture* (pp. 175–186). CRC Press.

Kroetsch, D. J., Geng, X., Chang, S. X., & Saurette, D. D. (2011). Organic soils of Canada: Part 1. Wetland organic soils. *Canadian Journal of Soil Science*, *91*(5), 807–822.

Lafond, J. A., Piette, É. B., Caron, J., & Rancourt, G. T. (2014). Evaluating fluxes in Histosols for water management in lettuce: A comparison of mass balance, evapotranspiration and lysimeter methods. *Agricultural Water Management*, *135*, 73–83.

Laine, J., Minkkinen, K., & Trettin, C. (2009). Direct human impacts on the peatland carbon sink. *Carbon Cycling in Northern Peatlands*, *184*, 71–78.

Lal, R. (2020). Soil organic matter and water retention. *Agronomy Journal*, *112*(5), 3265–3277.

Libohova, Z., Seybold, C., Wysocki, D., Wills, S., Schoeneberger, P., Williams, C., Lindbo, D., Stott, D., & Owens, P. R. (2018). Reevaluating the effects of soil organic matter and other properties on available water-holding capacity using the National Cooperative Soil Survey Characterization Database. *Journal of Soil and Water Conservation*, *73*(4), 411–421.

Mobilian, C., & Craft, C. B. (2022). Wetland soils: Physical and chemical properties and biogeochemical processes. In *Reference module in earth systems and environmental sciences* (pp. 157–168). https://doi.org/10.1016/B978-0-12-819166-8.00049-9

Mukherjee, A., & Lal, R. (2015). Tillage effects on quality of organic and mineral soils under on-farm conditions in Ohio. *Environmental Earth Sciences*, *74*, 1815–1822.

Murphy, B. W. (2015). Impact of soil organic matter on soil properties—A review with emphasis on Australian soils. *Soil Research*, *53*(6), 605–635.

Naasz, R., Michel, J. C., & Charpentier, S. (2008). Water repellency of organic growing media related to hysteretic water retention properties. *European Journal of Soil science*, *59*(2), 156–165.

Novak, J. M., Lima, I., Xing, B., Gaskin, J. W., Steiner, C., Das, K. C., & Schomberg, H. (2009). Characterization of designer biochar produced at different temperatures and their effects on a loamy sand. *Annals of Environmental Science*, *3*, 195–206.

Oleszczuk, R., Regina, K., Szajdak, L., Höper, H., & Maryganova, V. (2008). Impacts of agricultural utilization of peat soils on the greenhouse gas balance. *Peatlands and Climate Change*, 70–97.

Rasse, D. P., Smucker, A. J., & Santos, D. (2000). Alfalfa root and shoot mulching effects on soil hydraulic properties and aggregation. *Soil Science Society of America Journal*, *64*(2), 725–731.

Rawls, W. J., Ahuja, L. R., & Brakensiek, D. L. (1992). Estimating soil hydraulic properties from soils data. *Indirect Methods for Estimating the Hydraulic Properties of Unsaturated Soils*, *32*(9), 329–340.

Rawls, W. J., Nemes, A. T. T. I. L. A., & Pachepsky, Y. A. (2004). Effect of soil organic carbon on soil hydraulic properties. *Developments in Soil Science*, *30*, 95–114.

Rezanezhad, F., Price, J. S., Quinton, W. L., Lennartz, B., Milojevic, T., & Van Cappellen, P. (2016). Structure of peat soils and implications for water storage, flow and solute transport: A review update for geochemists. *Chemical Geology*, *429*, 75–84.

Rycroft, D. W., Williams, D. J. A., & Ingram, H. A. P. (1975). The transmission of water through peat: I. Review. *The Journal of Ecology*, 535–556.

Silva, M. L. D., Silva, A. C., Silva, B. P. C., Barral, U. M., Soares, P. G., & Vidal-Torrado, P. (2013). Surface mapping, organic matter and water stocks in peatlands of the Serra do Espinhaço Meridional-Brazil. *Revista Brasileira de Ciência do Solo*, *37*, 1149–1157.

Silva Neto, E. C., Coelho-Junior, M. G., Horák-Terra, I., Gonçalves, T. S., Anjos, L. H. C., & Pereira, M. G. (2023). Organic soils: Formation, classification and environmental changes records in the highlands of Southeastern Brazil. *Sustainability*, *15*(4), 3416.

Skaggs, R. W., Amatya, D., Evans, R. O., & Parsons, J. E. (1994). Characterization and evaluation of proposed hydrologic criteria for wetlands. *Journal of Soil and Water Conservation*, *49*(5), 501–510.

Strudley, M. W., Green, T. R., & Ascough II, J. C. (2008). Tillage effects on soil hydraulic properties in space and time: State of the science. *Soil and Tillage Research*, *99*(1), 4–48.

Surridge, B. W., Baird, A. J., & Heathwaite, A. L. (2005). Evaluating the quality of hydraulic conductivity estimates from piezometer slug tests in peat. *Hydrological Processes: An International Journal*, *19*(6), 1227–1244.

Védère, C., Lebrun, M., Honvault, N., Aubertin, M. L., Girardin, C., Garnier, P., & Rumpel, C. (2022). How does soil water status influence the fate of soil organic matter? A review of processes across scales. *Earth-Science Reviews*, *234*, 104214.

Verry, E. S., Boelter, D. H., Päivänen, J., Nichols, D. S., Malterer, T., & Gafni, A. (2011). Physical properties of organic soils. *Peatland Biogeochemistry and Watershed Hydrology at the Marcell Experimental Forest*, 135–176.

Wallor, E., Rosskopf, N., & Zeitz, J. (2018). Hydraulic properties of drained and cultivated fen soils: Part I-Horizon-based evaluation of van Genuchten parameters considering the state of moorsh-forming process. *Geoderma*, *313*, 69–81.

Wang, M., Liu, H., Zak, D., & Lennartz, B. (2020). Effect of anisotropy on solute transport in degraded fen peat soils. *Hydrological Processes*, *34*(9), 2128–2138.

Xu, J., Morris, P. J., Liu, J., & Holden, J. (2018). PEATMAP: Refining estimates of global peatland distribution based on a meta-analysis. *Catena*, *160*, 134–140.

Zhou, X., Lin, H. S., & White, E. A. (2008). Surface soil hydraulic properties in four soil series under different land uses and their temporal changes. *Catena*, *73*(2), 180–188.

15 Design of Different Structures for Composting

Nifa Mehraj, Rohitashw Kumar, and Munjid Maryam

15.1 INTRODUCTION

Worldwide waste generation has increased due to urbanization and the continuous increase in human population. Numerous environmental, social, and economic issues have arisen as a result of these waste streams, particularly in developing nations (Awasthi et al., 2014; Sukholthaman & Sharp, 2016). Human health suffers from poor waste management. In addition to being undesirable, it pollutes the air, damages water bodies when disposed of in the water, and, when burned, destroys the ozone layer, intensifying the effects of climate change. Frequently, wastes are handled incorrectly while employing traditional techniques (Aruna et al., 2018). Wastes are burned, thrown into rivers and oceans, and piled up along the sides of roadways (Ogwueleka, 2009). In addition to being unappealing these activities create disagreeable odours, attract flies and other pests, and (during burning) contribute to global warming. Adopting efficient waste management techniques is necessary to allay these worries (Chiee & Manaf, 2017).

Composting has been more popular among the several waste treatment solutions because of its related benefits, which include waste hygienization, cost efficiency, and waste conversion into value-added goods (Qian et al., 2014). Composting is a secure way to handle waste. Microorganisms break down complex degradable materials during the aerobic process of composting, producing both organic and inorganic byproducts (Toledo et al., 2018). The process of composting is biochemical and heterogeneous, involving the mineralization of organic materials to CO_2, NH_3, and H_2O as well as partial humification. The end product is stabilized and has a lower toxicity and harmful organism count (Das et al., 2011). It is a set of methods for treating organic waste that completely aligns with sustainable agriculture (Albrecht et al., 2011). Reducing the potential harmful impacts of applying organic waste to the soil may be achieved by composting, a dependable waste management method. This is because it offers items that have been cleaned and stabilized, which may be used in soil amendments or as a possible source of organic fertilizers (Qian et al., 2014). The process of breaking down organic materials into smaller amounts is continuous (Raut et al., 2008). This happens in controlled or natural environments when

DOI: 10.1201/9781003491460-15

microbial activity breaks down organic materials to produce compost, a beneficial product (Gigliotti et al., 2005). Thus, physico-chemical factors including temperature, pH, particle size, moisture content, aeration, and electrical conductivity affect the composting process (Juárez et al., 2015; Li et al., 2013).

Composting is applied to improve soil organic matter content and agricultural output because composted materials contain adequate nutrients and are home to organisms that promote plant development (Luo et al., 2017). This greatly contributes to ensuring food security. Compost can be used for bioremediation (Ventorino et al., 2019), plant disease management (Pane et al., 2019), weed control (Coelho et al., 2019), pollution prevention (Uyizeye et al., 2019), erosion control, landscaping, and wetland restoration in addition to being used as fertilizer. Additionally, composting lowers the environmental dangers resulting from synthetic fertilizer and improves soil biodiversity (Pose-Juan et al., 2017). Despite the fact that composting has many of the previously mentioned benefits that have been reported, this waste management technique has drawbacks as well. These include the method's effects on climate change, its release of carbon dioxide into the atmosphere, its depletion of oxygen, and the production of offensive odours due to the release of hydrogen sulphide from anaerobic activity, a lengthier preparation time, a lengthy mineralization period, and the potential for some pathogens—known as thermo tolerant pathogens—as well as low nutritional content. Because of these health effects, various organizations in different countries have implemented different rules regarding the usage of this method.

15.2 COMPOSTING: MICROBIOLOGY OF COMPOSTING

The end product of carefully regulated biological degradation of organic materials is compost. More precisely, compost is the stable, humus-like residue of carefully regulated biological breakdown of organic waste. Composting differs from the organic matter's natural breakdown in that it involves human control over the biological decomposition process. Whether or not we compost organic waste, they are still recycled; nevertheless, controlling and optimizing the environment guarantees a quicker process and the production of a high-quality final product.

Organic wastes naturally degrade; the process starts immediately after the wastes are produced. The majority of the first breakdown of organic materials into smaller particles occurs naturally by earthworms, nematodes, and soil insects such mites, springtails, ants, and beetles, which increases their susceptibility to microbial degradation. Composting operators use grinding or chopping to break down big waste particles under regulated circumstances. The resident microbial population is primarily in charge of the biodegradation and conversion process during composting (Hafeez et al., 2018). A diverse microbial community's activity is what causes composting to happen. The most abundant microorganisms among those reported to be present during composting are fungi and bacteria (Galitskaya et al., 2017). Mesophilic and thermophilic aerobic bacteria are the two types of microorganisms that are engaged in composting different types of waste. These organisms, which control various stages of composting, may be bacteria, actinomycetes, moulds, or yeasts. The three critical stages of the composting process are the thermophilic, initial mesophilic, and final mesophilic stages (sometimes referred to as the cooling). A mesophilic stage,

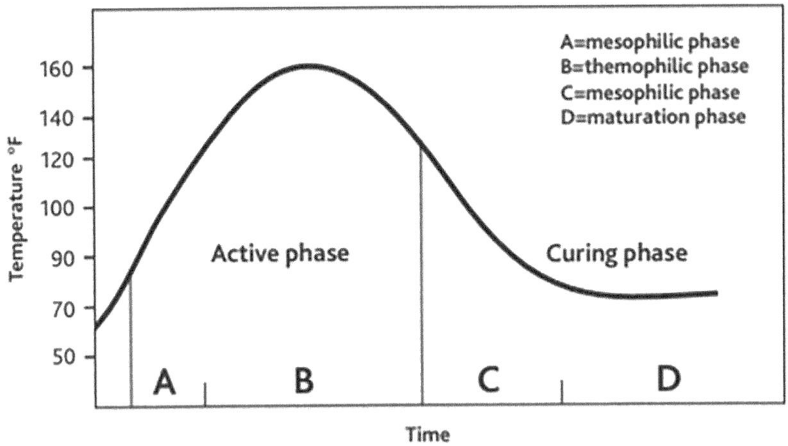

FIGURE 15.1 Temperature changes in composting.

with temperatures between 20 and 40°C, might precede the composting process. Following the mesophilic stage is the thermophilic stage. Compared to the mesophilic stage, active breakdown occurs in the thermophilic stage (40–70°C) (Hafeez et al., 2018). Mesophilic species are eliminated or rendered inactive during this phase, while the variety and population of actinomycetes, bacteria, and fungi that are thermophiles or thermotolerant rise (Chennaoui et al., 2018). Following the thermophile stage, the compost reaches its second mesophilic stage, sometimes referred to as the curing phase, at which point it has reached maturity (Figure 15.1).

The carbon (C) molecules found in organic materials are converted into carbon dioxide (CO_2) and discharged into the environment by microbes as a source of energy. The compost gets denser and the air holes inside the pile shrink as C is removed from the pile. The remaining O_2 in the pile is rapidly used up by the local bacteria, therefore it needs to be added back in to keep the system from being anaerobic. The first materials to oxidize are those that break down the quickest, including sugars and starches. The main ingredients of the final compost product are substances like cellulose and lignin that are resistant to breakdown.

15.3 BENEFITS OF COMPOSTING

Growing interest has been shown in composting as an eco-friendly method of using and discarding organic waste (Zhang & Sun, 2014). According to Raut et al., (2008), composting recycles organic materials into valuable products and aids in the sustainable management of huge volumes of organic wastes. Relatively persistent organic chemicals, such as veterinary medicines, have been demonstrated to be effectively reduced by composting (Li et al., 2015). It has also been discovered that compost, the end result of composting, has several advantages. It has recently been demonstrated that composts made from organic wastes are of higher quality than commercial inorganic fertilizers (Chowdhury et al., 2015). Hermann et al. (2011) state that composts can take the role of soil conditioners in order to promote the creation of humus, which is an advantage that cannot be produced artificially. According to Traversa et al. (2010), the humic-like proportion of composts promotes healthy plant development and also acts as a biocontrol agent against several soil-borne phyto-pathogens, including fungus. Composting successfully rendered Parascaris equorum eggs non-viable, despite the fact that these eggs are commonly recognized to have pathogenic consequences when consumed by animals, including horses (Gould et al., 2013). Pathogenic organisms have been eliminated from trash as a consequence of temperature buildup during composting (Pandey et al., 2016). According to Mohammad et al. (2012), compost can help increase the soil's ability to store water. They went on to say that compost offers superior tilts.

With the recent push against the use of synthetic fertilizers, compost in the form of compound fertilizer is a welcome concept right now. Majbar et al. (2018) state that compost enhances plant production and soil fertility. Another method of applying compost for plant development is to combine it with synthetic fertilizer. Research has indicated that synthetic fertilizer may be a more effective tool for promoting plant development than compost (Pampuro et al., 2017). For this reason, it is recommended to apply both fertilizers in the proper amounts together. Additionally, microorganisms that support plant growth are found in composts and contribute to increased soil fertility and plant growth. The fertility of the soil is diminished by erosion. Erosion results in a significant loss of potassium, phosphorus, and nitrogen. It has been shown that using organic amendments put to the surface works extremely well to stop erosion. Compost improves the soil's ability to store water as well as its structure and aggregate stability (Gonawala & Jardosh, 2018). The reason for this is the existence of humus, which is a stable residue left over from a high degree of organic matter decomposition. Humus binds to the soil and is sometimes referred to as the "glue" that holds the contents of the soil together (Epelde et al., 2018).

A biological control for plant diseases is provided by compost. The bacteria found in compost fight pathogenic counterparts using various strategies. These comprise competition for resources, parasitism, predation, lytic production, antibiotic synthesis, and the synthesis of additional extracellular chemicals or enzymes (Olanrewaju et al., 2017). For example, *Bacillus* sp. in compost has been found to counteract damping-off diseases and plant-wilt (Lin et al., 2014). Composting is an effective way to remediate heavy metal-polluted soil. Chlorinated and non-chlorinated hydrocarbons, solvents, heavy metals, pesticides, petroleum products, explosives, and wood-preserving compounds have all been degraded in soil by compost. By absorbing or breaking down some chemical contaminants, compost might lessen their toxicity (Huang et al., 2017). Redox processes, complexation, adsorption, and precipitation can all render heavy metals inaccessible. Managing degradable organic waste safely may be achieved by composting. Composting is a viable option for waste that would otherwise be burnt, roadside dumps, or disposed of in aquatic bodies. These composted waste materials' byproducts are put to many beneficial applications (Khater, 2015).

15.4 FACTORS INFLUENCING THE COMPOSTING PROCESS

The process of composting is driven by microbes. For survival and growth, bacteria require the correct conditions, just like any other living thing. Microbes require nutrient-rich "food," as well as the right amount of moisture, pH, temperature, and oxygen, for composting to be successful. The factors influencing composting include pH, composting temperature, texture of raw materials, particle size, moisture content, oxygen, C/N ratio, and turning frequency.

15.4.1 pH

The pace of composting is influenced by the pH of the materials being composted, which is a measurement of the acidity or alkalinity of the materials in the compost pile. The development and activity of microorganisms as well as the fate of N compounds are influenced by the pH. It has been suggested that alkaline pH is ideal for composting. Composting proceeds very slowly at an acidic pH because the bacteria are killed (Ameen et al., 2016). For bacteria, the ideal pH range is 6.0 to 7.5, whereas for fungi, it is 5.5 to 8.0. The likelihood of ammonia gaseous losses increases when the pH of the compost rises over 7.5. Food processing wastes can drop pH, whereas other items, including dairy manure and paper mill wastes, can boost pH. Nonetheless, maintaining pH within an ideal range is challenging and seldom done. Both inside the pile and throughout the composting process, the pH fluctuates. The pH of finished compost typically ranges from 6.5 to 7.5 (Ameen et al., 2016).

15.4.2 Composting Temperature

The pace at which the raw materials disintegrate is influenced by microbial development and activity, which are influenced by the temperature within compost heaps. Increased temperatures cause diseases to die, weed seeds to be destroyed, and organic materials to break down more quickly. On the other hand, temperatures over 70°C have the potential to suppress microbial activity. Temperatures between 40° and 70°C, known as thermophilic temperatures, are optimal for composting. In order to expedite the composting process and get rid of pathogenic organisms that are bad for soil organisms, plants, animals, and people, temperature is a crucial component in composting (Hafeez et al., 2018). The temperature at which microorganisms reside during the composting process determines their classification. Mesophilic microorganisms are those that grow between 20 and 40°C, while thermophilic bacteria are those that thrive between 40 and 70°C (Chennaoui et al., 2018). The composting process is initiated by mesophilic microbes, which decompose the readily degradable components of the waste. The compost heats up quickly as a result of their metabolism. Even if the temperature doesn't always reach 45°C throughout the composting process, bacteria can still perish when the nutrients in the composting materials run out and when competing organisms release enzymes that can kill the pathogens.

15.4.3 Texture of Raw Materials

Organic materials that are degradable and have a hard texture, high lignin content, or a rough texture often decompose slowly. Hard-textured leaves, for example, tend to decompose more slowly than soft-textured leaves. Because of their physical barrier, leaves with thorns may also take longer to decompose. A high lignin concentration may be the cause of leaves that have a leathery or hard feel (Wilson, 1989).

15.4.4 Particle Size

The ideal conditions for composting are often reached when the diameter of the material falls between one and two inches (Misra et al., 2003). This size results in a larger surface area, which speeds up the composting process and microbial activity. As the particle size gets smaller, aerobic breakdown happens more quickly. Nonetheless, minuscule particles have the potential to hinder the flow of oxygen in the pile, thus reducing the pace of composting. Additionally, small particle size promotes moisture retention and decreases airspace, which slows down the composting process (Zhao et al., 2017).

15.4.5 Moisture Content

One essential component that keeps bacteria's metabolic processes going is moisture. Materials intended for composting should have a moisture level of 40% to 60% (Ameen et al., 2016). It has been suggested that the moisture in compost comes from the water that was introduced initially or from the water that the microbe produces during metabolism. A decrease in oxygen diffusion caused by excess water lowers the metabolic activity of the organisms. Water is the only substance that microbial cells need for their metabolic processes. Therefore, only when organic molecules have been dissolved in water may bacteria begin to metabolize them. As the composting process progresses, moisture content falls (Chennaoui et al., 2018).

15.4.6 Oxygen

During the composting process, oxygen must be present. Gases are created when carbon is oxidized by organisms to provide energy. This exhausts the oxygen that is already there. Insufficient oxygen will cause the composting process to turn anaerobic, producing gases like ammonia, carbon dioxide, and methane that will cause the compost to smell bad (Gonawala & Jardosh, 2018).

15.4.7 Carbon to Nitrogen (C:N) Ratio

For a composting process, a C:N ratio of 30 is excellent (Gonawala & Jardosh, 2018; Yan et al., 2015). Composting materials with a low carbon to nitrogen ratio will prevent air from penetrating the pile, leading to anaerobic conditions and the creation of odours in addition to nitrogen loss in the form of ammonia gas. Furthermore, if the C/N ratio is very high, microbial activity will be diminished and the pace of breakdown would be slowed down (Méndez-Matías et al., 2018).

15.4.8 Turning Frequency

The most popular method of aeration throughout the composting process is by turning the compost material. This causes gas emissions by making the composting material more accessible for microbial utilization (Parkinson et al., 2004). The time it takes for waste to stabilize can be shortened by improving aeration during the early phases of organic matter breakdown; on the other hand, excessive aeration or stirring may cause the loss of essential composting components (Awasthi et al., 2014). Thus, it is necessary to optimize the turning regime to maintain the pertinent nutrients or to accomplish other particular goals like a better rate of hygienization (pathogen reduction) (Kalamdhad & Kazmi, 2009).

15.5 COMPOSTING SYSTEMS

The aerated static-pile system, the enclosed system, and the windrow system are the three primary methods for centralized composting.

15.5.1 Aerated Static-Pile System

In this approach, organic waste is piled up into heaps, which are occasionally covered with screened compost to keep the pile's temperature high and minimize odours. With the employment of blowers and air diffusers, aeration is achieved. These piles work best with homogenous materials (like sludge) and should not be used with heterogeneous materials.

15.5.2 Enclosed System

There are two types of enclosed systems: agitated beds and silos. Temperature, oxygen concentration, and odour control are all provided by the enclosed container. This technology is only utilized when the compost needs to be applied to the soil due to the high expense of enclosed systems. The vessel may take the shape of a trench lined with concrete or a silo. The agitated bed system combines the benefits of both the windrow and the aerated static-pile system by moving the composting material through the vessel by mixing, whereas the silo-type system enables the material to flow through the vessel with gravity.

15.5.3 Windrow System

A windrow is just an accumulation of garbage that is allowed to break down. It's the standard method for turning organic waste into compost. It is necessary to properly manage the quantity of organic wastes produced as a result of the rapidly growing population. However, improving the composting method's efficiency can result in cost savings. A windrow's cross-sectional form can range from rectangular to trapezoidal to triangular, mostly dependent on the properties of the composting material and the turning tools utilized. Turning the composting heap frequently allows for aeration. Aeration and turning machines spin windrows in big systems on a regular basis. Occasionally, windrows are rotated when a certain temperature range—typically between 55°C and 60°C—is attained by using the temperature of the composting material as a turning indication (Kumar, 2011). Since mechanical equipment is used to aerate the composting heap (Figure 15.2), windrow composting is comparatively inexpensive. The kind of input material and the turning apparatus

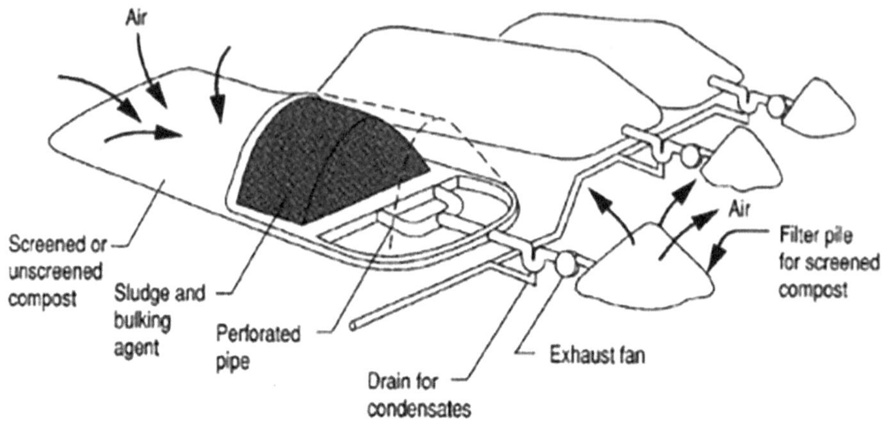

FIGURE 15.2 Windrow arrangement for forced aeration (Kreith & Tchobanoglous, 2002).

employed determine the windrows' height, breadth, and length. Temperature control is achieved by rotating the apparatus periodically.

15.6 METHODS OF COMPOSTING USING DIFFERENT STRUCTURES AND THEIR DESIGN CONSIDERATIONS

The availability of land and the location are the two most crucial site factors. The composting location should ideally be close to a wastewater treatment facility or, in the case of sludge, to a landfill. This will simplify processes and lessen the handling of materials. Living or working close to homes affects the amenities needed and how much they cost. The site of operations, the climate, the properties of the sludge and solid waste, and the kinds of bulking material that are accessible all have a role in the composting system selection and facility design (Kreith & Tchobanoglous, 2002). The methods of composting using a wide variety of structures include the following:

15.6.1 Holding Units

Bins, crates, or other structures used to hold organic waste in place while they decompose are known as holding units. Permanent units can be constructed from landscape timbers, concrete blocks, or bricks, while portable units can be manufactured from lightweight materials like chicken wire on a wood frame, hardware cloth, or welded wire.

There are additional holding units that are manufactured. Turning units are designed to make turning easier, but holding units often do not turn organic materials. This is how holding units vary from turning units.

Whatever kind of structure is decided, one needs to be sure to have adequate ventilation to allow air to circulate throughout the composting materials. Additionally, the construction should be 3 feet × 3 feet × 3 feet in size. The composting process

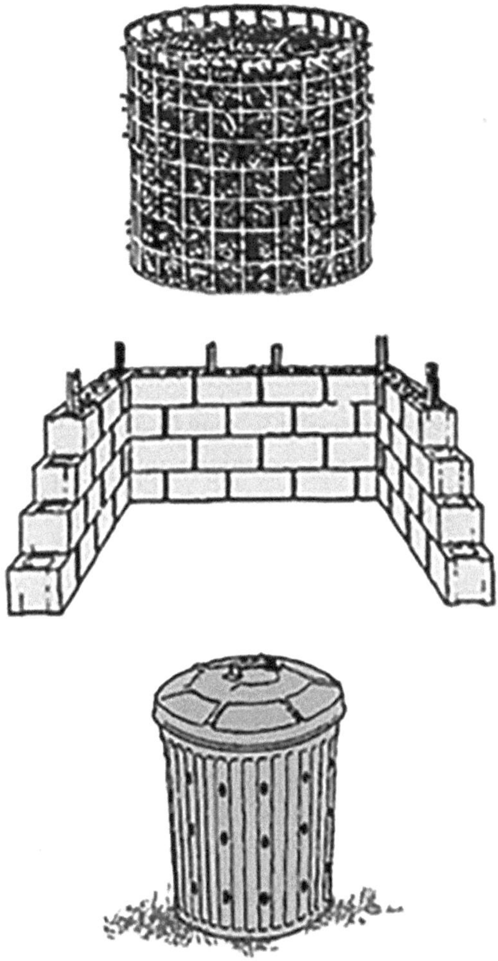

FIGURE 15.3 (a) Wire-mesh holding unit, (b) concrete-block holding unit, and (c) garbage-can composter.

happens more slowly and aeration becomes problematic if the container is taller or broader than three feet. Some of the commonly used holding units include wire-mesh holding unit, wooden pallet holding unit, concrete-block holding unit, and garbage-can composter (Figure 15.3).

15.6.2 Turning Units

Turning units, as their name suggests, are made to enable the periodic mixing of organic materials throughout the composting process. Barrel composters and three-bin turning units are the two most widely utilized varieties of turning units. Similar to holding units, a three-bin turning unit's bins should measure around 3 feet × 3 feet × 3 feet in order to provide enough aeration of the composting organics.

15.6.2.1 Wood and Wire Three-Bin Turning Unit

In a three-bin rotating unit (Figure 15.4), more room is available for composting with it than with a single unit. It is possible to convert organics from one unit to the next, with the three units being in distinct phases of development. Put otherwise, bin 1 can be utilized for recently received organic materials, bin 2 for partially completed compost, and bin 3 for completed compost. When the materials decompose to the appropriate degree, they are moved from one bin to the next and rotated inside each bin on a regular basis.

15.6.2.2 Wooden Three-Bin Turning Unit

The wooden three-bin turning unit (Figure 15.5) is a permanent, sturdy structure made of wooden posts and slats.

15.6.2.3 Concrete-Block Three-Bin Turning Unit

Three concrete-block holding units arranged in a row form a concrete-block turning unit (Figure 15.6). It is affordable to build and robust, provided that old concrete blocks are available.

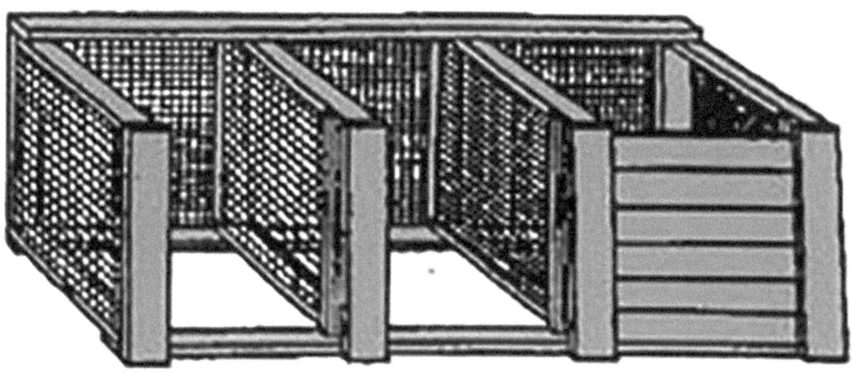

FIGURE 15.4 Wood and wire three-bin turning unit.

FIGURE 15.5 Wooden three-bin turning unit.

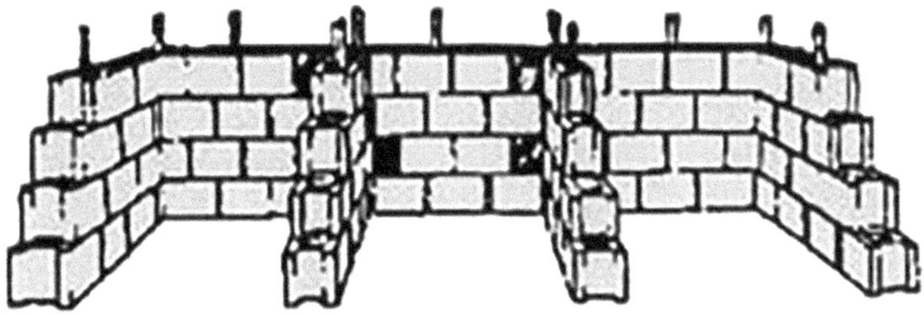

FIGURE 15.6 Concrete-block three-bin turning unit.

15.6.2.4 Barrel Composter

A barrel composter has an advantage over other types of structures since it is simple to turn organic materials; all that has to be done is rotate the barrel. Building a barrel composter is quite simple. Commercially, there are also a number of manufactured barrel composters available.

15.6.3 Heaps

Composting without a framework is known as heap composting. As organic materials become accessible, they are added to the pile. The heap's height should not be more than three feet for optimal decomposition; however, the length might vary based on the materials available. It is advisable to start a new heap with fresh materials once it reaches a certain size, allowing the current heap to break down with periodic rotation.

15.6.4 Pit Composting

Pit composting is the process of composting without using above-ground buildings, in shallow pits excavated into the earth. The hole is dug to a depth of around two feet, a width of three feet, and any length. If the surrounding soil packs tightly and stays solid, the pit's walls may be left unlined or lined with brick, masonry, or treated wood. If building materials are employed, the walls can go higher than the surface. Pit composting is mostly anaerobic (occurs without adequate air movement) and does not produce effective aeration. As a result, pit composting takes longer than above-ground composting.

15.6.5 Plastic Bag Composting

Composting using plastic bags is perhaps the easiest way because all you need is a black plastic rubbish bag and no extra equipment. The bags need to be at least 3 millilitres thick and between 30 and 40 gallons in capacity. Three bushels' worth of organic stuff ought to fit within a sack this size. A mix of organic materials that are both green and brown are preferably filled in the bag for optimal decomposition. One

FIGURE 15.7 Plastic bag composting.

cup of ammonium nitrate and roughly one-fourth cup of hydrated lime are added to each bag (lime is used in this instance to offset the acidity that will arise within the bag under anaerobic circumstances). After that, roughly a quart of water is added, and the bag is closed firmly, and left for half a year to a year. After sealing the bag, there is no need to stir the mixture or add more water. The bags are placed in a sunny area so they can absorb heat, if at all feasible. Composting in plastic bags (Figure 15.7) is handy, however it happens considerably more slowly and primarily in anaerobic circumstances than composting in well-ventilated structures, much like pit composting.

15.7 CONCLUSION

Composting is a safer alternative to improper waste management, which is a prevalent practice that is not safe. Human and environmental health are generally improving worldwide. Composting is an effective method of obtaining this objective. The usage of chemical fertilizers will decline as composting becomes more popular. Because fewer hazardous substances will be discharged into the environment as a result of this change, both human and environmental health will inevitably improve. The main reason influencing the use of compost in crop production has been its capacity to enhance soil structure and increase nutrient availability by adding to the nutrients already present. Moreover, plant-derived viricides, bactericides, fungicides, and anti-nematodes can be added to compost to create an efficient compost. This will discourage the use of chemicals and promote organic farming exclusively. Slowly decomposing items should be composted separately to prevent them from prolonging the time that other materials take to compost. Prior to being applied in the field, composts should always be examined for diseases and maturity. This will help to avoid any potential risks to the surrounding ecosystem and other living creatures. Lastly, further study has to be done to figure out how to extend the composting process. The composting process will be sustained by the development of speedier technologies. The aerated static-pile system, the enclosed system, and the windrow system are the three primary methods for centralized composting. The site of operations, the climate, the properties of the sludge and solid waste, and the kinds of bulking material that are accessible all have a role in the composting system selection and facility design. There is still more to be done to raise awareness about the possibilities of this technology in order for farmers to fully embrace it.

REFERENCES

Albrecht, R., Le Petit, J., Terrom, G., & Périssol, C. (2011). Comparison between UV spectroscopy and nirs to assess humification process during sewage sludge and green wastes co-composting. *Bioresource Technology*, *102*(6), 4495–4500.

Ameen, A., Ahmad, J., & Raza, S. (2016). Effect of pH and moisture content on composting of municipal solid waste. *International Journal of Scientific and Research Publications*, *6*(5), 35–37.

Aruna, G., Kavitha, B., Subashini, N., & Indira, S. (2018). An observational study on practices of disposal of waste garbages in Kamakshi Nagar at Nellore. *International Journal of Applied Research*, *4*(3), 392–394.

Awasthi, M. K., Pandey, A. K., Khan, J., Bundela, P. S., Wong, J. W., & Selvam, A. (2014). Evaluation of thermophilic fungal consortium for organic municipal solid waste composting. *Bioresource Technology*, *168*, 214–221.

Chennaoui, M., Salama, Y., Aouinty, B., Mountadar, M., & Assobhei, O. (2018). Evolution of bacterial and fungal flora during in-vessel composting of organic household waste under air pressure. *Journal of Materials and Environmental Science*, *9*(2), 680–688.

Chiee, M. Y., & Manaf, L. A. (2017). Solid waste management transformation and future challenges of source separation and recycling practice in Malaysia. *Resources, Conservation and Recycling*, *116*, 1–14. https://doi.org/10.1016/j.resconrec.2016.09.012

Chowdhury, A. K. M. M. B., Konstantinou, F., Damati, A., Akratos, C. S., Vlastos, D., Tekerlekopoulou, A. G., & Vayenas, D. V. (2015). Is physicochemical evaluation enough to characterize olive mill waste compost as soil amendment? The case of genotoxicity and cytotoxicity evaluation. *Journal of Cleaner Production*, *93*, 94–102.

Coelho, L., Osório, J., Beltrão, J., & Reis, M. (2019). Organic compost effects on Stevia rebaudiana weed control and on soil properties in the Mediterranean region. *Revista de Ciências Agrárias*, *42*(1), 109–121.

Das, M., Uppal, H. S., Singh, R., Beri, S., Mohan, K. S., Gupta, V. C., & Adholeya, A. (2011). Co-composting of physic nut (Jatropha curcas) deoiled cake with rice straw and different animal dung. *Bioresource Technology*, *102*(11), 6541–6546.

Epelde, L., Jauregi, L., Urra, J., Ibarretxe, L., Romo, J., Goikoetxea, I., & Garbisu, C. (2018). Characterization of composted organic amendments for agricultural use. *Frontiers in Sustainable Food Systems*, 2, 44.

Galitskaya, P., Biktasheva, L., Saveliev, A., Grigoryeva, T., Boulygina, E., & Selivanovskaya, S. (2017). Fungal and bacterial successions in the process of co-composting of organic wastes as revealed by 454 pyrosequencing. *PloS One*, *12*(10), e0186051.

Gigliotti, G., Valentini, F., Erriquens, F. G., & Said-Pullicino, D. (2005). Evaluating the efficiency of the composting process: A comparison of different parameters. In *Geophysical research abstracts* (Vol. 7, p. 09606). European Geosciences Union.

Gonawala, S. S., & Jardosh, H. (2018). Organic waste in composting: A brief review. *International Journal of Current Engineering and Technology*, *8*(1), 36–38.

Gould, J. C., Rossano, M. G., Lawrence, L. M., Burk, S. V., Ennis, R. B., & Lyons, E. T. (2013). The effects of windrow composting on the viability of Parascaris equorum eggs. *Veterinary Parasitology*, *191*(1–2), 73–80.

Hafeez, M., Gupta, P., & Gupta, Y. P. (2018). Rapid composting of different wastes with Yash Activator Plus. *International Journal of Life Sciences and Scientific Research*, *4*(2), 1670–1674.

Hermann, B. G., Debeer, L., De Wilde, B., Blok, K., & Patel, M. K. (2011). To compost or not to compost: Carbon and energy footprints of biodegradable materials' waste treatment. *Polymer Degradation and Stability*, *96*(6), 1159–1171.

Huang, J., Yu, Z., Gao, H., Yan, X., Chang, J., Wang, C., & Zhang, L. (2017). Chemical structures and characteristics of animal manures and composts during composting and assessment of maturity indices. *PloS One*, *12*(6), e0178110.

Juárez, M. F. D., Prähauser, B., Walter, A., Insam, H., & Franke-Whittle, I. H. (2015). Co-composting of biowaste and wood ash, influence on a microbially driven-process. *Waste Managementement*, *46*, 155–164.

Kalamdhad, A. S., & Kazmi, A. A. (2009). Effects of turning frequency on compost stability and some chemical characteristics in a rotary drum composter. *Chemosphere*, *74*(10), 1327–1334.

Khater, E. S. G. (2015). Some physical and chemical properties of compost. *International Journal of Waste Resources*, *5*(1), 72–79.

Kreith, F., & Tchobanoglous, G. (2002). *Handbook of solid waste management*. McGraw-Hill.

Kumar, S. (2011). Composting of municipal solid waste. *Critical Reviews in Biotechnology*, *31*(2), 112–136.

Li, Y., Liu, B., Zhang, X., Gao, M., & Wang, J. (2015). Effects of Cu exposure on enzyme activities and selection for microbial tolerances during swine-manure composting. *Journal of Hazardous Materials*, *283*, 512–518.

Li, Z., Lu, H., Ren, L., & He, L. (2013). Experimental and modeling approaches for food waste composting: A review. *Chemosphere*, *93*(7), 1247–1257.

Lin, Y., Du, D., Si, C., Zhao, Q., Li, Z., & Li, P. (2014). Potential biocontrol Bacillus sp. strains isolated by an improved method from vinegar waste compost exhibit antibiosis against fungal pathogens and promote growth of cucumbers. *Biological Control*, *71*, 7–15.

Luo, X., Liu, G., Xia, Y., Chen, L., Jiang, Z., Zheng, H., & Wang, Z. (2017). Use of biochar-compost to improve properties and productivity of the degraded coastal soil in the Yellow River Delta, China. *Journal of Soils and Sediments*, *17*, 780–789.

Majbar, Z., Lahlou, K., Ben Abbou, M., Ammar, E., Triki, A., Abid, W., & Rais, Z. (2018). Co-composting of olive mill waste and wine-processing waste: An application of compost as soil amendment. *Journal of Chemistry*, *2018*(1), 7918583.

Méndez-Matías, A., Robles, C., Ruiz-Vega, J., & Castañeda-Hidalgo, E. (2018). Composting agroindustrial waste inoculated with lignocellulosic fungi and modifying the C/N ratio. *Revista Mexicana de Ciencias Agrícolas*, *9*(2), 271–280.

Misra, R. V., Roy, R. N., & Hiraoka, H. (2003). *On-farm composting methods*. UN-FAO.

Mohammad, N., Alam, M. Z., Kabbashi, N. A., & Ahsan, A. (2012). Effective composting of oil palm industrial waste by filamentous fungi: A review. *Resources, Conservation and Recycling*, *58*, 69–78.

Ogwueleka, T. (2009). Municipal solid waste characteristics and management in Nigeria. *Journal of Environmental Health Science & Engineering*, *6*(3), 173–180.

Olanrewaju, O. S., Glick, B. R., & Babalola, O. O. (2017). Mechanisms of action of plant growth promoting bacteria. *World Journal of Microbiology and Biotechnology*, *33*, 1–16.

Pampuro, N., Bertora, C., Sacco, D., Dinuccio, E., Grignani, C., Balsari, P., & Bernal, M. P. (2017). Fertilizer value and greenhouse gas emissions from solid fraction pig slurry compost pellets. *The Journal of Agricultural Science*, *155*(10), 1646–1658.

Pandey, P. K., Cao, W., Biswas, S., & Vaddella, V. (2016). A new closed loop heating system for composting of green and food wastes. *Journal of Cleaner Production*, *133*, 1252–1259.
Pane, C., Spaccini, R., Piccolo, A., Celano, G., & Zaccardelli, M. (2019). Disease suppressiveness of agricultural greenwaste composts as related to chemical and bio-based properties shaped by different on-farm composting methods. *Biological Control*, *137*, 104026.
Parkinson, R., Gibbs, P., Burchett, S., & Misselbrook, T. (2004). Effect of turning regime and seasonal weather conditions on nitrogen and phosphorus losses during aerobic composting of cattle manure. *Bioresource Technology*, *91*(2), 171–178.
Pose-Juan, E., Igual, J. M., Sánchez-Martín, M. J., & Rodríguez-Cruz, M. S. (2017). Influence of herbicide triasulfuron on soil microbial community in an unamended soil and a soil amended with organic residues. *Frontiers in Microbiology*, *8*, 378.
Qian, X., Shen, G., Wang, Z., Guo, C., Liu, Y., Lei, Z., & Zhang, Z. (2014). Co-composting of livestock manure with rice straw: Characterization and establishment of maturity evaluation system. *Waste Management*, *34*(2), 530–535.
Raut, M. P., William, S. P., Bhattacharyya, J. K., Chakrabarti, T., & Devotta, S. (2008). Microbial dynamics and enzyme activities during rapid composting of municipal solid waste–a compost maturity analysis perspective. *Bioresource Technology*, *99*(14), 6512–6519.
Sukholthaman, P., & Sharp, A. (2016). A system dynamics model to evaluate effects of source separation of municipal solid waste management: A case of Bangkok, Thailand. *Waste Management*, *52*, 50–61.
Toledo, M., Siles, J. A., Gutiérrez, M. C., & Martín, M. A. (2018). Monitoring of the composting process of different agroindustrial waste: Influence of the operational variables on the odorous impact. *Waste Management*, *76*, 266–274.
Traversa, A., Loffredo, E., Gattullo, C. E., & Senesi, N. (2010). Water-extractable organic matter of different composts: A comparative study of properties and allelochemical effects on horticultural plants. *Geoderma*, *156*(3–4), 287–292.
Uyizeye, O. C., Thiet, R. K., & Knorr, M. A. (2019). Effects of community-accessible biochar and compost on diesel-contaminated soil. *Bioremediation Journal*, *23*(2), 107–117.
Ventorino, V., Pascale, A., Fagnano, M., Adamo, P., Faraco, V., Rocco, C., & Pepe, O. (2019). Soil tillage and compost amendment promote bioremediation and biofertility of polluted area. *Journal of Cleaner Production*, *239*, 118087.
Wilson, G. B. (1989). Combining raw materials for composting. *BioCycle*, *30*(8), 82–85.
Yan, Z., Song, Z., Li, D., Yuan, Y., Liu, X., & Zheng, T. (2015). The effects of initial substrate concentration, C/N ratio, and temperature on solid-state anaerobic digestion from composting rice straw. *Bioresource Technology*, *177*, 266–273.
Zhang, L., & Sun, X. (2014). Changes in physical, chemical, and microbiological properties during the two-stage co-composting of green waste with spent mushroom compost and biochar. *Bioresource Technology*, *171*, 274–284.
Zhao, G. H., Yu, Y. L., Zhou, X. T., Lu, B. Y., Li, Z. M., & Feng, Y. J. (2017). Effects of drying pretreatment and particle size adjustment on the composting process of discarded flue-cured tobacco leaves. *Waste Management & Research*, *35*(5), 534–540.

16 Sensor-Based Structure for Vermicomposting

Noureen Khursid, Rohitashw Kumar, and Tanzeel Khan

16.1 INTRODUCTION

Composting technology has advanced from static piles and windrows to smart. More recent research has focused on finding novel process-enhancing additives, whereas past efforts have focused primarily on identifying appropriate organic waste streams and potential co-composting. However, there is still more to be done to mitigate the environmental impact and maximize the goals of the circular economy. This chapter showcases current achievements in the integration of vermicomposting methods and proposes an innovative Internet of Things-based system for monitoring agricultural compost. The system consists of edge/cloud platforms, an Internet of Things sensor probe, integrated networking, monitoring, logging, reporting, and visualization solutions offering a forecast for upcoming changes.

Enormous amounts of waste have led to several environmental issues, such as soil, water, and air pollution, depletion of biodiversity, etc. Solid waste dumps release various greenhouse gasses, which aid in global warming. The synthesis of nanomaterials from various wastes has been reported in certain papers, although there are currently no viable uses for environmental advantages (Griffin et al., 2018). Researchers constantly look for environmentally friendly ways to manage solid waste without impacting the ecosystem. Biodegradable waste makes up a sizable portion of the waste generated by numerous industries. It can be a vital resource for several biological processes, such as vermicomposting, composting, and biomethanation.

The past few decades have seen a worldwide spike in attention to vermicomposting due to its efficiency and ease of use (Hussain & Abbasi, 2018). Vermicomposting has been identified as a sustainable strategy for managing organic waste (Sim & Wu, 2010). Many other types of waste can be used as raw materials, including animal excrement, commercial waste, waste from agriculture, and municipalities for the process of vermicomposting.

Vermicomposting is a mesophilic, bio-oxidative natural decomposition mechanism wherein earthworms and microbes combine to mineralize organic waste substrates and transform them into nutrient-rich organic manure (Pramanik & Chung, 2011). It is a collaborative endeavour between microbes and earthworms (Figure 16.1). Organic waste types include agricultural, residential, and animal manure. Rich in nutrients such as NPK and some micronutrients, the resulting dark brown, granular, fine, homogenous vermicompost resembles peat. Its numerous advantages include

DOI: 10.1201/9781003491460-16

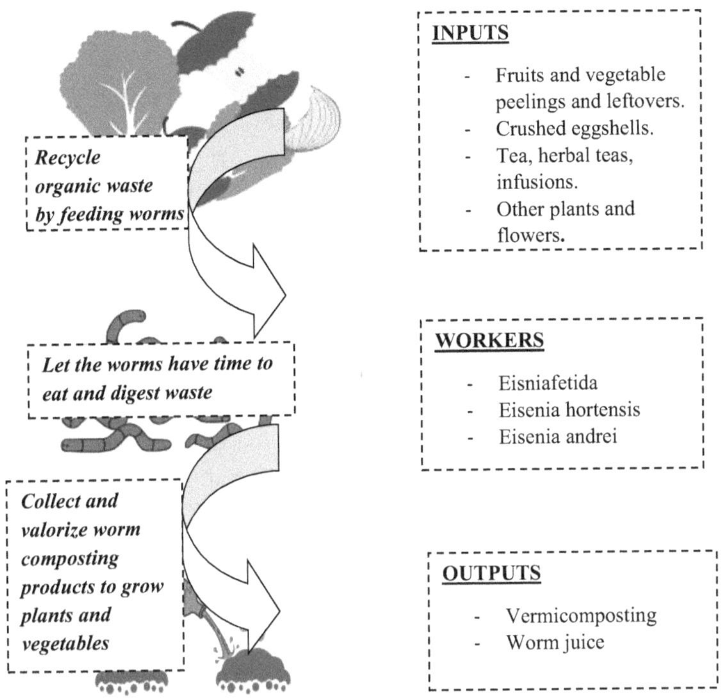

FIGURE 16.1 Vermicomposting principle.

efficient drainage, aeration, and water-holding capacity (Edwards et al., 2010). It improves the soil's microbial composition, which is good for plant development, health, and yield. It is a biofertilizer in agriculture and is safe for the environment. Research suggests that using vermicompost can improve seed germination and crop growth and yield without affecting soil health. Other applications for vermicomposting include energy production, soil remediation, and wastewater treatment (Dey et al., 2017). Recognizing the importance of recycling garbage, the Indian government has also encouraged using biological waste recycling methods like sensor-based composting and vermicomposting. The potential benefits of integrating automation and sensor technology in the field of smart agriculture are substantial for society (Srisruthi et al., 2016).

Abbasi et al.'s (2019) research study on the Internet of Things (IoT) in the agriculture domain provides an overview of the area and potential applications. The study illustrated how contemporary technologies are permeating formerly closed systems by examining several angles, such as IoT hardware, cloud platforms, and application domains. These so-called smart systems rely on connectivity. Improved connectivity for IoT devices and the cloud is made possible by 5G and next technologies like 6G, which offer quicker and more dependable data transport. Composter probe data are transported to the cloud over these networks via smart agriculture and Internet of Things-based compost management systems, allowing for real-time processing, analysis, and visualization.

While the benefits of smart agriculture systems have been extensively researched, Ammar et al. (2018) point out that there are still several difficult gaps in the design of high- and low-complexity IoT system frameworks, a range of communication protocols, hardware, and software layers that must be accomplished to boost systems' effectiveness. Thirty million devices have received and had improved IoT systems provided since 2015 (Elijah et al., 2018). IoT applications may enable sophisticated artificial intelligence (AI) devices that meet user needs for sufficient flexibility and adaptability (Sastra & Wiharta, 2016). The Internet of Things is attracting remote sensor-equipped and Internet-connected devices to its development of smart connectivity (Lindqvist & Neumann, 2017) for information analytics, centralized management, and data collection.

To address IoT infrastructural gaps, using the sensor-based vermiculture farming structure is used. One crucial method of treating trash is vermiculture-sensor technology. Vermicomposting uses earthworms to convert organic waste into nutrient-rich compost. It produces a useful resource for agriculture and is an eco-friendly method of getting rid of organic waste. Nevertheless, the procedure can take a while, and it needs close supervision to ensure that the compost is of the highest caliber. The worms' growth and well-being are dependent on many environmental factors, such as temperature and moisture, which are tracked by sensors installed in the bin. The sensors are linked to a central control system so that it may optimize the composting process by adjusting the conditions inside the bin as needed. An additional feature of the Internet of Things-based vermicomposting container is a data logging system that records the temperature, moisture content, and other environmental parameters over time. The composting process can be examined using this data to find areas that need improvement. All things considered, the Internet of Things vermicompost bin is a potent instrument for raising the efficacy and efficiency of the composting process. It reduces waste and helps to manufacture high-quality compost with little effort on the part of agricultural producers and urban dwellers alike, paving the way for a more sustainable future.

16.2 EXISTING LITERATURE WORK

Among the more contemporary pieces, an overview presented by Casas et al. (2014) was the development of a complicated wireless sensor network for remotely monitoring compost humidity and temperature.

An IoT-based architectural approach has been put out by Bharadwaj et al. (2016) to address the problems with the existing approach of solid waste management systems. They used IoT technologies to automate and monitor the tracking, collection, and management of solid trash.

With unsupervised and continuous composting processes, Jordão et al. (2017) proposed a reasonably priced station for sensing the air temperature at two independent spots in a compost pile. The writers claim that the station survived extremely high humidity and low-temperature circumstances while guaranteeing autonomy for seven days. The authors employed a DS18B20 temperature sensor and an 8-bit ATmega328 microprocessor in an Arduino/Genuino-based prototype platform. For seven days, the local memory stored the sensor readings every five minutes. It was not possible to classify this system as an integrated IoT solution because data were solely handled locally. After more thought, our objective was to create a method that could be applied to open compost piles or any other type of infrastructure, eliminating the need to create or purchase specialist equipment.

"COMPosting as a Service (COMPaaS)," a cloud service "in composition with specialized Internet of Things (IoT)-based composting machinery that allows for unsupervised composting," is introduced by Nikoloudakis et al. (2018). This complex system uses sensors both inside and outside the composting frames, as well as sensors located in the center of the compost pile, to collect and communicate real-time data (temperature, humidity, and solar intensity) to a cloud server. In addition to analyzing data in real time, the service can give composting hardware instructions. The system operates entirely autonomously, and the creators assert that their method provides 30% faster post-production.

Barcelon et al. (2019) proposed an automated vermicast production system in an enlarged application domain. Here, the sensor data readings are recorded using a Raspberry Pi, two Arduino microcontrollers, and this configuration. The data gathered are stored on the Raspberry Pi, which also acts as an independent controller for the sprinkler system. To preserve the ideal living conditions for worms, the vermicompost must have a constant moisture level.

Building upon the idea of composting-as-a-service, Daliakopoulos et al. (2019) detailed the development of a system consisting of sensors, an app for a smartphone, and a cloud service. The compost pile is filled with the sensor, which is housed in a watertight capsule. After that, data are sent over Bluetooth Low Energy to the smartphone app. Decisions on manual interventions are influenced by the composting process insights provided by a cloud service.

Ahmadi et al. (2020) presented a prototype reactor for the Fusagasuga municipality's agricultural waste composting program. A rotary mechanism with a set of rotating blades is used in the construction of this 20-litre closed-system reactor. The compost materials can be oxygenated with the help of a PI controller that has three humidity and temperature sensors. Remote access users can monitor and control the system thanks to its integrated Human-Machine Interface. Its small footprint and limited capacity may make it impractical to deploy the technology at large-scale compost facilities that already exist.

"Smart composting" is the necessity of the hour, as it allows for IoT application and automation, according to Balaganesh et al. (2022), who provided a comprehensive review of the most current advancements in the field of Internet of Things-enabled composters. The research's specifics were arranged according to technological elements such as hardware parts, protocols, data mining, IoT gateways, data collecting and storage systems, and gateways.

Further IoT integration, lower costs, more mobility, and simpler setup and operation of the IoT compost system prototype are among the possible advantages of the sector's present research work.

16.3 ADVANCEMENTS IN DESIGN AND TECHNOLOGICAL DEVELOPMENT

Technological progress driven by research and innovation endeavours will enhance the efficacy of composting and vermicomposting methods in transitioning toward a circular economy. This will result in the production of high-quality natural fertilizers, promoting sustainable agricultural practices (Awasthi et al., 2022). Zheng et al. (2020) noted that China had the largest percentage of global patents (44%) about compost technology, with a concentration on innovative composting equipment.

Sustainable rural development may be achieved by using urban and agricultural wastes in vermicomposting and composting processes to produce useful natural fertilizers for organic farming (Pereira et al., 2020). For family farmers, composting might be made easier with the use of new tools, bio-waste mixes, and financially viable technologies. To keep bio-waste out of landfills and open burning methods, home composting is essential in rural areas (Awasthi et al., 2021). However, both the emissions generated during the process of composting and the quality of the natural fertilizers obtained from home composting require improvement. By combining four bio-waste combinations and employing a straightforward centroid design to speed up the composting process, potential solutions for managing various rural waste sources in China and beyond can be illustrated (Guo, 2021). Indeed, such integrated techniques ought to be encouraged in future studies on small-scale composting processes.

Large-scale composting facilities are frequently used to service metropolitan agglomerations that generate a lot of bio-waste from their residents, visitors, and agri-food enterprises. According to Sakarika et al. (2019), major urban regions may find a dependable substitute for increasing capacity in centralized composting plants by pre-treating food waste (FW) by reducing its volume and mass using a decentralized bioreactor. The resultant frass from composting or vermicomposting may be used directly on agricultural land or may require further processing (Awasthi et al., 2020; Klammsteiner et al., 2020; Lopes et al., 2022). According to the tenets of the circular economy, more investigation is needed into the connection between bioenergy production and composting or vermicomposting.

In large-scale vermicomposting systems, one of the main issues is that earthworm extraction is frequently done by hand. Manual extraction is a lot of work, takes a long time, and is inefficient. Regarding this, Walling et al. (2020) recommended a novel technique that uses centrifugation to separate earthworms, resulting in an approximate 84% worm recovery rate. It is recommended to use a combination approach, such as room drying for ten days and vermicomposting for ten days, to enhance the stabilization process of dewatered sewage sludge and achieve good maturity (Huang et al., 2022). The vermicomposting reactor is another technologically advanced way to enhance working conditions and speed up the biodegradation process (Ramprasad & Alekhya, 2021). Ghorbani and Sabour (2021) proposed that a smart vermicompost reactor may produce compost in half the time compared to a conventional container, while simultaneously accelerating worm growth by 30%. Research indicates that reactor systems for composting and vermicomposting produce high-quality nutrients faster and more sustainably. Additionally, they highlight the advantages and viewpoints of cutting-edge technology (vermicomposting and reactor composting) over traditional outdoor choices for both small- and large-scale composting processes. It is anticipated that additional research will be conducted in the development of sensors to improve many aspects of composting and vermicomposting, including artificial intelligence.

16.3.1 Materials Needed

The IoT vermicompost probe is designed with four main requirements: simplicity, robustness, holistic approach, and mobility. The physical parts of the suggested Internet of Things system are joined by bespoke programming, common network protocols, and APIs connected to pertinent cloud services. The authors rank soil moisture as the most important factor, followed by pH levels, temperature, and light intensity. The components are chosen according to the specifications and standards that determine the probe design. (Figure 16.2)

The only switches on the probe are one for turning it on and another for transferring data regarding the first need to the network. Furthermore, there are a pair

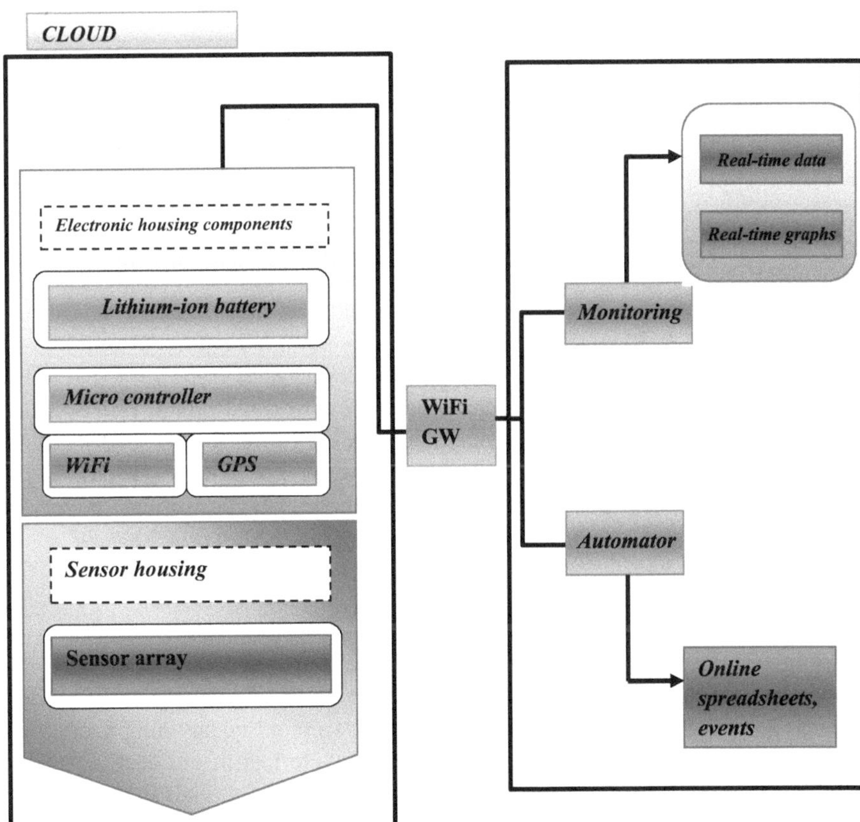

FIGURE 16.2 Framework of the suggested prototype.

of LEDs: one indicates whether the probe is powered on, and the other indicates whether it is ready to provide data. This readiness includes setting up sensors, locking onto GPS signals, and connecting to all relevant services. As such, using the probe is very easy. The operator pushes the button to send the readings to the network when every indicator is green. Confirmation of the successful transfer of all pertinent data is provided by a blinking green LED. Although automation is an option, this manual procedure enables the operator to correct any positioning issues before transmitting the data. In response to the second need, the penetrating portion of the probe is made of aluminium or inox, both of which can withstand the harsh characteristics of compost. The IP-certified electronics casing is designed to withstand environmental influences and is waterproof. Because of their strength and resistance to wear, the switches and buttons are guaranteed to last in demanding compost conditions. As for the third prerequisite, the energy required for the probe's operation comes from a lithium-ion battery. Deep sleep is a state in which the selected microcontroller can drastically cut power consumption to microamperes. By doing this, battery life is maximized. Convenient portability is made possible by the probe's easy manipulation and transfer by hand design. The IoT prototype may work effectively in any environment without requiring significant adjustments or additional hardware because it is designed to be location agnostic overall.

16.3.2 Soil Moisture and Soil Temperature

Sensors for soil temperature and moisture levels keep an eye on the condition of the vermicompost in the worm bin. The sensors will provide data to the Arduino microcontroller for condition checking if the soil temperature hits 33°C and the soil moisture falls below 60%. The earthworms won't perish because the specified condition will then cause the water pump to turn on the sprinkler, maintaining the soil's wetness and temperature. Decomposition is slowed between 40% and 50% moisture content (Bhargava & Bhattacharya, 2004). To promote high microbial activity and simple feeding of the food matter, the moisture level needs to be kept at roughly 50% (Gajalakshmi & Abbasi, 2004). Stocking more worms leads to faster casting production when the substrate is entirely decomposed in four to five weeks (25–35 days), allowing for more vermicompost to be recovered. For monitoring reasons, the Arduino microcontroller processes the data that the sensors pick up and processed data are sent via a WiFi hotspot to generate results.

16.3.3 Worm Migration System

To facilitate faster worm migration, two worm bins are used. For the first vermicomposting, the first bin is ready. For the next vermicomposting, the second worm bin is prepared for worm migration. If earthworms cannot find anything to eat, they leave the area. In this instance, the decomposed materials in the next worm bin will draw worms in once they have eaten the decomposed materials in the first bin. Worms were found to take 20 days to eat the disintegrated material. After 20 days, the Arduino microcontroller will cause the gate to open, and a lamp will illuminate to guarantee that the worms are migrating between the bins. The lamp will turn off when the gate is closed. The created vermicast in the evacuated worm bin first is ready to be harvested.

16.3.4 Sensor Network

Sensor types include soil moisture, ambient and soil temperature, light intensity, and water level sensors. Two electrodes make up the soil moisture sensor. These electrodes are covered in plaster of Paris before being inserted into the vermicompost, allowing it to absorb water from the vermicompost before exposing the sensors to the moisture in the soil. This removes variations in the value that the soil moisture sensor reads. The sensor's reading range was set at 0% to 100% soil moisture. The Arduino module is attached to thermocouple nodes, which are the sensors for the ambient temperature and the soil. Two dissimilar conductors connected at one end to create a hot junction that generates voltage are the components of a thermocouple. The voltage fluctuates and is converted when the junction heats or cools, returning the temperature. When the light-dependent resistor (LDR) is linked to the supply voltage, it produces an analog voltage whose magnitude changes in direct proportion to the amount of light it receives. The Arduino's analog input pin receives this LDR's output analog voltage after that. The water level ultrasonic sensor is used to gauge the water's height within the tank. Figure 16.3 represents the sensor-based framework of vermicomposting.

16.3.5 Irrigation System

Solenoids were employed as the sprinkler system's control valves. The pressure is created by the water tank's elevation used in place of a pump to provide pressure, and solenoid water valves control whether water flows or not. When the system detects that the compost needs to be kept at the ideal temperature and moisture level, the sprinkler system is in charge of watering it.

16.3.6 Display Unit

In this case, the display device is an LCD (Liquid Crystal Display). Temperature, moisture content, and gas are just a few of the vermicomposting data that it displays. Additionally, it can be utilized to show various messages on the LCD, fan pump, and lamp status.

16.3.7 Application

The capacity to monitor or manipulate data over the Internet is essential for any application that is enabled by the Internet of Things. The controlling operation in the suggested system is carried out automatically. When internet connectivity is present, the system sends the current state of the application so that it can be monitored and controlled by taking the necessary action. In addition, if any parameter surpasses its matching threshold value, an email is also sent to the provided email address.

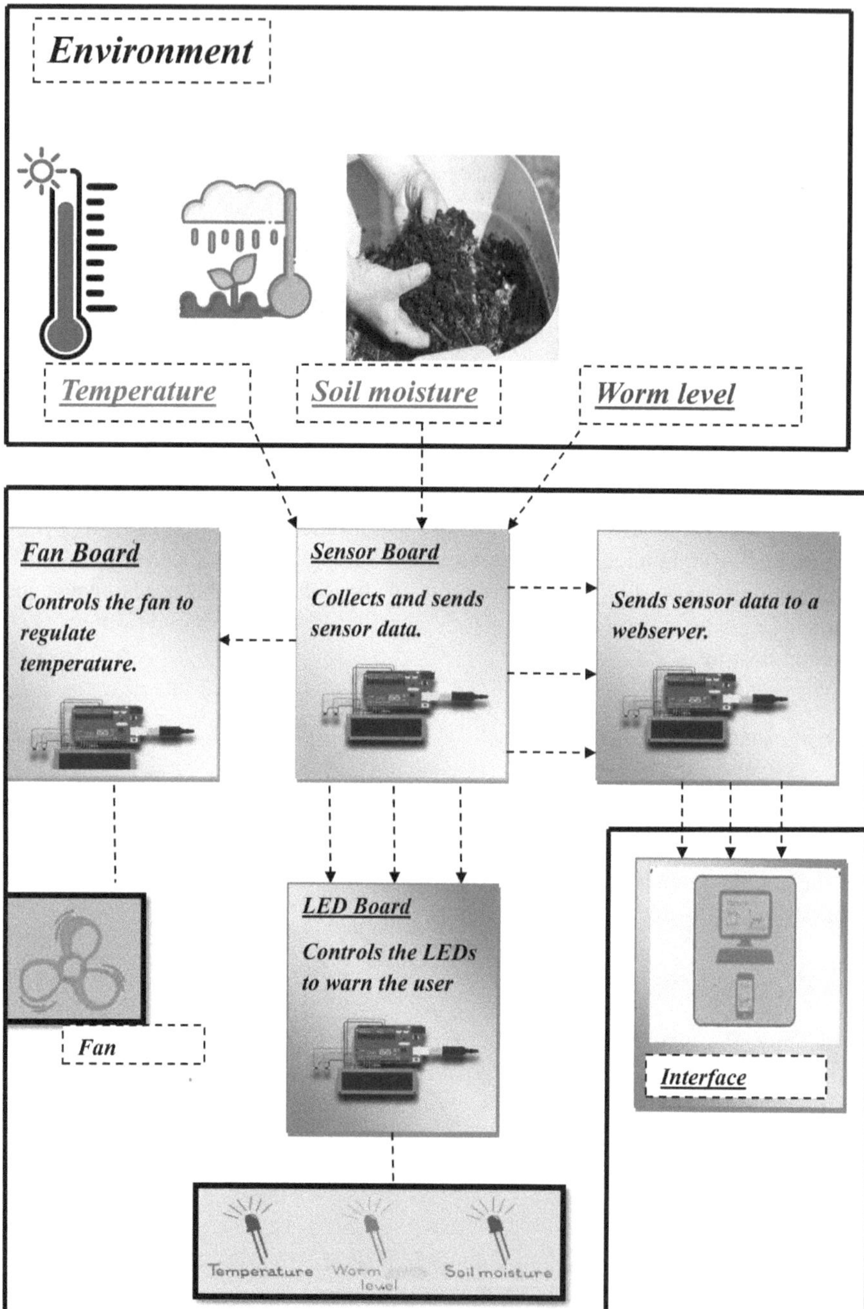

FIGURE 16.3 Sensor-based vermicomposting framework.

The following steps could be used to summarize the prototype's functionality:

- The sensor housing and electrical component housing make up the portable IoT probe. It is powered on. GPS tries to lock onto the position, the sensors are initializing, and the network and service connections are all being initialized. The probe is inserted into the designated compost pile.
- Sensors read physical values, which are processed within the probe. An LED provides a visible indicator when the data are prepared to be sent over the network.
- The LED on the IoT probe flashes many times to indicate that the data transmission was successful.

16.4 CONCLUSION

Vermicomposting is thought to be a zero-waste, sustainable, and clean method of managing organic waste, but there are still certain obstacles in the way of its widespread adoption. It is necessary to focus more on the practical use of vermicomposting rather than expanding research in the sector. For small-scale vermicomposting operations, IoT-based bins provide several advantages. Utilizing IoT technology makes it possible to remotely monitor and regulate vital environmental parameters, which might enhance the composting process' efficacy and efficiency. The chapter comes to two conclusions: First of all, it presents a cutting-edge Internet of Things (IoT) based vermicompost monitoring system. Numerous functions are included in this system, such as data recording, monitoring, logging, reporting, visualization, and alerting for important compost parameters. It lessens a number of the shortcomings of current solutions, including the requirement for significant financial resources, vendor-specific infrastructure or equipment, specialist training, complex procedures, or pricey sensors and tools. Second, the analysis identifies the optimal sensor technology.

REFERENCES

Abbasi, M., Yaghmaee, M. H., & Rahnama, F. (2019). Internet of things in agriculture: A survey. In *3rd international conference on internet of things and applications (IoT)* (pp. 1–12). IEEE.

Ahmadi, T., Casas Díaz, C. A., García Vera, Y. E., & Escobar Escobar, N. (2020). *A prototype reactor to compost agricultural wastes of Fusagasuga Municipality, Colombia.* https://dspace.emu.ee/items/ad175578-02e5-4bf7-ae06-8fc8a5162e82

Ammar, M., Russello, G., & Crispo, B. (2018). Internet of things: A survey on the security of IoT frameworks. *Journal of Information Security and Applications*, *38*, 8–27.

Awasthi, M. K., Liu, T., Awasthi, S. K., Duan, Y., Pandey, A., & Zhang, Z. (2020). Manure pretreatments with black soldier fly Hermetiaillucens: A study to reduce pathogen content. *Science of the Total Environment*, *737*, 139842.

Awasthi, M. K., Singh, E., Binod, P., Sindhu, R., Sarsaiya, S., Kumar, A., Chen, H., Duan, Y., Pandey, A., Kumar, S., & Taherzadeh, M. J. (2022). Biotechnological strategies for bio-transforming biosolid into resources toward circular bio-economy: A review. *Renewable and Sustainable Energy Reviews*, *156*, 111987.

Awasthi, S. K., Duan, Y., Liu, T., Zhang, Z., Pandey, A., Varjani, S., Awasthi, M. K., & Taherzadeh, M. J. (2021). Can biochar regulate the fate of heavy metals (Cu and Zn) resistant bacteria community during the poultry manure composting? *Journal of Hazardous Materials*, *406*, 124593.

Balaganesh, P., Vasudevan, M., Rameswari, R., & Natarajan, N. (2022). Recent trends in IoT-enabled composter for organic wastes. In *Sustainable Cities and Resilience: Select Proceedings of VCDRR 2021* (pp. 445–457). Springer Singapore.

Barcelon, M. S., Orilla, A. A., Mahilum, J. A., & Adtoon, J. J. (2019). Automated vermiculture monitoring and compost segregating system using microcontrollers. *Proceedings of the 6th International Conference on Bioinformatics Research and Applications*, 38–42.

Bharadwaj, A. S., Rego, R., & Chowdhury, A. (2016). IoT based solid waste management system: A conceptual approach with an architectural solution as a smart city application. In *2016 IEEE annual India conference (INDICON)* (pp. 1–6). IEEE.

Bhargava, N. S. A. K. D., & Bhattacharya, S. (2004). Optimum moisture requirement during vermicomposting using Perionyx excavatus. *Applied Ecology and Environmental Research*, *2*(1), 53–62.

Casas, O., López, M., Quílez, M., Martinez-Farre, X., Hornero, G., Rovira, C., Pinilla, M. R., Ramos, P. M., Borges, B., Marques, H., & Girão, P. S. (2014). Wireless sensor network for smart composting monitoring and control. *Measurement*, *47*, 483–495.

Daliakopoulos, I., Daskalakis, G., Markakis, N., Papastefanakis, N., & Manios, T. (2019). Automatic monitoring of a community backyard composting program. *Multidisciplinary Digital Publishing Institute Proceedings*, *30*(1), 45.

Dey, M. D., Das, S., Kumar, R., Doley, R., Bhattacharya, S. S., & Mukhopadhyay, R. (2017). Vermiremoval of methylene blue using Eisenia fetida: A potential strategy for bioremediation of synthetic dye-containing effluents. *Ecological Engineering*, *106*, 200–208.

Edwards, C. A., Arancon, N. Q., & Sherman, R. L. (Eds.). (2010). *Vermiculture Technology: Earthworms, Organic Wastes, and Environmental Management*. CRC Press.

Elijah, O., Rahman, T. A., Orikumhi, I., Leow, C. Y., & Hindia, M. N. (2018). An overview of Internet of Things (IoT) and data analytics in agriculture: Benefits and challenges. *IEEE Internet of Things Journal*, *5*(5), 3758–3773.

Gajalakshmi, S., & Abbasi, S. A. (2004). Earthworms and vermicomposting. *Indian Journal of Biotechnology*, *3*, 486–494.

Ghorbani, M., & Sabour, M. R. (2021). Global trends and characteristics of vermicompost research over the past 24 years. *Environmental Science and Pollution Research*, *28*, 94–102.

Griffin, S., Sarfraz, M., Farida, V., Nasim, M. J., Ebokaiwe, A. P., Keck, C. M., & Jacob, C., (2018). No time to waste organic waste: Nanosizing converts remains of food processing into refined materials. *Journal of Environmental Management*, *210*, 114–121.

Guo, C. (2021). December. Application of Fuzzy Control Technology in Aerobic Compost Heat Recovery Device. In *Journal of Physics: Conference Series*, *2138*(1), 012014.

Huang, K., Zhang, Y., Xu, J., Guan, M., & Xia, H. (2022). Feasibility of vermicomposting combined with room drying for enhancing the stabilization efficiency of dewatered sludge. *Waste Management*, *143*, 116–124.

Hussain, N., & Abbasi, S. A. (2018). Efficacy of the vermicompost of different organic wastes as "clean" fertilizers: State-of-the-art. *Sustainability*, *10*(4), 1205.

Jordão, M. D., Paiva, K. D., Firmo, H. T., Inácio, C. T., Rotunno, O. C., & Lima, T. D. A. (2017). Low-cost automatic station for compost temperature monitoring. *Revista Brasileira de EngenhariaAgrícola e Ambiental*, *21*, 809–813.

Klammsteiner, T., Turan, V., Fernandez-Delgado Juarez, M., Oberegger, S., & Insam, H., (2020). Suitability of black soldier fly frass as soil amendment and implication for organic waste hygienization. *Agronomy*, *10*(10), 1578.

Lindqvist, U., & Neumann, P. G. (2017). The future of the Internet of Things. *Communications of the ACM*, *60*(2), 26–30.

Lopes, I. G., Yong, J. W., & Lalander, C. (2022). Frass derived from black soldier fly larvae treatment of biodegradable wastes. A critical review and future perspectives. *Waste Management*, *142*, 65–76.

Nikoloudakis, Y., Panagiotakis, S., Manios, T., Markakis, E., & Pallis, E. (2018). Composting as a service: A real-world IoT implementation. *Future Internet, 10*(11), 107.

Pereira, M. M. A., Morais, L. C., Martins, A. D., Luz, J. M. Q., Pasqual, M., Oliveira, R. C., & Doria, J. (2020). Challenges of organic agriculture for the production of composts and vermicompost for the production of medicinal plants-a socioeconomic demand. *Bioscience Journal, 36*.

Pramanik, P., & Chung, Y. R. (2011). Changes in fungal population of fly ash and vinasse mixture during vermicomposting by Eudriluseugeniae and Eisenia fetida: Documentation of cellulase isozymes in vermicompost. *Waste Management, 31*(6), 1169–1175.

Ramprasad, C., & Alekhya, D. (2021). Design and optimisation of high rate decentralised vermicomposting reactor for the household organic waste. *International Journal of Environment and Waste Management, 27*(4), 420–439.

Sakarika, M., Spiller, M., Baetens, R., Donies, G., Vanderstuyf, J., Vinck, K., Vrancken, K. C., Van Barel, G., Du Bois, E., & Vlaeminck, S. E. (2019). Proof of concept of high-rate decentralized pre-composting of kitchen waste: Optimizing design and operation of a novel drum reactor. *Waste Management, 91*, 20–32.

Sastra, N. P., & Wiharta, D. M. (2016). October. Environmental monitoring as an IoT application in building smart campus of Universitas Udayana. In *International Conference on Smart Green Technology in Electrical and Information Systems (ICSGTEIS)* (pp. 85–88). IEEE.

Sim, E. Y. S., & Wu, T. Y. (2010). The potential reuse of biodegradable municipal solid wastes (MSW) as feedstocks in vermicomposting. *Journal of the Science of Food and Agriculture, 90*(13), 2153–2162.

Srisruthi, S., Swarna, N., Ros, G. S., & Elizabeth, E. (2016). Sustainable agriculture using eco-friendly and energy efficient sensor technology. In *2016 IEEE international conference on recent trends in electronics, information & communication technology (RTEICT)* (pp. 1442–1446). IEEE.

Walling, E., Trémier, A., & Vaneeckhaute, C. (2020). A review of mathematical models for composting. *Waste Management, 113*, 379–394.

Zheng, X., Aborisade, M. A., Liu, S., Lu, S., Oba, B. T., Xu, X., Cheng, X., He, M., Song, Y., & Ding, H. (2020). The history and prediction of composting technology: A patent mining. *Journal of Cleaner Production, 276*, 124232.

17 Application of Plastics in Promotion of Organic Farming

Munjid Maryam and Rohitashw Kumar

17.1 INTRODUCTION

Plastic is a broad term for a material based on one or more organic polymers with added substances (additives) to achieve desired properties, such as flame retardants, antistatics, coloring agents, stabilizers, reinforcing materials, UV-adsorbents, and plasticizers. Conventional polymers like polyethylene (PE), polypropylene (PP), and polyvinyl chloride (PVC) are environmentally persistent due to their chemically stable polymer backbones. Agricultural plastics made from these polymers can break down into micro- and nanoplastics (MNP), accumulating in soils, being taken up by organisms, or moving into surrounding environments. Biodegradable polymers, on the other hand, are designed to be metabolized by microbes, converting the polymer carbon into CO_2 and microbial biomass in oxygen-rich conditions. The biodegradability of a polymer depends on its physicochemical properties and the environmental conditions where it degrades. Although biodegradable polymers are intended to fully mineralize to CO_2, they may also temporarily form smaller MNP fragments. Both biodegradable and conventional polymers can be sourced from non-renewable petrochemicals or renewable bio-based materials. For example, polybutylene adipate-co-terephthalate is made from petrochemicals, while starch, cellulose, chitosan, and poly(lactic acid) come from renewable sources. While this review mainly focuses on structural polymers, non-structural polymers (e.g., polymers in liquid formulations or water-soluble polymers) are also widely used in agriculture for purposes like encapsulating or dispersing pesticides, herbicides, and nutrients, as well as in seed coatings and soil conditioners. Structural polymers can often be collected, reused, or recycled after use, whereas non-structural polymers dissipate into the environment. However, the key concepts discussed also apply to non-structural polymers (Rosenboom et al., 2022).

There are multiple applications of plastics in agriculture and their use has increased over the last 70 years to an estimated 12.5 million tons annually, which has led to the accumulation of large quantities of macro-, micro-, and nanoplastics in soils and other receiving environments. The pervasive use of plastics in agriculture poses a growing risk to soil functions, and the wider natural environment, highlighting the urgency for a more sustainable use of plastics in agricultural food production (FAO, 2021). In March 2022, 175 countries agreed to negotiate a global and legally binding

 DOI: 10.1201/9781003491460-17

Plastics Treaty (UNEA-5.2) to end plastic pollution by 2024. This international policy instrument aims to address the ecological and human health risks posed by the entire plastic lifecycle, including those linked to agriculture. It is anticipated that regular reporting will be required by governments and nongovernment stakeholders on the implementation and performance of their goals to reduce plastic pollution (Bauchmüller et al., 2021; Bergmann et al., 2022; Thompson et al., 2022; Walker, 2022).

17.2 APPLICATION AND BENEFITS OF PLASTIC IN ORGANIC AGRICULTURE

Plastics in organic agriculture offer numerous environmental and societal benefits. Plastic mulch films, which represent about 50% of the total mass of agricultural plastics, are extensively used in crop production (Figure 17.1). They provide several agronomic advantages, such as weed and pest control, soil moisture conservation, temperature regulation of soil and air, and enhanced nutrient uptake. These benefits result in increased yields, improved water and nutrient use efficiency, and reduced pesticide usage. For instance, in China, an additional 3.9 million hectares of arable

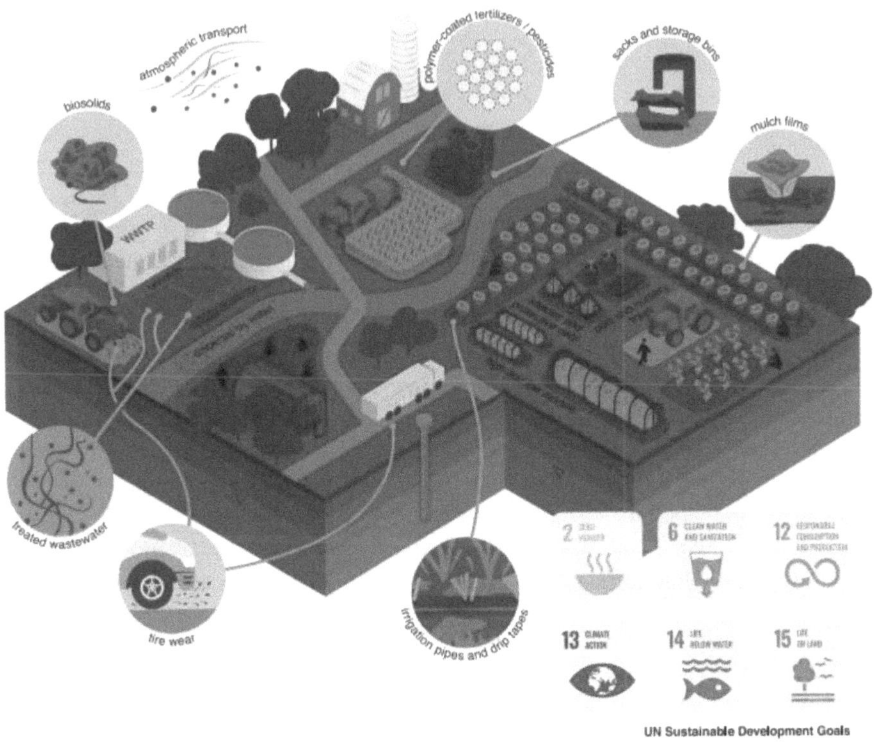

FIGURE 17.1 Applications of plastics in organic agriculture (USDG, 2024).

land would be needed to produce the same amount of food without mulch film. The higher soil temperatures under plastic mulch films enable farmers to plant and harvest crops earlier, providing a market advantage. Additionally, plastic mulch films are used in organic agriculture to suppress weeds and insect infestations without the need for synthetic pesticides (Andrady & Neal, 2009; Kasirajan & Ngouajio, 2012; Sun et al., 2020).

Plastics are utilized in various ways within plant agriculture, aiding in the reduction of irrigation water, pesticide, and fertilizer usage. This, in turn, lowers greenhouse gas emissions and boosts crop yields, thereby contributing to several UN Sustainable Development Goals. Additionally, incidental plastic pollution on agricultural lands can originate from various sources, such as the dispersal of wastewater biosolids or treated wastewater from wastewater treatment plants (WWTP), which may contain tire wear particles.

Conventional plastic mulch films are usually made from low-density PE but can also consist of other polymers like PVC or ethylene-vinyl acetate copolymers. Usage rates are highest in Europe and Asia at 40–50 kg/ha, compared to slightly lower rates of 10–20 kg/ha in North and South America. Incomplete collection after use results in persistent plastic residues accumulating in soils. Additionally, chemical additives can leach from the mulch films. Repeated applications can thus lead to harmful effects on soil productivity and health due to the buildup of plastic residues and additives. Certified soil-biodegradable mulch films are promoted as alternatives to conventional thin (<20–25 μm) mulch films. These biodegradable films can be ploughed into the soil after harvest and are designed to fully biodegrade into CO_2 and microbial biomass under oxic conditions (FAO, 2021; Hayes et al., 2019).

Plastic sheets are commonly used as covering materials in greenhouses, as well as in high and low tunnels (Figure 17.2), offering thermal insulation, radiant energy capture, and protection from weather and pests. These sheets are made from various polymers and include additives that give them a wide range of optical and energy-related properties. Additionally, plastics are utilized as shade and protective nets, such as sunshades, anti-hail, and anti-bird covers, or as seedling plugs (Maraveas, 2019, 2020).

Food security relies heavily on nitrogen fertilizers, but their production and usage contribute to 5% of global GHG emissions. Enhancing nitrogen-use efficiency is the most effective way to reduce these emissions, and polymer coatings in fertilizers help regulate nutrient release to align with the plant life cycle. For instance, the Japanese Association of Agriculture and Forestry Statistics estimates that between 1976 and 2018, Japan used 2.3 million tons of fertilizers coated with polyurethane and PE. These coatings helped manage nutrient release, improve nutrient use efficiency, and reduce nutrient loss. The European Chemicals Agency (ECHA) has proposed a ban on non-biodegradable polymer coatings for controlled-release fertilizers, with the European Commission discussing a draft regulation in September 2022, including a five-year transition period. Recent advancements have focused on biodegradable, non-toxic biopolymers to enhance efficiency and ensure precise nutrient release. Using this approach, 25% of the recommended nutrient load has shown to provide the same photosynthetic performance in soybean and wheat as the full application rate of conventional fertilizers (Hayes et al., 2019; Gao & Cabrera Serrenho, 2023; Jariwala et al., 2022; Katsumi et al., 2021; Xu et al., 2022).

FIGURE 17.2 Low tunnels.

Mulching involves covering the soil to prevent its degradation over time (Figure 17.4). Plastic films are employed to maintain soil moisture and reduce water evaporation. They also improve thermal conditions for plants, inhibit weed growth, and help preserve soil nutrients. Water scarcity is a significant constraint on crop growth and yield. Besides mulch films, irrigation pipes and drip tapes are used to deliver precise amounts of water to plant roots, thereby enhancing water use efficiency. Irrigation equipment (Figure 17.3) is typically made from conventional polymers like high-density PE or PVC. Seedling plugs and nursery pot trays are common in crop production; seedling plug trays facilitate efficient germination and optimize plant growth. Additionally, some crops, particularly climbing species and vines, need support during cultivation. Nursery pot trays are also essential for transporting plants, and plastics are used to manufacture agricultural packaging materials, including fertilizer and storage bags, flexible bulk containers, crates, and pesticide containers (Jia et al., 2018; Rillig & Lehmann, 2020; Zhang et al., 2017).

Plastics play a crucial role in silage production by enhancing the preservation, quality, and storage of forage crops used for animal feed. They are employed in several key ways: plastic covers are used to seal the top of silage piles, preventing air and moisture from causing spoilage; plastic wrap encases bales, creating an airtight seal that promotes proper fermentation and protects the silage from external contaminants; plastic bags offer a compact and controlled storage solution for small-scale operations, maintaining anaerobic conditions; and specialized plastic films and

FIGURE 17.3 Laying of plastic drip lines.

FIGURE 17.4 Plastic mulch.

sheeting provide durable, weather-resistant barriers that further safeguard the silage. These plastic materials collectively help in extending the storage life of silage, reducing spoilage, and preserving the nutritional content of the feed, making them indispensable in modern silage management.

In aquaculture, plastics are essential for optimizing fish and shellfish production, providing both practical benefits and enhanced operational efficiency. They are used in various applications, including the construction of tanks, ponds, and cages, which offer controlled environments for aquatic species. Plastic materials such as high-density polyethylene (HDPE) and polyvinyl chloride (PVC) are favored for their durability, resistance to corrosion, and ease of maintenance. In addition, plastics are employed in the manufacturing of filtration systems, aeration devices, and water treatment equipment, which are crucial for maintaining water quality and ensuring the health of aquatic organisms. Plastic nets and screens are also used to contain and protect fish, while flexible plastic piping and tubing facilitate the efficient delivery of water and nutrients. The versatility of plastics in aquaculture supports improved resource management, better environmental control, and enhanced productivity, making them a vital component of modern aquaculture systems.

A variety of plastics are also employed in hydroponic systems and indoor vertical farming for tubing, clamps, grow trays, grids, packaging, and net pots. Plastics play a crucial role in hydroponics and indoor farming, offering numerous advantages that enhance plant growth and resource management. In hydroponic systems, plastics are used extensively in growing containers, nutrient delivery systems, and water management. Plastic containers and trays support plant roots and hold nutrient solutions, providing a durable and lightweight alternative that resists corrosion. Flexible plastic pipes and tubing transport nutrients efficiently, while plastic tanks and reservoirs store water and nutrient solutions. Additionally, plastics are used in growth media like expanded clay pellets, perlite, and vermiculite, which come in plastic bags or containers to support root systems and retain moisture. In indoor farming, plastics contribute significantly to creating optimal growing conditions. They are used to construct growth chambers, greenhouses, and tunnels, which control temperature, humidity, and light. Plastic fixtures and covers for lighting systems ensure even distribution of light, while plastic structures like hydroponic towers and vertical racks maximize space utilization. Irrigation systems incorporate plastic pipes and drip systems for precise water and nutrient delivery, reducing waste and enhancing resource efficiency. Plastics are also essential in climate control systems, including ventilation, fans, and humidifiers, which maintain ideal conditions for plant growth. Seedling trays and pots made of plastic facilitate the starting and growth of young plants before they are transferred to the main system. The benefits of using plastics in these farming methods include improved resource efficiency, precise climate control, and optimized space usage. Plastics offer durability and ease of maintenance, making them suitable for the demanding environments of hydroponics and indoor farming. Their cost-effectiveness and long-term durability contribute to reduced operational costs, supporting higher yields and sustainability in controlled agricultural settings (Figure 17.5).

FIGURE 17.5 Application of plastic for making rows in indoor farming.

Plastics thus play a multifaceted role in organic farming, offering several benefits that support sustainable agricultural practices including:

- **Weed Control:** Plastic mulch films are laid over soil to create a barrier that blocks sunlight, preventing weed seeds from germinating and growing. By suppressing weeds, these films reduce the need for mechanical weeding and herbicides, which aligns with organic farming principles that avoid synthetic chemicals. Common types include black, clear, and colored mulches, each offering different benefits based on the crop and environmental conditions.
- **Soil Moisture Conservation:** Plastic mulch films help retain soil moisture by reducing evaporation. This is especially beneficial in regions prone to drought or during hot periods. Conserving water is crucial for sustainable farming. The reduced need for irrigation helps maintain water resources and supports organic practices where water conservation is emphasized.
- **Temperature Regulation:** Plastic mulches can regulate soil temperature by trapping heat. In cooler climates, they help warm the soil, while in hotter climates, they can keep the soil cooler. Maintaining optimal soil temperatures promotes better plant growth and development, contributing to higher yields and better-quality crops in organic systems.
- **Pest Control:** Plastic covers (like row covers) and nets protect crops from pests, birds, and harsh weather conditions. They create a physical barrier

between pests and plants. Reducing reliance on chemical pesticides aligns with organic farming principles. These plastic solutions provide non-chemical methods for pest management.

- **Crop Protection:** Plastic tunnels (hoop houses) and cloches (mini-greenhouses) protect crops from frost, wind, and heavy rain. They also extend the growing season by providing a controlled environment. By extending the growing season and protecting crops, these plastic structures help increase productivity and reduce the risk of crop loss, which supports sustainable and resilient farming practices.
- **Seedling Growth:** Plastic seedling trays and pots are used to germinate and grow young plants before transplanting them into the field. They provide a controlled environment for seedling development. Efficient seedling growth leads to healthier plants and better crop yields. The use of plastic trays helps manage plant growth and soil health effectively, which is important for maintaining organic standards.
- **Efficient Use of Resources:** Plastic irrigation pipes and drip tapes deliver precise amounts of water directly to the plant roots. This targeted irrigation reduces water waste and improves efficiency. Efficient water use is critical for sustainability. Drip irrigation minimizes water runoff and ensures that plants receive adequate moisture, supporting organic practices that emphasize resource conservation.
- **Reduced Soil Erosion:** Plastic mulch helps stabilize the soil, preventing erosion caused by wind and rain. It also helps reduce soil compaction and maintain soil structure. Preventing soil erosion helps preserve soil fertility and structure, which is essential for long-term soil health and productivity in organic farming.

17.3 SUSTAINABLE USE OF PLASTICS IN ORGANIC AGRICULTURE

The sustainable use of plastics in organic agriculture must align with the "3 R" waste hierarchy: reducing, reusing, and recycling plastics rather than disposing of them after use (Figure 17.6). More attention is needed for plastic applications that lack circular end-of-life treatment options. Additionally, two criteria can guide sustainable use strategies and identify after-use treatment options for agricultural plastics: (i) the collectability of plastics after use and (ii) the duration of plastic use. For applications where plastics can be completely collected after use, reuse and recycling are the preferred treatment options, regardless of how long the plastics were used. For these applications, advancing sustainability involves using plastics that resist chemical weathering and fragmentation, and deploying technologies that ensure complete collection after use, especially for below-ground plastics like irrigation pipes. When plastics cannot be fully collected after use, or if collected plastics are too degraded or contaminated for reuse or recycling, less toxic and biodegradable polymers should replace conventional persistent ones, particularly for short-duration uses like mulch films or non-biodegradable coatings for controlled-release fertilizers. These biodegradable plastics must perform effectively throughout their application period and completely convert to CO_2 and microbial biomass in the soil within a specified time

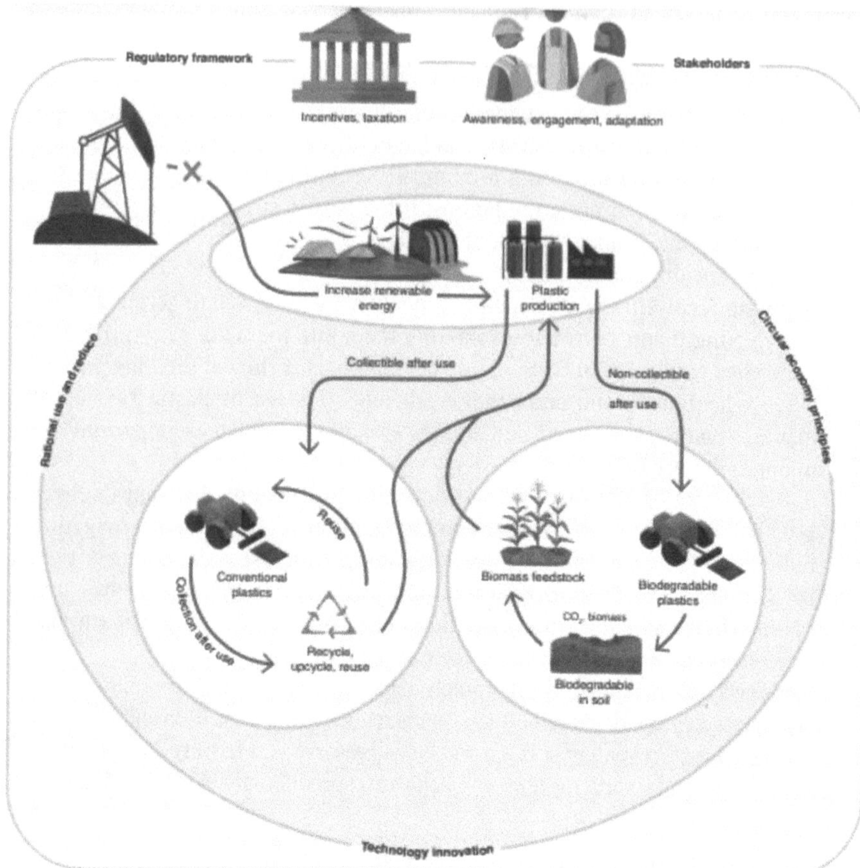

FIGURE 17.6 Sustainable use of plastics.

frame at the end of their lifecycle. Applications include seed coatings, slow-release fertilizers, thin mulch films, and geotextiles.

17.3.1 Collection After Use, Reuse, and Recycling Should Be Prioritized

The rates of reuse and recycling for agricultural plastic waste are currently very low, particularly with polyvinyl chloride usage and the phasing out of intentionally added microplastic applications. Therefore, plastics that are not significantly damaged during normal use should be prioritized for reuse on farms or through collection programs like extended user responsibility and reverse logistics. After collection, non-reusable agricultural plastics should undergo thorough cleaning to remove soil contamination and other debris before being sent to materials resource facilities (MRF) for recycling. Some agricultural plastics, such as agrichemical containers, can be effectively decontaminated and reused. Collecting bale wrap films, which are typically less soiled than ground-collected mulch films, can offer economic feasibility and environmental benefits. Additionally, plastic mulch films can be recycled through pyrolysis to produce fuels or polymers, as this thermal treatment process is unaffected by soil or other contaminants. To enhance agricultural plastic recycling, dedicated programs need to be established, either through legislation or voluntary initiatives, with well-coordinated and widely accessible collection facilities. If agricultural plastic waste is too contaminated or physically degraded, it should be properly disposed of in a landfill, although this is the least preferred option. Materials resource facilities (MRF) sort and process plastics mechanically, such as by shredding, melting, and re-extrusion, to produce downcycled plastic resins or asphalt (Cassou, 2022; De Lucia & Pazienza, 2019; Braga Marsola et al., 2021).

Chemical processing techniques like pyrolysis, dissolution, re-extrusion, chemical or enzymatic methanolysis, and gasification can generate a variety of upcycled products. These include feedstock chemicals such as aromatic chemicals and olefins, high-quality plastic resins and original plastic monomers, methanol, syngas, and transportation fuels. Extensive research has been conducted on various upcycling technologies for waste plastics, such as hydrogenolysis to produce lubricants and oils, and enzymatic or enzyme-assisted chemical treatments of PET to yield primary monomers. Newly developed zirconium-catalyzed C-H and C-C alumination processes may simplify the recycling and upcycling of plastics or render them biodegradable. Many of these technologies are already in commercial operation, with some at the industrial demonstration stage, and are capable of recycling the diverse types of plastics used in agriculture. Although processing techniques vary for different plastics and their mixtures, agricultural plastics are generally recyclable. For example, plastic film waste from greenhouses has been directly used in laboratory demonstrations of catalytic cracking for feedstock recycling. Additionally, a closed-loop supply chain analysis for bale wrap collection, sorting, and recovery in Finland demonstrated significant economic savings and a reduced global warming potential. This closed-loop recycling approach could also be applied to self-immolative polymers. Although large-scale feasibility of breaking down and reconstituting these polymers repeatedly has not been fully explored, recent reviews have highlighted their potential and importance. The types of chemical additives in plastic products can significantly impact recyclability. In mechanical recycling, certain additives and their transformation products can affect the safety and marketability of secondary plastics. For example, mixing plastics with different pigments can result in colored secondary products, limiting their further use. Moreover, toxic by-products can be produced if recycling is performed at unsuitable temperatures, leading to the accumulation of these substances and reducing the suitability of plastics in a circular economy. Thus, integrating the management of chemical additives into the global Plastics Treaty will be crucial for simplifying plastic mixtures. Technologies like dissolution-based recycling can be developed to selectively eliminate problematic additives and components, though these methods tend to be more costly and energy-intensive. While agricultural plastic wastes offer the benefit of a relatively uniform and compositionally suitable stream of materials for recycling, especially in concentrated geographic areas, there are still technical challenges to address. Thin films that become too brittle or weathered cannot be fully collected after use. Exposure to heat, solar UV radiation, soil, and agricultural chemicals can deteriorate some plastics, complicating recovery and often limiting it to down-cycling through mechanical processing. Additionally, the seasonal variability in agricultural plastic waste generation poses challenges for recycling facilities in establishing stable secondary markets. For plastic mulch films, the primary recycling challenge is the soil contamination that adheres to the plastics when they are collected from fields post-harvest. Soil and plant residue contamination can constitute up to 80% of the total weight of the plastic materials, whereas existing recycling facilities require contamination levels to be below 5%. It is essential to develop effective methods for removing soil and plant residues from plastic films. These methods may include mechanical or manual cleaning processes and strategies to minimize soil adhesion. Ideally, such techniques should be applied directly at the farm to reduce transportation costs, using tools like mechanical rotary brooms or manual methods on dry days to aid in collection and recyclability. Therefore, there is a strong need for improved techniques and standardized procedures for post-use collection, decontamination, and handling to enhance recycling efforts (European Commission, 2016; Kaabel et al., 2021; Kanbur et al., 2023; Mayanti & Helo, 2022; Madrid et al., 2022; Picuno et al., 2020; Serrano, 2004; Tricker et al., 2022; Yardley et al., 2019).

17.3.2 Plastics That Cannot Be Fully Collected After Use Should Be Biodegradable

Biodegradation is a viable end-of-life option for agricultural plastics that cannot be completely retrieved from the field, as well as for cases where collected plastic fragments are too weathered or contaminated for reuse and recycling. Examples of such applications include thin agricultural mulch films, abraded plastic from string trimmers and plant clips, and geotextiles. In these situations, ensuring complete biodegradation of the plastics into CO_2 and biomass prevents residue accumulation in soils. Using polymers with cleavable bonds, such as esters or amides, can enable biodegradation to be controlled to occur at the end of the product's life. Altering the backbone chemistry of existing polymers to make them biodegradable is an emerging research area. For instance, a new bio-based polyester-2, 18 has properties similar to high-density PE but can be easily hydrolyzed by naturally occurring enzymes. On the other hand, using prooxidant (or "oxo") additives that initiate polyolefin breakdown through thermal or photochemical activation has proven ineffective in ensuring plastic biodegradability (Bauchmüller et al., 2021; Eck et al., 2023; von Vacano et al., 2022).

It will be essential to thoroughly evaluate the increased use of biodegradable plastics to confirm that they truly biodegrade in situ within a reasonable and specified timeframe. Currently, the biodegradability of plastics in soil is tested through laboratory soil incubation studies conducted under aerobic conditions. These studies involve respirometric analyses to measure the conversion of polymers to CO_2 under controlled humidity and constant temperatures, typically between 20 and 25 °C. These conditions are anticipated to promote biodegradation more effectively than field conditions, and further research is needed to assess how well laboratory biodegradation data apply to real-world field environments. Additionally, increased efforts are necessary to monitor the levels of biodegradable plastics in situ over extended periods (approximately years) and after multiple applications. This will help establish steady-state concentrations in soils and identify the main factors influencing biodegradation rates. There is also a critical need for analytical methods that can accurately measure low concentrations of polymers in soils. Strict biodegradability standards that define testing conditions and specify the required mineralization timeframe are essential for ensuring biodegradability. For example, the standard for biodegradable mulch films (EN 17033:2019) mandates 90% polymer mineralization in soils within two-year laboratory soil incubation at 20–25 °C, as determined by established methods (ASTM-D5988). Scientific guidance will be crucial for future regulations concerning both plastics and their additives. Additives are likely to be released during biodegradation, potentially leading to higher concentrations in soil porewaters compared to those leached from conventional plastics, underscoring the need for environmentally benign additives (Nelson et al., 2020; Svechnikov et al., 2008).

Currently, biodegradable plastics are 20–80% more expensive to produce. While these higher costs could significantly affect their acceptance, the calculations often exclude the costs associated with collecting, recycling, or disposing of conventional plastics. Additionally, the negative effects on soil productivity from the long-term accumulation of non-biodegradable plastics in agricultural soils are not typically factored into these assessments.

17.3.3 Mandating the Use of Environmentally Friendly Additives Is Crucial to Minimizing Toxicity

The degradation of plastics and the subsequent leaching of additives into soils are significant concerns. Plasticizers, which are commonly used in high concentrations in many plastic materials (e.g., about 30–35 wt% in PVC), have been particularly scrutinized. DEHP, one of the most prevalent plasticizers, and its monoester metabolite have been identified as endocrine disruptors, leading to their ban in children's toys. Research into the degradation pathways of phthalates has spurred the development of alternative plasticizers, such as non-phthalate di(2-ethylhexyl) terephthalate and di(isononyl) cyclohexane-1,2-dicarboxylate (DINCH). These alternative plasticizers are categorized as "green plasticizers" and are considered non-toxic along with their metabolites. However, recent discussions have raised concerns about DINCH potentially having metabolic toxicity to aquatic organisms, and several reviews have emphasized the difficulties in determining whether a plasticizer is genuinely "green." Although some plasticizers are derived from renewable feedstocks, their production often involves hazardous chemicals, highlighting the need for life cycle analysis to prevent undesirable substitutions (Bocqué et al., 2016; Jamarani et al., 2018; Jia et al., 2018; Qadeer et al., 2022; Qadeer et al., 2022).

17.4 CONCLUSION

Plastics have become integral to modern agricultural practices, offering numerous benefits that enhance productivity and resource efficiency. However, their pervasive use has led to significant environmental challenges, particularly the accumulation of macro-, micro-, and nanoplastics in soils. Conventional plastics such as polyethylene, polypropylene, and polyvinyl chloride are highly persistent in the environment due to their stable polymer backbones, posing long-term ecological risks. The transition to biodegradable polymers, designed to degrade into CO_2 and microbial biomass under specific conditions, represents a promising solution. Nevertheless, biodegradable plastics must be rigorously evaluated to ensure they effectively degrade in field conditions without leaving harmful residues. The sustainable use of plastics in agriculture hinges on improving collection, reuse, and recycling processes, as well as developing environmentally benign additives to minimize toxicity. With global initiatives like the Plastics Treaty aiming to address plastic pollution, the agricultural sector must adopt comprehensive strategies to mitigate the environmental impact of plastics while maintaining their agronomic benefits. This balanced approach will be critical for advancing sustainable agricultural practices and safeguarding soil health and ecosystem integrity.

REFERENCES

Andrady, A. L., & Neal, M. A. (2009). Applications and societal benefits of plastics. *Philosophical Transactions of the Royal Society B: Biological Sciences, 364*, 1977–1984.

Bauchmüller, V., Carus, M., Chinthapalli, R., Dammer, L., Hark, N., Partanen, A., Ruiz, P., & Lajewski, S. (2021). *BioSinn: Products for which biodegradation makes sense* (Report) Nova-Institute and IKT-Stuttgart 25.

Bergmann, M., Almroth, B. C., Brander, S. M., Dey, T., Green, D. S., Gundogdu, S., Krieger, A., Wagner, M., & Walker, T. R. (2022). A global plastic treaty must cap production. *Science, 376*, 469–470.

Bocqué, M., Voirin, C., Lapinte, V., Caillol, S., & Robin, J. J. (2016). Petro-based and bio-based plasticizers: Chemical structures to plasticizing properties. *Journal of Polymer Science Part A: Polymer Chemistry, 54*, 11–33.

Braga Marsola, K., Leda Ramos de Oliveira, A., Filassi, M., Elias, A. A., & Andrade Rodrigues, F. (2021). Reverse logistics of empty pesticide containers: Solution or a problem? *International Journal of Sustainable Engineering, 14*, 1451–1462.

Cassou, E. (2022). *Plastics: Agricultural pollution*. World Bank.

De Lucia, C., & Pazienza, P. (2019). Market-based tools for a plastic waste reduction policy in agriculture: A case study in the south of Italy. *Journal of Environmental Management, 250*, 109468.

Eck, M., Schwab, S. T., Nelson, T. F., Wurst, K., Iberl, S., Schleheck, D., Link, C., Battagliarin, G., & Mecking, S. (2023). Biodegradable high-density polyethylene-like material. *Angewandte Chemie International Edition, 62*, e202213438.

European Commission. (2016). *The impact of the use of 'oxo-degradable' plastic on the environment: Final report*. Publications Office.

FAO. (2021). Assessment of agricultural plastics and their sustainability. *A call for action*. https://www.fao.org/3/cb7856en/cb7856en.pdf (Rome)

Gao, Y., & Cabrera Serrenho, A. (2023). Greenhouse gas emissions from nitrogen fertilizers could be reduced by up to one-fifth of current levels by 2050 with combined interventions. *Nature Food, 4*, 170–173.

Hayes, D. G., Anunciado, M. B., DeBruyn, J. M., Bandopadhyay, S., Schaeffer, S., English, M., Ghimire, S., Miles, C., Flury, M., & Sintim, H. Y. (2019). Biodegradable plastic mulch films for sustainable specialty crop production. In *Polymers for Agri-Food Applications* (pp. 183–213). Springer International Publishing.

Jamarani, R., Erythropel, H., Nicell, J., Leask, R., & Marić, M. (2018). How green is your plasticizer? *Polymers*, *10*, 834.

Jariwala, H., Santos, R. M., Lauzon, J. D., Dutta, A., & Wai Chiang, Y. (2022). Controlled release fertilizers (CRFs) for climate-smart agriculture practices: A comprehensive review on release mechanism, materials, methods of preparation, and effect on environmental parameters. *Environmental Science and Pollution Research*, *29*, 53967–53995.

Jia, P., Xia, H., Tang, K., & Zhou, Y. (2018). Plasticizers derived from biomass resources: A short review. *Polymers*, *10*, 1303.

Kaabel, S., Therien, J. D., Deschênes, C. E., Duncan, D., Friščić, T., & Auclair, K. (2021). Enzymatic depolymerization of highly crystalline polyethylene terephthalate enabled in moist-solid reaction mixtures. *Proceedings of the National Academy of Sciences*, *118*, e2026452118.

Kanbur, U., Paterson, A. L., Rodriguez, J., Kocen, A. L., Yappert, R., Hackler, R. A., Wang, Y. Y., Peters, B., Delferro, M., LaPointe, A. M., & Coates, G. W. (2023). Zirconium-catalyzed C–H alumination of polyolefins, paraffins, and methane. *Journal of the American Chemical Society*, *145*, 2901–2910.

Kasirajan, S., & Ngouajio, M. (2012). Polyethylene and biodegradable mulches for agricultural applications: A review. *Agronomy for Sustainable Development*, *32*, 501–529.

Katsumi, N., Kusube, T., Nagao, S., & Okochi, H. (2021). The input–output balance of microplastics derived from coated fertilizer in paddy fields and the timing of their discharge during the irrigation season. *Chemosphere*, *279*, 130574.

Madrid, B., Wortman, S., Hayes, D. G., DeBruyn, J. M., Miles, C., Flury, M., Marsh, T. L., Galinato, S. P., Englund, K., Agehara, S., & DeVetter, L. W. (2022). End-of-life management options for agricultural mulch films in the United States—A review. *Frontiers in Sustainable Food Systems*, *6*.

Maraveas, C. (2019). Environmental sustainability of greenhouse covering materials. *Sustainability*, *11*, 6129.

Maraveas, C. (2020). Environmental sustainability of plastic in agriculture. *Agriculture*, *10*, 310.

Mayanti, B., & Helo, P. (2022). Closed-loop supply chain potential of agricultural plastic waste: Economic and environmental assessment of bale wrap waste recycling in Finland. *International Journal of Production Economics*, *244*, 108347.

Nelson, T. F., Remke, S. C., Kohler, H.-P. E., McNeill, K., & Sander, M. (2020). Quantification of synthetic polyesters from biodegradable mulch films in soils. *Environmental Science & Technology*, *54*, 266–275.

Picuno, C., Alassali, A., Sundermann, M., Godosi, Z., Picuno, P., & Kuchta, K. (2020). Decontamination and recycling of agrochemical plastic packaging waste. *Journal of Hazardous Materials*, *381*, 120965.

Qadeer, A., Kirsten, K. L., Ajmal, Z., Jiang, X., & Zhao, X. (2022). Alternative plasticizers as emerging global environmental and health threat: Another regrettable substitution? *Environmental Science & Technology*, *56*, 1482–1488.

Qadeer, A., Kirsten, K. L., Ajmal, Z., & Xingru, Z. (2022). Rebuttal to comment on "alternative plasticizers as emerging global environmental and health threat: Another regrettable substitution?" Focus on DINCH as an example. *Environmental Science & Technology*, *56*, 5294–5297.

Rillig, M. C., & Lehmann, A. (2020). Microplastic in terrestrial ecosystems. *Science*, *368*, 1430–1431.

Rosenboom, J. G., Langer, R., & Traverso, G. (2022). Bioplastics for a circular economy. *Nature Reviews Materials*, *7*, 117–137.

Serrano, D. (2004). Feedstock recycling of agriculture plastic film wastes by catalytic cracking. *Applied Catalysis B: Environmental*, *49*, 257–265.

Sun, D., Li, H., Wang, E., He, W., Hao, W., Yan, C., Li, Y., Mei, X., Zhang, Y., Sun, Z., & Jia, Z. (2020). An overview of the use of plastic-film mulching in China to increase crop yield and water-use efficiency. *National Science Review*, *7*, 1523–1526.

Svechnikov, K., Svechnikova, I., & Söder, O. (2008). Inhibitory effects of monoethylhexyl phthalate on steroidogenesis in immature and adult rat Leydig cells in vitro. *Reproductive Toxicology*, *25*, 485–490.

Thompson, R. C., Pahl, S., & Sembiring, E. (2022). Plastics treaty—Research must inform action, *Nature*, *608*, 472.

Tricker, A. W., Osibo, A. A., Chang, Y., Kang, J. X., Ganesan, A., Anglou, E., Boukouvala, F., Nair, S., Jones, C. W., & Sievers, C. (2022). Stages and kinetics of mechanochemical depolymerization of poly(ethylene terephthalate) with sodium hydroxide. *ACS Sustainable Chemistry & Engineering*, *10*, 11338–11347.

von Vacano, B., Mangold, H., Vandermeulen, G. W., Battagliarin, G., Hofmann, M., Bean, J., & Künkel, A. (2022). Sustainable design of structural and functional polymers for a circular economy. *Angewandte Chemie International Edition*, *62*, e202210823.

Walker, T. R. (2022). Calling for a decision to launch negotiations on a new global agreement on plastic pollution at UNEA5.2. *Marine Pollution Bulletin*, *176*, 113447.

Xu, T., Wang, Y., Aytac, Z., Zuverza-Mena, N., Zhao, Z., Hu, X., Ng, K. W., White, J. C., & Demokritou, P. (2022). Enhancing agrichemical delivery and plant development with biopolymer-based stimuli responsive core–shell nanostructures. *ACS Nano*, *16*, 6034–6048.

Yardley, R. E., Kenaree, A. R., & Gillies, E. R. (2019). Triggering depolymerization: Progress and opportunities for self-immolative polymers. *Macromolecules*, *52*, 6342–6360.

Zhang, G., Liu, C., Xiao, C., Xie, R., Ming, B., Hou, P., Liu, G., Xu, X., Shen, D., Wang, K., & Li, S. (2017). Optimizing water use efficiency and economic return of super high yield spring maize under drip irrigation and plastic mulching in arid areas of China. *Field Crops Research*, *211*, 137–146.

Index

Note: Page numbers in *italics* indicate a figure and page numbers in **bold** indicate a table on the corresponding page.

D

E

F

G

H

I

J

K

L

M

N

O

P

R

S

FSC
www.fsc.org
MIX
Papier | Fördert gute Waldnutzung
FSC® C083411

Kolibri 360 GmbH
Ferdinand-Jühlke-Straße 7
99095 Erfurt, Deutschland
produktsicherheit@kolibri360.de